NEURORECEPTOR MECHANISMS IN BRAIN

ADVANCES IN EXPERIMENTAL MEDICINE AND BIOLOGY

Recent Volumes in this Series

NEURORECEPTOR MECHANISMS IN BRAIN

Edited by

Shozo Kito

University of the Air
Chiba, Japan

Tomio Segawa

Hiroshima University School of Medicine
Hiroshima, Japan

and

Richard W. Olsen

University of California–Los Angeles School of Medicine
Los Angeles, California

PLENUM PRESS • NEW YORK AND LONDON

Library of Congress Cataloging in Publication Data

International Symposium on Neurotransmitter Receptors (3rd: 1990: Hiroshima-shi, Japan)
    Neuroreceptor mechanisms in brain / edited by Shozo Kito, Tomio Segawa, and Richard W. Olsen.
        p.   cm. — (Advances in experimental medicine and biology; v. 287)
    "Proceedings of the Third International Symposium on Neurotransmitter Receptors, held February 5–8, 1990, in Hiroshima, Japan"—T.p. verso.
    Includes bibliographical references.
    Includes index.
    ISBN-13: 978-1-4684-5909-8
    1. Neurotransmitter receptors—Congresses. 2. Brain—Congresses. I. Kitō, Shōzō, date. II. Segawa, Tomio, 1927 –   . III. Olsen, Richard W. IV. Title. V. Series.
    [DNLM: 1. Brain—physiology—congresses. 2. Receptors, Synaptic—physiology—congresses. W1 AD559 v. 287 / WL 300 I61243n 1990]
QP364.7.I59   1990
612.8′22—dc20
DNLM/DLC   91-3019
for Library of Congress   CIP

Proceedings of the Third International Symposium on Neurotransmitter Receptors, held February 5–8, 1990, in Hiroshima, Japan

ISBN-13: 978-1-4684-5909-8   e-ISBN-13: 978-1-4684-5907-4
DOI: 10.1007/ 978-1-4684-5907-4

© 1991 Plenum Press, New York
Softcover reprint of the hardcover 1st edition 1991
A Division of Plenum Publishing Corporation
233 Spring Street, New York, N.Y. 10013

FOREWORD

The Third International Symposium on Neurotransmitter Receptors was held in Hiroshima at a time when the entire field of neurotransmitter receptors in the brain is progressing at an unprecedented pace. The symposium also marked my retirement as Professor and Chairman of the Third Department of Internal Medicine, Hiroshima University School of Medicine, and a new beginning as a Professor of the University of the Air.

The symposium was remarkably successful, and there were enthusiastic responses from scientists all over the world, proving that the meeting was timely. The selected papers contained in this volume constitute a state-of-the-art survey of the most advanced aspects of neurotransmitter receptor mechanisms in the brain.

I owe thanks for the great success of the symposium to Prof. Richard Olsen of UCLA, Prof. Tomio Segawa of Hiroshima University, Prof. Kinya Kuriyama of Kyoto Prefectural University of Medicine, and Prof. Masaya Tohyama of Osaka University. I express my sincere gratitude to many friends for making this publication possible. I especially thank Dr. Rie Miyoshi, whose devoted efforts as secretary-general were vital to the success of the symposium. Dr. Miyoshi is currently an instructor in the Department of Pharmacology at Tokyo Women's Medical College. I would also like to acknowledge the excellent secretarial work of Misses Ritsuko Sato and Yuko Wakita.

My appointment to the University of the Air affords me access to a well-equipped laboratory for basic biological research. I hope that my concentration on receptor studies over the next several years will make it possible for me to organize the Fourth International Symposium on Neurotransmitter Receptors, probably not at Hiroshima, but at the University of the Air, which is located close to the Tokyo metropolitan area.

Finally, this symposium was financially supported by the Japanese Ministry of Science, Culture, and Education.

Shozo Kito

PREFACE

The Third International Symposium on Neurotransmitter Receptors, held in Hiroshima, Fabruary 5-8, 1990, continues the tradition of the first two such symposia held in 1983 and 1987, also published by Plenum Press. However, the fame of these meetings has obviously been growing, as indicated by both the large number of abstracts received and the great number of countries represented by the attending neuroscientists.

The topics covered in this publication of the proceedings include structural and functional studies on some well-characterized receptor proteins, and characterization of receptor-coupling mechanisms, receptor interactions, and receptor regulation, with concentration on the central nervous system. Techniques described include now-classical molecular pharmacology through imaging of second messengers in cells and in intact brain to the most sophisticated use of recombinant DNA technology. Readers will no doubt find these representative chapters at the cutting edge of neuroscience .

Speaking on behalf of the participants at the Hiroshima symposium, I would like to thank Professor Kito sincerely and warmly for providing this excellent scientific program in the midst of an extremely friendly and enjoyable visit to Japan. I hope that there will be more such occasions.

Richard W. Olsen

CONTENTS

EFFECT OF NEUROPEPTIDES ON CLASSIC TYPES OF NEUROTRANSMISSION IN THE RAT CENTRAL NERVOUS SYSTEM

Shozo Kito and Rie Miyoshi

Third Department of Internal Medicine, Hiroshima University
School of Medicine, 1-2-3 Kasumi, Minamiku, Hiroshima 734
Japan

INTRODUCTION

Neuromodulator is a term for which there is no clear definition.
For the past several years, the authors have been studing how
neuropeptides modulate classic types of neurotransmission. We observed
two examples of peptide action, that is, an effect of somatostatin on
muscarinic acetylcholine receptors (mAChR) and that of cholecystokinin
(CCK) on dopamine receptors in the rat brain, and obtained results
suggesting that these two peptides modulated classical neurotransmitter
receptors in a similar manner.

Both acetylcholine (Coyle et al., 1983; Perry et al., 1977) and
somatostatin (Beal et al., 1986; Roberts et al., 1985) were definitely
diminished in the brain of Alzheimer's disease. Morphologically, it has
been reported that septo-hippocampal cholinergic neurons project to
somatostatin-containing neurons in the hippocampus (Yamano and Luiten,
1989). In addition, it has been reported that somatostatin affects the
turnover rate of acetylcholine in this brain area (Wood et al., 1981).
Within the hippocampus, there are abundant M1 subtype mAChR (Vickroy et
al., 1984). The authors investigated a modulatory effect of somatostatin
on mAChR and an intracellular response after somatostatin receptor
activation in the rat hippocampus from the viewpoint of pharmacology.

CCK receptors are observed   richly in the striatum where
dopaminergic nerve terminals ascending from the substantia nigra are
observed with high density (Gaudreau et al., 1985).   There has been
strong evidence for dopamine and CCK interaction in the basal ganglia.
For instance, CCK coexists with dopamine in the midbrain region (Hommer
et al., 1986), and the peptide modulates dopamine release in the
striatum (Hokfelt et al., 1985; Kovacs et al., 1981; Starr, 1982).
Dopamine receptors have been classified into two major subtypes on the
basis of their pharmacological and biochemical characteristics (Leff and
Creese, 1983; Stoof and Kebabian, 1984). The functional significance of
D2 receptors in the central nervous system has been well discussed from
the viewpoints of pharmacotherapy of psychosis and movement disorders.
Recently, availability of D1 selective ligands has made it possible to
study physiological roles of D1 receptors in the central nervous system
(Iorio et al., 1983). In the present paper, the effect of CCK on D1

receptor binding and the intracellular transduction system of CCK receptors were examined in the rat striatum.

MATERIALS AND METHODS

Effect of Somatostatin on mAChR in the Hippocampus

Wistar strain male rats weighing 200-250 g were used in these experiments. After decapitation, the hippocampus was rapidly removed and P_2 fractions of the tissue were prepared. Both ^{3}H-QNB and ^{3}H-N-methyl-scopolamine (NMS) were used as mAChR antagonists. For observation of mAChR agonist binding sites, oxotremorine/^{3}H-NMS inhibition experiments were performed. As assay medium, Krebs-Henseleit solution containing bovine serum albumin, bacitracin and pepstatin was used. Non-specific binding was defined as the binding in the presence of 1 μM atropine. Aliquots of the tissue preparation were incubated at 30°C for 15 min. Membrane bound ^{3}H-ligands were trapped by the rapid vacuum filtration method.

The phosphatidylinositol (PI) turnover is one of the important second messenger systems involving an increase of intracellular Ca^{2+} concentration ($[Ca^{2+}]i$) and activation of protein kinase C (Berridge and Irvine, 1984; Nishizuka, 1986). There has been strong evidence that mAChR activate this signal transduction system in various tissues (Brown et al., 1984; Janowsky et al., 1984). In this experiment, the synergistic effect of carbachol, a mAChR agonist, and somatostatin was investigated using hippocampal slices according to Berridge's method.

In addition, changes of $[Ca^{2+}]i$ in response to somatostatin were examined by fura-2 fluorometry on a single cell basis. Hippocampal neurons were obtained from 18-day rat embryos and dissociated cell cultures were made. Cultures were maintained for 12-14 days. After 3 days in vitro, cultures were treated with 10 μM cytosine arabinofuranoside for 24 hrs in order to suppress growth of fibroblasts and glial cells. The level of $[Ca^{2+}]i$ was measured in combination of fura-2, a fluorescence microscope, a video camera and photometrical devices as previously described (Kudo et al., 1986). Drugs used in this experiment were applied to cultured neurons by perfusion. During measurement, 1 μM tetrodotoxin was added to a buffer in order to block spontaneous Ca^{2+} flux. The amount of $[Ca^{2+}]i$ was quantitatively determined by a dual beam excitation method.

Effect of CCK on Dopamine D1 Receptors in the Striatum

For dopamine D1 receptor binding, the P_2 fraction of the striatum was used. ^{3}H-SCH23390 was used as a D1 selective antagonist. For observation of D1 antagonist binding sites, saturation experiments with ^{3}H-SCH23390 were performed. Non-specific binding was defined as binding in the presence of 10^{-5} M SCH23390. For D1 agonist binding sites, dopamine/^{3}H-SCH23390 inhibition experiments were done. As assay medium, 50 mM Tris HCl buffer, pH7.4 containing 5 mM KCl, 1 mM $MgCl_2$, 2 mM $CaCl_2$, bovine serum albumin and bacitracin was used. Aliquots of the tissue preparation were incubated at 22°C for 30 min.

To study intracellular responses of CCK receptors, changes of the PI turnover-Ca^{2+} signalling system in response to either CCK-8 or ceruletide, an octapeptide analog of CCK-8, were examined with use of the rat striatum.

2

RESULTS

Effect of Somatostatin on mAChR in the Hippocampus

Somatostatin had no effect on both ^{3}H-QNB and ^{3}H-NMS binding in the rat hippocampus (data not shown). However, as shown in Fig. 1, the peptide affected the affinity of agonist binding sites. An oxotremorine/^{3}H-NMS inhibition curve exhibited heterogeneous characteristics with a Hill coefficient much less than 1. The binding data were fitted best by a two-site model, characterized by two dissociation constants: a K_H value of 6.6×10^{-9} M and a K_L value of 2.6×10^{-6} M. Percentages of high and low affinity binding sites to the total binding capacity were 36.3 % and 63.7 %, respectively. The inhibition curve after adding 1 μM [D-trp^8]somatostatin in the incubation medium exhibited a Hill slope factor close to 1, and the curve fitted best to a binding model consisting of a single homogeneous binding site whose Ki value was consistent with the K_L value.

As previously reported by various groups (Brown et al., 1984; Janowsky et al., 1984), carbachol caused an enhancement of ^{3}H-inositol-1-phosphate (IP$_1$) accumulation. Somatostatin significantly augmented the effect of carbachol (p<0.01), while somatostatin itself had no effect on the basal accumulation of ^{3}H-IP$_1$ (Fig. 2).

As the next step, to elucidate the mechanism of the above-mentioned effects of somatostatin on mAChR, changes of [Ca^{2+}]i in response to somatostatin were examined in cultured hippocampal neurons. Fig. 3 shows the response to either [D-trp^8]somatostatin or SMS201-995 at the concentration of 10^{-5} M in the same cultured hippocampal neuron. The basal level of [Ca^{2+}]i in neurons was about 100 nM. By perfusing with [D-trp^8]somatostatin for 30 sec., a monophasic increase of [Ca^{2+}]i was observed. The cell population which responded to [D-trp^8]somatostatin stimulation was 26.5 % of the total neurons. SMS201-995, a very potent octapeptide analog of somatostatin, had about 2 times more potency on elevation of [Ca^{2+}]i. This effect of somatostatin was found to be dose-dependent when concentrations ranging from 10^{-8} to 10^{-4} M were used. As next experiments, the Ca^{2+} source for the effect of somatostatin was determined. Fig. 4 shows changes of [Ca^{2+}]i in response to [D-trp^8]somatostatin in cases of perfusing with either normal buffer or Ca^{2+}-depleted medium in the same neuron. The effect of somatostatin was completely blocked in Ca^{2+}-depleted medium. The

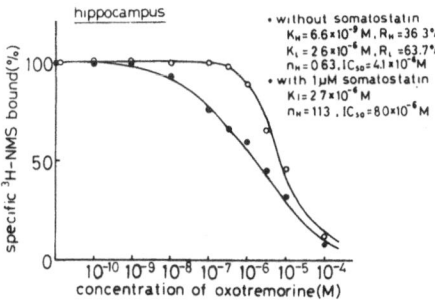

Fig. 1 Oxotremorine binding to P$_2$ fraction of the rat hippocampus measured by competition with ^{3}H-NMS in the presence and absence of 1 μM [D-trp^8]somatostatin. n_H: Hill coefficient. The concentration of ^{3}H-NMS was 0.2 nM. These values were means in 3 repeated experiments.

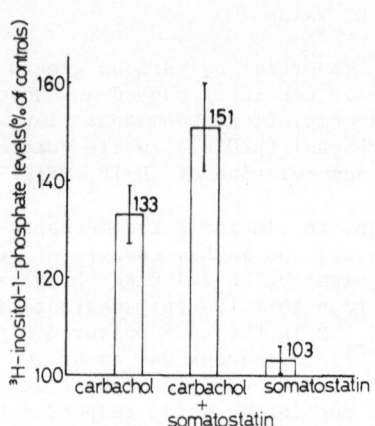

Fig. 2   Effect of [D-trp[8]]somatostatin on carbachol-stimulated
inositol phospholipid hydrolysis in rat hippocampal
slices. The concentrations of carbachol and somatostatin
were 1 mM and 50 μM, respectively.   Somatostatin
augmented carbachol-induced accumulation of ^{3}H-IP$_1$ in a
dose-dependent manner (data not shown).   Values
represent % of those in experiments in which 10 mM LiCl
alone was added to the incubation medium. p<0.05, vs
the data with 1 mM carbachol. n=3.

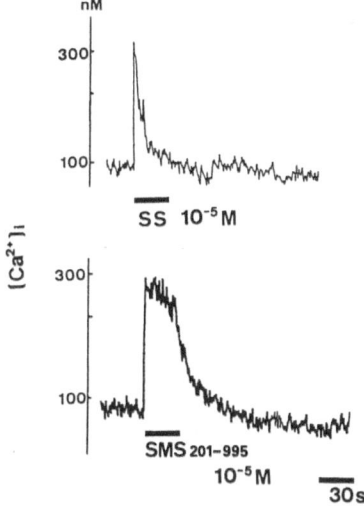

Fig. 3 Effect of 10^{-5} M [D-trp^8]somatostatin (SS) or SMS201-995 on [Ca^{2+}]i in the same cultured rat hippocampal neuron. Drugs were applied to the neuron by perfusion for 30 sec.

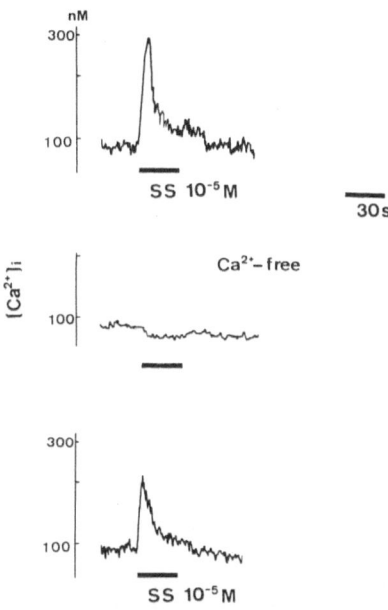

Fig. 4 Determination of the source of Ca^{2+} for [D-trp^8]somatostatin (SS)-induced elevation of [Ca^{2+}]i in cultured rat hippocampal neurons. The upper plate shows a result of an experiment perfusing with normal buffer and the middle Ca^{2+}-depleted medium. In the latter case, the neuron was perfused for 1 min with Ca^{2+}-depleted medium in which CaCl$_2$ was removed and 0.1 mM EGTA was added, prior to SS stimulation. The lower plate shows recovery of the response by re-perfusing with normal buffer.

5

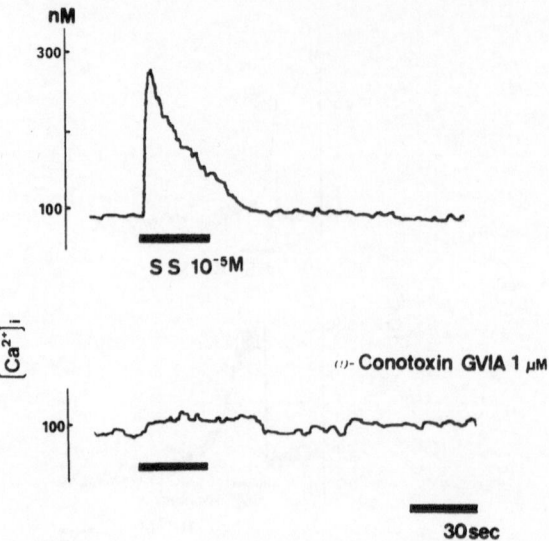

Fig. 5 Effect of ω-conotoxin GVIA on $[D-trp^8]$somatostatin (SS)-induced increase of $[Ca^{2+}]i$ in cultured rat hippocampal neurons. The upper plate shows a result of an experiment perfusing with normal buffer and the lower that of pretreatment with 1 μM ω-conotoxin GVIA for 1 min. A set of experiments were performed using the same neuron.

binding sites without changing dissociation constants, independently. In the presence of NaCl, CCK-8 caused further conversion to a low affinity state. However, when both GTP and CCK-8 were added, CCK-8 did not convert anymore.

Effect of CCK on Dopamine D1 Receptors in the Striatum

Scatchard analysis revealed that ^{3}H-SCH23390 had a single high affinity binding site whose Kd and Bmax values were 1.57 nM and 907.6 fmol/mg protein, respectively. CCK-8 at the concentration of 1 μM had no effect on ^{3}H-SCH23390 binding (data not shown).

Table 1 shows results obtained from dopamine/^{3}H-SCH23390 inhibition experiments in the rat striatum. Curves without and with CCK-8 exhibited heterogeneous characteristics with Hill coefficients much less than 1. In the absence of CCK-8, the binding data were fitted best by a two-site model, characterized by two dissociation constants: a K_H value of 7.76×10^{-7} M and a K_L value of 2.45×10^{-5} M. Percentages of high (R_H) and low (R_L) affinity binding sites to the total binding capacity were 74.6 and 25.4 %, respectively. After adding 1 μM CCK-8, R_H decreased to 51.3 % and R_L increased to 48.7 %, with their affinity constants unchanged. Guanine nucleotides and sodium ion have been known to convert agonist high affinity binding sites into low affinity ones in striatal dopamine receptors (Grigoriadis and Seeman, 1985). The mechanism of the effect of guanine nucleotides on agonist binding has been well investigated and accepted to be the result of the dissociation of a GTP binding protein from a receptor protein. The mechanism of sodium ion is not yet clear, but it is suggested that sodium ion acts like an allosteric inhibitor of dopamine receptors. In our experiments, both 50 μM GTP and 120 mM NaCl caused a reduction of high affinity

Table 1. Effect of CCK-8, NaCl and GTP on dopamine binding to the rat striatum measured by competition with ^{3}H-SCH23390

	n_H	$-\log[K_H]$	$-\log[K_L]$	R_H (%)	R_L (%)
	0.64 ± 0.10	6.11 ± 0.53	4.61 ± 0.17	74.60 ± 1.44	25.40 ± 1.44
CCK-8	0.75 ± 0.06	7.02 ± 0.09	5.67 ± 0.14	51.27 ± 5.83	48.73 ± 5.83
NaCl	0.67 ± 0.04	6.17 ± 0.01	4.73 ± 0.10	55.61 ± 0.81	44.39 ± 0.81
NaCl + CCK-8	0.73 ± 0.02	6.91 ± 0.26	5.05 ± 0.04	27.61 ± 4.81	72.39 ± 4.81
GTP	0.71 ± 0.20	6.99 ± 1.47	4.93 ± 0.49	48.13 ± 7.53	51.87 ± 7.53
GTP + CCK-8	0.70 ± 0.04	6.97 ± 0.43	5.41 ± 0.14	45.94 ± 9.29	54.06 ± 9.29

Synergistic effects of CCK-8 and NaCl or GTP on dopamine/^{3}H-SCH23390 inhibition experiments were studied in rat striatal homogenates. These drugs were added to the incubation medium. These values were means of 3 repeated experiments.

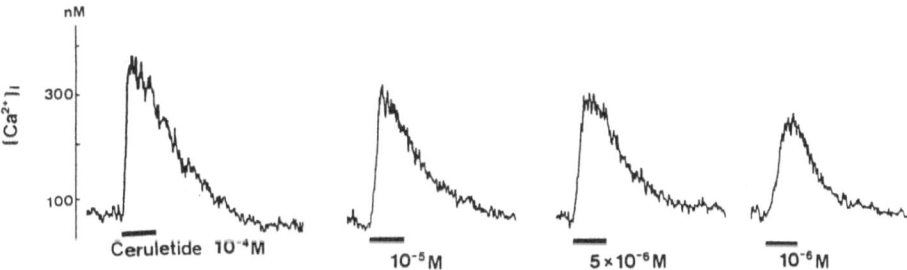

Fig. 6 Effect of various concentrations of ceruletide on $[Ca^{2+}]i$ in the same cultured rat striatal neuron. Ceruletide was applied to the neuron by perfusion for 30 sec.

blockage was recovered by re-perfusing with the normal buffer. After 1 min pretreatment with 1 μM ω-conotoxin GVIA, a neurotoxic peptide from the venom of a marine snail, the effect of somatostatin was completely inhibited (Fig. 5). Nifedipine, one of the dihydropyridine Ca^{2+} channel blockers, had no effect on the $[Ca^{2+}]i$ rise induced by somatostatin.

Using the same experimental process as in the case of somatostatin, the authors investigated the effect of CCK on the PI turnover-Ca^{2+} signalling system in the rat striatum. Neither CCK-8 nor ceruletide changed basal PI turnover. However, ceruletide caused a monophasic increase of $[Ca^{2+}]i$ in a dose-dependent manner as shown in Fig. 6. The response to ceruletide was completely blocked by perfusing with Ca^{2+}-depleted medium or pretreatment with ω-conotoxin GVIA.

DISCUSSION

Since the effect of somatostatin on mAChR in the rat hippocampus was limited to agonist binding, it seems that somatostatin dose not bind mAChR directly, but the peptide affects the functional state of receptors through some intracellular responses.

Somatostatin has been considered to be an inhibitory neuropeptide throughout the endocrine and nervous systems. As for the intracellular response, somatostatin inhibits adenylate cyclase activity and Ca^{2+}

mobilization in the pituitary, pancreas and so on (Catalan et al., 1979; Chneiweiss et al., 1987; Diez and Tamargo, 1987; Luini et al., 1986). In our experiments, in the presence of somatostatin, the muscarinic function seems to be activated, that is, a reduction of the affinity of receptors and augmentation of PI turnover elicited by a mAChR agonist in the hippocampus occur. In previous reports, an excitatory effect of somatostatin has been described in the hippocampus and cerebral cortex from viewpoints of electrophysiology (Dodd and Kelly, 1978; Ioffe et al., 1978; Olpe et al., 1980). In this paper, as a novel excitatory response of somatostatin, the authors obtained results that the peptide increased $[Ca^{2+}]i$, probably via N-type Ca^{2+} channels. It is considered that an increase of $[Ca^{2+}]i$ induced by somatostatin is a key event for modulating the muscarinic function.

Galanin is a neuropeptide which is considered to have some relation to the septo-hippocampal cholinergic system (Dutar et al., 1989; Melander et al., 1985; Senut et al., 1989). Consolo et al. reported tnat galanin reduced PI turnover elicited by carbachol probably because of its lowering action of $[Ca^{2+}]i$ in the rat hippocampus (Consolo et al., 1989). The authors confirmed that galanin decreased $[Ca^{2+}]i$ stimulated by high K^+ (unpublished data). A possible mechanism on modulation of hippocampal mAChR by neuropeptides is schematically shown in Fig. 7.

Our indirect agonist binding studies revealed that CCK-8 converted a part of D1 agonist high affinity binding sites into low affinity ones in the rat striatum. It was noticed that the effect of CCK-8 was similar to that of GTP rather than that of NaCl. Therefore, it is considered that CCK-8 modulates D1 agonist binding sites in a GTP-like manner. It is noteworthy that somatostatin and CCK exert the same kind of effect on classical neurotransmission, that is, these two peptides reduce the affinity of agonist binding for muscarinic and D1 receptors,

Regulation of mAChR-coupled PI turnover by peptides in the hippocampus

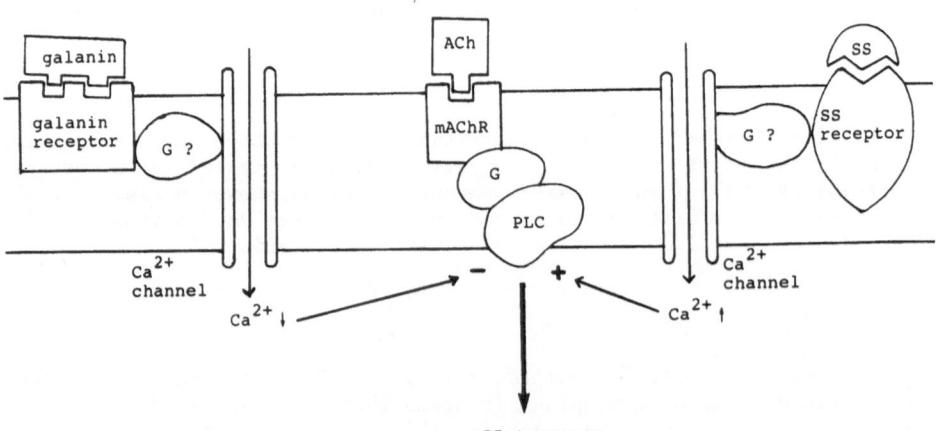

PI turnover

Fig. 7 A possible scheme for regulation of mAChR-induced stimulation of PI turnover by neuropeptides. It seems that the hippocampal mAChR function is regulated by at least two neuropeptides, that is, somatostatin and galanin, in an opposite manner.

respectively. Both mAChR and dopamine receptors are classified into the same group of receptor, a GTP binding protein-coupled receptor family. The authors demonstrated that somatostatin and CCK caused the same type of intracellular response in the central nervous system. From these facts, it is suggested that an increase of $[Ca^{2+}]i$ is a key point to the reduction of affinity for agonist binding. It is also likely that the modulatory effect of neuropeptides on classical neurotransmission is something like a common pathway and there is not much difference among the peptides.

ACKNOWLEDGEMENTS

This study was supported in part by a Grant-in-Aid for Science Research from the Ministry of Education, Science and Culture, Japan.

REFERENCES

Beal, M.F., Mazurek, M.F., Svendsen, C.N., Bird, E.D., and Martin, J.B., 1986, Widespread reduction of somatostatin-like immunoreactivity in the cerebral cortex in Alzheimer's disease, Ann Neurol., 20: 489-495.

Berridge, M.J., and Irvine, R.F., 1984, Inositol trisphosphate, a novel second messenger in cellular signal transduction, Nature, 312:315-321.

Brown, E., Kendall, D.A., and Nahorski, S.R., 1984, Inositol phospholipid hydrolysis in rat cerebral cortical slices: I. receptor characterisation, J. Neurochem., 42:1379-1387.

Catalan, R.E., Aragones, M.D., and Martinez, A.M., 1979, Somatostatin effect on cyclic AMP and cyclic GMP levels in rat brain, Biochim. Biophys. Acta, 586:213-216.

Chneiweiss, H., Bertrand, Ph., Epelbaum, J., Kordon, C., Glowinski, J., Premont, J., and Enjalbert, A., 1987, Somatostatin receptors on cortical neurons and adenohypophysis: comparison between specific binding and adenylate cyclase inhibition, Eur. J. Pharmacol., 138:249-255.

Consolo, S., Fisone, G., Bartfai, T., and Palazzi, E., 1989, Voltage-dependent Ca^{2+} channels modulate galanin inhibition of the muscarinic stimulation of phosphoinositide turnover in rat ventral hippocampus, Society for Neuroscience, Abstract, 1277.

Coyle, J.T., Price, D.L., and DeLong, M.R., 1983, Alzheimer's disease: a disorder of cortical cholinergic innervation, Science, 219:1184-1190.

Diez, J., and Tamargo, J., 1987, Effect of somatostatin on ^{45}Ca fluxes in guinea-pig isolated atria, Br. J. Pharmacol., 90:309-314.

Dodd, J. and Kelly, J.S., 1978, Is somatostatin an excitatory transmitter in the hippocampus? Nature, 273:674-675.

Dutar, P., Lamour, Y., and Nicoll, R.A., 1989, Galanin blocks the slow cholinergic EPSP in CA1 pyramidal neurons from ventral hippocampus, Eur. J. Pharmacol., 164:355-360.

Gaudreau, P., St-Pierre, S., Pert, C.B., and Quirion, R., 1985, Cholecystokinin receptors in mammalian brain, Ann. N. Y. Acad. Sci., 198-219.

Grigoriadis, D., and Seeman, P., 1985, Complete conversion of brain D_2 dopamine receptors from the high- to the low-affinity state for dopamine agonists, using sodium ions and guanine nucleotide, J. Neurochem., 44:1925-1935.

Hokfelt, T., Sharp, T., Ungerstedt, U., and Zetterstrom, T., 1985, Effect of CCK peptides on dopamine release and metabolism in rat striatum in vivo, Br. J. Pharmacol., 7P.

Hommer, D.W., Stoner, G., Crawley, J.N., Paul, S.M., and Skirboll, L.R., 1986, Cholecystokinin-dopamine coexistence: electrophysiological actions corresponding to cholecystokinin receptor subtype, J. Neurosci., 10:3039-3043.

Ioffe, S., Havlicek, V., Friesen, H., and Chernick, V., 1978, Effect of somatostatin (SRIF) and L-glutamate on neurons of the sensorimotor cortex in awake habituated rabbits, Brain Res., 153:414-418.

Iorio, L.C., Barnett, A., Leitz, F.H., Houser, V.P., and Korduba, C.A., 1983, SCH 23390, a potential benzazepine antipsychotic with unique interactions on dopaminergic systems, J. Pharmacol. Exp. Ther., 226:462-468.

Janowsky, A., Labarca, R., and Paul, S.M., 1984, Characterization of neurotransmitter receptor-mediated phosphatidylinositol hydrolysis in the rat hippocampus, Life Sci., 35:1953-1961.

Kudo, Y., Ozaki, K., Miyakawa, A., Amano, T., and Ogura, A., 1986, Monitoring of intracellular Ca^{2+} elevation in a single neural cell using a fluorescence microscope/video-camera system, Japan. J. Pharmacol., 41:345-351.

Kovacs, G.L., Szabo, G., Penke, B., and Telegdy, G., 1981, Effects of cholecystokinin octapeptide on striatal dopamine metabolism and on apomorphine-induced stereotyped cage-climbing in mice, Eur. J. Pharmacol., 69:313-319.

Leff, S.E., and Creese, I., 1983, Dopamine receptors reexplained, Trends Pharmacol. Sci., 463-467.

Luini, A., Lewis, D., Guild, S., Schofield, G., and Weight, F., 1986, Somatostatin, an inhibitor of ACTH secretion, decreases cytosolic free calcium and voltage-dependent calcium current in a pituitary cell line, J. Neurosci., 6:3128-3132.

Melander, T., Staines, W.A., Hokfelt, T., Rokaeus, A., Eckenstein, F., Salvaterra, P.M., and Wainer, B.H., 1985, Galanin-like immunoreactivity in cholinergic neurons of the septum-basal forebrain complex projecting to the hippocampus of the rat, Brain Res., 360:130-138.

Nishizuka, Y., 1986, Studies and perspectives of protein kinase C, Science, 233:305-312.

Olpe, H.R., Balcar, V.J., Bittiger, H., Rink, H., and Sieber, P., 1980, Central actions of somatostatin, Eur. J. Pharmacol., 63:127-133.

Perry, E.K., Perry, R.H., Blessed, G., and Tomlinson, B.E., 1977, Necropsy evidence of central cholinergic deficits in senile dementia, The Lancet, 22:189.

Roberts, G.W., Crow, T.J., and Polak, J.M., 1985, Location of neuronal tangles in somatostatin neurones in Alzheimer's disease, Nature, 314:92-94.

Senut, M.C., Menetrey, D., and Lamour, Y., 1989, Cholinergic and peptidergic projections from the medial septum and the nucleus of the diagonal band of Broca to dorsal hippocampus, cingulate cortex and olfactory bulb: a combined wheatgerm agglutinin-apohorseradish peroxidase-gold immunohistochemical study, Neurosci., 30:385-403.

Starr, M.S., 1982, Influence of peptides on ^{3}H-dopamine release from superfused rat striatal slices, Neurochem. Int., 4:233-240.

Stoof, J.C., and Kebabian, J.W., 1984, Two dopamine receptors: biochemistry, physiology and pharmacology, Life Sci., 35:2281-2296.

Vickroy, T.W., Watson, M., Yamamura, H.I., and Roeske, W.R., 1984, Differential regulation of putative M1/M2 muscarinic receptors: implications for different receptor-effector coupling mechanisms, in "Neurotransmitter Receptors: Mechanisms of Action and Regulation", S. Kito, T. Segawa, K. Kuriyama, H.I. Yamamura, and R.W. Olsen, eds., Plenum Press, New York, 99-114.

Wood, P.L., Cheney, D.L., and Costa, E., 1981, Interaction of neuropeptides with cholinergic septal-hippocampal pathway: indication for a possible trans-synaptic regulation, in "Cholinergic Mechanisms", G. Perpru, and H. Ladinsky, eds., Plenum Press, New York, 715-722.

Yamano, M., and Luiten, P.G., 1989, Direct synaptic contacts of medial septal efferents with somatostatin immunoreactive neurons in the rat hippocampus, Brain Res. Bull., 22:993-1001.

SOLUBILIZATION AND CHARACTERIZATION OF SUBSTANCE P RECEPTORS

IN THE CENTRAL NERVOUS SYSTEM

Tomio Segawa and Yoshihiro Nakata

Department of Pharmacology
Institute of Pharmaceutical Sciences
Hiroshima University School of Medicine
Kasumi 1-2-3, Minami-ku, Hiroshima 734, Japan

INTRODUCTION

Substance P (SP) is one of the best characterized neuro-peptides and acts as a neurotransmitter or a neuromodulator in central and peripheral nervous systems. In recent years, two other mammalian tachykinins, neurokinin A (NKA, or substance K) and neurokinin B (NKB, or neuromedin K) have been identified (Nawa et al., 1983; Kimura et al., 1983; Kangawa et al., 1983). Many radioligand binding studies and autoradiographic experiments have provided the evidence for the existence of three distinct types of tachykinin receptors (Quirion, 1983; Buck et al., 1985). One of these receptor types (NK-1 or SP-P) exhibits preferential affinity for SP, another (NK-2 or SP-K) for NKA, and the third (NK-3 or SP-E) type for NKB.

Although SP receptors are sensitive to monovalent and divalent cations and guanine nucleotides (Cascieri and Liang, 1983, Lee et al., 1983; Cascieri et al., 1985; Bahouth et al., 1985), little is known about the molecular properties, effectors and regulatory mechanisms of SP receptors. On the other hand, solubilization of the receptor components from membranes is the first step to purify the receptor protein and a useful method to study the molecular mechanisms of receptor regulation. Solubilization of SP receptors has been reported only by Maruyama (1985) who used rat brain and Payan et al. (1986) who used human IM-9 lymphoblast. Previously, we have reported the successful solubilization of SP binding protein from bovine brainstem with 10 mM 3-[(3-cholamidopropyl) dimethylammonio]-1-propane sulfonate (CHAPS) and 100 mM NaCl in an active form retaining sensitivity to GTP (Nakata et al., 1988). Many effects evoked by tachykinins have been reviewed but the effect of SP on lipid metabolism has not been studied in detail and this particular area is far from being fully understood. In this respect, some investigators have documented that SP is able to induce hydrolysis of inositol phospholipids in some tissues. The phosphatidylinositol (PtdIns) cycle is one major second messenger system in which receptors stimulate the phosphatidylinositol biphosphate to inositol triphosphate and diacylglycerol (Berridge and Irvine, 1984). Pertussis toxin (islet activating protein, IAP) which ADP-ribosylates several GTP-binding proteins, also inhibits the PtdIns cycle (Okajima and Ui, 1984; Smith et al., 1985). These data encouraged us to investigate whether or not IAP inhibits the PtdIns cycle stimulated by SP (Watson et al., 1983; Hunter et al., 1985). To achieve this idea, we have developed the new method in prelabeling the PtdIns cycle in rats and stimulating it in vivo. Inositol phosphates were measured by high performance liquid chromatography (HPLC) using an anionic exchange column.

In this study, we tried to analyze SP binding profiles by nonlinear least-square regression using a computer program (Simplex and modified Marquardt methods) and investigated the effects of $MgCl_2$, NaCl, GTP and the changes induced by IAP on tritiated SP binding activity in intact rat cerebral membranes and in a CHAPS-solubilized fraction. We also examined a functional interaction between SP binding states and PtdIns cycle.

MATERIALS AND METHODS

Membrane preparation

Male Wistar rats (180-250 g) were used. Cerebral cortical membranes were prepared by the method of Arima et al. (1986). Briefly, after decapitation, the cerebral cortices were homogenized in ice-cold 50 mM Tris-HCl buffer (pH 7.4) with a Polytron. The homogenates were centrifuged at 20,000 g for 10 min. and the pellets were washed three times. The preparation was used immediately for radioligand binding assay.

Preparation of the solubilized fraction

The solubilized fraction was prepared as described previously (Nakata et al., 1988). Membranes were homogenized in ice-cold solubilization buffer (10 mM CHAPS plus 100 mM NaCl in 50 mM Tris-HCl buffer, pH 7.4) to make 10 mg protein/ml. The homogenate was kept on ice for 30 min. with frequent mixing, then centrifuged at 100,000 g for 60 min. The supernatant was concentrated with Amicon Centriflo CF 25 (about 25,000 daltons cut off), then supplemented with 50 mM Tris-HCl buffer (pH 7.4) to make about 2.5 mg protein/ml. The mixture was centrifuged at 100,000 g for 60 min. once more, and the supernatant was referred to as a CHAPS-solubilized fraction.

[^{3}H]SP binding assay

The binding assays were performed as described previously (Nakata et al., 1988). For the membrane fractions, the fresh membrane preparations were suspended in 50 mM Tris-HCl buffer (pH 7.4) containing 0.02 % bovine serum albumin, 40 µg/ml bacitracin, 4 µg/ml leupeptin, 2 µg/ml chymostatin, 10 µM captopril and 10 mM $MgCl_2$. A 50 µl aliquot of [^{3}H]SP was incubated at room temperature for 30 min. with 1 mg of protein per tube in a final volume of 500 µl. Binding reactions were terminated by rapid filtration of the incubation mixture through Whatman GF/B glass filter (presoaked with 0.1 %

14

polyethyleneimine). The filters were washed four times with 10 ml of ice-cold 50 mM Tris-HCl buffer. Nonspecific binding of labeled ligand was determined by 2 µM unlabeled SP. In the case of the CHAPS-solubilized fractions, the method was modified slightly with the incubation time and temperature being 120 min. and 0°C, respectively. Filtrations were carried out using GF/B glass filters presoaked with 0.3 % poly-ethyleneimine according to the method of Bruns et al.(1983). For competition experiments using both preparations, [^{3}H]SP (1.8 nM of final concentration in the membrane fraction and 2.2 nM of final concentration in the solubilized fraction, respectively) was incubated with increasing concentrations of unlabeled SP (7 pM- 2 µM), as described above.

In vivo pretreatment with IAP

IAP solution was microinjected intracerebroventricularly (i.c.v.) according to a slight modification of the method of Nomura et al. (1987). In brief, the animals were anesthetized with pentobarbital, (40 mg/kg, i.p.) and the head was placed in a horizontal position. A midsagittal incision was made from eyes to ears and the bregma was exposed. Small holes were made bilaterally at points 1.7 mm lateral to midline and 1.0 mm posterior to the bregma. IAP solution (2.5 µg/5 µl of saline) or saline was drawn into a Hamilton syringe having a stop at 4.0 mm from its top. The syringe was held in the ventricular positions and the IAP or the vehicle solution was administered bilaterally. The syringe was withdrawn and the holes were stopped up with bone wax immediately, then the incision was sewn closed. The locomotor activity was measured with an AUTOMEX activity meter (Columbus Instrument) for 10 min. every 10 min. for 70 min. on the day 10 after the treatment with IAP. The rats were used for experiments on day 10 after the treatment with IAP.

Assay of ^{3}H-Phosphatidylinositol break down

Male Wistar rats weight 200g were used. Animals were maintained at 23 °C under the normal daylight cycle. Rats were anesthetized with pentobarbital sodium (40 mg/kg, i.p.) and placed in a stereotaxic holder. One or two µCi of ^{3}H-inositol dissolved in artificial cerebro spinal fluid was injected in the bilateral i. c. v. as described above.

After i.c.v. administration of ^{3}H-inositol, guide cannulae were inserted to the injection site of the brain. The cannulae were fastened to the skill with dental cement and acrylic resin.

Two days before i. c. v. administration of ^{3}H-inositol, IAP was injected in a total volume of 5 µl/ a lateral ventricle over a 60s period to rats that were taken by the same opera-tion of ^{3}H-inositol administration. 48 hours after i.c.v. administration of ^{3}H-inositol, neuropeptide dissolved at a final concentration of 1.0 µg/ one µl in a saline containing 100 mM LiCl were injected through an injection cannula. The certain period after neuropeptide i.c.v. injection, the rats were decapitated and the brain removed. The injected hemi-whole brain excluding cerebellum were sliced and homogenized in 1 ml of ice cold 5% trichloroacetic acid. The trichloro-acetic acid supernatants were obtained by centrifugation, and washed four times with one ml aliquots of diethyl ether to remove the trichloroacetic acid. 1 ml of chloroform/methanol/HCl (100:200:2, v/v) was added and centrifuged. The aqueous phase was submitted to analysis of inositol phosphates(IP_s).

The trichloroacetic acid precipitate was homogenized in chloroform/ methanol/ HCl (100:200:2, v/v). The lower organic phase was transferred and the radioactivity in lipids was determined by a liquid scintillation counting. The water soluble inositol phosphates were characterized by HPLC. A Shimazu LC-4A solvent delivery system was used with a TSK Gel DEAE-5PW (the flow rate was 1.35 ml/min) and fractions were collected at 30 seconds intervals using two linear ammonium formate gradients.

Data analysis

The data obtained from competition experiments were analyzed by nonlinear least-square regression using the Simplex and the modified Marquardt methods. The program provided pseudo-Hill co-efficient (nH), the Kd values and the percent ratios of the binding sites. Each inhibition plot was routinely fit to both single- and multiple-sites models. In order to determine whether the goodness of the fit for a model with additional parameters was significantly better, F-test was used as described by Munson and Rodbard (1980). Significance of differences between control and test values was determined by Student's t-test.

Protein determination

Protein content was assayed by the method of Lowry et al. (1951) with bovine serum albumin as the standard.

Materials

[3H]Substance P (40-60 Ci/mmol) and myo-(2-3H)Inositol were purchased from Amersham. Unlabeled SP was obtained from Peptide Institute, Inc. (Osaka, Japan). Bovine serum albumin, bacitracin, leupeptin, chymostatin, and Tris(hydroxymethyl)-aminomethane were purchased from Sigma (St. Louis, MO). Guanosine 5'-triphosphate disodium salt was obtained from Boehringer Mannheim. Polyethyleneimine was from Sigma or Nakarai Chemicals (Kyoto, Japan). Islet-activating protein was from Funakoshi Chemicals (Tokyo, Japan). Pentobarbital sodium salt from Pitman-Moore. Captopril was a gift from Sankyo Pharmaceutical Co Ltd. (Tokyo, Japan) and all other reagents were purchased from Nakarai Chemicals (Kyoto, Japan). TSK gel DEAE-5PW (7.5 mm x 750 mm) was from Tosoh (Tokyo, Japan).

RESULTS

$MgCl_2$ produced a concentration dependent increase in the [3H] SP specific binding to both the membrane and the CHAPS-solubilized fractions. At 10 mM $MgCl_2$ (the highest concentration tested), the increase in [3H]SP binding was 163.3$\pm$5.1 % above $MgCl_2$ free control (19.0$\pm$3.5 fmol/mg protein) in the membrane fraction and 378$\pm$15.2 % above $MgCl_2$ free control (6.23$\pm$0.26 fmol/mg protein) in the CHAPS-solubilized fraction. NaCl and GTP decreased [3H]SP specific binding in a concentration-dependent manner. At 100 mM NaCl, the decrease in [³H]SP binding was 37.3±4.0 % of control in the membrane fraction (in the presence of 10 mM MgCl₂, 30.8±3.5 fmol/mg protein). In the case of the solubilized fraction, at 50 mM NaCl, the decrease was 70.4±3.6 % of control (in the presence of 10 mM MgCl₂, 23.5±3.6 fmol/mg protein). At 100 µM GTP in the membrane fraction, the decrease in [³H]SP binding was 26.2±0.58 % of control. On the other hand, at 100 µM GTP in CHAPS-solubilized fraction, the decrease in [³H]SP binding was 80.0±6.1 % of control. These results indicated that the effects of MgCl₂, NaCl and GTP on [³H]SP binding were greater in the CHAPS-solubilized fraction than in the membrane fraction.

In IAP-pretreated membranes (prepared from the cerebral cortex of rats 10 days after i.c.v. injection of 5 µg IAP), [³H]SP binding (25.0±2.1 fmol/mg protein in the presence of 10 mM MgCl₂) was reduced in the absence of GTP compared with that of control membrane fraction (30.8±3.5 fmol/mg protein). Further significant reduction (P < 0.05) was observed in the presence of 100 µM GTP (19.6 ±1.1 fmol/mg protein binding).

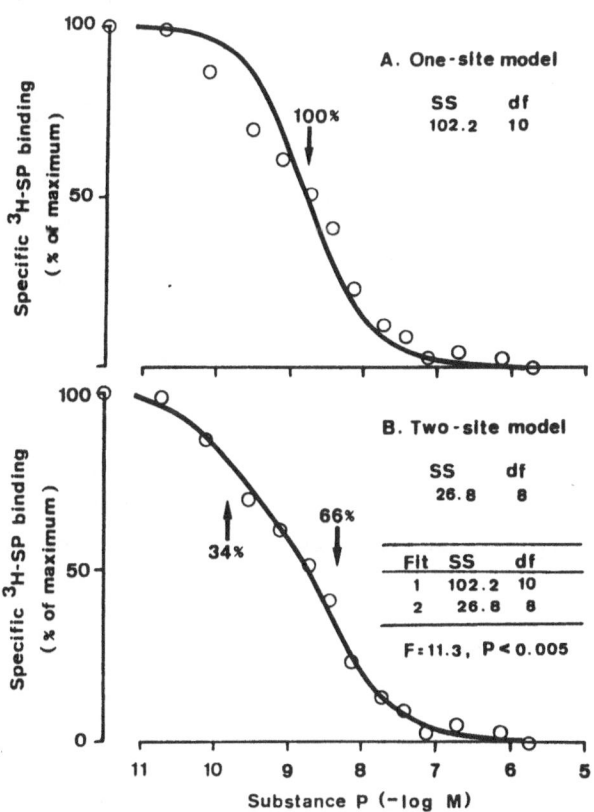

Fig. 1. Analysis of data by a nonlinear least-squares computer fitting program. A: The data fit to a one site model. B: The data fit to a two-site model. Each point represents the mean of duplicate determinations from a typical experiment. The two-site model shows a significantly better goodness of fit (p < 0.005). Arrows denote the Kd values, and percent ratios of sites present are indicated. SS, df, and F represent sum of squares, degrees of freedom, and F value, respectively.

Fig. 1 shows a typical inhibition curve of [3H]SP binding by unlabeled SP in the presence of 10 mM $MgCl_2$ (control). The data was fitted to a one-site model (A) and a two-site model (B). Computer analysis of the data indicated that nH was less than unity (0.69) and that the curve was best fitted to a two site-model; a high affinity site (H, Kd=0.14 nM, 34 %) and a low affinity site (L, Kd=4.47 nM, 66 %) in the membrane fraction. In contrast to the membrane fraction, the inhibition curve under control conditions (in the presence of 10 mM $MgCl_2$) in the CHAPS-solubilized fraction was steeper than in the case of the membrane fraction and best fitted to a one-site model with nH close to unity (0.86 $\pm$ 0.05). The Kd value (11.1 nM) was relatively close to that of the low affinity state in the membrane fraction.

Omission of $MgCl_2$ from assay buffer or addition of 100 mM NaCl in the presence of 10 mM $MgCl_2$ induced a decrease in specific binding, and probably affected the affinity of SP binding. In the absence of $MgCl_2$, the nH was less than unity (0.44 $\pm$ 0.26) and the curve was best fitted to a two site model. But the H site disappeared and the L site (Kd=2.15 nM, 70%) and a new super-low site (SL; 1174 nM, 30%) were observed. In the presence of 100 mM NaCl and 10 mM $MgCl_2$, the nH (0.44 $\pm$0.26), Kd values and percent ratios of the L and SL sites (1.03 $\pm$0.08 nM, 62 $\pm$10 %, 424 $\pm$146 nM, 38 $\pm$10 %) were similar to those observed in the absence of $MgCl_2$.

In the presence of 100 μM GTP and 10 mM $MgCl_2$, the competition curve and the binding parameters estimated from computer analysis (the nH 0.69 $\pm$0.08, 1.51 $\pm$0.06 nM (73 %), 355 $\pm$39 nN (27 %)) were similar to those in the absence of $MgCl_2$ or in the presence of 100 mM NaCl and 10 mM $MgCl_2$. In IAP treated membranes, the specific binding in the presence of 10 mM $MgCl_2$ was approximately 80 % of that found with nontreated membranes. The nH, Kd values and percent ratios of the L and SL sites were similar to those in the presence of 100 μM GTP in non-treated membranes. Further addition of 100 μM GTP in IAP-treated membranes induced a slight decrease in the specific binding, but the Kd values and percent ratios of the L and SL sites were little affected.

Table 1. Binding parameters estimated from inhibition curves of [3H]SP binding to the CHAPS-solubilized fractions from rat cerebral cortex by unlabeled SP

	nH	Kd (nM)
Control(10 mM $MgCl_2$)	0.86$\pm$0.05	11.1$\pm$1.5
1 mM $MgCl_2$	0.92$\pm$0.06	17.1$\pm$1.3
50 mM NaCl + 10 mM $MgCl_2$	0.90$\pm$0.02	18.7$\pm$3.5
10 μM GTP	1.03$\pm$0.08	15.8$\pm$2.0

In the case of the CHAPS-solubilized fraction, the reduction of $MgCl_2$ concentration to 1 mM, or the addition of 50 mM NaCl or 10 μM GTP, decreased the specific binding of [3H]SP. But the curves kept a fitting to the one site model and the Kd values were changed only slightly (Table 1).

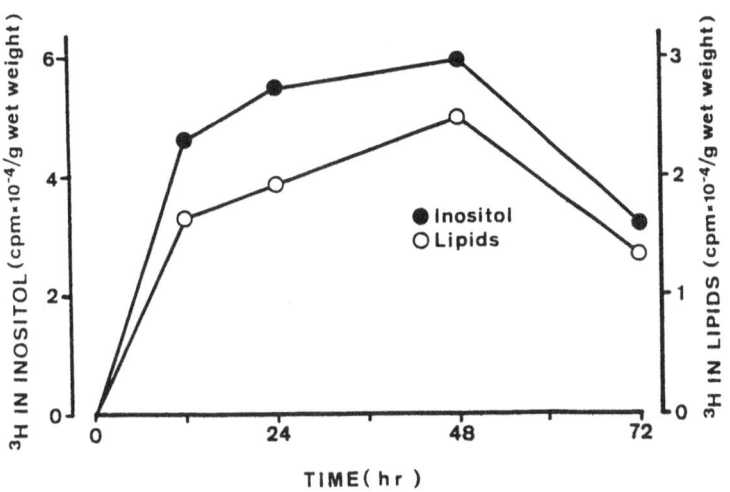

Fig. 2. Time course of [³H]inositol uptake and its incorporation into rat brain membranes lipids.

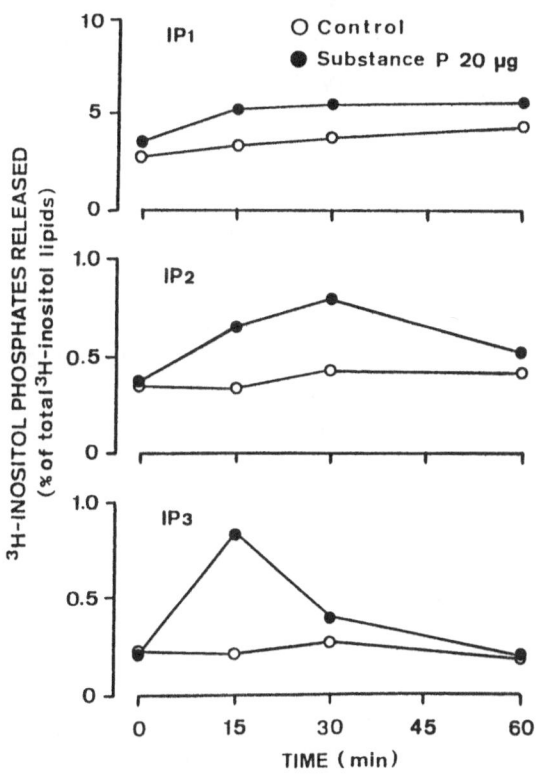

Fig. 3. Stimulation of [³H]inositol phosphates(IP$_S$) formation
by SP.

PtdIns cycle experiments employed our new routine technique, where rat brain tissue was labeled in vivo for 48 hours with ^{3}H-inositol. The time course of ^{3}H-inositol uptake and incorporation into the intracellular inositol and membrane lipid was essentially linear for the first 12 hours after which time maximal incorporation was reached and maintained up to 48 hours after ^{3}H-inositol i.c.v. injection (Fig. 2). Time courses of inositol phosphates formation after i.c.v. injection of SP were investigated. The radioactivity of the IP_3 fraction reached maximal at 15 min. after an injection and that of IP_2 fraction was maximal at 30 min. IP_1 formation was linear over the 60 min. After these experiments, we decided on a period of stimulating by SP of 15 min. (Fig. 3).

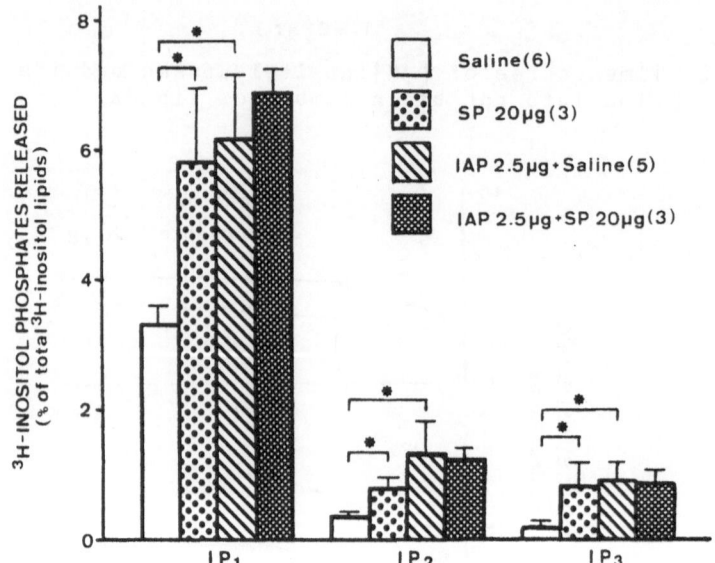

Fig. 4. Effect of IAP pretreatment on SP induced PtdIns cycle. Each values shows the mean $\pm$ S.E. of the number of rats indicated in parentheses. * P < 0.05 vs. normal saline.

The water soluble fraction obtained 15 min. after i.c.v. injection of SP was analyzed. SP elicits a rapid significant increase in the amount of radioactivity present in all three inositol phosphates (Fig. 4). The effect of SP on IP_1 was linear for at least 60 min. due to the presence of LiCl, which inhibits IP_s catabolism (Berridge et al., 1982).

However, there was no significant changes in terms of the three inositol phosphates between saline and SP in IAP treated rats. SP does not stimulate the PtdIns cycle any more in IAP treated rats (Fig. 4). It was not unexpected that the basal activity of the three inositol phosphates is significantly increased in IAP treated rats compared to that of normal rats (control), because guanine nucleotide regulatory proteins could be ADP ribosylated by an injected IAP. The locomotor activity of IAP treated rats in measuring with an ANIMEX activity meter after 4 days injection is significantly increased compared to that of sham (saline) injected rats. We also observed that the locomotor activity of the IAP treated rats increased about 133 % of sham rats during one hour measurement from 12:00 to 13:00.

DISCUSSION

The present results may suggest the existence of multiple "SP (NK-1)" receptors or multiple affinity states of SP receptors in rat cerebral cortex membranes. The possibility that SP binds to more than one site or state is tenable as curvilinear Scatchard plots have been observed in guinea pig small intestine (Buck et al., 1984), rat submaxillary gland (Bahouth et al., 1985) and rat spinal cord (Charlton and Helke, 1985). The Kd values of the H and L sites (states) reported here are in good agreement with those reported by Charlton and Helke (1985), and that of the SL site presently found is very close to that of a low affinity site in low ionic strength media reported by Bahouth and Musacchio (1985). In our laboratory, the existence of high and low states of SP receptors in rat cerebral cortical membranes was also demonstrated by Scatchard analysis using [^{3}H] SP (data not shown). Three distinct affinity states for one agonist were also observed for muscarinic acetylcholine receptors (Uchda et al., 1984; McMahon et al., 1985).

In contrast, Cascieri and Liang (1983) have shown that only one single site for [^{125}I] Bolton Hunter SP was present in rat cerebral cortical membranes. This discrepancy between our and their results might be due to the use of different radioligands. Furthermore, in contrast to the present conditions, Cascieri and Liang (1983) used a centrifugation method to separate bound and free radioligand, and store the membranes at - 70 °C. Our preliminary study indicated that freezing and thawing of tissues affected SP binding (data not shown), so it is also possible that the discrepancy noted above reflects technical differences. But we cannot rule out the possibility that SP interacts with other binding sites, such as NK-2 or NK-3 receptors.

In this study, we demonstrated also in the CHAPS-solubilized fraction that $MgCl_2$ increased whereas NaCl and GTP decreased [^{3}H] SP binding activity. Since similar effects were observed in the membrane fraction, these results suggest that CHAPS could solubilize SP receptors while leaving intact regulatory mechanisms, presumably the GTP-binding protein coupling activity to SP receptors (binding sites). The computer analyses of the inhibition data indicate that a single affinity state for SP binding exists in the CHAPS-solubilized fraction, and that cations and GTP cause changes in the specific binding not in the affinity. The effects of these cations and GTP were greater in the CHAPS-solubilized fraction than in the membrane fraction. There could be two possibilities for these phenomena. One possibility is that these discrepancies may be due to the differences in the incubation temperature and time. Another possibility may be the loss of some important factor(s), probably other than GTP-binding protein, upon membrane solubilization or concentration with Amicon Centriflo CF 25. We suppose that unknown factor(s) may regulate SP receptor function. The existence of some factor(s) has also been speculated in serotonin$_{1A}$ and serotonin$_{1B}$ receptors (Asarch et al., 1987) and in muscarinic acetylcholine receptors (Wang et al., 1985).

Since the effects of Mg^{2+}, Na^+ and guanosine triphosphate on the affinity of SP receptors are similar to several receptors which mediate inhibition of adenylate cyclase activity, such as α_2 adrenergic receptors (Kitamura et al., 1986), and D_2 dopamine receptors (Grigoriadis et al., 1985), it is speculated that SP receptors may couple with inhibitory GTP-binding protein(s). Then, we investigated the effect of IAP pretreatment on SP binding in the membrane fraction. In IAP-treated membranes, [^{3}H]SP binding was decreased and the effect of GTP was less pronounced than in control membranes. Similar results were obtained when P_2 fraction from bovine brainstem was treated with IAP in vitro (Tanaka et al., 1986). The advantage of in vivo treatment is that much lower concentrations of IAP are required than for in vitro treatment (Katada et al., 1982). In competition experiments, the effect of IAP pretreatment was similar to that of GTP in non-treated membranes suggesting that GTP-binding proteins are ADP-ribosylated by IAP and uncoupled from SP receptors. However, even in the IAP-treated membranes, 100 μM GTP caused a significant reduction of [^{3}H]SP binding. This may be due to incomplete ADP-ribosylation of GTP-binding proteins. Although we did not determine how the GTP-binding proteins were ADP-ribosylated by IAP under our conditions, we confirmed the behavioural excitement and the increase in locomotor activity. Another possibility is the participation of other GTP-binding proteins in addition to the IAP-substrate.

The present study has confirmed the results of Watson and Downes, (1983) and Hunter et al. (1985) that SP induces a hydrolysis of inositol lipids in rat brain. Our method used in this study has therefore enabled us to detect a large and reproducible breakdown of ligand stimulating inositol phospholipid hydrolysis in rat brain. This method is also useful in measuring the PtdIns cycle in vivo.

An injection (i.c.v.) of IAP into rats increased the basal levels of accumulation of inositol phosphates due to the ADP-ribosylation of guanine nucleotide regulatory protein, and blocked any accumulation of inositol phosphates induced by SP, as seen in control rats. If SP coupled to an another guanine nucleotide inhibitory binding protein that is not an IAP-substrate, these inositol phosphates induced by SP injection should be increased even in IAP treated rats. However it is not ruled out that more than one GTP-binding protein may be capable of interacting with SP receptors.

The "high-affinity" was no longer observed in IAP-pretreated membranes. Taken together, these data strongly suggested that an inhibitory guanine nucleotide regulatory protein(s) couples to SP "high-affinity" state (site) in brain to regulate the binding, and that the PtdIns cycle stimulated by SP could be linked to GTP binding protein(s).

ACKNOWLEDGMENTS

This work was supported by Grant-in-Aid for Special Project Research 60126005, Grant-in-Aid for Scientific Research on Priority Areas 63641005 from the Ministry of Education, Science and Culture, Japan and Grant from Uehara Memorial Foundation.

The skillful assistance of Mr. Y. Morishima is gratefully acknowledged. We also are grateful to Drs. Y. Kitamura (Hokkaidou University) and A. Seo (Hiroshima University) for computer programming and Prof. T. Masujima of Hiroshima University for his technical advise.

REFERENCES

Arima, T., Segawa, T., and Nomura, Y., 1986, Influence of pertussis toxin on the effects of guanine nucleotide on adenylate cyclase in rat striatal membranes, Life Sci., 39: 2429-2434.

Asarch, K. B., and Shih, J. C., 1987, Solubilization of serotonin$_{1a}$ and serotonin$_{1b}$ binding sites from bovine brain, J. Neurochem., 48: 1494-1501.

Bahouth, S. W., and J. M. Musacchio, 1985, Specific binding of [^{3}H]substance P to the rat submaxillary gland. The effects of ions and guanine nucleotides, J. Pharmacol. Exp. Ther., 234: 326-336.

Berridge, M. J., Downers, P. C., and Hanley, M. R., 1982, Lithium amplifies agonist-dependent phosphatidylinositol responses in brain and salivary grands, Biochem. J., 206: 587-595.

Berridge, M. J., and Irvine, R. F., 1984, Inositol triphosphate, a novel second messenger in cellular signal transduction, Nature, 312: 315-321.

Bruns, R. F., Lawson-Wendling, K., and Pugsley, T. A., 1983, A rapid filtration assay for soluble receptors using polyethylenimine treated filters. Anal. Biochem. 132: 74-82.

Buck, S. H. and Burcher, E., 1986, The tachykinins: a family of peptides with a brood of 'receptors', Trends Pharmacol. Sci. 7: 65-68.

Buck, S. H., Maurin, Y., Burks, T. F., and Yamamura, H. I., 1984, High-affinity ^{3}H-substance P binding to longitudinal muscle membranes of the guinea pig small intestine, Life Sci., 34: 497-507.

Cascieri, M. A., and Liang, T., 1983, Characterization of the substance P receptor in rat brain cortex membranes and the inhibition of radioligand binding by guanine nucleotides, J. Biol. Chem., 258: 5158-5164.

Cascieri, M. A., Chicchi, G. G., and Liang, T., 1985, Demonstration of two distinct tachykinin receptors in rat brain cortex, J. Biol. Chem., 260: 1501-1507.

Charlton, C. G., and Helke, C. J., 1985, Characterization and segmental distribution of ^{125}I-Bolton-Hunter-labeled substance P binding sites in rat spinal cord, J. Neurosci., 5: 1293-1299.

Grigoriadis, D., and Seeman, P., 1985, Complete conversion of brain D_2 dopamine receptors from the high- to the low affinity states for dopamine agonists, using sodium ions and guanine nucleotide, J. Neurochem., 44: 1925-1935.

Hunter, J. C., Goedert, M., and Pinnock, R. D., 1985, Mammalian tachykinin-induced hydrolysis of inositol phospholipids in rat brain slices, Biochem. Biophy. Res. Comm., 127: 616-622.

Kangawa, K., Minamino, N., Fukuda A., and Matsuo, H., 1983, Neuromedin K: a novel mammalian tachykinin identified in porcine spinal cord, Biochem. Biophys. Res. Commun., 114: 533-540.

Katada, T., and Ui, M., 1982, Direct modification of the membrane adenylate cyclase system by islet-activating protein due to ADP-ribosylation of a membrane protein, Proc. Natl. Acad. Sci. USA., 79: 3129-3133.

Kimura, S., Okada, M., Sugita, Y., Kanazawa, I., and Munekata, E., 1983, Novel neuropeptides: neurokinin A and B isolated from porcine spinal cord, Proc. Jpn. Acad. Ser. B Phys. Biol Sci., 59: 101-104.

Kitamura, Y., Tanaka, H., and Nomura, Y., 1986, [^{3}H]Clonidine and [^{3}H]yohimbine binding to solubilized $_2$-adrenoceptors from rat cerebral cortex, Eur. J. Pharmacol., 123: 263-270.

Lee, C.-M., Javitch, J. A., and Snyder, S. H., 1983, ^{3}H-substance P binding to salivary gland membranes. Regulation by guanyl nucleotides and divalent cations, Mol. Pharmacol., 23: 563-569.

Lowry, O. H., Rosebrough, N. J., Farr, A. L., and Randall, R. J., 1951, Protein measurement with the Folin phenol reagent, J. Biol. Chem., 193: 265-267.

Maruyama, M., 1986, Selective solubilization of physalaemin-type substance P binding sites from rat brain membranes by glycodeoxycholate and NaCl, Brain Res., 370: 186-190.

McMahon, K. K., and Hosey, M. M., 1985, Agonist interactions with cardiac muscarinic receptors. Effects of Mg^+, guanine nucleotides, and monovalent cations, Mol. Pharmacol., 28: 400-409.

Munson, P. J., and Rodbard, D., 1980, LIGAND: a versatile computerized approach for characterization of ligand-binding systems, Anal. Biochem., 107: 220-239.

Nawa, H., Hirose, T., Takashima, H., Inayama, S., and Nakanishi, S., 1983, Nucleotide sequences of cloned cDNAs for two types of substance P precursor, Nature, 306: 32-36.

Nakata, Y., Tanaka, H., Morishima, Y., and Segawa, T., 1988, Solubilization and characterization of substance P binding protein from bovine brainstem, J. Neurochem., 50: 522-527.

Nomura, Y., Kawata, K., Kitamura, Y., and Watanabe, H., 1987, Effects of pertussis toxin on the α_2 adrenoceptor-inhibitory GTP- binding protein-adenylate cyclase system in rat brain: pharmacological and neurochemical studies, Eur. J. Pharmacol., 134: 123-129.

Okajima, F., and M, Ui., 1984, ADP-ribosylation of the specific membrane protein, pertussis toxin, associated with inhibitory of a chemotactic peptide induced arachidonate release in neutrophils. J. Biol. Chem., 259: 13868-13781.

Payan, D. G., McGillis, J. P., and Organist, M. L., 1986, Binding characteristics and affinity labeling of protein constituents of the human IM-9 lymphoblast receptor for substance P, J. Biol. Chem., 261: 14321-14329.

Quirion, R., 1985, Multiple tachykinin receptors, Trends Neurosci., 8: 183-185.

Smith, C.D., Lane, B. C., Kusaka, I., Verghese, M. V., and Snyderman, R., 1985, Chemoattractant receptor induced hydrolysis of phosphatidylinositol 4, 5 bisphosphate in human polymorphonuclear leukocyte, J. Biol. Chem., 260: 5875-5878.

Tanaka, H., Nakata, Y., and Segawa, T., 1986, Substance P receptors in bovine brain membranes are coupled to GTP inhibitory binding protein, Eur. J. Pharmacol., 125: 157-158.

Uchida, S., Matsumota, K., Mizushima, A., Osugi, T., Higuchi H., and Yoshida, H., 1984, Effects of guanine nucleotide and sulfhydryl reagent on subpopulations of muscarinic acetylcholine receptors in mammalian hearts: possible evidence for interconversion of super-high and low affinity agonist binding sites, Eur. J. Pharmacol., 100: 291-298.

Wang, J. -X., Mei, L., Yamamura, H.I., and Roeske, W. R., 1987, Solubilization with digitonin alters the kinetics of pirenzepine binding to muscarinic receptors from rat forebrain and heart, J. Pharmacol. Exp. Ther., 242: 981-900.

Watson, S. P., and Downes, C. P., 1983, Substance P induced hydrolysis of inositol phosholipids in guinea-pig ileum and rat hypothalamus, Eur. J. Pharmacol., 93: 245-253.

EVIDENCE FOR CARDIOVASCULAR ROLES OF TACHYKININ PEPTIDES IN THE BRAIN OF THE RAT

Hiro-o Kamiya, Yukio Takano and Akira Nagashima

Department of Pharmacology, Faculty of Pharmaceutical
Sciences, Fukuoka University, Fukuoka 814-01, Japan

INTRODUCTION

In addition to substance P (SP), 4 other tachykinin peptides have so far been isolated from the mammalian nervous system, namely neurokinin A (NKA), neuropeptide K (NPK), neuropeptide γ (NPγ) and neurokinin B (NKB) (Maggio, 1988; Krause et al., 1989). These tachykinin peptides have a variety of pharmacological effects, indicating smooth muscle contraction, salivation, endothelium-dependent vasodilator actions, and depolarization of central neurons (Erspamer, 1981; Pernow, 1983). These peptides are extensively distributed and play important physiological roles in the mammalian central nervous system (CNS). Several lines of evidence have proposed the existence of multiple tachykinin receptor subtypes in peripheral tissues and the CNS, called NK-1, NK-2 and NK-3 (TABLE I). SP, NKA and NKB are considered to be endogenous ligands of the NK-1, NK-2 and NK-3 subtypes, respectively (Henry, 1987).

On the basis of cardiovascular and neurochemical evidence, SP has become of interest as a neuropeptides regulating the blood pressure in the brain and spinal cord (Helke et al., 1980; Unger et al., 1981; Loewy and Sawyer, 1982; Takano et al., 1984; Takano et al., 1985; Takano et al., 1988; Nagashima et al., 1989a, Nagashima et al., 1989b).

Here we report studies on the distribution of tachykinin peptides (SP, NKA and NKB) in the CNS in spontaneously hypertensive rats (SHRs) and normotensive Wistar Kyoto rats (WKYs), and adults Wistar rats, and tests on the cardiovascular effects of these peptides injected into the lateral brain ventricle (i.c.v.) and into the nucleus tractus solitarii (NTS).

I. DISTRIBUTION OF TACHYKININ PEPTIDES IN THE CNS

Extraction and radioimmunoassay (RIA) were described previously (Takano et al., 1986; Nagashima et al., 1989a, Nagashima et al., 1989b). The anti-SP antiserum (FP03) used showed very low cross-reactivity with NKA (0.19%). High specific anti-NKA and anti-NKB antisera were obtained by a novel immunization procedure described previously (Nagashima et al., 1987; Tateishi et al., 1989). The anti-NKA antiserum (FK05) did not cross-reacted with SP (<0.01%) and cross-reacted only slightly with NKB (5.4%). The anti-NKB antiserum (R707) was highly specific for NKB and its cross-reactivity with NKA, neuropeptide K, kassinin and other tachykinins were all less than 0.001%.

TABLE I. Classification of mammalian tachykinin receptor

Subtype	Model Tissue	Biological Potency	Selective Agonist
NK-1	Porcine Coronary Artery	$SP \gg NP\gamma \geqq NKA = NPK > NKB$	SP-methy Ester
	Guinea Pig Aorta	or	Dimeric-SP_{3-11}[**]
	Dog Carotid Artery	$NP\gamma \geqq SP \gg NKA > NPK > NKB$	$[Sar^9, Met(O_2)^{11}]SP$
	Rat Salivary Gland		$[Cys^{3,6}, Tyr^8, Pro^9]SP$
	Rat Spinal Cord (IML)[*]		$[Cys^{3,6}, Tyr^8, Pro^{10}]SP$
	CNS		$Ac-[Arg^6, Sar^9, Met(O_2)^{11}]SP_{6-1}$
NK-2	Rat Vas Deferens	$NPK = NP\gamma > NKA > SP > NKB$	$[Nle^{10}]NKA_{4-10}$
	Rabbit Pulmonary Artery		
NK-3	Rat Portal Vein	$NKB > NKA > SP$	senktide
	Rat Cerebral Cortex		$[Pro^7]NKB$
			$[MePhe^7]NKB$

[*]IML; intermediolateral cell column, [**]Higuchi et al., 1989.

1) Comparison between SP and NKA contents in male Wistar rats

The level of NKA-like immunoreactivity (NKA-LI) was high in the sybstantia nigra, NTS and nucleus habenula, moderate in the lateral hypothalamus (LH), interpedunculus nucleus and dorsal horn of spinal cord, low in the nucleus accumbens, striatum, ventral tegmental area (VTA), raphe and intermediolateral cell column (IML) of the spinal cord. Thus the distribution of NKA-LI in rat brain was similar to that of SP-like immunoreactivity (SP-LI) (Table II).

Although the ventral medulla oblongata, which is thought to play a critical role in maintaining vasomotor tone (Nagashima et al., 1989b), contained a comparatively high level of SP-LI, NKA-LI was not detected in this area. The SP-LI/NKA-LI ratio was in the range of 1.15-3.94 in the brain, and was high in the interpedunclus nucleus (SP-LI/NKA-LI; 3.35) and raphe (SP-LI/NKA-LI; 3.94).

2) Comparison between SP and NKB contents in WKYs and SHRs

The systolic blood pressure of WKYs and SHRs of 12 weeks were 120 ± 3 and 184 ± 3 mmHg (p<0.001), respectively. The level of NKB-like immunoreactivity (NKB-LI) was high in the hypothalamus, amygdala, NTS and spinal cord in both WKYs and SHRs (Table III). Thus the distribution of NKB-LI in the CNS was not proportional to that of SP-LI, although the levels of a tachykinin peptide, NKA were similar to that of SP-LI (Table II) (Takano et al., 1986). For instance, the SP-LI/NKB-LI ratio in various regions of the CNS varied over the wide range of 1.4-54.2 in WKYs and of 1.8-64.7 in SHRs (Table III). The SP-LI/NKA-LI ratio, on the other hand, varied only between 1.2 and 3.9 (Table II).

Table III compares the SP-LI and NKB-LI contents in the brain and spinal cord of WKYs and SHRs. The SP-LI contents in 12-week-old SHRs were higher than those of the age-matched WKYs in the striatum, substantia nigra, VTA, locus ceruleus, nucleus ambiguus and the caudal part of the NTS. On the other hand, SHRs showed significantly higher NKB-LI contents than age-matched WKYs in the striatum, globus pallidus,

TABLE II.   Distributions of SP-LI and NKA-LI in various regions of rat brain

Values in pg/μg protein are means ± S.E.M. for number of rats shown in parentheses. n.d., not detectable.

	Substance P	*Neurokinin* A	*SP/NKA*
Frontal cortex	0.52 ± 0.07 (9)	n.d.	
Nucleus accumbens	1.97 ± 0.06 (12)	0.95 ± 0.15 (8)	2.07
Striatum			
Central	1.48 ± 0.08 (11)	1.15 ± 0.35 (7)	1.29
Dorsolateral	1.00 ± 0.04 (12)	0.49 ± 0.10 (5)	2.04
Globus pallidus	1.84 ± 0.15 (10)	0.93 ± 0.17 (8)	1.98
Hypothalamus			
LH	3.23 ± 0.23 (9)	2.31 ± 0.44 (11)	1.40
VMH	2.52 ± 0.14 (9)	1.35 ± 0.20 (10)	1.87
Nucleus habenula	6.51 ± 0.66 (11)	3.24 ± 1.10 (5)	2.01
Hippocampus	0.17 ± 0.02 (3)	n.d.	
Substantia nigra (SN)	9.64 ± 0.43 (9)	5.77 ± 0.71 (12)	1.67
Ventral tegmental area (VTA)	1.85 ± 0.29 (9)	0.93 ± 0.20 (8)	1.99
Interpedunculus nucleus (IP)	7.61 ± 0.76 (7)	2.27 ± 0.32 (6)	3.35
Raphe	4.41 ± 0.54 (4)	1.12 ± 0.22 (5)	3.94
Nucleus tractus solitarius (NTS)	7.36 ± 0.80 (6)	5.25 ± 0.74 (4)	1.40
Ventral medulla oblongata	3.63 ± 0.57 (4)	n.d.	
Spinal cord			
Dorsal horn	3.12 ± 0.35 (11)	2.17 ± 0.28 (7)	1.44
IML	1.95 ± 0.26 (5)	1.70 ± 0.13 (5)	1.15
Ventral horn	1.27 ± 0.16 (6)	n.d.	

supraoptic nucleus of the hypothalamus and caudal part of the NTS (Table III). This evidence is interesting because these areas of brain have been proposed to play important roles in blood pressure control.

II.CARDIOVASCULAR RESPONSES INDUCED BY TACHYKININ PEPTIDES

1) Cardiovascular effects of tachykinin peptides in the NTS

Experiments on cardiovascular responses were described previously (Nagashima et al., 1989 a, b). Rats were anesthetized with urethane (1g/kg i.p., with supplementary intravenous injections as required). Then one catheter was inserted into the right femoral artery for recording the blood pressure and heart rate, and another into the left femoral vein for adminstration of drugs. The body temperature was maintained at 37-38 ℃ with a heating pad. The systolic blood pressure was maintained above 90 mmHg by infusion of 4% Ficoll 70 (Nakarai Chemicals, Japan) in saline solution.

For experiments of cardiovascular effects of tachykinin peptides injected into the NTS, each rat was placed in a stereotaxic apparatus with the head fixed to 15 ℃. The dorsal surface of the lower brainstem was exposed by limited occipital craniotomy. The caudal tip of the area postorema in the midline was used as the rostrocaudal zero. A glass pipette (outer diameter <50μm) and a double-barrel glass pipette (outer diameter <70μm) were inserted into the NTS (0.5-1.0mm rostral and

TABLE III   *SP-LI and NKB-LI contents of the brain and spinal cord of WKYs and SHRs*

Values in pg/μg protein are means ± S.E.M. for the numbers of rats shown in parentheses. n.d., not detectable.

Regions	WKY		SP/NKB in WKY	SHR		SP/NKB in SHR
	Substance P	Neurokinin B		Substance P	Neurokinin B	
Cortex						
prefrontal	0.26 ± 0.02 (8)	0.078 ± 0.009 (6)	3.3	0.36 ± 0.05 (8)	0.072 ± 0.006 (4)	5.0
cingulate	0.12 ± 0.02 (7)	0.069 ± 0.009 (6)	1.7	0.14 ± 0.02 (8)	0.057 ± 0.002 (6)	2.5
piriform	0.12 ± 0.02 (8)	0.088 ± 0.005 (6)	1.4	0.19 ± 0.04 (4)	0.105 ± 0.022 (4)	1.8
parietal	n.d.	0.051 ± 0.006 (5)		n.d.	0.067 ± 0.011 (6)	
Septum	1.36 ± 0.19 (8)	0.071 ± 0.005 (6)	19.2	1.07 ± 0.06 (7)	0.075 ± 0.010 (6)	14.3
Striatum						
central	0.99 ± 0.05 (8)	0.089 ± 0.009 (6)	11.1	1.21 ± 0.13 (8)	0.100 ± 0.007 (6)	12.1
dorsolateral	0.76 ± 0.05 (8)	0.061 ± 0.007 (6)	12.5	1.16 ± 0.08 (8)***	0.097 ± 0.007 (6)*	12.0
Globus pallidus	1.01 ± 0.10 (8)	0.120 ± 0.010 (6)	8.4	1.31 ± 0.16 (8)	0.216 ± 0.015 (6)***	6.1
Hypothalamus						
paraventricular nucleus	4.85 ± 0.39 (8)	0.378 ± 0.016 (6)	12.8	5.05 ± 0.62 (6)	0.430 ± 0.041 (6)	11.7
supraoptic nucleus	3.08 ± 0.20 (8)	0.248 ± 0.022 (6)	12.4	3.59 ± 0.38 (8)	0.334 ± 0.031 (6)*	10.7
AH	4.79 ± 0.45 (8)	0.432 ± 0.050 (6)	11.1	5.37 ± 0.61 (8)	0.488 ± 0.065 (6)	11.0
LH	2.45 ± 0.30 (8)	0.453 ± 0.032 (5)	5.4	3.20 ± 0.20 (8)	0.434 ± 0.056 (6)	7.4
VMH	1.96 ± 0.16 (8)	0.293 ± 0.018 (6)	6.7	2.65 ± 0.33 (8)	0.302 ± 0.013 (6)	8.8
Amygdala	1.48 ± 0.19 (8)	0.253 ± 0.022 (6)	5.9	2.43 ± 0.47 (8)	0.303 ± 0.045 (6)	8.0
Hippocampus	n.d.	0.056 ± 0.005 (5)		n.d.	0.049 ± 0.003 (4)	
Substantia nigra	8.57 ± 0.57 (8)	0.158 ± 0.019 (6)	54.2	12.3 ± 1.05 (8)**	0.190 ± 0.014 (6)	64.7
Ventral tegmental area	1.30 ± 0.11 (8)	0.248 ± 0.031 (6)	5.2	2.23 ± 0.24 (8)**	0.293 ± 0.023 (6)	7.6
Locus ceruleus	0.63 ± 0.18 (6)	0.113 ± 0.013 (6)	5.6	1.34 ± 0.20 (8)*	0.151 ± 0.013 (4)	8.9
Nucleus tractus solitarii						
intermediate	7.96 ± 0.74 (8)	0.345 ± 0.033 (6)	23.1	8.55 ± 0.96 (8)	0.448 ± 0.050 (6)	19.1
caudal	5.27 ± 0.58 (8)	0.312 ± 0.032 (6)	16.9	8.34 ± 0.75 (8)**	0.452 ± 0.032 (6)*	18.5
Nucleus ambiguus	3.25 ± 0.27 (8)	0.183 ± 0.024 (5)	17.8	4.65 ± 0.58 (7)*	0.190 ± 0.022 (6)	24.5
Ventral medulla oblongata	3.27 ± 0.58 (8)	0.171 ± 0.025 (4)	19.1	2.46 ± 0.39 (8)	0.179 ± 0.015 (6)	13.7
Spinal cord						
dorsal horn	4.40 ± 0.76 (8)	0.510 ± 0.024 (6)	8.6	3.69 ± 0.52 (8)	0.471 ± 0.047 (6)	7.8
IML	1.77 ± 0.19 (8)	0.455 ± 0.018 (6)	3.9	2.26 ± 0.26 (8)	0.411 ± 0.038 (6)	5.5

*$P < 0.05$, **$P < 0.01$, ***$P < 0.001$ vs the corresponding value for WKYs.

0.5-0.8mm lateral to the bregma as the zero point and 0.5-1.0mm beneath the dorsal surface of the brainstem).

Peptides were dissolved in artificial cerebrospinal fluid (ACSF). Peptide solutions were injected manually in a volume of $0.1\mu l$ over 5 sec for injection into the NTS and in that of $10\mu l$ over a period of 1 min for i.c.v. injection, using a Hamilton microsyringe, respectively. Injection of ACSF (i.c.v.) did not produce any detectable changes in blood pressure and heart rate. After each experiment, malachite green was injected for confirmation of the injection site.

The primary afferent fibers of the grossopharyngeal and vagal nerves (the IXth and Xth cranial nerves) transmit baro- and chemoreceptor signals to the NTS (Crill and Reis, 1968; Lipski et al., 1975; Panenton and Loewy, 1980). In order to examine the effect of unilateral deafferentation of the IXth and Xth cranial nerves (nodose ganglionectomy) on the content of tachykinin peptides, the levels of tachykinin peptides in tissues from the NTS and the nucleus ambiguus were assayed. The levels of SP-LI and NKA-LI in the NTS 7 or 8 days after the right nodose ganglionectomy were decreased significantly by about 22% and 18%, respectively. No significant differences were found in their levels in the nucleus ambiguus after the denervation. The decrease in the NKA-LI content after removal of the nodose ganglion indicates the existence of NKA-LI pathways in vagal afferents, as well as SP-LI pathways. These results are

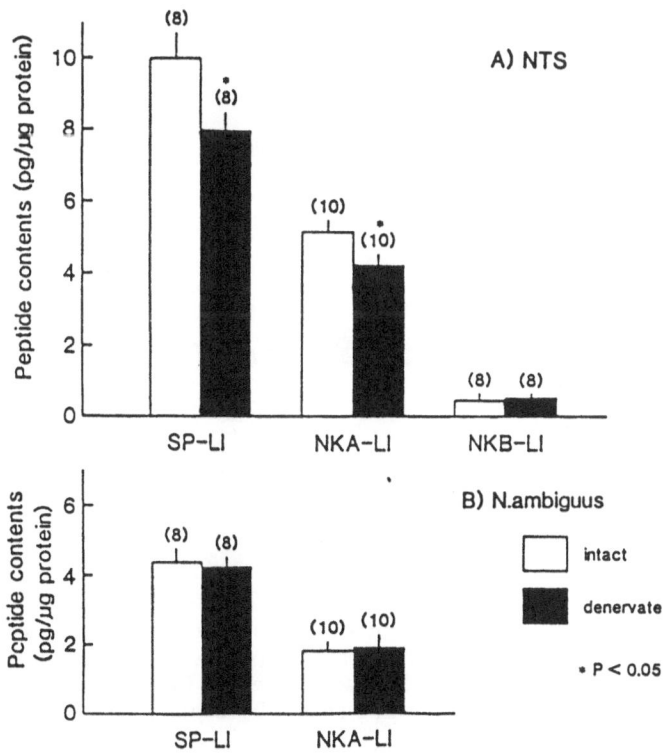

Fig.1. Effects of unilateral nodose ganglionectomy on the SP-LI, NKA-LI and NKB-LI contents in the NTS (A) and the Nucleus ambiguus. No NKB-LI was detectable in the Nucleus ambiguus. Values are means ± S.E.M., and asterisks indicate significant difference from corresponding values on the control side calculated by the two-tailed Student's t-test.

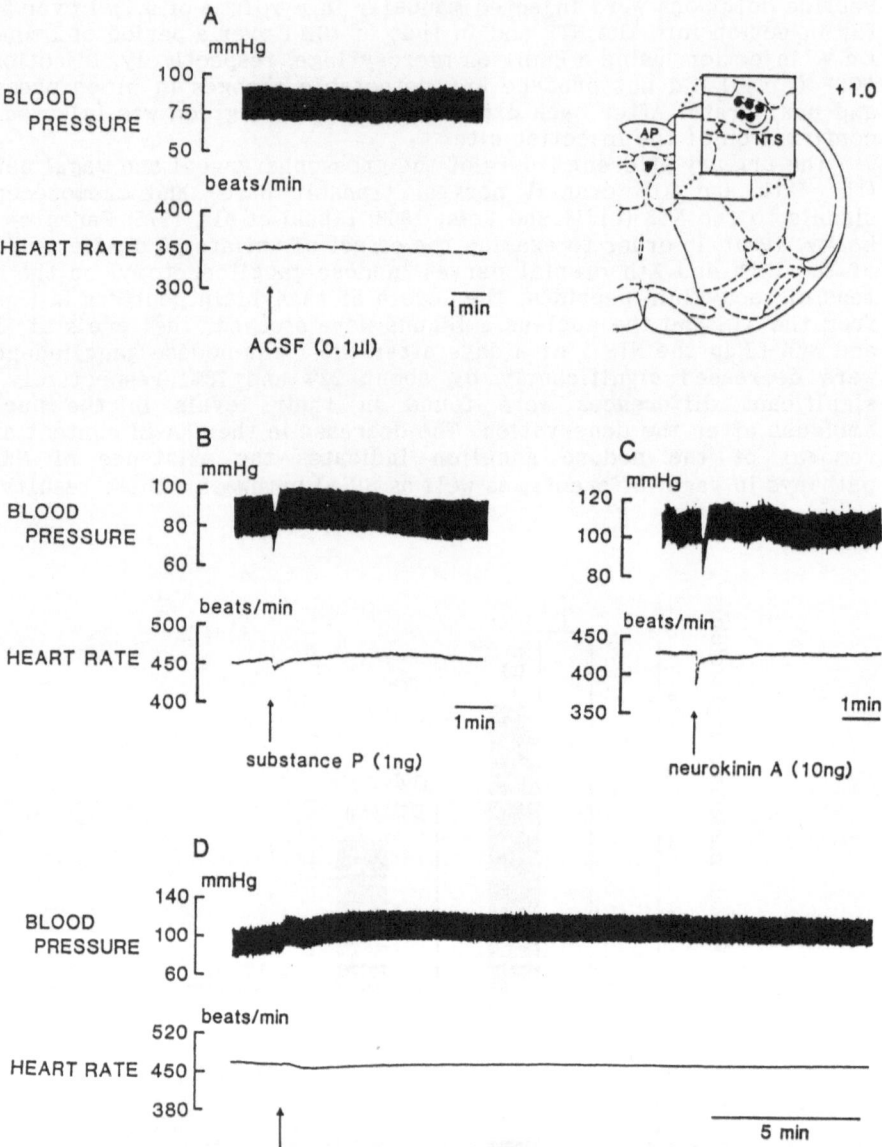

Fig. 2. Typical recordings of blood pressure and heart rate after microinjections of vehicle (ACSF, A), substance P (B), neurokinin A (C) and senktide (D) into the NTS. The sites of microinjections are shown by black circles.

similar to our previous finding that thermal lesions of the striatum of rats markedly decreased the contents of both SP-LI and NKA-LI in the substantia nigra, indicating the existence of both a striatonigral NKA-LI pathway and an SP-LI pathway (Nagashima et al., 1987). These findings suggest that SP and NKA may be co-transmitters that are involved in pathways for cardiovascular regulation in vagal afferents.

The content of NKB-LI was lower than those of SP-LI and NKA-LI in the NTS (Fig. 1A), and no NKB-LI was detectable in the nucleus ambiguus (Fig. 1B). Interestingly, the NKB-LI content of NTS was not affected by unilateral nodose ganglionectomy. These results suggest that in the NTS the system for biosynthesis of NKB-LI is different from that for biosynthesis of SP-LI and NKA-LI.

Fig. 2 shows typical recordings of the blood pressure and heart rate after microinjections of vehicle and peptides into the NTS. The microinjections of 1ng of SP and 10ng of NKA into the caudolateral area of the NTS resulted in decreases in the blood pressure and heart rate during the first minute (Fig. 2B,C). These results indicated that NKA as well as SP may be involved in the baroreceptor reflex in the NTS. In contrast, senktide (suc-[Asp^6,Me-Phe^8]-SP$_{6-11}$), which is thought to be a selective NKB analogue (Wormer et al., 1986), caused long-lasting hypertension (Fig. 2D).

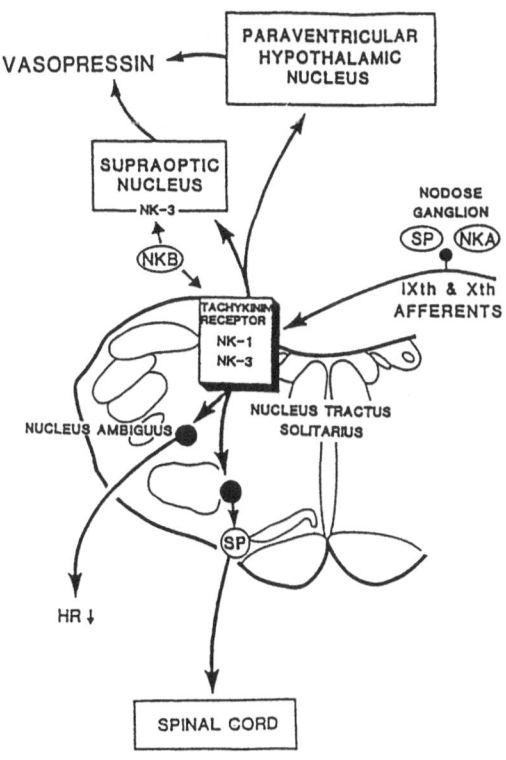

Fig. 3. Proposed model of the cardiovascular roles of tachykinin peptides.

Fig. 3 is a diagram summarizing our proposal on the cardiovascular roles of tachykinin peptides in the NTS. The present results provide physiological evidence supporting the idea that NKA as well as SP may be a neurotransmitter of the baroreceptor afferent nerve in the NTS, because the contents of SP-LI and NKA-LI in the NTS decrease after denervation of the baroreceptor afferents. On the other hand, NKB neurons in the NTS may be interneurons and NKB may stimulate the ascending pathway from the NTS to the hypothalamus.

2) Cardiovascular effects of tachykinin peptides (i.c.v.)

There are several reports that the central pressor response to SP seems to be mediated by the sympathetic nervous system, without participation of vasopressin in conscious rat (Unger et al., 1981; Unger et al., 1988). However, there are few reports of studies on the central roles of other tachykinin peptides, probably partly because of the poor solubility of NKB in water and the absence of selective ligands for the tachykinin receptor subtype. Recently, selective NKB receptor ligand, senktide have been developed (Wormser et al., 1986). Therefore, we tested on the cardiovascular effects of NKB analogue, senktide, injected into the lateral brain ventricle (i.c.v.).

Fig.4 shows typical recordings of the arterial blood pressure and heart rate after injection of senktide (i.c.v.). Injection of 10μg of senktide increased the blood pressure about 20 mmHg and the heart rate with a latency of 60 sec and a peak after about 5 min, and these effects persisted for 40 min at least. This central pressure response to senktide was dose-dependent. The injections (i.c.v.) of [Pro[7]]-NKB, another NK-3 agonist, SP and NKA also caused increase in blood pressure and heart rate.

For determination of whether vasopressin (AVP) receptors in the peripheral vascular system are involved in the central pressor responses observed after injections (i.c.v.) of tachykinin peptides, the AVP antagonist, d(CH₂)₅OMe(Tyr)AVP, was administered 10 min before injections of tachykinin peptides. Intravenous administration of the AVP antagonist (10μg/kg) did not cause any detectable change in either the blood pressure or the heart rate (Fig. 5A). Pretreatment with the AVP antagonist inhibited the sustained hypertension induced by senktide (10μg, i.c.v.) and [Pro[7]]-NKB (50μg, i.c.v.) (Fig. 5C,E), did not influence the pressor response induced by SP(Fig.5B) and NKA. The AVP antagonist had no significant effect on the increases in heart rate induced by SP and senktide (Fig. 5B,C).

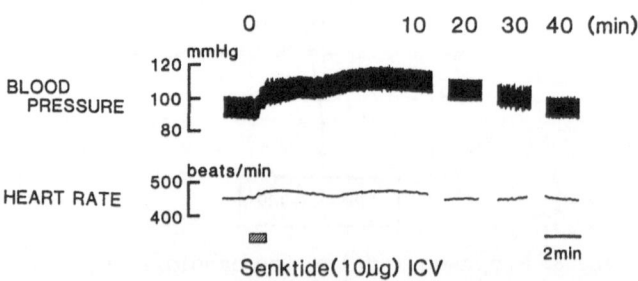

Fig. 4. Typical recordings of changes in blood pressure and heart rate after injection of the NKB analogue, senktide, into the lateral brain ventricle (i.c.v.) of a normotensive rat.

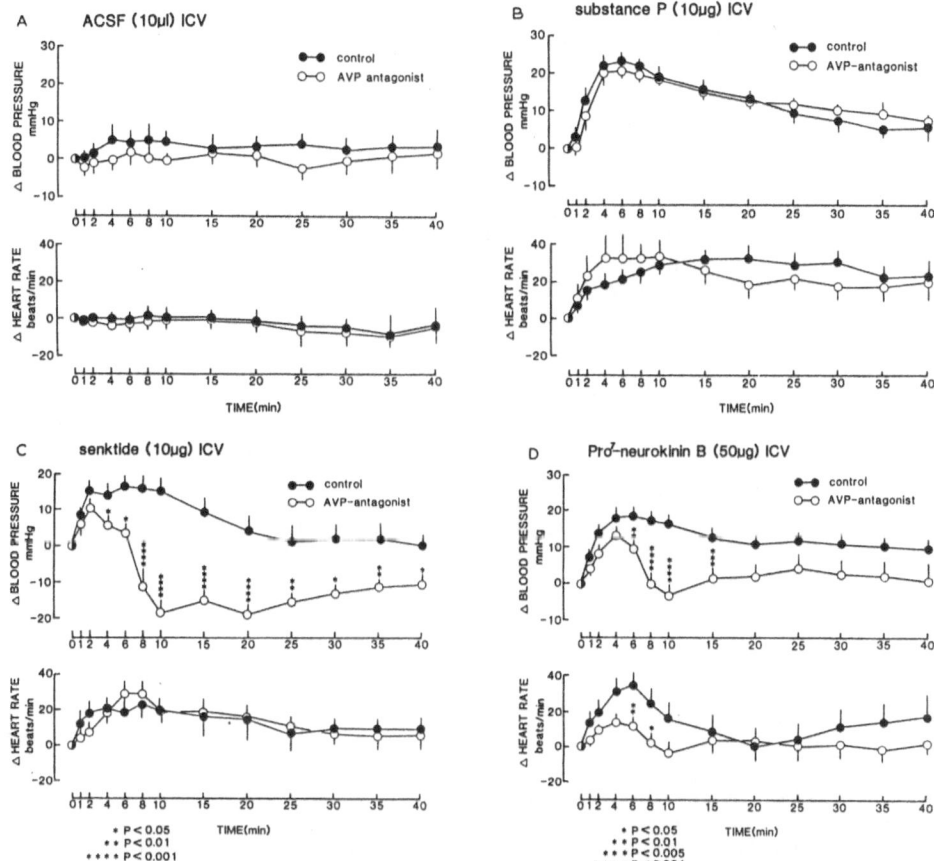

Fig. 5. Effects of blockade of peripheral vascular AVP receptors on cardiovascular responses after injections of (A) vehicle (ACSF), (B) substance P (10 μ g), (C) senktide (10 μ g), (D) [Pro[7]]-NKB (50 μ g) into the lateral brain ventricle (i.c.v.) of the normotensive rats. Values are means $\pm$ S.E.M. for 5-11 rats and asterisks indicate significant differences from corresponding values on the control rats calculated by the two-tailed Student's t-test.

The findings of differences between SHRs and WKYs in their NKB-LI contents in the supraoptic nucleus of the hypothalamus and the caudal part of the NTS (Table Ⅲ) suggests that NKB may participate in the central cardiovascular control mechanisms. In addition, the results shown in Fig.5 suggest that a part of the NKB analogue-induced hypertensive response is mediated by the peripheral vascular AVP receptor. On the other hand, the central pressor response to SP appears to be mediated exclusively by the sympathetic nervous system, because injection of SP (i.c.v.) did not increase the plasma AVP level (Unger et al., 1981; Unger et al., 1988) and the pressor responses to SP and NKA were blocked by the sympathetic blocking agents (unpublished data).

The findings that the NKB-LI contents in the supraoptic nucleus of the hypothalamus and the NTS are higher in SHRs than in WKYs is interesting because these areas of the brain have been proposed to play important roles in blood pressure control. In addition, autoradiographic binding studies have demonstrated that the distributions of [^{3}H]SP and [^{3}H]NKB binding sites (NK-1 and NK-3) are markedly different, and that high densities of [^{3}H]NKB binding sites are present in the cerebral cortex, paraventricular nucleus and supraoptic nucleus of the hypothalamus and the NTS (Bergstorm et al., 1987; Saffroy et al., 1988).

These findings suggest that NKB may increase the blood pressure by stimulating AVP release from the paraventricular nucleus and/or supraoptic nucleus of the hypothalamus through the NKB receptor (NK-3 subtype).

Acknowledgements

We are grateful to Drs. K. Tateishi and R. Saito (Fukuoka University) for helpful discussions. We thank T. Hagio for technical assistance.

REFERENCES

Bergstrom, L., Torrens, Y., Saffroy, M.., Beaujouan, J.C., Lavielle, S., Chassaing, G., Morgat, J.L., Glowinski, J. and Marquet, A., 1987, [^{3}H]neurokinin B and ^{125}I-Bolton Hunter eledoisin label identical tachykinin binding sites in the rat brain, J. Neurochem., 48: 125-133.

Crill, N.E. and Reis, D.J., 1968, Distribution of carotid sinus and depressor nerves in the cat brainstem, Am. J. Physiol., 214: 269-276.

Erspamer, V., 1981, The tachykinin peptide family, Trends Neurosci., 4: 267-269.

Helke, C.J., O'Donohue, T.L. and Jacobowitz, D.M, 1980, Substance P as a baro- and chemoreceptor afferent neurotransmitter: immunocyto-chemical and neurochemical evidence in the rat, Peptide, 1:1-9.

Henry, J.L., 1987, "Substance P and Neurokinins", Springer, New York.

Higuchi, Y. Takano, Y. Shimazaki, H., Shimohigashi, Y., Kodama, H., Matsumoto, H., Sakaguchi, K., Nonaka, S., Saito, R., Waki, M. and Kamiya, H., 1989, Dimeric substance P analogue shows a high potent activity of the in vivo salivary secretion in the rat, Eur. J. Pharmacol., 160: 413-416.

Krause, J.E., MacDonald, M.R. and Takeda, Y., 1989, The polyprotein nature of substance P precursors, BioEssays, 10: 62-69.

Lipski, J., MacAllen, R.M. and Spyer, K.M., 1975, The sinus nerve and baroreceptor input to the medulla of cat, J.Physiol. (Lond.), 251: 61-78.

Loewy, A.D. and Sawyer, W.B., 1982, Substance P antagonist inhibits vasomotor responses elicited from ventral medulla in rat, Brain Res., 245: 379-383.

Maggio, J.E., 1988, Tachykinins, Annu. Rev. Neurosci., 11: 13-28.

Nagashima, A., Takano, Y., Masui, H. and Kamiya, H., 1987, Evidence that neurokinin A (substance K) neurons project from the striatum to the substntia nigra in the rats, Neurosci. Lett., 77: 103-108.

Nagashima, A., Takano, Y., Tateishi, K., Matsuoka, Y., Hamaoka, T. and Kamiya, H., 1989a, Cardiovascular roles of tachykinin peptides in the nucleus tractus solitarii of rats, Brain Res., 487: 392-396.

Nagashima, A., Takano, Y., Tateishi, K., Matsuoka, Y., Hamaoka, T. and Kamiya, H., 1989b, Central pressor actions of neurokinin B:increases in neurokinin B contents in discrete nuclei in spontaneously hypertensive rats, Brain Res., 499: 198-203.

Panenton, W.M. and Loewy, A.D., 1980, Projections of the cartid sinus nerve to the nucleus of the solitary tract in the cat, Brain Res., 191: 239-244.

Pernow, B., 1983, Substance P, Pharmacol. Rev., 5: 85-141.

Saffroy, M., Beaujouan, J.C., Torrens, Y., Besseyre, J., Bergstrom, L. and Glowinski, J., 1988, Localization of tachykinin binding sites (NK_1, NK_2, NK_3 ligands) in the rat brain, Peptides, 9: 227-241.

Takano, Y., Martin, J.E., Leeman, S.E. and Loewy, A.D., 1984, Substance P immunoreactivity released from rat spinal cord after kainic acid excitation of the ventral medulla oblongata: a correlation with increased blood pressure, Brain Res., 291:168-172.

Takano, Y., Sawyer,W.B. and Loewy, A.D., 1985, Substance P mechanisms of the spinal cord related to vasomotor tone in the spontaneously hypertensive rat, Brain Res., 334: 106-116.

Takano, Y., Nagashima, A., Masui, H., Kuromizu, K. and Kamiya, H., 1986, Distribution of substance K (neurokinin A) in the brain and peripheral tissues of rats, Brain Res., 369: 400-404.

Takano, Y., Nagashima, A., Kamiya, H., Kurosawa, M. and Sato, A., 1988, Well-maintained reflex responses of sympathetic nerve activity tostimulation of baroreceptor, chemoreceptor and cutaneous mechanoreceptors in neonatal capsaicin treated rats, Brain Res., 455:188-192.

Tateishi, K., Matsuoka, Y. and Hamaoka, T., 1989, Establishment of highly specific radioimmunoassay for neurokinin A and neurokinin B and de-termination of tissue distribution of these peptides in the rat central nervous system, Regul. Pept., 24: 245-257.

Unger, T., Rascher, W., Schuster, C., Pavlovitch, R., Schomig, A., Dietz, R. and Ganten, D., 1981, Central blood pressure effects of substance P and angiotensin II : Role of the sympathetic nervous system and vasopressin, Eur. J. Pharmacol., 71:33-42.

Unger, T., Carolus, S., Demmert, G., Ganten, D., Lang, R.E., Maser-Gluth, C., Steinberg, H. and Veelken, R., 1988, Substance P induces a cardiovascular defense reaction in the rat: Pharmacological characterization, Cir. Res., 63: 812-820.

Wormser, U., Laufer, R., Hart, Y., Chorev, M., Gilon, C. and Selinger, Z., 1986, Highly selective agonists for substance P receptor subtypes, The EMBO J., 5: 2805-2808.

FUNCTIONS OF A CO-TRANSMITTER, CALCITONIN GENE-RELATED PEPTIDE,

ON THE NEUROMUSCULAR JUNCTION

SHUJI UCHIDA, KENJI TAKAMI, HIDEYUKI KOBAYASHI,
KAZUYA HASHIMOTO AND NAOKO MATSUMOTO

Department of Pharmacology 1, School of Medicine, Osaka
University, Nakanoshima 4-3-57, Kitaku, Osaka 530, Japan

INTRODUCTION

Calcitonin gene-related peptide (CGRP) is a peptide that is synthesized in nervous tissues by a different processing of the messenger RNA for calcitonin (Amara et al.,1982; Rosenfeld et al.,1983) . A wide, but non-random distribution of CGRP-like immunoreactive structures in the central and peripheral nervous system has been demonstrated by immunohistochcmistry and radioimmunoassay (Rosenfeld et al.,1983; Gibson et al.,1984; Kawai et al.,1985). CGRP has been shown to cause physiological responses and to have specific binding sites, suggesting that it is a neurotransmitter or neuromodulator.

Among these nervous systems, motor neurons and motor nerve terminals have been shown by immunohistochemistry and in situ hybridization to contain CGRP with acetylcholine, so we have been interested in the function of CGRP in signal transduction at neuromuscular junctions (Takami et al.,1985A; Fontaine et al.,1986; New and Mudge,1986; Gibson et al.,1988).

Many neuropeptides including CGRP are known to co-exist with classical neurotransmitters and are called "co-transmitters". However, the physiological significance of these co-transmitters is not yet clear. Neuromuscular junctions in which acetylcholine and CGRP are co-existent seemed a very good system for studying the significance of co-transmitters because the innervation of skeletal muscles is simple, the tissue is homogeneous and it is easy to detect physiological and biochemical responses.

Another reason that we are interested in this subject is that CGRP may be a trophic factor. Destruction of a motor nerve is known to cause drastic changes in the structure and metabolism of the skeletal muscle cells that it innervates. The main changes observed are atrophy of the muscle cells, decrease in insulin-dependent glucose uptake, increase in the amount of nicotinic acetylcholine receptors and changes in the properties of ion channels, etc., most of which are thought to be due to disuse of the denervated muscles. However, some of these changes may be caused by disappearance of a tropic factor. So, it seemed very interesting to study the effects of CGRP on these changes after denervation.

Neuroreceptor Mechanisms in Brain, Edited by S. Kito *et al.*
Plenum Press, New York, 1991

In this report, we describe the release of CGRP by nerve stimulation and the short- and long-term effects of CGRP on contraction and the adenylate cyclase system in skeletal muscle cells.

I.EFFECTS OF CGRP ON CONTRACTION AND THE ADENYLATE CYCLASE SYSTEM IN SKELETAL MUSCLE

In mouse and rat phrenic nerve-diaphragm preparations, CGRP caused dose-dependent increase in twitch contraction by nerve and direct stimulations (Fig.1,2) (Takami et al.,1985B; Uchida et al.,1990). The ED_{50} value of CGRP was about 3×10^{-8} M and the increase in contraction at 10^{-6} M CGRP was about 30% more than the basal contraction. This stimulatory effect of CGRP was observed even in the presence of curare and propranolol, suggesting a direct action of CGRP on skeletal muscle cells.

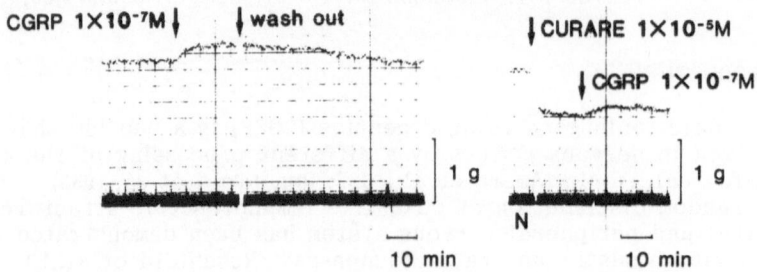

Fig.1 Effect of CGRP on striated muscle contraction under nerve stimulation (left) and on direct stimulation (right) of the phrenic nerve-diaphragm preparation of mice.

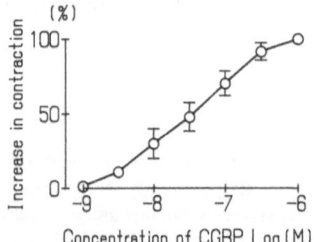

Fig.2 Dose-dependence of effect of CGRP on twitch contraction of rat diaphragm induced by direct stimulation.
CGRP was added to the medium cumulatively and increase of the contraction at 1 μM CGRP was taken as 100 % increase.

Fig. 3 shows that CGRP increased the cyclic AMP content of diaphragm muscle cells dose-dependently, which is consistent with its effect on twitch contraction (Takami et al.,1986). CGRP elevated the cyclic AMP content within 1 min after its addition, which is fast enough to cause the increase in twitch contraction. The elevation of cyclic AMP content was due to activation of adenylate cyclase in the membranes by CGRP. Muscle cell membranes had a specific binding site for CGRP with a similar affinity as the ED_{50} values for activation of adenylate cyclase and stimulation of twitch contraction (Kobayashi et al.,1987).

These results suggest that CGRP binds to specific receptors on cell membranes and then activates adenylate cyclase. Activation of twitch contraction by CGRP seemed to be mainly due to elevation of the cyclic AMP content because dibutyryl-cyclic AMP, 8-bromo-cyclic AMP or phosphodiesterase inhibitors also caused activation of twitch contraction.

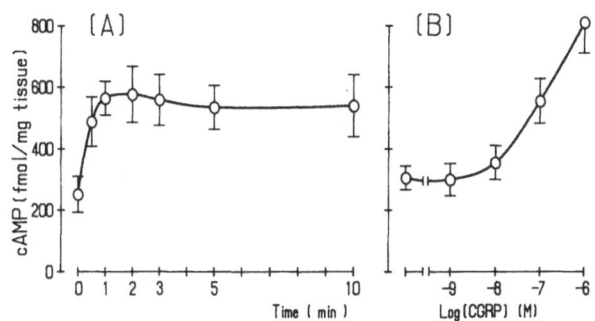

Fig.3 Time course (A) and dose response curve (B) of CGRP-induced elevation of cyclic AMP levels in mouse diaphragm.
A) After 5 min preincubation, 0.1µM CGRP was added at 0 min. B) The diaphragm was incubated for 3 min in Krebs-Ringer solution containing various concentrations of CGRP.

II.RELEASE OF CGRP FROM MOTOR NERVE TERMINAL BY NERVE EXCITATION

For confirmation of the physiological significance of CGRP as a neurotransmitter or neuromodulator, its release on nerve excitation must be demonstrated. Electrical stimulation of the phrenic nerve or membrane depolarization by high K^+ medium increased the content of CGRP-like immunoreactive substance (CGRP-LIS) in the medium in rat phrenic nerve-hemidiaphragm preparations.(Fig.4,A and B). This increased release was not observed in Ca^{++}-free medium (Fig.4, C and D), indicating that release of CGRP-LIS from nerve terminals on tetanic stimulation was Ca^{++}-dependent.

Fig.5A shows the Ca^{++}-dependent increase in the cyclic AMP level after phrenic nerve stimulation in the presence of a phosphodiesterase inhibitor. Addition of antiserum against human CGRP to the medium prevented the increase in cyclic AMP content without affecting the cyclic AMP content in the resting state (Fig.5B).

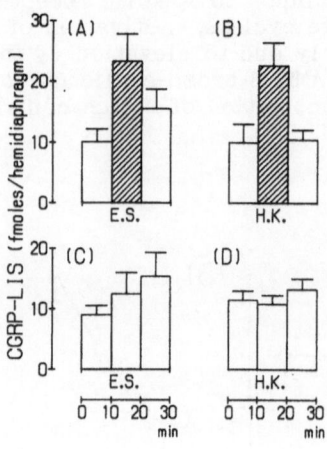

Fig.4 Release of CGRP-LIS from phrenic nerve-hemidiaphragm preparations induced by electrical stimulation of the phrenic nerve or high K⁺ medium.

Tissues were incubated for 30 min after equilibration. The incubation media after the first 10 min and last 10 min were used as control samples. In the second 10 min, the phrenic nerve was stimulated electrically (A,C) or with high K⁺ medium (B,D). The incubation media for (A) and (B) contained 1.0 mM $CaCl_2$ and those for (C) and (D) were Ca^{++}-free. Hatched columns indicate significant difference from values in the first 10 min (control) ($p<0.02$).

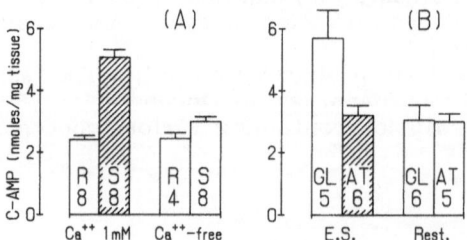

Fig.5 Elevation of the cyclic AMP content in phrenic nerve-hemidiaphragm preparations by electrical stimulation of the phrenic nerve and its inhibition by anti-CGRP serum.
(A) Preparations were incubated for 10 min with (S) or without (R) electrical stimulation in 1 mM $CaCl_2$ or Ca^{++}-free medium.
(B) Preparations were incubated for 10 min with anti-CGRP antiserum (AT) or with bovine gamma-globlin (GL) during electrical stimulation (E.S.) or in the resting condition (Rest.).
Data are means ± S.E.M. from the numbers of experiments indicated in columns. Hatched columns indicate a significant difference from the left adjacent (control) column (A:$p<0.002$, B:$p<0.02$).

These results show that excitation of the motor nerve caused Ca^{++}-dependent release of CGRP-LIS, which increased the content of cyclic AMP in skeletal muscle cells, suggesting that CGRP has a physiological role as a co-transmitter in this tissue.

III.CHANGES IN THE CONTRACTION AND ADENYLATE CYCLASE SYSTEM OF SKELE-TAL MUSCLE AFTER DENERVATION

We demonstrated that CGRP activated adenylate cyclase via CGRP receptors. Therefore, we studied the effect of denervation on adenylate cyclase and contraction in skeletal muscle with the idea that denervation may induce supersensitivity of CGRP-activated adenylate cyclase and CGRP-induced enhancement of twitch contraction.

A) Adenylate Cyclase System in Rat Gastrocnemius Muscle

Rat gastrocnemius muscle was denervated by unilateral removal of the sciatic nerve. The weight of the denervated muscle decreased gradually and reached 28% of that of the control side 30-40 days after denervation (Hashimoto et al.,1989).

The stimulations by CGRP and forskolin of the adenylate cyclase activity of muscle membranes 30 days after denervation were compared with those of membranes from the contralateral normal muscle. Although the denervation caused increase in reactivity to CGRP, the reactivity to forskolin and the basal activity also increased (Fig.6). These results indicate that denervation caused supersensitivity of adenylate cyclase itself, but not that of CGRP-activation.

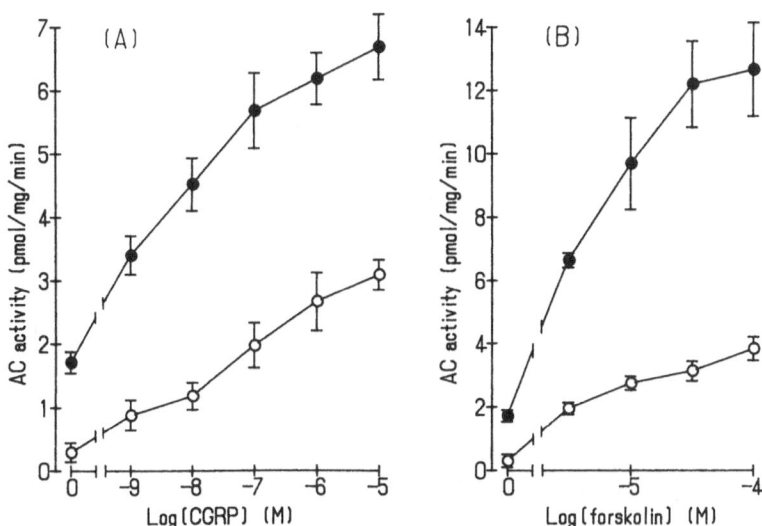

Fig.6 Effects of CGRP (A) and forskolin (B) on adenylate cyclase activity in denervated (●) and nondenervated (○) gastrocnemius muscle 30 days after unilateral denervation.

Then we examined the basal activity of adenylate cyclase and ^{3}H-forskolin binding as an indicator of the amount of adenylate cyclase molecules. Fig.7 shows the change in the ratio of the basal adenylate cyclase activity on the denervated side to that on the control side with time after denervation. The activity is expressed as total adenylate cyclase activity in whole muscle to avoid the effect of muscle atrophy.

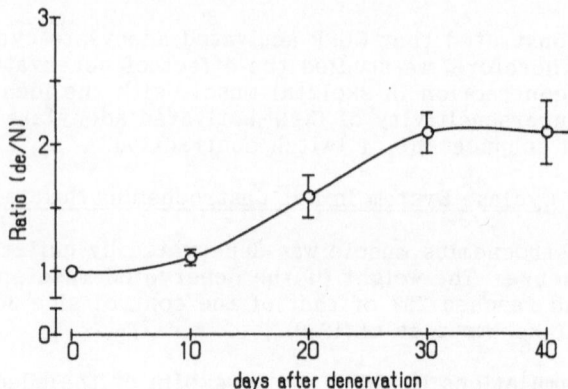

Fig.7 Change in ratio of total adenylate cyclase activities in denervated and control whole muscles with time after denervation. de; denervated muscle N; control muscle.

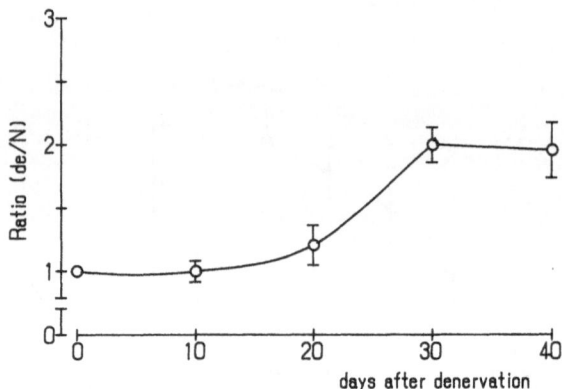

Fig.8 Change in ratio of specific ^{3}H-forskolin binding per cell of denervated and control muscle with time after denervation.

As shown in Fig.8, the ratio of specific ^{3}H-forskolin binding sites on denervated muscle to that on the control side also increased with time after denervation. This change was in good agreement with that of the basal adenylate cyclase activity. It is unknown whether forskolin binds to adenylate cyclase molecules directly. There are reports that forskolin acts on and bind to other molecules besides adenylate cyclase (Wadzinski et al.,1987; Hoshi et al.,1988; Wagoner and Pallotta,1988). But most,if not all, the the change in forskolin binding may reflect change in a component of adenylate cyclase system after the receptor.

These results indicate that the supersensitivity of the adenylate cyclase system after denervation is heterologous, resulting from increase in the amount of adenylate cyclase or a factor related to adenylate cyclase.

There are two possible mechanism for this heterologous supersensitivity. One is loss of continuous stimulation by CGRP, resulting in heterologous supersensitivity of the adenylate cyclase system. The other is that muscle disuse due to denervation may result in supersensitivity of adenylate cyclase for some unknown reason.

B) Changes in CGRP-Induced Enhancement of Twitch Contraction in Rat Diaphragm

Similar experiments were made on the left hemidiaphragm of rats by cutting the phrenic nerve. Two weeks after denervation, basal adenylate cyclase activity of the membrane was increased about 1.7 fold. But, activation ratios and ED_{50} values for CGRP and isoproterenol were not affected by denervation. These results were consistent with those in gastrocnemius muscles.

Twitch contraction by transmural stimulation was increased about 2 fold by denervation reflecting the increase of adenylate cyclase activity. However, CGRP-induced enhancement of twitch contraction was abolished by denervation, as shown in fig.9.

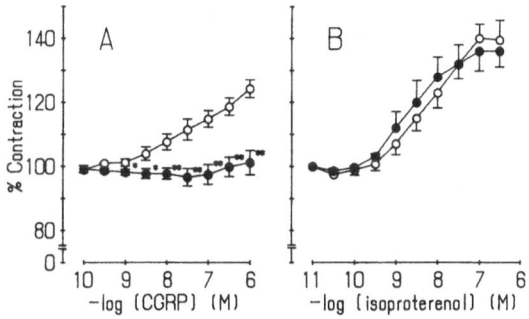

Fig.9 Enhancement of twitch contraction by CGRP(A) and isoproterenol(B) in normal and denervated diaphragms. Data are expressed as percentage increases in isometric tension induced by drugs over the steady level in normal (O) and denervated (●) muscles. Values are means ± SEM's.

CGRP at concentration of 3 x 10^{-8}M to 1 x 10^{-6}M caused dose-dependent potentiation of the contraction of normal diaphragm. In denervated muscle, no potentiation by CGRP was observed. On the other hand, isoproterenol potentiated the contractile responses of normal and denervated diaphragms in similar concentration-dependent manners and to similar extent.

In general, denervated tissues become hypersensitive to the neurotransmitters present in the nerves that are destroyed, as the result of increase in neurotransmitter receptors. After denervation, the adenylate cyclase system became hypersensitive to CGRP, although this hypersensitivity was heterologous resulting from an increase in adenylate cyclase molecules. Therefore, we expected that the muscle would become hypersensitive in CGRP-induced enhancement of contraction. But, on the contrary, the response to CGRP in twitch contraction was abolished by phrenic nerve destruction.

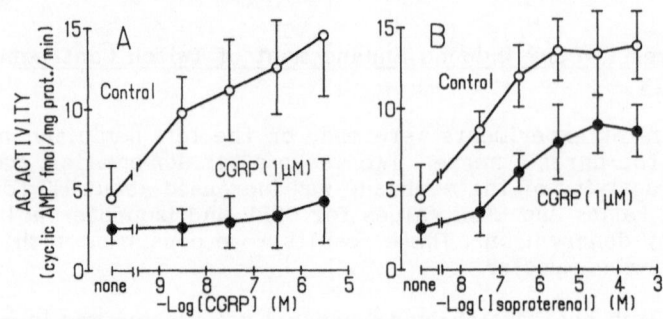

Fig.10 Effect of exposure of cultured cells to CGRP (1μM) for 24 hours on CGRP- (A) and isoproterenol- (B) stimulated adenylate cyclase activities.

To clarify this discrepancy, we measured cyclic AMP content in the muscle. The basal cyclic AMP content in the tissue was also elevated by denervation. This increase corresponds to the increase in the basal contractile force and suggests that the cyclic AMP level regulates the contractile force of skeletal muscle. But, the stimulation by CGRP and isoproterenol of the cyclic AMP content over basal level were decreased by denervation. This may be one reason of the decrease in CGRP-induced enhancement of twitch contraction.

Destruction of a motor nerve causes very unique heterologous supersensitization of the adenylate cyclase system in skeletal muscle, probably due to lack of CGRP release. The elevation of the basal twitch contraction by denervation may result from an increase in the basal cyclic AMP level. However, disappearance of CGRP-enhanced contraction cannot be explained entirely by change in the cyclic AMP system, and may have other underlying mechanisms.

IV. ADENYLATE CYCLASE SYSTEM IN CULTURED MUSCLE CELLS AND EFFECT OF CGRP

Isolated cells from gastrocnemius muscle of newborn rats were cultured in the presence of CGRP (Hashimoto et al.,1989B). The muscle cells in culture are free from control by the nerve, so that we thought they were a good model for denervated muscles. The adenylate cyclase activity of muscle cells stimulated by isoproterenol, CGRP, NaF and forskolin increased with time during culture until 4 days. These changes may reflect those by denervation in vivo although the time courses were different from each other.

Exposure of the cell cultures to CGRP (1μM) for 24 hr depressed their CGRP-stimulated adenylate cyclase activity to about one fourth of that of control (Fig.10,A). This treatment also depressed isoproterenol-stimulated adenylate cyclase activity as shown in Fig.10,B.

The effects of exposure of the cells to CGRP for 24 hr on their adenylate cyclase activities stimulated by drugs are summarized in Table 1. All drug-stimulated adenylate cyclase activities were depressed by exposure to CGRP, but the depression of CGRP-stimulated activity was the greatest.

Next, exposure to CGRP was prolonged to 3 days in the same conditions and the results are shown in Table 2. Exposure to CGRP for 3 days again depressed the stimulations by various drugs, but its effects on the adenylate cyclase activities stimulated by CGRP, NaF and forskolin were not significantly different.

Table 1

Comparison of adenylate cyclase activities stimulated by various drugs after exposure of cultured cells to CGRP (1µM) for 24 hr.

Stimulator	Control	CGRP-exposure	% Ratio
CGRP (5 µM)	10.1 ± 1.3	2.1 ± 1.0	20.8 ± 10.5
Isoproterenol (0.3 mM)	9.4 ± 0.9	6.3 ± 0.6	67.0 ± 11.7
NaF (10 mM)	11.2 ± 2.0	5.9 ± 1.0	52.7 ± 15.6
Forskolin (1 µM)	13.2 ± 1.9	9.7 ± 1.0	73.5 ± 15.9
None(Basal Activity)	4.5 ± 2.0	2.0 ± 0.9	57.7 ± 21.6

Adenylate cyclase activities in the presence of stimulators are shown as total minus basal activity (cyclic AMP fmol/mg prot./min). Values are means ± S.D. Percentage ratios mean CGRP-exposure/Control. * Significantly different from other values ($p < 0.01$).

Table 2

Comparison of adenylate cyclase activities stimulated by various drugs after exposure of cultured cells to CGRP (1 µM) for 3 days.

Stimulator	Control	CGRP-exposure	% Ratio
CGRP (5 µM)	16.7 ± 3.6	9.4 ± 2.6	56.3 ± 22.8
Isoproterenol (0.3 mM)	16.4 ± 0.9	14.3 ± 3.8	87.2 ± 10.8 *
NaF (10 mM)	23.8 ± 3.1	8.6 ± 4.3	36.1 ± 20.1
Forskolin (1 µM)	27.2 ± 3.1	10.7 ± 3.6	39.3 ± 15.9
None(Basal Activity)	6.6 ± 2.1	2.9 ± 1.8	43.7 ± 31.3

* Significantly different from other values ($p < 0.01$).
Explanations are as for Table 1.

These results suggest that exposure of cultured skeletal muscle cells to CGRP for 24 hr caused homologous desensitization mainly at the level of the CGRP receptor, whereas exposure to CGRP for a longer period caused heterologous desensitization mainly at some level after the receptor.

To determine whether these desensitization were mediated by an increased concentration of cyclic AMP, muscle cells were cultured in medium containing dibutyryl cyclic-AMP (dbc-AMP). Table 3 shows the CGRP-stimulated adenylate cyclase activity in muscle cells after exposure to dbc-AMP (1 mM) for 24 hr.

Table 3

Effect of 24 hr exposure of cultured cells to dbc-AMP (1 mM) on CGRP-stimulated adenylate cyclase activity.

Adenylate cyclase activity	Control	Exposure dbc-AMP (1mM)	CGRP(1μM)
CGRP-stimulated	12.1 ± 2.7	6.3 ± 3.4	4.3 ± 2.6
Basal	4.5 ± 2.0	1.8 ± 1.0	2.6 ± 0.9

Values are means ± S.D. (cyclic-AMP, fmol/mg prot./min).

We also examined the time courses of change in adenylate cyclase activity stimulated by CGRP and forskolin in cultures with 1 mM dbc-AMP for 3 days(data not shown). The results showed that the adenylate cyclase activities stimulated by CGRP and forskolin were inhibited by exposure to dbc-AMP and that the time course was very similar to that of the desensitization by CGRP-exposure.

The results for cultured muscle cells indicated that desensitization of the adenylate cyclase system in muscle cells by exposure to CGRP was at first mainly homologous but later became heterologous with time of culture. These desensitizations seem to be mediated by elevation of the intracellular cyclic-AMP level.

V. DISCUSSION

We demonstrated that CGRP, like other neurotransmitters, is released Ca^{++}-dependently from nerve terminal by nerve excitation. The released CGRP activated adenylate cyclase via CGRP receptors and enhanced twitch contraction mediated by elevation of the cyclic-AMP level.

Affinities of CGRP for activation of adenylate cyclase and binding to membranes in this tissue were much lower than those reported in other tissues. The reason for this difference is unknown. However, it is possible that the concentration of CGRP released in the synaptic cleft can reach more than 10^{-7} M and cause a physiological response because the synaptic cleft of the neuromuscular junction is larger and narrower than other synapses. Our findings that elevation of cyclic AMP by nerve stimulation was blocked by antiserum against CGRP supports this idea. The CGRP receptor in this tissue may be a different subtype with lower affinity for CGRP from those in other tissues.

CGRP inhibited adenylate cyclase via inhibitory GTP binding protein in breast cancer cells, in which the affinity was higher than that of activation of adenylate cyclase (Barsony and Marx,1988). However, we could not detect any inhibition of adenylate cyclase by CGRP in skeletal muscles.

CGRP may regulate acetylcholine release from nerve terminals by binding to presynaptic receptors. But, this possibility seems unlikely from our analysis of the effects of CGRP on twitch contraction by nerve stimulation. Further studies are necessary on this problem.

As the enhancement of twitch contraction by CGRP was at most 30 % in rat diaphragm, we thought that CGRP might have other physiological roles, such as a neuronal trophic factor regulating the homeostasis of muscle cells. In this report, we demonstrated that denervation caused heterologous supersensitivity of the adenylate cyclase system in the innervated muscles and that CGRP induced the heterologous desensitization in cultured muscle cells. These findings suggest that CGRP regulate the adenylate cyclase system in skeletal muscle cells.

Some properties of muscle have been demonstrated to be regulated trophically by unknown molecular mechanisms (Fambrough,1970; Grampp et al.,1972). Lentz (1972) suggested that the actions of a presumed 'trophic hormone' might be mediated by cyclic-AMP. Additional evidence that the adenylate cyclase system in skeletal muscle is neurotrophically regulated is as follows : First, the changes in cyclic-AMP concentration resulting from muscle denervation and simple disuse are different. Second, the onset of increase in cyclic-AMP concentration after denervation depends on the length of the residual nerve stump. Third, reinnervation of a denervated muscle leads to reduction in the cyclic-AMP concentration (Carlson,1975).

Some long-term effects of CGRP on skeletal muscle have been also reported. Culture in the presence of CGRP increased the number of nicotinic acetylcholine receptors on skeletal muscle cells (Fontaine et al.,1986; New and Mudge,1986). Daily injection of CGRP prevented disuse-induced sprouting of motor nerve terminals in rat skeletal muscle (Tsujimoto and Kuno,1988).

Our results and these findings suggest that CGRP, which is located in motor nerve terminal, may regulate the adenylate cyclase system in skeletal muscle physiologically and may be one of the trophic factors derived from motor neurons. To establish that CGRP is a trophic factor, more work on the effect of continuous treatment with CGRP in the denervated state is necessary.

REFERENCES

Amara, S. G., Jones, V., Rosenfeld, M. G., Ong, E. S., and Ewans, R. M., 1982, Alternative RNA processing in calcitonin gene expression generate mRNA encoding different polypeptide products, Nature, 298:240-244.
Barsony, J., and Marx, S. J., 1988, Dual effects of calcitonin and calcitonin gene-related peptide on intracellular cyclic AMP in a human breast cancer cell line. Endocrinology, 122:1218-1223.
Carlsen, R. C., 1975, The possible role of cyclic AMP in the neurotrophic control of skeletal muscle. J.Physiol., 247:343-361.
Fambrough, G. H., 1970, Acetylcholine sensitivity of muscle fiber membrane: Mechanism of regulation by motoneurons. Science, 168:372.
Fontaine, B., Klarsfeld, A., Hökfelt, T., and Changeux, J. -P., 1986, Calcitonin gene-related peptide, a peptide present in spinal cord motoneurons, increases the number of acetylcholine receptors in primary cultures of chick embryo myotubes. Neurosci. Lett., 71:51-65.
Gibson, S. J., Polak, J. M., Bloom, S. R., Sabate, I. M., Mulderry, P. M., Chatei, M. A., McGregor, G. P., Morrison, J. F. B., Kelly, J. S., Evans, R. M., and Rosenfeld, M. G., 1984, Calcitonin gene-related peptide immunoreactivity in spinal cord of man and of eight other species. J. Neuroscience, 4:3101-3111.
Gibson, S. J., Polak, J. M., Giaid, A., Hamid, Q. A., Kar, S., Jones, P. M., Denny, P., Legon, S., Amara, S. G., Craig, R. K., Bloom, S. R., Penketh, R. J. A., Rodek, C., Ibrachim, N. B. N., and Dawson, A., 1988, Calcitonin gene-related peptide messenger RNA is expressed in sensory neurones of dorsal root ganglia and also in spinal motoneurons in man and rat. Neurosci. Lett., 91:283-288.

Grampp, W., Harrus, J. B., and Thesleff, S., 1972, Inhibition of denervation changes in skeletal muscle by blockers of protein synthesis. J. Physiol. 221:743-745.

Hashimoto, K., Watanabe, Y., Uchida S., and Yoshida, H., 1989A, Increase in the amount of adenylate cyclase in rat gastrocnemius muscle after denervation. Life Sci., 44:1887-1895.

Hashimoto, K., Uchida, S., and Yoshida, H., 1989B, Effects of calcitonin gene-related peptide on the adenylate cyclase system in cultured rat skeletal muscle cells. Life Sci. 45:2183-2193.

Hoshi, T., Garber, S. S., and Aldrich, R. W., 1988, Effect of forskolin on voltage-gated potassium channels is independent of adenylate cyclase activation. Science, 240:1652-1655.

Kawai, Y., Takami, K., Shiosaka, S., Emson, P. C., Hillyard, C. J., McIntyre, I., and Tohyama, M., 1985, Topographic localization of calcitonin gene-related peptide in rat brain : An immunohistochemical analysis. Neuroscience, 15:747-763.

Kobayashi, H., Hashimoto, K., Uchida, S., Sakuma, J., Takami, K., Tohyama, M., Izumi, F., and Yoshida, H., 1987, Calcitonin gene-related peptide stimulates adenylate cyclase activity in rat striated muscle. Experientia, 43:314-316.

Lentz, T. L., 1972, A role of cyclic AMP in a neurotropic process. Nature (New Biology), 238:154-155.

New, H. V., and Mudge, A.V., 1986, Calcitonin gene-related peptide regulates muscle acetylcholine receptor synthesis. Nature, 323:129-135.

Rosenfeld, M. G., Mermod, J. J., Amara, S. G., Swanson, L. W., Sawchenko, P. E., River, J., Vale, W. W., and Evans, R. M., 1983, Production of novel neuropeptide encoded by the calcitonin gene via tissue specific RNA processing. Nature, 304:129-135.

Takami, K., Kawai, Y., Shiosaka, S., Lee, Y., Girgis, S., Hillyard, C. J., MacIntyre, I., Emson P. C., and Tohyama, M., 1985A, Immunohistochemical evidence for coexistence of calcitonin gene-related peptide- and choline acetyltransferase-like immunoreactivity in neurons of the rat hypoglossal, facial and ambiguus nuclei. Brain Res., 328:386-389.

Takami, K., Kawai, Y., Uchida, S., Tohyama, M., Shiotani, Y., Yoshida, H., Emson, P. C., Girgis, S., Hillyard, C. J., and MacIntyre, I., 1985B, Effect of calcitonin gene-related peptide on contraction of striated muscle in the mouse. Neurosci. Lett., 60:227-230.

Takami, K., Hashimoto, K., Uchida, S., Tohyama M., and Yoshida, H., 1986, Effect of calcitonin gene-related peptide on the cyclic AMP level of isolated mouse diaphragm. Jpn. J. Pharmacol., 42:345-350.

Tsujimoto, T., and Kuno, M., 1988, Calcitonin gene-related peptide prevents disuse-induced sprouting of rat motor nerve terminals. J. Neurosci., 8:3951-3957.

Uchida, S., Yamamoto, H., Iio, S., Matsumoto, N., Wang, X. -B., Yonehara, N., Imai, Y., Inoki, R., and Yoshida, H., in press, Release of CGRP-like immunoreactive substance from neuromuscular junction by nerve excitation and its action on striated muscle. J. Neurochem.,

Wadzinski, B. E., Shanahan, M. F,, and Ruoho, A. E., 1987, Derivatization of the human erythrocyte glucose transporter using a novel forskolin photoaffinity label. J. Biol. Chem., 262:17683-17689.

Wagoner, P. K., and Pallotta, B. S., 1988, Modulation of acetylcholine receptor desensitization by forskolin is independent of cyclic AMP. Science 240:1656-1657.

PHARMACOLOGICAL CHARACTERIZATION OF SIGMA BINDING SITES

IN GUINEA PIG BRAIN MEMBRANES

E.W. Karbon and S.J. Enna

Nova Pharmaceutical Corporation
6200 Freeport Centre
Baltimore, MD 21224-2788

INTRODUCTION

The pharmacological, biochemical and behavioral characterization of sigma binding sites is currently the focus of intense investigation. First described by Martin and his colleagues (1976) to account for the behavioral effects of selected benzomorphan opiates, sigma binding sites were originally classified as an opioid receptor subtype. Subsequent work suggested this designation was inaccurate since, unlike opioid receptors, the sigma site is insensitive to blockade by naloxone (Itzhak et al., 1985; Katz et al., 1985). In addition, the (+) isomers of opiates such as pentazocine and N-allylnormetazocine (SKF 10,047) are more active at sigma binding sites, while opiate receptors are typically more responsive to (-) isomers (Weber et al., 1986).

Sigma binding sites were first identified using [^{3}H] ($\pm$)-SKF 10,047 in the presence of etorphine to mask opiate receptors (Su, 1982), and subsequently with [^{3}H] (+)-SKF 10,047 or [^{3}H] haloperidol (Tam and Cook, 1984). The pharmacological selectivity of (+)-SKF 10,047 binding was similar to that found for [^{3}H] phencyclidine (PCP), suggesting the two binding sites may be the same (Mendelsohn et al., 1985). This hypothesis was supported by the apparent similarity between the behavioral effects elicited by PCP, (+)-cyclazocine and (+)-SKF 10,047 (Holtzman, 1985), and by the fact that PCP generalizes to (+)-SKF 10,047 in drug discrimination tests (Shannon, 1982). However, the identification of sigma and PCP-selective ligands indicates the two sites represent distinct entities. For example, [^{3}H] di-o-tolylguanidine ([^{3}H] DTG), [^{3}H] (+)-3-[3-hydroxyphenyl]-N-(1-propyl)piperidine ([^{3}H] (+)-3-PPP), and [^{3}H] (+)-pentazocine selectively label sigma binding sites having high affinity for haloperidol, DTG, (+)-3-PPP and (+)-benzomorphans, moderate to low affinity for (-)-benzomorphans, some antipsychotics and antidepressants, and negligible affinity for TCP, a PCP derivative which selectively labels the phencylidine binding site (Largent et al., 1986; Loo et al., 1986; Weber et al., 1986; Bowen et al., 1989). In contrast, [^{3}H] TCP, which binds to a component of the N-methyl-D-aspartate (NMDA) receptor/ionophore complex, is potently displaced by MK-801 and PCP analogs but is less influenced by (+)-3-PPP and DTG. Although PCP is more potent at PCP than at sigma sites and, conversely, (+)-SKF 10,047 is more potent at sigma binding sites, the fact that both display some affinity for each site accounts for the earlier conclusion that PCP and sigma sites are identical (Table 1).

Neuroreceptor Mechanisms in Brain, Edited by S. Kito *et al.*
Plenum Press, New York, 1991

Table 1.   Substrate Selectivity of Sigma and PCP Binding Sites

| Drug | Binding Site IC$_{50}$ (nM) | |
	[^{3}H] (+)-3-PPP	[^{3}H] TCP
(+)-SKF 10,047	365	405
(-)-SKF 10,047	1440	821
(±)-Pentazocine	25	2820
(+)-3-PPP	32	> 100,000
Haloperidol	2	> 50,000
PCP	758	66
TCP	3720	11

Adapted from Largent et al. (1986). Binding assays were performed using rat brain membranes.

Autoradiographic studies have provided further evidence for the existence of distinct sigma and PCP binding sites. Thus, the regional distribution of high affinity binding sites labeled with either [^{3}H] (+)-3-PPP or [^{3}H] (+)-SKF 10,047 is similar, but distinct from, that found for [^{3}H] TCP (Largent et al., 1986). The density of sigma binding sites is high in the spinal cord, particularly the ventral horn and dorsal root ganglia, cerebellar Purkinje cells, midbrain, and the hippocampal pyramidal cell layer. In contrast, PCP binding sites are most abundant in the thalamus, cerebral cortex and non-pyramidal cell regions of the hippocampus. A separation between PCP and sigma binding sites is also suggested from drug discrimination studies revealing that PCP does not generalize to (+)-pentazocine, a selective sigma ligand (Steinfels et al., 1988). Thus, it now appears that sigma and PCP binding sites are distinct molecular entities. Although the functional role of sigma binding sites has yet to be established, the fact that psychotomimetics and antipsychotics have high affinity for these sites has led to the suggestion that some sigma ligands may be useful in the treatment of schizophrenia (Snyder and Largent, 1989).

Sigma binding sites are not limited to the central nervous system, having also been detected in the pituitary, adrenal gland, testis and ovary, tissues devoid of PCP binding sites (Wolfe et al., 1989). In addition, sigma binding has been noted in guinea pig splenocytes which, like brain, is sensitive to haloperidol and progesterone, suggesting a possible link between endocrine, nervous, and immune systems (Su et al., 1988). Such a relationship has been proposed to explain steroid-induced mental disturbances and immune dysfunction (Maggi & Perez, 1985), and PCP-induced immunosuppression (Khansari et al., 1985). The existence of sigma binding sites in brain and peripheral tissues suggests a possible association with a common membrane protein, such as an ion channel. Given these findings, work has been undertaken to define more precisely the sigma binding site as it relates to the pharmacological effects of some centrally-active drugs.

DIFFERENTIATION BETWEEN [^{3}H]DTG AND [^{3}H](+)-3-PPP BINDING SITES

Experiments were undertaken to compare the binding of [^{3}H] DTG and [^{3}H] (+)-3-PPP in extensively washed guinea pig brain membranes. Preliminary studies indicated that the same tissue preparation and incubation conditions could be employed for binding of the two ligands, allowing for a direct comparison of the effects of various drugs on the binding sites labeled by these substances. The results indicated that [^{3}H]

(+)-3-PPP binds to a single population of sites having an equilibrium dissociation constant (K_d) of 43 nM as determined by non-linear regression analysis of Scatchard plots (Table 2). In contrast, Scatchard plots of [³H] DTG binding were biphasic and were best described as two populations of sites having K_d's of 13 and 186 nM (Table 2). This contrasts with a report by Weber et al. (1986) that [³H] DTG bound to a homogeneous population of sites in guinea pig brain membranes. However, in the Weber study the concentration of [³H]DTG was less than 90 nM, making it difficult to detect the lower affinity site. In the present study, the maximum number of binding sites (B_{max}) labeled by [³H] DTG was 3 times greater than that found for [³H] (+)-3-PPP (Table 2). Taken together these findings suggest a multiplicity of sigma binding sites in guinea pig brain tissue.

Support for this conclusion was provided by displacement studies aimed at defining the substrate selectivity of [³H]DTG and [³H](+)-3-PPP binding sites (Table 3). In agreement with previous reports (Weber et al., 1986), haloperidol exhibited very high affinity for sigma sites and failed to distinguish between [³H] DTG and [³H] (+)-3-PPP binding. Likewise, unlabeled DTG displaced [³H] DTG and [³H] (+)-3-PPP with nearly equal affinity, whereas (+)-3-PPP was 3-fold more potent against [³H] (+)-3-PPP than against [³H] DTG binding. In contrast, dextromethorphan, a non-opioid antitussive and anticonvulsant, was substantially more potent at displacing [³H] (+)-3-PPP than [³H] DTG binding. Moreover, dextromethorphan inhibited [³H] DTG binding in a biphasic manner, yielding a displacement curve having a pseudo-Hill coefficient of 0.5, compared to 0.9 for inhibition of [³H] (+)-3-PPP binding (data not shown). The high affinity component of dextromethorphan displacement of [³H] DTG was 30 nM, whereas the lower affinity component was approximately 10 μM, potencies similar to those reported by Musacchio et al. (1988) for high and low affinity [³H] dextromethorphan binding to guinea pig brain membranes. It was also found that (+)-SKF 10,047 is highly selective for [³H] (+)-3-PPP binding, exhibiting a nearly 40-fold greater affinity for these sites than for [³H] DTG binding sites (Table 3).

INTERACTIONS OF PHENYTOIN AND DEXTROMETHORPHAN

It has been reported that phenytoin, an anticonvulsant, increases [³H] (+)-3-PPP and [³H] dextromethorphan binding to guinea pig brain membranes by enhancing the affinity of the binding site (Musacchio et al., 1988; Musacchio et al., 1989). Experiments in the current study revealed that phenytoin also increased [³H] (+)-SKF 10,047 binding, but failed to augment either [³H] DTG or [³H] haloperidol binding to guinea pig brain membranes (Table 4). This effect is consistent with the finding that the substrate

Table 2. Characteristics of [³H] DTG and [³H] (+)-3-PPP
Binding to Sigma Sites in Guinea Pig Brain Membranes

| | | Ligand | |
		[³H] (+)-3-PPP	[³H] DTG
K_d (nM)	High	43	13
	Low	--	186
B_{max} (fm/mg)	High	750	210
	Low	--	1906

Adapted from Naper et al. (1989).

Table 3. Effects of Various Compounds on [^{3}H] DTG and
[^{3}H] (+)-3-PPP Binding to Sigma Sites
in Guinea Pig Brain Membranes

| Compound | Inhibition of Ligand Binding IC_{50} (nM) | |
	[^{3}H] DTG	[^{3}H] (+)-3-PPP
Haloperidol	7	4
DTG	55	73
(+)-3-PPP	145	43
Dextromethorphan	2093	143
(+)-SKF 10,047	9800	265

Adapted from Naper et al. (1989).

selectivities of [^{3}H] (+)-3-PPP and [^{3}H] (+)-SKF 10,047 binding sites are
similar, as are those for [^{3}H] DTG and [^{3}H] haloperidol binding (data not
shown). Of particular interest is the finding that phenytoin, while not
affecting [^{3}H] DTG binding directly, increases the potency of dextromethorphan to displace [^{3}H] DTG without influencing the potency of haloperidol.
These findings might indicate that phenytoin modulates agonist, but not
antagonist, binding to sigma sites, making it possible to distinguish these
two classes of sigma ligands. Taken together, these data indicate that the
sigma binding component may represent a site of action for anticonvulsants
such as phenytoin and dextromethorphan.

HALOPERIDOL ADMINISTRATION AND SIGMA BINDING

Chronic administration of haloperidol causes an increase in the number
of dopamine (D_2) binding sites in rat striatal membranes, an effect thought
to be responsible for neuroleptic-induced tardive dyskinesia (Jenner &
Marsden, 1983). Because of the possible involvement of sigma binding sites
in mediating the symptoms of psychosis, experiments were undertaken to
assess the effects of haloperidol administration on these sites. Previous
studies have shown that repeated (14 day) administration of haloperidol

Table 4. Effect of Phenytoin on Ligand Binding to Sigma Sites in
Guinea Pig Brain Membranes

Ligand	Specific Binding (% Control)
[^{3}H] DTG	88 $\pm$ 6
[^{3}H] Haloperidol	81 $\pm$ 7
[^{3}H] (+)-3-PPP	175 $\pm$ 20
[^{3}H] (+)-SKF 10,047	165 $\pm$ 10

Adapted from Naper et al. (1989). The results are expressed as
the percentage of specifically bound radioligand obtained in the
presence of phenytoin (300 μM) relative to that observed in the
absence of phenytoin, and represent the mean $\pm$ SEM of four
separate determinations.

(2 mg/kg, i.p., b.i.d.) to mice causes a reduction in $[^3H]$ $(+)$-SKF 10,047 binding in brain membranes (Itzhak and Alerhand, 1989). In the present work, both $[^3H]$ DTG and $[^3H]$ $(+)$-3-PPP binding to guinea pig brain membranes were examined following 14 days of haloperidol administration (1 mg/kg, i.p.). Whereas there was a 40% reduction in $[^3H]$ DTG binding in the treated animals, there was nearly a complete loss of $[^3H]$ $(+)$-3-PPP binding sites in these subjects (Table 5). This finding suggests the existence of a heterogeneous population of sigma binding sites, and indicates that while haloperidol is capable of displacing $[^3H]$ DTG *in vitro*, only a subpopulation of $[^3H]$ DTG binding sites are affected by the neuroleptic *in vivo*.

SIGMA BINDING SITES AND THE NMDA RECEPTOR

Little is known about the biochemical and physiological properties of sigma binding sites. Sigma ligands have been reported to influence neurotransmitter-stimulated phospholipid turnover and smooth muscle contractility (Vaupel and Su, 1987; Bowen et al., 1988). However, these effects require concentrations much higher than those needed to inhibit sigma binding *in vitro*. Recently, it was reported that ifenprodil, an anti-ischemic agent and NMDA antagonist (Carter et al., 1988), potently displaced $[^3H]$ $(+)$-3-PPP from its binding sites, exhibiting an IC_{50} of 42 nM (Karbon et al., 1990). This value is similar to the K_d reported by Schoemaker et al. (1989) for $[^3H]$ ifenprodil binding, suggesting that the binding sites labeled by $[^3H]$ $(+)$-3-PPP and $[^3H]$ ifenprodil may be the same. The finding that ifenprodil may be a potent sigma ligand suggests a functional association between this binding site and the NMDA receptor/ionophore complex. Moreover, these results indicate that sigma agents might be useful in the treatment of cerebral ischemia.

SUMMARY AND CONCLUSIONS

The discovery of sigma binding sites has prompted investigation into the functional role of these sites. Binding studies have revealed that sigma sites exhibit a unique pharmacological profile, and have provided evidence favoring the existence of a multiplicity of sigma binding sites in central nervous system. However, the findings that chemicals having diverse structures and therapeutic applications are all potent sigma agents, and that sigma binding sites are present in peripheral tissues, have raised concerns about the physiological and pharmacological relevances of this site. Furthermore, an endogenous ligand for the sigma binding site has not yet been identified. Finally, there is a lack of data regarding the

Table 5. Effect of Haloperidol Administration on Sigma Binding Sites in Guinea Pig Brain Membranes

Treatment	Specific Binding (fm/mg protein)	
	$[^3H]$ DTG	$[^3H]$ $(+)$-3-PPP
Vehicle	62 ± 9	50 ± 6
Haloperidol	38 ± 4	4 ± 1

Adapted from Karbon and Naper (1990). Guinea pigs were treated with either vehicle or haloperidol (1 mg/kg, i.p.) for 14 consecutive days followed by a 4 day washout period. The values represent the mean $\pm$ SEM of 8 separate determinations.

functional coupling of sigma binding sites, although recent studies have provided some clues as to a possible signalling mechanism (Karbon et al., 1990). Besides raising questions about the relevance of sigma sites, the paucity of information on their functional properties has made it difficult to distinguish sigma agonists from and antagonists. While haloperidol is assumed to be a sigma antagonist, the finding that prolonged administration of this neuroleptic decreases the number of sigma sites (see Table 5) would seem to argue in favor of it being an agonist for this site.

Regardless of the precise nature of the sigma binding site, studies have suggested that it may represent the site of action for a number of important drugs. For example, haloperidol, a butyrophenone antipsychotic, exhibits high affinity for sigma binding sites, and several psychotomimetics, including PCP and (+)-benzomorphans, also bind this site. This suggests that sigma agents might be useful in the treatment of schizophrenia, and, if lacking D_2 antagonist properties, would be less likely to produce the extrapyramidal symptoms which limit the utility of most antipsychotics.

Several sigma compounds have been developed as antipsychotics. Rimcazole (BW 234U), a weak ($IC_{50} \sim 1$ μM) sigma agent with negligible affinity for dopamine (D_2) sites, has exhibited some antipsychotic potential in acute schizophrenia (Chouinard and Annable, 1984). Gevetroline (WY 47384) has high affinity ($IC_{50} \sim 100$ nM) for sigma binding sites, but also interacts with several other neurotransmitter receptors (Snyder and Largent, 1989). Other sigma agents currently in clinical trials are remoxipride, which is 15 times more potent at sigma ($IC_{50} \sim 60$ nM) than at dopamine D_2 sites (Snyder and Largent, 1989), and BMY 14802, which is approximately 60 times more potent at sigma ($IC_{50} \sim 75$ nM) than D_2 sites (Taylor and Dekleva, 1987; Snyder and Largent, 1989). In preclinical trials, BMY 14802 resembles conventional antipsychotics by inhibiting apomorphine-induced stereotypy and pole climbing, in reducing amfonelic acid-induced hyperlocomotion, and in producing inhibition of conditioned avoidance responding (Taylor et al., 1989). However, unlike traditional antipsychotics, BMY 14802 does not produce catalepsy in laboratory animals and, following chronic administration, induces depolarization blockade in mesolimbic (A10) dopaminergic cells, but not in nigrostriatal (A9) neurons (Wachtel and White, 1988). This profile suggests that BMY 14802 is a promising antipsychotic which may lack the potential for producing extrapyramidal side effects.

Additional therapeutic targets for sigma compounds include epilepsy and brain ischemia. The recent finding that ifenprodil, an NMDA antagonist with anti-ischemic properties, potently interacts with sigma binding sites, suggests that sigma agents might influence excitatory amino acid neurotransmission. In fact, several sigma agents appear to interact with the NMDA receptor/ionophore complex since they inhibit drug-stimulated [^{3}H] TCP binding (unpublished observation). In addition, dextromethorphan has displayed neuroprotective activity in both cultured neurons and in a rat hypoxia model (Goldberg et al., 1987; Prince and Feeser, 1988), and is able to inhibit NMDA-induced convulsions in mice (Ferkany et al., 1988). Furthermore, phenytoin, which augments sigma binding, acts synergistically with dextromethorphan to inhibit maximal electroshock-induced seizures (Tortella and Musacchio, 1986). Finally, BMY 14802 is effective in preventing hypoxic lethality in mice, and attenuates NMDA-induced convulsions (D. Taylor, personal communication).

In summary, there appear to be at least two pharmacologically distinct sigma binding sites in guinea pig brain membranes. The continued discovery of potent and selective sigma agents will help to characterize further the pharmacological properties of these sites, and will ultimately reveal their physiological function. While much remains to be discovered about sigma binding sites, current data suggest their pharmacological manipulation may have therapeutic relevance, making the development of sigma agents a promising area for further research.

References

Bowen, W.D., DeCosta, B., Hellewell, S.B., Thurkauf, A., Rice, K.C. and Walker, J.M., 1989, Characterization of [³H] (+)-pentazocine, a highly selective sigma ligand, Soc. Neurosci. Abstr., 15:984.

Bowen, W.D., Kirschner, B.N., Newman, A.H. and Rice, K.C., 1988, Sigma receptors negatively modulate agonist-stimulated phosphoinositide metabolism in rat brain, Eur. J. Pharmacol., 149:399-400.

Carter, C., Benavides, J., Legendre, P., Vincent, J.D., Noel, F., Thuret, F., Lloyd, K.G., Arbilla, S., Zivkovic, B., Mackenzie, E.T., Scatton, B. and Langer, S.Z., 1988, Ifenprofil and SL 82,0715 as cerebral anti-ischemic agents. II. Evidence for N-methyl-D-aspartate receptor antagonist properties, J. Pharmacol. Exper. Ther., 247: 1222-1232.

Chouinard, G. and Annable, L., 1984, An early phase II clinical trial of BW234U in the treatment of acute schizophrenia in newly admitted patients, Psychopharmacol., 84:282-284.

Ferkany, J.W., Borosky, S.A., Clissold, D.B. and Pontecorvo, M.J., 1988, Dextromethorphan inhibits NMDA-induced convulsions, Eur. J. Pharmacol., 151:151-154.

Goldberg, M., Pham, P. and Choi, D., 1987, Dextrorphan and dextromethorphan block anoxic neuronal injury in culture, Neurology (suppl. 1), 37:250.

Holtzman, S.G., 1985, Discriminative stimulus properties of opioids that interact with mu, kappa and PCP/sigma receptors, in: "Behavioral Pharmacology: The Current Status," L.S. Seiden and R.L. Balster, eds., Alan R. Liss, Inc., New York.

Itzhak, Y. and Alerhand, S., 1989, Differential regulation of sigma and PCP receptors after chronic administration of haloperidol and phencyclidine in mice, FASEB J., 3:1868-1872.

Itzhak, Y., Hiller, J.M. and Simon, E.J., 1985, Characterization of specific binding sites for [³H] (d)-N-allylnormetazocine in rat brain membranes, Mol. Pharmacol., 27:46-52.

Jenner, P. and Marsden, C.D., 1983, Neuroleptics and tardive dyskinesia, in: "Neurolpetics: Neurochemical, Behavioral, and Clinical Perspectives," J.T. Coyle and S.J. Enna, eds., Raven Press, New York.

Karbon, E.W., Patch, R.J., Pontecorvo, M.J. and Ferkany, J.W., 1990, Ifenprodil potently interacts with [³H] (+)-3-PPP labeled sigma binding sites in guinea pig brain membranes, Eur. J. Pharmacol., in press.

Karbon, E.W. and Naper, K., 1990, Chronic haloperidol administration differentially regulates [³H] DTG and [³H] (+)-3-PPP labeled sigma binding sites in guinea pig brain membranes. manuscript in preparation.

Katz, J.L., Spealman, R.D. and Clark, R.D., 1985, Stereoselective behavioral effects of N-allylnormetazocine in pigeons and squirrel monkeys, J. Pharmacol. Exper. Ther., 232:452-461.

Khansari, N., Whitten, H.D. and Fudenberg, H.H., 1985, Phencyclidine-induced immunodepression, Science, 225:76-78.

Largent, B.L., Gundlach, A.L. and Snyder, S.H., 1986, Pharmacological and autoradiographic discrimination of sigma and phencyclidine receptor binding sites in brain with (+)-[³H]-SKF 10,047, (+)-[³H]-3-[3-hydroxyphenyl]-N-(1-propyl)piperidine and [³H]-1-[1-(2-thienyl)cyclohexyl]piperidine, J. Pharmacol. Exper. Ther., 238:739-748.

Loo, P., Braunwalder, A., Lehmann, J. and Williams, M., 1986, Radioligand binding to central phencyclidine recognition sites is dependent on excitatory amino acid receptor agonists, Eur. J. Pharmacol., 123:467-468.

Maggi, A. and Perez, J., 1985, Role of female gonadal hormones in the CNS: clinical and experimental aspects, Life Sci., 37:893-906.

Martin, W.R., Eades, C.G., Thompson, J.A., Huppler, R.E. and Gilbert, P.E., 1976, The effects of morphine- and nalorphine-like drugs in the nondependent and morphine-dependent chronic spinal dog, J. Pharmacol. Exper. Ther., 197:517-532.

Mendelsohn, L.G., Kalra, V., Johnson, B.G. and Kerchner, G.A., 1985, Sigma opioid receptor: characterization and coidentity with the phencyclidine receptor, J. Pharmacol. Exper. Ther., 233:597-602.

Musacchio, J.M., Klein, M. and Santiago, L.J., 1988, High affinity dextromethorphan binding sites in guinea pig brain: further characterization and allosteric interactions, J. Pharmacol. Exper. Ther., 247:424-431.

Musacchio, J.M., Klein, M. and Paturzo, J.J., 1989, Effects of dextromethorphan site ligands and allosteric modifiers on the binding of (+)-[^{3}H]3-(-3-hydroxyphenyl)-N-(1-propyl)piperidine, Mol. Pharmacol., 35:1-5.

Naper, K., Pontecorvo, M.J. and Karbon, E.W. 1989, Dextromethorphan distinguishes between [^{3}H] DTG and [^{3}H] (+)-3-PPP labeled sigma receptor binding sites in guinea pig brain membranes, Soc. Neurosci. Abstr., 15:1236.

Prince, D.A. and Feeser, H.R., 1988, Dextromethorphan protects against cerebral infarction in a rat model of hypoxia-ischemia, Neurosci. Lett., 85:291-296.

Schoemaker, H., Allen, J. and Langer, S.Z., 1989, Ifenprodil, a novel NMDA antagonist, binds to a polyamine-sensitive site. Soc. Neurosci. Abstr., 15:200.

Shannon, H.E., 1982, Phencyclidine-like discriminative properties of (+)- and (-)-N-allylnormetazocine in rats, Eur. J. Pharmacol., 84:225-228.

Snyder, S.H. and Largent, B.L., 1989, Receptor mechanisms in antipsychotic drug action: focus on sigma receptors, J. Neuropsychiatry, 1:7-15.

Steinfels, G.F., Alberici, G.P., Tam, S.W. and Cook, L., 1988, Biochemical, behavioral and electrophysiological actions of the selective sigma receptor ligand (+)-pentazocine, Neuropsychopharmacology, 4:321-327.

Su, T.-P., 1982, Evidence for sigma opioid receptor: binding of [^{3}H] SKF-10,047 to etorphine-inaccesible sites in guinea pig brain, J. Pharmacol. Exper. Ther., 223:284-290.

Su, T.-P., London, E.D. and Jaffe, J.H., 1988, Steroid binding at sigma receptors suggests a link between endocrine, nervous and immune systems, Science, 240:219-221.

Tam, S.W. and Cook, L., 1984, Sigma opiates and certain antipsychotic drugs mutually inhibit (+)-[^{3}H] SKF 10,047 and [^{3}H] haloperidol binding in guinea pig brain membranes, Proc. Natl. Acad. Sci. U.S.A., 80:5618-5621.

Taylor, D.P. and Dekleva, J., 1987, Potential antipsychotic BMY 14802 selectively binds to sigma sites, Drug. Dev. Res., 11:65-70.

Taylor, D.P., Eison, M.S., Moon, S.L. and Yoccam F.D., 1989, BMY 14802: a potential antipsychotic with selective affinity for sigma binding sites, in: "Schizophrenia," C.A. Tamminga and S.C. Schulz, eds., Advances in Neuropsychiatry, vol. 1, Raven Press, New York.

Tortella, F.C. and Musacchio, J.M., 1986, Dextromethorphan and carbetapentane: centrally acting non-opioid antitussive agents with novel anticonvulsant properties, Brain Res., 383:314-318.

Vaupel, D.B. and Su, T.-P., 1987, Guinea pig vas deferens may contain both sigma receptors and phencyclidine receptors, Eur. J. Pharmacol., 139:125-128.

Wachtel, S.R. and White, F.J., 1988, Electrophysiological effects of BMY 14802, a new potential antipsychotic drug, on midbrain dopamine neurons in the rat: acute and chronic studies, J. Pharmacol. Exper. Ther., 244:410-416.

Weber, E., Sonders, M., Quarum, M., McLean, S., Pou, S. and Keana, J.F.W., 1986, 1,3-Di(2-[5-^{3}H]tolyl)guanidine: A selective ligand that labels sigma-type receptors for psychotomimetic opiates and antipsychotic drugs, Proc. Natl. Acad. Sci. U.S.A., 83:8784-8788.

Wolfe, S.A., Jr., Culp, S.G. and De Souza, E.B., 1989, Sigma receptors in endocrine organs: identification, characterization and autoradiographic localization in rat pituitary, adrenal, testis and ovary, Endocrinology, 124:1160-1172.

TRANSMITTER-ACTIVATED ION CHANNELS AS THE TARGET OF CHEMICAL AGENTS

Toshio Narahashi

Department of Pharmacology
Northwestern University Medical School
Chicago, IL 60611
U.S.A.

ABSTRACT

The transmitter-activated ion channels are known to be important target
sites of a variety of therapeutic and toxic agents. The GABA-activated
chloride channel has been shown to be modulated by general anesthetics,
alcohols, and the pyrethroid, cyclodiene and lindane insecticides. The
general anesthetics halothane, enflurane and isoflurane greatly augmented
the GABA-activated current before desensitization took place, and suppressed
it after desensitization at clinically relevant concentrations equivalent to
1-2 minimum alveolar concentrations. The stimulating effect appears to be a
mechanism of general anesthesia. It seems that general anesthetics have a
specific affinity for the GABA receptor-channel complex. Ethanol also
augmented the GABA-activated peak chloride current with little or no effect
on the desensitized sustained current. Longer chain alcohols n-butanol,
n-hexanol, n-octanol, and n-decanol also exerted the same type of effect,
with the potency and efficacy increasing with lengthening of the carbon
chain.

The GABA receptor-channel complex has also been shown to be an
important target site of certain insecticides. The type II pyrethroids
deltamethrin and fenvalerate augmented the GABA-activated peak chloride
current when applied concurrently with GABA, but the effect was diminished
as the pyrethroids were applied for long periods of time prior to GABA
application. The latter effect might explain the controversy in the
literature regarding the pyrethroid action on the GABA system. The type I
pyrethroid allethrin suppressed the GABA-activated peak chloride current
when co-applied with GABA. Both types of pyrethroids suppressed the
N-methyl-d-aspartate-induced current. Lindane and the cyclodienes dieldrin,
endrin, heptachlor-epoxide, and isobenzan suppressed the GABA-activated
chloride current. These effects can account for the convulsant action of
lindane and the cyclodienes.

INTRODUCTION

A variety of chemical agents, including therapeutic drugs and
environmental toxicants, are known to act on the transmitter-activated ion
channels thereby exerting therapeutic or toxic effects. The ion channels
associated with acetylcholine receptors, GABA receptors and excitatory amino

acid receptors have been studied most extensively for their modulation by
various chemical agents. The classical prototype of such studies is
represented by voltage clamp experiments with skeletal muscle end-plate
membranes where acetylcholine activates ion channels permeable to sodium and
potassium ions. The development of patch clamp techniques during the past
decade has made it possible to use practically any type of cells for voltage
clamp analysis of channel behavior. Thus the studies of neuronal ion
channels activated by various neurotransmitters have advanced rapidly, and a
number of therapeutic drugs and toxicants which were known to act on the
transmitter receptor-channel system have been the subject of intense
investigation. In this chapter, some of the recent developments along this
line of studies in our laboratory will be presented. Most of these studies
deal with the GABA-activated chloride channels, but some of our recent
studies of the excitatory amino acid-activated channels are also discussed.

METHODS

     The rat dorsal root ganglia contain neurons which generate chloride
current in response to GABA application. They are convenient materials
as a model for the $GABA_A$ receptor-channel system. The ganglia were isolated
from the lumbo-sacral region of newborn rats, digested in calcium-free
Eagle's minimum essential medium containing 0.1% collagenase/dispase plus
0.025% trypsin. The ganglia were then dissociated by repeated trituration,
and the dissociated neurons were maintained in primary culture as described
previously (Ogata et al., 1988).

     The rat hippocampal neurons in primary culture were also used for the
experiments with the GABA-activated chloride current and the excitatory
amino acid-activated current.

     Membrane currents were recorded by the whole cell patch clamp
techniques (Hamill et al., 1981) at room temperature. The external and
internal solutions used were slightly different in different sets of
experiments. Basically, the external solutions contained NaCl or choline
chloride as the major component, and the internal solutions contained
Cs-glutamate or CsCl as the major component (Table 1).

     Drugs were applied either to the bath through the inlet tubing or by
puffer for 20-30 sec.

GABA-ACTIVATED CURRENTS

     Before studying the effects of chemical agents on the GABA-activated
current, the species of ions that carry the current must be identified.
Figure 1 illustrates the results of such experiments (Ogata et al., 1988).
Bath application of $1x10^{-5}$M GABA to the dorsal root ganglion neurons caused
an inward current to flow. The current after attaining a peak decreased to
a lower level and decayed slowly to the original baseline level (Fig. 1A).
During the current flow, the membrane conductance increased as shown by an
increase in current amplitude associated with short hyperpolarizing pulses.
Bicuculline blocked the current completely (Fig. 1B). Repeated application
of $1x10^{-5}$M GABA with a short interval (<2 min) generated a small slow
current only without a transient current (Fig. 1C1). A higher concentration
of GABA ($1x10^{-3}$M) applied immediately after $1x10^{-5}$M GABA generated a slow
current which was slightly larger than that associated with $1x10^{-5}$M GABA
(Fig. 1C2). These results indicated that the GABA-induced current was
desensitized by a prolonged or repeated application of GABA, and that the
current was sensitive to bicuculline suggesting $GABA_A$ type.

If the GABA-induced current is mediated by GABA$_A$ receptors, the current is expected to be carried by chloride ions. To prove the validity of this hypothesis, the GABA-induced currents were measured at different chloride concentrations, and the current-voltage curves were plotted (Fig. 1D). At 130 mM chloride in the external solution, the reversal potential was estimated to be -50.3 ± 2.0 mV. When the external chloride concentration was reduced to 20 mM, the reversal potential was shifted to -2.5 mV in a manner predicted by the Nernst potential for chloride. Thus it was concluded that the GABA-induced current was carried by chloride ions.

Table 1. Compositions of External and Internal Solutions

Component	External Solution		Internal Solution	
	I	II	I	II
	mM	mM	mM	mM
NaCl	120	10	--	--
CsCl	5	5	--	140
Cs Glutamate	--	115	--	--
Choline Chloride	--	--	136	--
CaCl$_2$	1.8	--	2	--
MgCl$_2$	1	2.5	1	1
Glucose	25	5	--	--
HEPES[a]	5	--	10	10
PIPES[b]	--	5	--	--
EGTA[c]	--	5	--	5
Tetrodotoxin	0.001	--	--	--
pH	7.4	7.0	7.3	7.3

[a]N-2-hydroxyethyl-piperazine-N'-2-ethanesulfonic acid
[b]Piperazine-N,N'-bis(2-ethanesulfonic acid)
[c]Ethyleneglycol-bis-(β-aminoethyl ether)N,N,N',N'7-tetraacetic
acid

The degree of desensitization of GABA-induced chloride current depends on the concentration of GABA (Nakahiro et al., 1989). At a low concentration (3×10^{-6}M), GABA induced a small and steady-state current without a sign of desensitization (Fig. 2A). At a medium concentration (3×10^{-5}M), GABA-induced current attained a larger peak and was desensitized to a low steady-state level (Fig. 2B). At a high concentration (3×10^{-4}M), the peak current amplitude did not increase further but was desensitized more quickly and to a lower steady-state level (Fig. 2C). The desensitization of the GABA-activated chloride current was also discovered in other preparations including frog dorsal root ganglion neurons (Akaike et al., 1986, 1987; Yasui et al., 1985). Although the mechanism by which the apparent two components of chloride current are produced remains to be seen, it is of interest to see three or four conductance states in single GABA-induced currents (Bormann et al., 1987).

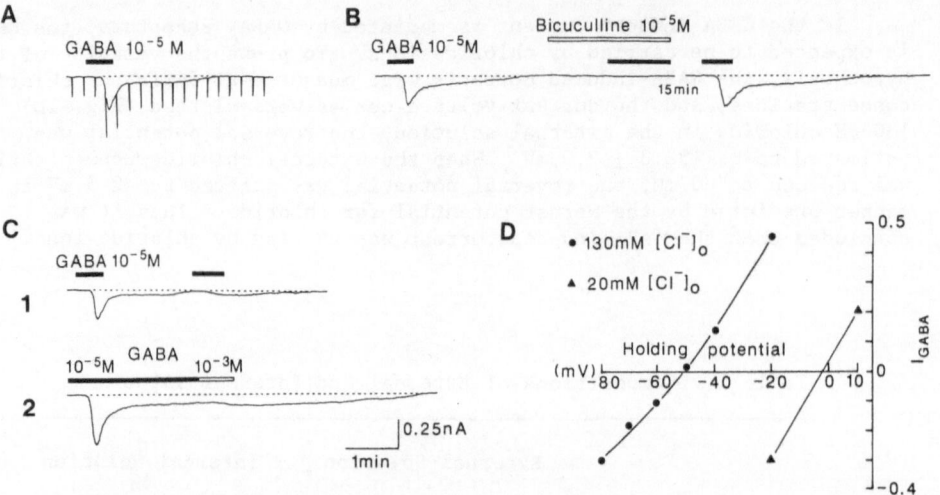

Fig. 1. GABA-induced currents in the rat dorsal root ganglion neurons in primary culture. The currents were recorded on a pen recorder. A. Inward current induced by bath application of 1×10^{-5}M GABA during the time indicated by a horizontal bar. Holding potential -70 mV. Hyperpolarizing 600-msec pulses to -110 mV were applied to monitor the change in membrane conductance. B. Bicuculline block of GABA-induced current. C. A second application of 1×10^{-5}M GABA within a short interval (record 1) failed to evoke a transient current while a small sustained current was still induced. An increase in GABA concentration to 1×10^{-3}M after the decay of the initial component (record 2) produced only an increased sustained component. D. Current-voltage relationships of the GABA-induced transient current in 130 mM and 20 mM chloride in external solutions (from Ogata et al., 1988).

EFFECTS OF GENERAL ANESTHETICS ON GABA-ACTIVATED CURRENTS

General anesthetics caused drastic changes in GABA-induced chloride current; they increased the current before desensitization took place and decreased it after desensitization (Nakahiro et al., 1989). Halothane, enflurane and isoflurane were tested at concentrations equivalent to 1-2 minimum alveolar concentrations (MACs), and all of them had essentially the same effects on the current.

An example of an experiment with halothane is illustrated in Fig. 3A. The chloride current induced by a low concentration of GABA (3×10^{-6}M) was increased substantially by 0.86 mM halothane. The effect was completely reversible upon washing out halothane, and could be repeated by a second application of halothane. When a medium concentration of GABA (3×10^{-5}M) was used, 0.96 mM isoflurane increased the current before desensitization and suppressed it after desensitization (Fig. 3B). It should be noted that isoflurane itself did not generate an inward current as can be seen in the middle record of Fig. 3B before GABA was introduced. The differential effects of anesthetics on two components of GABA-induced current are more

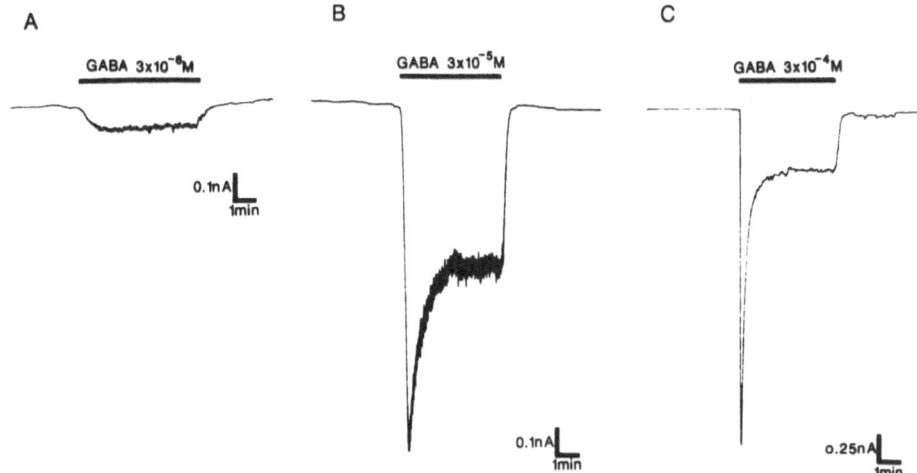

Fig. 2. Currents induced by bath application of various concentrations of
GABA. A. GABA at low concentration ($3x10^{-6}M$) generated a steady
inward current. B. At a medium concentration ($3x10^{-5}M$), GABA
induced a peak transient current that decayed to a low steady-state
level. C. At a high concentration ($3x10^{-4}M$), the current attained a
maximum peak level and decayed to a very low steady-state level. The
three records were obtained from different neurons (from Nakahiro et
al., 1989).

clearly seen in the experiment shown in Fig. 4. Enflurane 1.89 mM applied
during the falling phase of GABA-induced current caused a transient increase
in current and a sizable decrease following desensitization (Fig. 4A).
After completion of desensitization, enflurane simply decreased the current
(Fig. 4B).

The aforementioned experimental results raise a few important issues.
First of all, anesthetics exerted the effects on GABA-induced current at
clinically relevant concentrations equivalent to 1-2 MACs. Thus these
effects are likely to play an important role in general anesthesia. Second,
the increase in current before desensitization takes place appears to be
relevant to clinical situation, because GABA is normally released from
inhibitory nerve terminals for a short period of time. Thus the observed
augmentation of GABA-induced peak current would increase synaptic inhibition
leading to general anesthesia. Third, although various general anesthetics
have been shown to suppress other ion channel currents such as
voltage-activated sodium and potassium currents (Haydon and Urban, 1983,
1986) and acetylcholine-activated current (Lechleiter and Gruener, 1984),
the concentrations of general anesthetics required for these effects are in
many cases higher than the clinical concentrations to produce surgical
anesthesia. The fact that general anesthetics affect various ion channels
at different concentration levels is not compatible with the prevailing
notion that they produce effects through general membrane actions such as
changes in fluidity and expansion of the membrane (Franks and Lieb, 1987;
Janoff and Miller, 1982; Roth, 1979; Seeman, 1972). It appears that general
anesthetics have a specific affinity for the GABA receptor-channel complex,
and that the augmentation of its activity leads to surgical anesthesia.

A

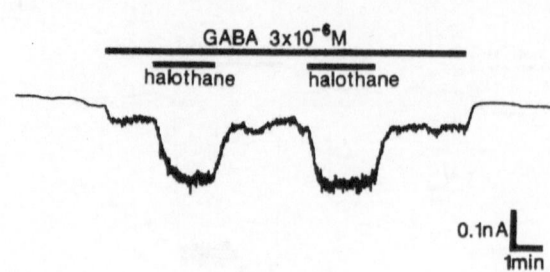

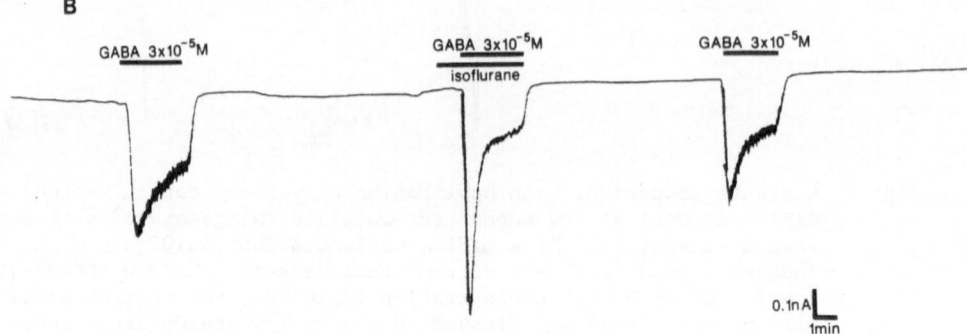

Fig. 3. Effects of halothane (0.86 mM) and isoflurane (0.96 mM) on the currents induced by 3×10^{-6}M (A) and 3×10^{-5}M (B) GABA. The current before desensitization was augmented by the anesthetics, while the current after desensitization was suppressed (from Nakahiro et al., 1989).

EFFECTS OF ALCOHOLS ON GABA-INDUCED CURRENTS

GABA receptor-channel complex has been suspected to be a target site of ethanol. For example, GABA-induced ^{36}Cl influx in neurons and synaptosomes was enhanced by ethanol (Mehta and Ticku, 1988; Suzdak et al., 1986; Tucku et al., 1986). However, electrophysiological experiments have yielded controversial results; in some experiments enhancement of GABA responses was observed whereas in other experiments no effect on GABA responses was noted (Celentano et al., 1988; Mancillas et al., 1986; Nestoros, 1980; Siggins et al., 1987). In order to clarify the point, patch clamp experiments were performed with the rat dorsal root ganglion neurons (Nishio and Narahashi, 1988; Nakahiro et al., 1990).

Ethanol increased the peak transient current induced by GABA while having little or no effect on the current after desensitization. The amount of increase in peak current was concentration dependent, being 20, 24.5 and 32.6% at 30, 100 and 300 mM ethanol, respectively. These increases are statistically significant.

In order to lend further support to the notion that ethanol-induced increase in GABA-activated chloride current is an important factor responsible for the depressant action of ethanol on animals and humans, longer-chain alcohols were tested for their effects on the GABA system. It

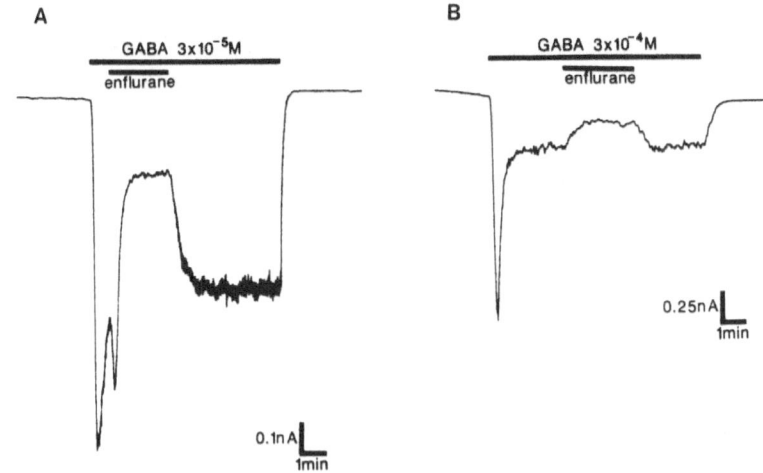

Fig. 4. The effects of enflurane (1.89 mM) on the currents induced by
$3x10^{-5}$M (A) and $3x10^{-4}$M (B) GABA. The currents were recorded from
different neurons. Enflurane increased the current before
desensitization took place, and decreased it after desensitization
(from Nakahiro et al., 1989).

is well known that the potency of alcohol-induced anesthesia increases with
lengthening the carbon chain of the alcohol molecule (Janoff and Miller,
1982; Miller et al., 1987). Therefore, if the ethanol-induced augmentation
of GABA-activated current is significant in ethanol intoxication, higher
alcohols are expected to augment the current even more strongly.

The results of patch clamp experiments with the rat dorsal root
ganglion neurons support the above notion (Nakahiro et al., 1990). Ethanol,
n-butanol, n-hexanol, n-octanol, and n-decanol enhanced the GABA-induced
peak current in a dose-dependent manner, and the potency increased with the
length of the carbon chain. n-Octanol was almost 10,000 times more potent
than ethanol. Furthermore, the efficacy also increased with lengthening of
the carbon chain. Thus, it was concluded that the enhancement of the
GABA-induced chloride peak current by alcohols including ethanol plays an
important role in intoxication.

ROLE OF GABA RECEPTOR-CHANNEL COMPLEX IN INSECTICIDAL ACTION

The GABA receptor-channel complex plays an important role in
intoxication with certain insecticides. Before introducing the results of
experiments, a brief description of the mechanisms of action of insecticides
on the nervous system will be given in order to provide the whole picture of
the field. Several review articles have been published (Narahashi, 1971,
1976, 1985, 1987, 1988, 1990; Ruigt, 1984; Wouters and van den Bercken,
1978; Woolley, 1981).

The major target sites of insecticides may be divided into three
groups, i.e. cholinesterases, voltage-activated sodium channels and
GABA-activated chloride channels. Organophosphate and carbamate
insecticides (e.g. parathion, malathion, fonofos, diazinon, carbaryl,
carbofuran) inhibit acetylcholinesterase in their original form or activated

form (e.g. paraoxon, malaoxon) thereby causing the symptoms of poisoning in animals. Pyrethroids and DDT modulate the voltage-activated sodium channel to cause a very prolonged opening which in turn results in a prolonged sodium current. This leads to hyperexcitation of the nervous system. Cyclodienes and lindane suppress the GABA-activated chloride current thereby causing hyperactivity. Whereas the inhibition of acetylcholinesterase by organophosphate and carbamate insecticides and the modulation of the sodium channel by pyrethroids and DDT are thoroughly studied and demonstrated, the role of the GABA system in insecticidal action remains largely to be seen. Furthermore there has been some controversy over its role in pyrethroid action. Therefore, we have performed extensive experiments to characterize and determine the action of insecticides on the GABA-activated chloride current.

Pyrethroids

Whereas the pyrethroid-induced modulation of voltage-activated sodium channel is well documented, there have been some observations to indicate that type II pyrethroids containing a cyano group in the α position suppress the activity of the GABA receptor-channel complex. The bindings of [^{35}S]t-butylbicyclophosphorothionate (TBPS) and Ro5-4864 to the GABA receptor-channel complex were inhibited by type II pyrethroids but not type I pyrethroids which are devoid of an α-cyano group (Crofton et al., 1987; Gammon and Sander, 1985; Lawrence and Casida, 1983; Lawrence et al., 1985; Lummis et al., 1987). GABA-induced chloride uptake by mouse brain vesicles was also inhibited by the type II pyrethroid deltamethrin (Bloomquist and Soderlund, 1985). However, the effective concentrations for these actions were high.

Patch clamp experiments with the rat dorsal root ganglion neurons have clearly demonstrated that the GABA-induced chloride current was not affected by deltamethrin while the sodium current was greatly prolonged (Ogata et al., 1988). An example of such an experiment is shown in Fig. 5. The same neuron was stimulated by GABA and a depolarizing pulse alternatively, and the effects of 1×10^{-5}M deltamethrin were examined. The sodium current induced by a depolarizing pulse was greatly prolonged 10 min after application of deltamethrin, while the GABA-induced chloride current remained unchanged. Thus it was concluded that the GABA system plays little or no role in intoxication with deltamethrin. Some other observations do not support the GABA hypothesis (Matsumura and Tanaka, 1984; Chalmers et al., 1985).

In an attempt to settle the controversy over the GABA hypothesis, we have performed more detailed patch clamp experiments with the rat hippocampal neurons in primary culture (Frey et al., 1989; Frey and Narahashi, 1989). One of the critical factors that influence the action of pyrethroids is the period of time of exposure to pyrethroids. When $1-5 \times 10^{-5}$M fenvalerate, a type II pyrethroid, was applied together with GABA or shortly before GABA by puffer, the GABA-induced chloride current was increased. This was also true when the chloride current was induced by pentobarbital instead of GABA. Fenvalerate or deltamethrin enhanced the pentobarbital-induced chloride current when applied concurrently with GABA. However, the enhancement became less when 1×10^{-5}M deltamethrin was continuously applied for a long time (5-15 min) prior to pentobarbital application. Thus, it is possible that in binding and other experiments in which type II pyrethroids were incubated with the preparation for long periods of time a decrease was observed. The type I pyrethroid allethrin (5×10^{-5}M) decreased rather than increased the pentobarbital-induced chloride current when applied concurrently with pentobarbital. The toxicological significance of these effects of pyrethroids remains to be seen. However, it is clear that the sodium channel modulation is the primary action of

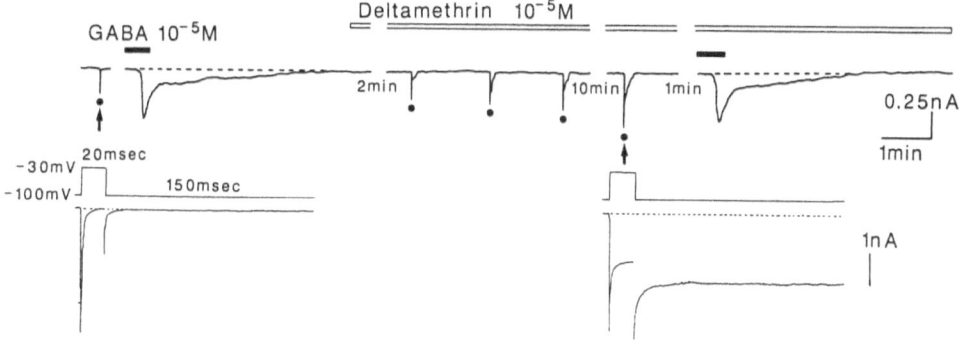

Fig. 5. Effects of deltamethrin on the voltage-activated sodium current and
the GABA-activated chloride current. At points indicated by dots,
20-msec depolarizing pulses to -30 mV were applied to generate
sodium currents. At the two points indicated by upward arrows, the
sodium currents were recorded on an oscilloscope. Whereas the
GABA-induced current remained unchanged by deltamethrin, the sodium
current underwent drastic changes being prolonged during and after
the depolarizing pulse (from Ogata et al., 1988).

pyrethroids in producing the symptoms of poisoning in animals such as
hyperexcitability, hypersensitivity, convulsions and tremors (Narahashi,
1988, 1990).

The pyrethroid insecticides were found to suppress the current induced
by N-methyl-d-aspartate (NMDA) in rat hippocampal and neocortical neurons
maintained in primary culture (Frey and Narahashi, 1990). Both type I and
type II pyrethroids including allethrin and deltamethrin suppressed the
NMDA-induced current by 35-60% at $1-5 \times 10^{-5}$M. The rate of desensitization
was often accelerated. Glycine attenuated the pyrethroid block of
NMDA-induced current. The toxicological significance of these effects of
pyrethroids remains to be seen.

Lindane and Cyclodienes

Lindane and cyclodiene insecticides are known to stimulate synaptic
transmission producing prolonged postsynaptic after-discharges in response
to a single presynaptic stimulus (Narahashi, 1971; Yamasaki and Ishii
(Narahashi), 1954; Yamasaki and Narahashi, 1958). More recently,
GABA-induced ^{36}Cl uptake by cockroach muscles and mouse brain vesicles and
the binding of [^{3}H]α-dihydropicrotoxinin and [^{35}S]TBPS to rat brain membrane
preparations were found to be antagonized by lindane and cyclodienes
(Ghiasuddin and Matsumura, 1982; Matsumura and Ghiasuddin, 1983; Bloomquist
and Soderlund, 1985). In order to demonstrate these actions in a more
straightforward manner, patch clamp experiments were performed with the rat
dorsal root ganglion neurons and rat hippocampal neurons (Ogata et al.,
1988; Frey et al., 1989; Narahashi and Frey, 1989).

Lindane at a concentration of 1×10^{-5}M suppressed the GABA-induced
chloride current (Fig. 6). The peak component of the current was completely
blocked, while the slow component was decreased only to a small extent or
not affected at all. The cyclodienes dieldrin, endrin, heptachlor-epoxide
and isobenzan also reduced the GABA-induced current with a relative potency

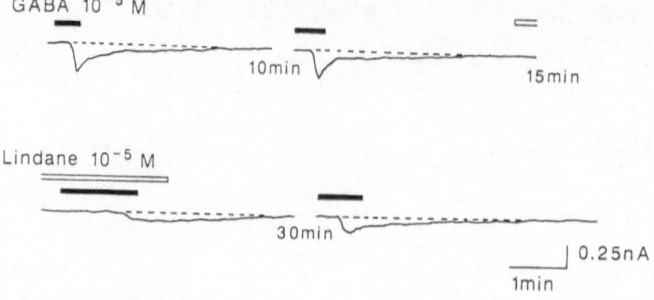

Fig. 6. Effects of lindane on the GABA-induced chloride current. The
transient current was completely blocked by lindane, while the
sustained current remained almost unchanged (from Ogata et al.,
1988).

of endrin = isobenzan > dieldrin = hiptachlor-epoxide. However, the onset
of action of the cyclodienes was much slower than lindane. These blocking
actions on the GABA-induced chloride current are compatible with the results
of ^{36}Cl uptake and binding experiments, and can account for the convulsant
action.

The studies cited in this paper were supported by NIH grants NS14143
and NS14144 and ADAMHA grant AA07836. I wish to thank Vicky James-Houff for
secretarial assistance.

REFERENCES

Akaike, N., Inoue, M., and Krishtal, O. A., 1986, 'Concentration-clamp'
   study of γ-aminobutyric-acid-induced chloride current. Kinetics in frog
   sensory neurones, J. Physiol. (London), 379:171-185.

Akaike, N., Yakushiji, T., Tokutomi, N., and Carpenter, D. O., 1987,
   Multiple mechanisms of antagonism of γ-aminobutyric acid (GABA)
   responses, Cell. Mol. Neurobiol., 7:97-103.

Bloomquist, J. R., and Soderlund, D. M., 1985, Neurotoxic insecticides
   inhibit GABA-dependent chloride uptake by mouse brain vesicles, Biochem.
   Biophys. Res. Comm., 133:37-43.

Bormann, J., Hamill, O. P., and Sakmann, B., 1987, Mechanism of anion
   permeation through channels gated by glycine and γ-aminobutyric acid in
   mouse cultured spinal neurones, J. Physiol (London)., 385:243-286.

Celentano, J. J., Gibbs, T. T., and Farb, D. H., 1988, Ethanol potentiates
   GABA- and glycine-induced chloride currents in chick spinal cord neurons,
   Brain Res., 455:377-380.

Chalmers, A. E., Miller, T. A., and Olsen, R. W., 1985, A pharmacological
   investigation of invertebrate GABA receptors, in: "Neurotox '85.
   Neuropharmacology and Pesticide Action," Univ. Bath, Abstr. pp. 41-42.

Crofton, K. M., Reiter, L. W., and Mailman, R. B., 1987, Pyrethroid
   insecticides and radioligand displacement from the GABA receptor chloride
   ionophore complex, Toxicol. Letters, 35:183-190.

Franks, N. P., and Lieb, W. R., 1987, What is the molecular nature of general anaesthetic target sites? Trends in Pharmacol. Sci., 8:169-174.

Frey, J., and Narahashi, T., 1990, Pyrethroid insecticides block calcium and NMDA-activated currents in cultured mammalian neurons, Biophys. J., in press.

Frey, J., Dichter, M., and Narahashi, T., 1989, Effects of lindane and fenvalerate on GABA-activated chloride currents in cultured hippocampal neurons, The Toxicologist, 9:149.

Gammon, D. W., and Sander, G., 1985, Two mechanisms of pyrethroid action: Electrophysiological and pharmacological evidence, NeuroToxicology 6(2):63-86.

Ghiasuddin, S. M., and Matsumura, F., 1982, Inhibition of gamma-aminobutyric acid (GABA)-induced chloride uptake by gamma-BHC and heptachlor epoxide, Comp. Biochem. Physiol., 73C:141-144.

Hamill, O. P., Marty, A., Neher, E., Sakmann, B., and Sigworth, F. J., 1981, Improved patch-clamp techniques for high-resolution current recording from cells and cell-free membrane patches, Pflügers Arch., 391:85-100.

Haydon, D. A., and Urban, B. W., 1983, The effects of some inhalation anaesthetics on the sodium current of the squid giant axon, J. Physiol. (London), 341:429-439.

Haydon, D. A., and Urban, B. W., 1986, The actions of some general anaesthetics on the potassium current of the squid giant axon, J. Physiol. (London), 373:311-327.

Janoff, A. S., and Miller, K. W., 1982, A critical assessment of the lipid theories of general anaesthetic action, in: "Biological Membranes," Vol. 4, D. Chapman, ed., Academic Press, London, pp. 417-476.

Lawrence, L. J., and Casida, J. E., 1983, Stereospecific action of pyrethroid insecticides on the γ-aminobutyric acid receptor-ionophore complex, Science, 221:1399-1401.

Lawrence, L. J., Gee, K. W., and Yamamura, H. I., 1985, Interactions of pyrethroid insecticides with chloride ionophore-associated binding sites, NeuroToxicology 6(2):87-98.

Lechleiter, J., and Gruener, R., 1984, Halothane shortens acetylcholine receptor channel kinetics without affecting conductance, Proc. Natl. Acad. Sci. USA, 81:2929-2933.

Lummis, S. C. R., Chow, S. C., Holan, G., and Johnston, G. A. R., 1987, γ-Aminobutyric acid receptor ionophore complexes: Differential effects of deltamethrin, dichlorodiphenyltrichloroethane, and some novel insecticides in a rat brain membrane preparation, J. Neurochem., 48:689-694.

Mancillas, J. R., Siggins, G. R., and Bloom, F. E., 1986, Systemic ethanol: Selective enhancement of responses to acetylcholine and somatostatin in hippocampus, Science, 231:161-163.

Matsumura, F., and Ghiasuddin, S. M., 1983, Evidence for similarities between cyclodiene type insecticides and picrotoxinin in their action mechanisms, J. Environ. Sci. Health, B18:1-14.

Matsumura, F., and Tanaka, K., 1984, Molecular basis of neuroexcitatory actions of cyclodiene-type insecticides, in: "Cellular and Molecular Neurotoxicology," T. Narahashi, ed., Raven Press, New York, pp. 225-240.

Mehta, A. K., and Ticku, M. K., 1988, Ethanol potentiation of GABAergic transmission in cultured spinal cord neurons involves γ-aminobutyric acid$_A$-gated chloride channels, J. Pharmacol. Exp. Ther., 246:558-564.

Miller, K. W., Firestone, L. L., and Forman, S. A., 1987, General anesthetic and specific effects of ethanol on acetylcholine receptors. New York Acad. Sci., 492:71-87.

Nakahiro, M., Yeh, J. Z., Brunner, E., and Narahashi, T., 1989, General anesthetics modulate GABA receptor channel complex in rat dorsal root ganglion neurons, FASEB J., 3:1850-1854.

Nakahiro, M., Arakawa, O., and Narahashi, T., 1990, Ethanol and longer chain alcohols enhance GABA-activated Cl⁻ current in rat dorsal root ganglion neurons, FASEB J., in press.

Narahashi, T., 1971, Effects of insecticides on excitable tissues, in: "Advances in Insect Physiology," Vol. 8, J. W. L. Beament, J. E. Treherne and V. B. Wigglesworth, eds., Academic Press, London and New York, pp. 1-93.

Narahashi, T., 1976, Effects of insecticides on nervous conduction and synaptic transmisssion, in: "Insecticide Biochemistry and Physiology," C. F. Wilkinson, ed., Plenum Press, New York, pp. 327-352.

Narahashi, T., 1985, Nerve membrane ionic channels as the primary target of pyrethroids, NeuroToxicology, 6(2):3-22.

Narahashi, T., 1987, Neuronal target sites of insecticides, in: "Sites of Action for Neurotoxic Pesticides," R. M. Hollingworth and M. B. Green, eds., American Chemical Society Symposium Series, No. 356, ACS, Washington, DC, pp. 226-250.

Narahashi, T., 1988, Molecular and cellular approaches to neurotoxicology: Past, present and future, in: "Neurotox '88: Molecular Basis of Drug and Pesticide Action," G. G. Lunt, ed., Elsevier, Amsterdam, pp. 269-288.

Narahashi, T., 1990, The role of ion channels in insecticide action, in: "Insecticide Action: From Molecule to Organism," T. Narahashi and J. E. Chambers, eds., Plenum Press, New York, in press.

Narahashi, T., and Frey, J. M., 1989, Lindane and cyclodiene insecticides block GABA-activated chloride current in cultured rat hippocampal neurons, Abstr. Soc. Neurosci., 15:1151.

Nestoros, J. N., 1980, Ethanol specifically potentiates GABA-mediated neurotransmission in feline cerebral cortex, Science, 209:708-710.

Nishio, M., and Narahashi, T., 1988, Ethanol enhances a component of GABA-gated Cl channel current in rat dorsal root ganglion neurons, Soc. Neurosci. Abstr., 14:642.

Ogata, N., Vogel, S. M., and Narahashi, T., 1988, Lindane but not deltamethrin blocks a component of GABA-activated chloride channels, FASEB J., 2:2895-2900.

Roth, S. H., 1979, Physical mechanisms of anesthesia, Ann. Rev. Pharmacol. Toxicol., 19:159-178.

Ruigt, G. S. F., 1984, Pyrethroids, in: "Comprehensive Insect Physiology, Biochemistry and Pharmacology," Vol. 12, Chapter 7, G. A. Kerkut and L. I. Gilbert, eds., Pergamon Press, Oxford, pp. 183-263.

Seeman, P., 1972, The membrane actions of anesthetics and tranquilizers, Pharmacol. Rev., 24:583-655.

Siggins, G. R., Pittman, Q. J., and French, E. D., 1987, Effects of ethanol on CA_1 and CA_3 pyramidal cells in the hippocampal slice preparation: An intracellular study, Brain Res., 414:22-34.

Suzdak, P. D., Schwartz, R. D., Skolnick, P., and Paul, S. M., 1986, Ethanol stimulates γ-aminobutyric acid receptor-mediated chloride transport in rat brain synaptoneurosomes, Proc. Natl. Acad. Sci. USA, 83:4071-4075.

Ticku, M. K., Lowrimore, P., and Lehoullier, P., 1986, Ethanol enhances GABA-induced ^{36}Cl-influx in primary spinal cord cultured neurons, Brain Res. Bull., 17:123-126.

Woolley, D. W., 1981, The neurotoxicity of DDT and possible mechanisms of action, in: "Mechanisms of Neurotoxic Substances," K. N. Prasal and A. Vernadakis, eds., Raven Press, New York, pp. 95-141.

Wouters, W., and van den Bercken, J., 1978, Action of pyrethroids, Gen. Pharmacol., 9:387-398.

Yamasaki, T., and Ishii (Narahashi), T., 1957, Studies on the mechanism of action of insecticides. X. Nervous activity as a factor of development of γ-BHC symptoms in the cockroach, Botyu-Kagaku (Scientific Insect Control), 19:106-112. English translation, Japanese Contributions to the Study of the Insecticide-Resistance Problem. Published by the Kyoto University for the W.H.O., pp. 176-183.

Yamasaki, T., and Narahashi, T., 1958, Nervous activity as a factor of development of dieldrin symptoms in the cockroach. Studies on the mechanism of action of insecticides, XVI, Botyu-Kagaku (Scientific Insect Control), 23:47-54.

Yasui, S., Ishizuka, S., and Akaike, N., 1985, GABA activates different types of chloride-conducting receptor-ionophore complexes in a dose-dependent manner, Brain Res., 344:176-180.

CHEMICAL KINETIC INVESTIGATIONS OF THE CHANNEL-OPENING PROCESS

OF NEUROTRANSMITTER RECEPTORS

Norio Matsubara, Andrew P. Billington, Hou Chang Chen, Anthony P. Guzikowski, Katherine W. Johnson, Doraiswamy Ramesh, Melody T. Sweet and George P. Hess

Section of Biochemistry, Molecular and Cell Biology, 217 Biotechnology Building, Cornell University, Ithaca, New York 14853-2703 U.S.A.

INTRODUCTION

The application of chemical kinetic investigations, using fast reaction techniques, to the studies of neuronal receptor-mediated reactions has been recently reported (Hess *et al.*, 1979, 1987). However, the use of such mixing techniques is limited to the studies of receptors in membrane vesicles (Hess *et al.*, 1979) with a time resolution of 5 msec, or to measurements with receptor-containing cells (Udgaonkar and Hess, 1987) with a 20 msec time resolution. In order to clarify the activation process of receptors, a new method which has a faster time resolution must be developed. Caged compounds pioneered by Professor Jack Kaplan, University of Pennsylvania, and Professor David Trentham, National Institute for Medical Research, London, are good candidates for this purpose. In contrast to their oxygen-bonded caged compounds, we protected nitrogen atoms included in many neurotransmitters. We have synthesized caged neurotransmitters bonded to a photo-dissociative *o*-nitrobenzyl group with the nitrogen atom. These compounds have shown relatively fast photolysis and have been useful for the study of neuronal receptors.

An inactive photolabile precursor of carbamoylcholine (caged carbamoylcholine) was equilibrated with mouse muscle cells held in the whole-cell current recording mode. A short laser pulse (318-319 nm, 600 ns) liberated carbamoylcholine within 200 µsec and the resulting current was recorded and analyzed. The method is also useful for the study of other neuronal receptors such as glycine receptors on mouse spinal neurons.

RESULTS

Caged Carbamoylcholine Derivatives

A number of photosensitive carbamoylcholine derivatives were synthesized and obtained in purified form by either HPLC or recrystallization. The proposed structures for these compounds (Table 1) were verified by proton NMR, uv absorption, and elemental analysis. Initially, we were interested only in the rates of photolysis. A comparison of the photolysis rates of compound I and III (Table 1) indicated that the trimethyl ammonium group did not alter the rate significantly. Therefore, the trimethyl ammonium was not included in compound III and IV. When it was noted that the photolysis rate of compound IV is 65 times slower than that of compound II, its further investigation was abandoned. Though the photolysis rate of compound II and V are both fast enough for the investigation of channel-opening kinetics of acetylcholine receptors, some of the properties of compound V made it

Neuroreceptor Mechanisms in Brain, Edited by S. Kito *et al.*
Plenum Press, New York, 1991

Table 1. Caged carbamoylcholine derivatives [a] (reprinted with permission from Milburn *et al.*, 1989)

compound	λ_{max} (nm)	ϵ (M^{-1} cm^{-1})	$t_{1/2}$ (ms)	k (s^{-1})
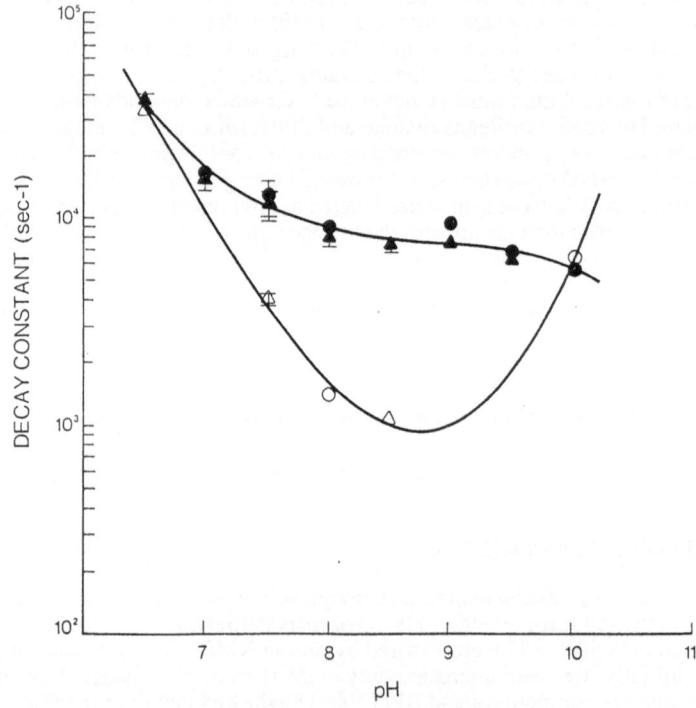 *N*-(2-nitrobenzyl)carbamoylcholine iodide (I)[b]	262	5200	1.7 ± 0.1	410
N-[1-(2-nitrophenyl)ethyl]carbamoylcholine iodide (II)[b]	262	5200	0.067 ± 0.002	10 000
2-bromoethyl *N*-(2-nitrobenzyl)carbamate (III)			0.82[c]	850[c]
2-bromoethyl *N*-(4-carboxy-2-nitrobenzyl)carbamate (IV)	260	5100	4.4	160
N-(α-carboxy-2-nitrobenzyl)carbamoylcholine trifluoroacetate (V)	266	5200	0.04 ± 0.001	17 000

[a] Measurements of $t_{1/2}$ and k were made in aqueous buffer at pH 7.0 and 23 °C. λ_{max} and ϵ_M were measured in aqueous solution and were independent of pH above 6. [b] Walker et al. (1986). [c] Solution contains 5% (v/v) methanol.

Figure 1 Rate constants for the decay of the transient intermediate in the photolysis of 2 mM-compound V as a function of pH measured at both 429 nm (▲) and 400 nm (●). The data for compound II, measured at 429 (△) and 406 nm (O), are shown for comparison.

more useful than compound II. As shown in Fig. 1, the photolysis rate of compound II is sensitive to pH around physiological conditions (pH 7-8) and slower than that of compound V in this region. Although compound V is inactive prior to photolysis, compound II binds to acetylcholine receptors and inhibits and desensitizes the receptors.

Photolysis Mechanism

The proposed mechanism (McCray *et al.,* 1980) for the photolysis of caged phosphates suggested that the photolytic cleavage of the 2-nitro benzyl moiety involves an intramolecular oxidation-reduction reaction between the nitro group and the adjacent benzylic carbon, resulting in the cleavage of the carbon-nitrogen bond to release free carbamate (Fig. 2). When spectral changes in the near uv were monitored by flash photolysis, an intermediate was detected with absorption properties different from those of the parent

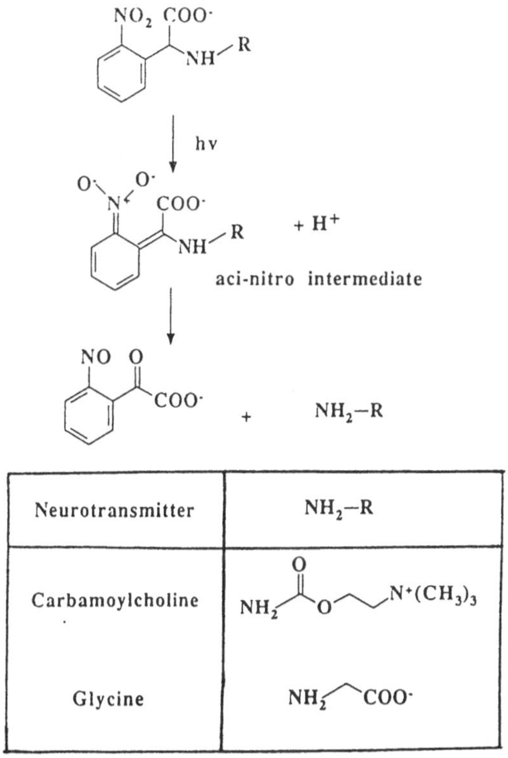

Figure 2 Proposed mechanism of caged neurotransmitter photolysis. A laser pulse (1 μsec duration) from a Candella (dye) laser at wavelengths between 308 and 355 nm was used to photolyze the caged neurotransmitters.

compound V. The intermediate was similar in its absorption spectrum to the *aci*-nitro intermediate originally identified in the flash photolysis of 2-nitrotoluene (Wettermark, 1962), caged ATP (McCray *et al.*, 1980) and more recently caged carbamoylcholine (compound II) (Walker *et al.*, 1986). The presumed *aci*-nitro species showed a pH-dependent exponential decay to products with time constants in the microsecond to millisecond time region.

Measurements of the Rising Phase of Channel-Opening Current

Flash photolysis was done using a Candella SLL-uv 600 dye-laser. 4×10^{-4} M Rhodamine 640 or Sulforhodamine 640 (Exciton) was used as the dye. A frequency-doubled secondary beam with a wavelength of 318-319 nm, energies up to 1 mJ, and a pulse length of 600 nsec was focused on the cell. The extracellular solution used in the measurements was 145 mM NaCl, 5 mM KCl, 1.8 mM $CaCl_2$, 1.7 mM $MgCl_2$ and 25 mM HEPES (pH 7.4). The intracellular solution was 140 mM KCl, 10 mM NaCl, 2 mM $MgCl_2$, 1 mM EGTA, and 25 mM HEPES (pH 7.4). Experiments were done at room temperature. The cells were equilibrated with caged carbamoylcholine dissolved in extracellular solution in the culture dish prior to photolysis. The concentration of carbamoylcholine produced by flash photolysis was calibrated by flowing a known concentration of carbamoylcholine over the cell (Udgaonkar and Hess, 1987) prior to the flash and comparing the current amplitudes.

Fig. 3a shows the current trace of the acetylcholine receptors of a BC3H1 cell observed by flash photolysis of caged carbamoylcholine at room temperature (22-23° C). At time 0, the BC3H1 cell equilibrated with the extracellular solution containing 200 μM caged carbamoylcholine, voltage-clamped at -60 mV, was exposed to a laser flash, which caused a small artifact in the recording and produced the carbamoylcholine, which opened and desensitized the acetylcholine receptors. This is shown on a separate time scale in Fig. 3a. The rising phase can be fitted by the following equation.

$$I_t = I_{max} - I_{max} \times e^{-kt}$$

where I_t is the current detected at time t, I_{max} is the maximum observed current, and k is the rate constant. The rising phase can be linearized by a logarithmic plot, shown in Fig. 3b. These traces show that, after an initial short period during the photolysis of caged carbamoylcholine, the current increases as a single exponential component, which proves that the rate-determining step is a single step of the opening mechanism.

The relationship between carbamoylcholine concentration and the rate constant, k_{obs}, calculated by fitting the current rising phase after flash photolysis of caged carbamoylcholine is shown in Fig. 4. The opening process is described in the following equation.

$$R \overset{K_1}{\underset{}{\rightleftharpoons}} AR \overset{K_1}{\underset{}{\rightleftharpoons}} A_2R \overset{k_{cl}}{\underset{k_{op}}{\rightleftharpoons}} \overline{A_2R}$$

The receptor binds to two molecules of agonist A, with the dissociation constant of K_1, and changes conformation to the open form, $\overline{A_2R}$, with the rate constant of k_{op}, and closes with the rate constant of k_{cl}. We can express the rate constant of the current rising phase, k_{obs}, with K_1, k_{op}, k_{cl}, and L,

$$k_{obs} = k_{cl} + k_{op} \times (L/(L + K_1))^2$$

as suggested previously (Udgaonkar and Hess, 1986). The data points were fitted according to this equation by a non-linear least square program with k_{cl}, k_{op}, and K_1 as parameters. The values of $k_{cl} = 580$ s^{-1}, $k_{op} = 9400$ s^{-1}, and $K_1 = 210$ μM were obtained from the fitting.

Similar experiments were performed with central nervous system receptors. Mouse spinal cord neurons dissociated from 12-day old embryos were cultured *in vitro* for 10 days. The culture media was replaced by recording medium which contained 145 mM NaCl, 1.8 mM $MgCl_2$, 1 mM $CaCl_2$ and 10 mM HEPES (pH 7.4). After the cells were equilibrated with recording medium containing 550 μM caged glycine (α-[(carboxymethyl)amino]-2-nitrobenzeneacetic acid) at room temperature, voltage-clamped (-40 mV) spinal cord cell in the whole-cell recording mode was exposed to a single laser pulse, and the current was recorded (Fig. 5). The results show the rapid activation process followed by desensitization of the receptors.

$$k_{obs} = k_{cl} + k_{op} \times (L/(L + K_1))^2$$

The values of $k_{cl} = 580$ s^{-1}, $k_{op} = 9400$ s^{-1}, $K_1 = 210$ μM were obtained as the best fit.

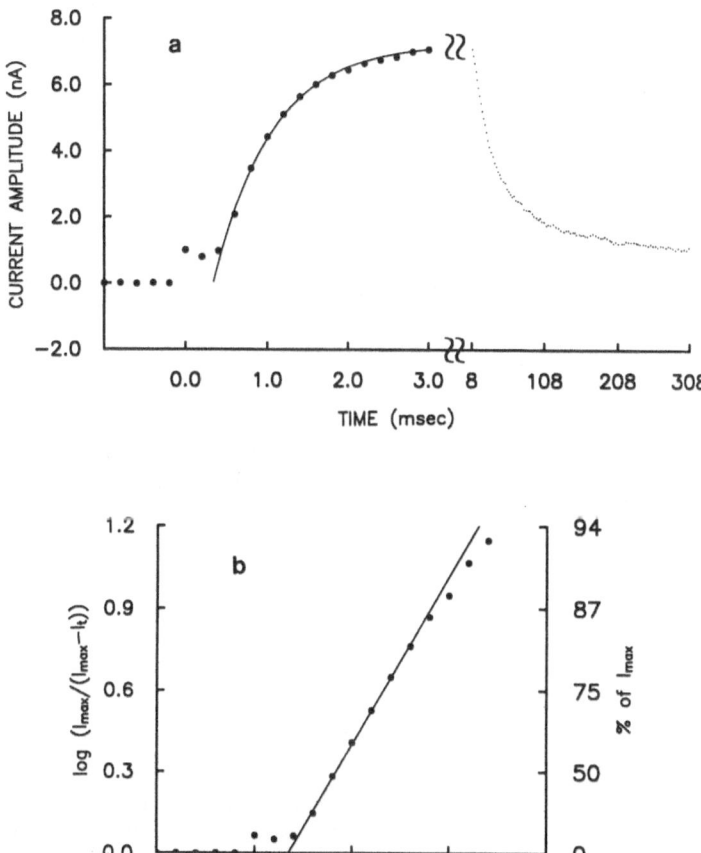

Figure 3(a) The whole-cell current generated by flash photolysis of 200 μM caged carbamoylcholine at pH 7.4, 22-23° C, and -60 mV. Excitation wavelength was 318 nm. **(b)** The first-order plot of the current rising phase in the upper Figure. From the slope, the rate constant of 1400 s^{-1} was obtained.

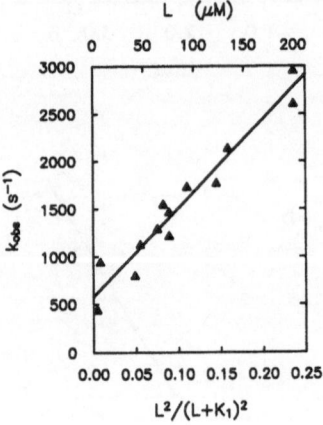

$$R \;\overset{K_1}{\rightleftharpoons}\; AR \;\overset{K_1}{\rightleftharpoons}\; A_2R \;\overset{k_{op}}{\underset{k_{cl}}{\rightleftharpoons}}\; \overline{A_2R}$$

$$k_{obs} \;=\; k_{cl} \;+\; k_{op} \;\times\; L^2/(L+K_1)^2$$

Figure 4 The relationship between concentration of photolyzed carbamoylcholine and observed rate constant (k_{obs}) of the current rising phase calculated by fitting. The values of k_{cl}, k_{op}, K_1 were obtained by the following equation using a non-linear least-square fitting program.

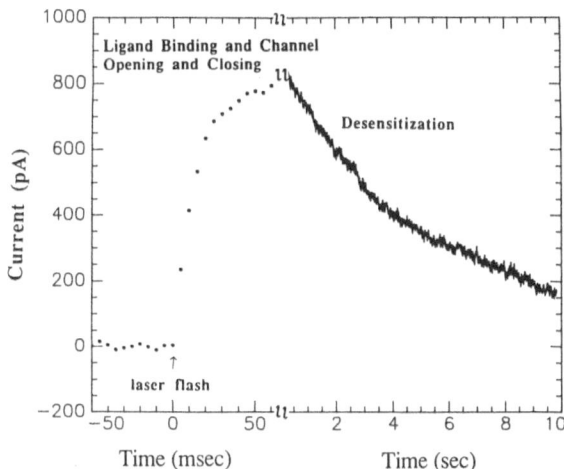

Figure 5 Spinal cord cell response to caged glycine. Caged glycine (550 μM) was photolyzed by an Eximer laser (308 nm) in the presence of a spinal cord cell under voltage clamp conditions.

FUTURE PROSPECTS

The application of this fast ligand releasing technique is not limited to small cells. We are now applying the method to exogenous receptors expressed in *Xenopus* oocytes. Though equilibration by a flow mixing technique is slow because of the large size oocytes, and problems from desensitization are serious, laser flash photolysis overcomes this defect.

We also hope to turn to another system, the nematode *C. elegans*. One unique feature of the *C. elegans* is that the locations of all its 302 neurons are known from the electron micrograph studies. However, the identity, the function of the neurotransmitters, and therefore, of receptors, at the single neuron level is not known. We believe caged compounds are useful for the identification of receptors combining with a laser ablation technique in *C. elegans* neurons developed by Professor Horvitz at M.I.T.

REFERENCES

Hess, G.P., Cash, D.J., and Aoshima, H. (1979). Acetylcholine receptor-controlled ion fluxes in membrane vesicles investigated by fast reaction techniques. *Nature (London)* **282**, 329-331.

Hess, G.P., Udgaonkar, J.B., and Olbricht, W.L. (1987). Chemical kinetic measurements of transmembrane processes using rapid reaction techniques: Acetylcholine receptor. *Ann. Rev. Biophys. Biophys. Chem.* **16**, 507-533.

McCray, J.A., Herbette, L., Kihara, T., and Trentham, D.R. (1980). A new approach to time-resolved studies of ATP-requiring biological systems: Laser flash photolysis of caged ATP. *Proc. Natl. Acad. Sci. U.S.A.* **77**, 7237-7241.

Milburn, T., Matsubara, N., Billington, A.P., Udgaonkar, J.B., Walker, J.W., Carpenter, B.K., Webb, W.W., Marque, J., Denk, W., McCray, J.A. and Hess, G.P. (1989). Synthesis, photochemistry, and biological activity of a caged photolabile acetylcholine receptor ligand. *Biochemistry* **28**, 49-55.

Udgaonkar, J.B., and Hess, G.P. (1986). Acetylcholine receptor kinetics: Chemical kinetics. *J. Membr. Biol.* **93**, 93-109.

Udgaonkar, J.B., and Hess, G.P. (1987). Chemical kinetic measurements of a mammalian acetylcholine receptor by a fast-reaction technique. *Proc. Natl. Acad. Sci. U.S.A.* **84**, 8758-8762.

Walker, J.W., McCray, J.A., and Hess, G.P. (1986). Photolabile protecting groups for an acetylcholine receptor ligand. Synthesis and photochemistry of a new class of *o*-nitrobenzyl derivatives and their effects on receptor function. *Biochemistry* **25**, 1799-1805.

Wettermark, G. (1962). Photochromism of *o*-nitrotoluenes. *J. Phys. Chem.* **66**, 2560-2562.

PRIMARY STRUCTURE AND FUNCTIONAL EXPRESSION OF THE INOSITOL 1,4,5-TRISPHOSPHATE RECEPTOR, P400

Katsuhiko Mikoshiba[1,2], Teiichi Furuichi[2], Nobuaki Maeda[1], Shingo Yoshikawa[2], Atsushi Miyawaki[2], Michio Niinobe[1] and Kentaro Wada[1]

[1]Division of Regulation of Macromolecular Function, Institute for Protein Research, Osaka University, 3-2 Yamadaoka, Suita, Osaka 565, Japan. [2]Division of Behavior and Neurobiology, National Institute for Basic Biology, 38 Nishigonaka, Myodaiji-cho, Okazaki 444, Japan

INTRODUCTION

There are several signal transduction pathways inside the cell, such as the C-kinase, A-kinase, and calmodulin dependent protein kinase systems. These systems exert physiological functions through phosphorylation. However, it is well known that calcium ions play an important role in physiological functions, and recently, it has been demonstrated that inositol trisphosphate, a counterpart of diacylglycerol (DG) hydrolysed from phosphatidyl inositol di-phosphates, mobilizes calcium ions from calcium storage sites inside the cell (Berridge, 1987). Although many researchers have been working on calcium mobilization, no detailed mechanism was known. Many questions have arisen, for example, do putative InsP3 receptors have both InsP3 binding sites and calcium channels or not, is the calcium channel a different molecule from the InsP3 receptor or not.

Quite recently, we succeeded in cloning and sequencing the cDNA of the InsP3 receptor as the result of the studying the development of the cerebellum using mutant mice (Maeda et al., 1988, 1989, 1990; Furuichi et al.,1989). We also analyzed the molecular mechanism of the InsP3 receptor.

Purkinje cell-deficient mutant mice and P400 protein

The mouse cerebellum contains five different types of neurons. Of these, the Purkinje cell is considered to play a central role in information processing, since it receives all the inputs and is the only neuron that sends outputs from the cerebellar cortex. There are many cerebellar ataxic mutants, and nervous and pcd mutants are Purkinje cell-deficient mutants. The high molecular weight protein referred to as P400 protein is greatly decreased in the cerebella of these mutants (Mallet et al., 1976; Mikoshiba et al.,1979; Maeda et al.,1989).

Characterization of P400 protein

P400 protein has an apparent molecular mass (Mr) of 250,000 (250K) and is phosphorylated (Mikoshiba et al.,1985; Maeda et al., 1988; Yamamoto et al., 1989). Endo-β-N-acetylglucosaminidase F digestion of P400 protein revealed that P400 protein has a small number of asparagine-linked oligosaccharide chains (Maeda et al.,1988). When the membrane fraction of the cerebellum was phosphorylated in the presence of ^{32}P-ATP, P400 protein was found to be highly phosphorylated (Yamamoto et al., 1989). It was one of the major proteins phosphorylated. Purified P400 protein was phosphorylated by both the purified catalytic subunit of cylcic AMP-dependent protein kinase (A-kinase) and calcium/calmodulin-dependent protein kinase II (CaM kinase II). Immunoprecipitation with monoclonal antibody to P400 protein confirmed that the P400 protein itself is definitely phosphorylated by the catalytic subunit of A-kinase and CaM kinase II. A-kinase phosphorylated only the seryl residue in P400 protein. Incubation of cultured cerebellar cells with [^{32}P]orthophosphate resulted in the labeling of P400 protein. These results suggest that P400 protein is one of the best substrates for A-kinase and CaM kinase II in the cerebellum and that the function of P400 protein is regulated by cyclic AMP- and calcium/calmodulin-dependent phosphorylation (Yamamoto et al.,1989).

Monoclonal antibodies against P400 protein

Three independent monoclonal antibodies against P400 protein, i.e., 4CII, IOA6, and I8AIO were prepared (Maeda et al., 1988, 1989). These antibodies were produced by injecting partially purified P400 protein from mouse cerebellum into a rat and fusing the spleen cells from the rat with Sp2 mouse myeloma cells (Maeda et al.,1988).

Table 1. Purification of InsP3 Receptor

Step	Protein	InsP3 binding	Specific activity	Recovery (%)	Purification (fold)
	mg	pmol	(pmol/mg)		
P2 + P3 fraction	500	538	1.1	100	1
Triton extract	260	615	2.4	114	2.2
DE52	44	423	9.6	79	8.7
Heparin agarose	9.9	361	36	67	33
Lentil lectin Sepharose	1.1	215	195	40	177
Hydroxyl-apatite	0.5	169	338	31	307

The first P400 protein-positive Purkinje cells were detected on postnatal day 3. The increase in the expression of P400 protein by Purkinje cells seems to occur around postnatal day 3 and continues until postnatal day 21. This period coincides with the stage of development of the dendritic arborization of Purkinje cells. Synapse formation between the Purkinje cells and the parallel fibers of granule cells also occurs vigorously during this period. The expression of P400 protein is abnormal in the Purkinje cells of the staggerer mutant mouse, which exhibits poor dendritic arborization and a lack of synapses between Purkinje cells and granule cells. No positive reaction was observed in any region of the Purkinje cell in the staggerer cerebellum, indicating that the Purkinje cells in staggerer cannot promote the biosynthesis of P400 protein (Maeda et al.,1989).

Electron microscopic observation showed that immunoreactivity was localized on the smooth endoplasmic reticulum. The Golgi complex, mitochondria and nuclei were not stained (Maeda et al.,1989). There was also a positive reaction in the cytoplasmic membrane region. It is necessary to determine whether the reaction in the cytoplasmic membrane region is located in the membrane or just beneath the membrane or the reaction on the endoplasmic reticulum adjacent to the plasma membrane.

Identification of P400 protein as an InsP3 receptor

Recently Crepel et al.(1984) reported that the calcium spike is absent in the Purkinje cells of the staggerer mutant. Since P400 protein is greatly decreased in the staggerer cerebellum, we presumed that calcium metabolism is abnormal, including the calcium spike in the staggerer Purkinje cells.

We first screened the various proteins reported to be enriched in the Purkinje cells of the cerebellum. However, most of them are soluble proteins and their molecular weight is much smaller than that of the P400 protein. Recently Supattapone et al.(1988a) purified an InsP3-binding protein from the rat cerebellum whose molecular weight was 260kDa. The binding protein is a glycoprotein capable of binding to concanavalin A and is a substrate for cAMP-dependent protein kinase (Supattapone et al.,1988ab). They also showed that $[^{3}\text{H}]$InsP3-binding activity is very much higher in the cerebellum than in any other brain area and is selectively localized in Purkinje cells (Worley et al.,1989). On the basis of these results, we presumed that InsP3 binding protein was very similar to P400 protein in its characteristics. In order to determine whether P400 protein is identical to InsP3 binding protein or not, we first compared the $[^{3}\text{H}]$InsP3 binding autoradiographic pattern with that of the immunohistochemical staining of the cerebellar sections by P400 antibody. In the cerebellar sections from normal mice, the InsP3 binding pattern was similar to that of the immunohistochemical staining. When we analyzed the cerebellum of the Purkinje cell deficient mutant, InsP3 binding was greatly decreased, indicating also similarity in P400 localication and InsP3 binding (Maeda et al., 1990).

Since P400 protein had no InsP3 binding activity due to the detergent treatment used to solubilize the protein, we first tried to purify P400 protein from mouse cerebellum under mild conditions. As we have reported previously, most of the P400 protein was recovered in the precipitate when cerebellar microsome fractions were treated with a 1% Triton X-100 solution and centrifuged at 105,000xg for 60 min (Maeda et al.,1988).   InsP3 binding activities were also precipitated under the same conditions.   Suppattapone et al. (1988) reported that InsP3 binding protein was solubilized in the 1% Triton X-100 solution from rat cerebellum at a critical tissue concentration and that the binding protein did not precipitate after centrifugation at 120,000xg for 2 hours.   In mouse cerebellum, most InsP3 binding activities in the microsome fraction of mouse cerebellum were recovered in the precipitate under these conditions.   Using CHAPs, octylglucoside or Zwittergent 3-14 yielded the same results. Most InsP3 binding activities and P400 protein were recovered in the supernatant when the microsomes were treated with 1% Triton X-100 and centrifuged at 20,000xg for 60 min.   We therefore have chosen Triton X-100 and this centrifugation to solubilize the InsP3 binding protein.

Purification of the InsP3 receptor

InsP3 binding activity was recovered in a crude mitochondrial fraction (P2 fraction) and in a microsomal fraction (P3 fraction). We used the P2 + P3 fraction from mouse cerebella as a starting material.   The P2 + P3 fraction was prepared from mouse

Table 2. Inhibition of Specific $[^3H]$ Ins-1,4,5-P3 Binding by Various Inositol Phosphates, Heparin and Calcium

Displacing agent		Binding of ^3H Ins-1,4,5-P3 (% of control)	
		P2 + P3	Purified receptor
None		100	100
Ins-1-P1	1 µM	105	101
	10 µM	93	85
Ins-1,4-P2	1 µM	102	95
	10 µM	87	97
Ins-1,3,4-P3	1 µM	104	92
	10 µM	90	92
Ins-1,4,5-P3	1 µM	14	3
	10 µM	0	0
Ins-2,4,5-P3	1 µM	47	27
	10 µM	0	4
Ins-1,3,4,5-P4	1 µM	98	59
	10 µM	22	12
Ins-P6	1 µM	92	95
	10 µM	25	46
Heparin	20 µg/ml	0	0
calcium ion	1 mM	0	99

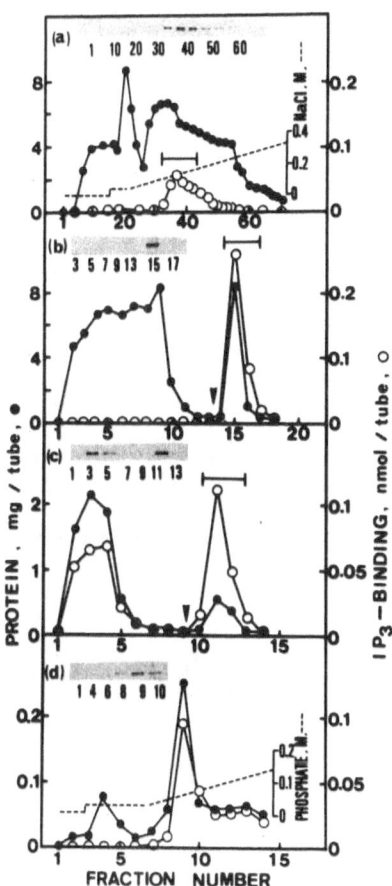

Fig. 1. Purification of P400/InsP3 receptor protein. Triton X-100
extract was applied to the column of DE52 and proteins
were eluted with a linear NaCl gradient (0.05-0.5 M NaCl
indicated by dashed line) (a). The collected fractions
indicated by the horizontal bar in (a) were applied to a
column of heparin-agarose. After washing the column with
the buffer containing 0.25 M NaCl, the receptors were
eluted with 0.5 M NaCl (the position of the buffer
change is indicated with the arrowhead)(b). The
peak fractions of InsP3 binding (indicated by the horizontal bar in (b) were collected and were applied to a
column of lentil lectin-Sepharose. After the washing, the
receptors were eluted with 0.8 M α-methyl-D-mannoside
(the position of the application of α-methyl-D-mannoside
is indicated by the arrowhead) (c). The receptor fractions indicated by the horizontal bar in (c) were applied to a column of hydroxylapatite. The receptors
were eluted with the linear sodium phosphate gradient
(0.025-0.15 M sodium phosphate indicated by the dashed
line)(d). The aliquots of the each fraction were assayed
for protein concentration (●) and for InsP3 binding
activity (○). Insets show the results of the immunoblot-
ing of the fractions indicated by fraction number with a
monoclonal antibody against P400 protein, 18A10.
(from Maeda et al.1990).

cerebella and treated with 1% Triton X-100. The solubilized proteins were applied to a column of DE52 and eluted as a single peak by increasing the NaCl concentration. The InsP3 receptor was further purified on a heparin-agarose column. The InsP3 binding activity was adsorbed on heparin-agarose at 0.25 M NaCl and eluted by increasing the concentration of NaCl to 0.5 M. The sample was then applied to a column of lentil lectin-Sepharose. The InsP3 binding activities adsorbed on the lentil lectin-Sepharose was eluted by 0.8 M α-methyl-D-mannoside. Finally, the sample was applied to a column of hydroxylapatite, and the bound proteins were eluted with a gradient of sodium phosphate (Fig. 1). SDS-PAGE analysis of the InsP3 binding fraction revealed that the 250 kD protein is only one protein band (Fig. 2).

Table 1 summarizes the purification of InsP3 binding protein. Approximately 0.5 mg of binding protein can be obtained from 10 g of mouse cerebellum. Suppattapone et al.(1988a) obtained 0.02 mg of binding protein from 16 rat cerebella (about 4 g of tissue). About 10 times more receptor protein can be obtained by our method than by the procedure of Suppattapone et al..

Scatchard analysis of InsP3 binding to the purified protein indicated that the Kd was 83 nM and the Bmax was 2.1 pmol/μg of protein (Fig. 2). The Kd value of the original P2 + P3 fraction was 34 nM, close to that of the purified binding protein (Maeda et al.,1990).

Inhibition of specific [3H]InsP3 binding by various inositol phosphates heparin and calcium. Ins2,4,5-P3 was a relatively potent competitor for [3H]InsP3 binding and Ins-1,3,4,5-P4 was

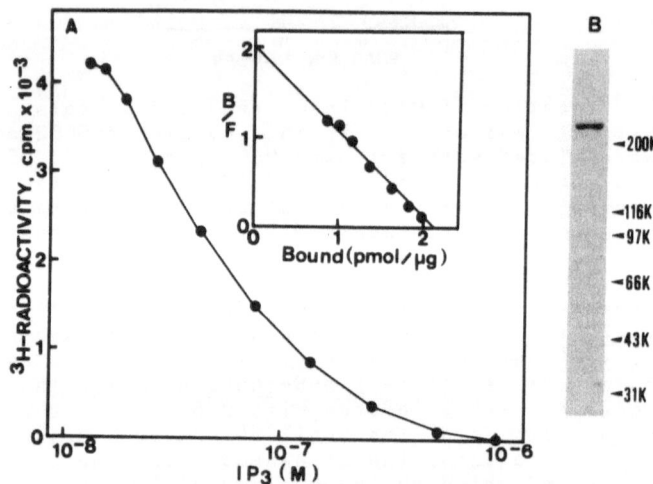

Fig. 2. Characterization of the InsP3 receptor protein. (A) The binding assays contained 0.42 μg of purified receptor, 11.2 nM [3H]InsP3 and various concentrations of cold InsP3 in 50 μl of 1 mM EDTA, 10 μM pepstatin A, 10 μM leupeptin, 0.1 mM PMSF, 50 mM Tris-HCL, pH8.0. Samples were incubated for 10 min and then processed for PEG precipitation. (B) Purified receptor (1 μg) was analyzed by SDS-PAGE (5-20% polyacrylamide gradient gel). Positions of M.W. markers are shown on the right. (from Maeda et al.,1990)

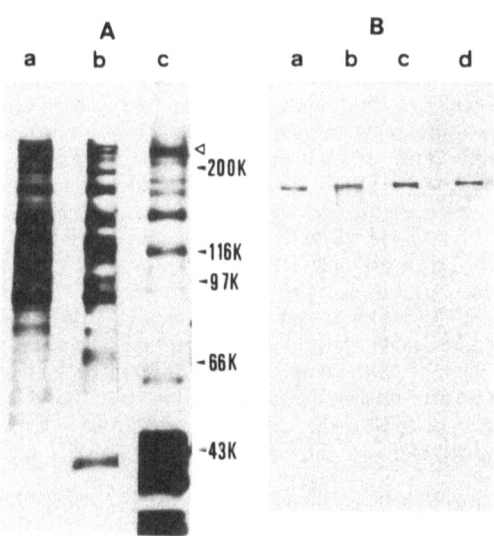

Fig.3. Analysis with the monoclonal antibodies. (A) The purified P400 protein was partially digested with V8 protease. The proteins were electrophoresed (5-12.5% SDS-PAGE) and immunodetected with 4C11 (a), 10A6 (b) and 18A10 (c) after the transfer of the proteins to a nitrocellulose sheet. Positions of M.W. markers (in Kd) are shown on the right. The triangle indicates the position of the intact P400 protein. (B) The purified InsP3 receptor protein was electrophoresed (5% SDS-PAGE) and transferred to a nitrocellulose sheet. The blots were stained with amido black (a), or immunostained with 4C11 (b), 10A6 (c) and 18A10 (d). (from Maeda et al., 1990)

less effective. Ins-1,4-P2 and Ins-1-P1 were inactive at the concentration of 10μM. These specificity is comparable to that of InsP3-sensitive calcium mobilization. Heparin inhibited the [^{3}H]InsP3 binding to both the P2 + P3 fraction and purified binding protein.

Identification of P400 protein as an InsP3 binding protein

InsP3 binding protein purified as a single band reacted with three monoclonal antibodies (4C11, 10A6, 18A10). Figure 3 shows the reactions of monoclonal antibodies with P400 protein partially digested with Staphylococcus aureus V8 protease. The 4C11 and 10A6 reacted with the common peptide in the higher molecular weight portion, but their reaction patterns were quite different from each other in the lower molecular weight portion. 18A10 monoclonal antibody reacted with the band which are not recognized by the other two antibodies. It is therefore apparent that these monoclonal antibodies recognize the three different epitopes on P400 protein.

Each fraction on the chromatograms shown in Figure 1 was analyzed by immunoblotting using a monoclonal antibody against P400 protein, 18A10. The immunoblot analysis revealed that P400 protein and InsP3 binding activity co-migrate completely at each purification step.

A Triton X-100 extract of the cerebellar P2 + P3 fraction was immunoprecipitated in the presence of InsP3 with a monoclonal antibody against P400 protein. InsP3-binding activity was immunoprecipitated dose-dependently by 18A10 monoclonal antibody (Maeda et al.,1990). Based on these results, we concluded that P400 protein is identical to InsP3 binding protein.

Structure of the P400 protein

Two mouse cerebellum cDNA libraries, synthesized by priming with random hexamer or oligo(dT), were constructed in phage-λ gt 11 expression vector. The library was screened with the three anti-P400 monoclonal antibodies. Two types of complete cDNA clones were obtained differing in their polyadenylation sites by 800 bases. The complete cDNA sequence was determined by sequencing the 12 clones listed in Figure 4. There is a single large open reading frame. Although the neighbouring sequence (CGGACATGT) of the first ATA shows only modest similarity to the initiation consensus sequence (CCA/GCCATG(G)), it is closer to the consessus sequence than the sequence (ACAAAATGT) surrounding the second ATG. The open reading frame continues for 8,247 bases when an in-frame TAG termination codon occurs at nucleotide position 8,248 from the first ATG. This frame was confirmed by sequencing the fusion points between the β-galactosidase gene and cDNA inserts in the immunopositive clones tested. The sequence of the long type cDNA is 9,871 nucleotides, consisting of a 328 base pair 5' untranslated region, a 8,247 bp coding region and a 1,296 bp 3' untranslated region which includes a polyadenylation signal (AATAAA) 14 bp upstream of a 23 base poly(A) tract. The long type cDNA is 9,071 nucleotides long, differing only in the 3' untranslated region (496 nucleotides, including AATAAA 12 bp upstream of a 23 base poly (A0) tract).

On the basis of the cDNA sequence, the P400 protein is predicted to comprise 2,749 amino acids and have a molecular weight of 313kd. From sequence analysis of immunopositive clones, the

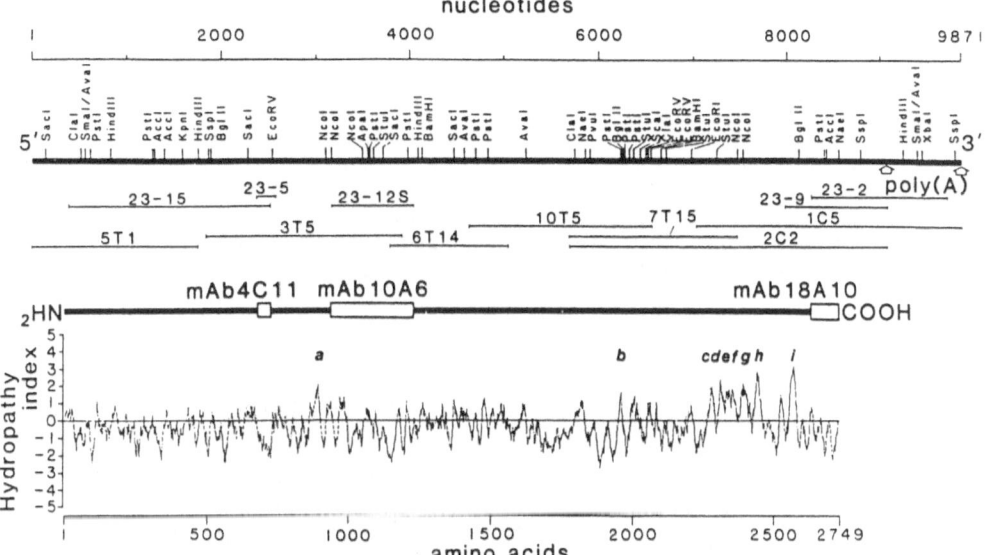

Fig.4. a, Restriction map of P400 cDNA clones. Position of 12
overlapping clones are indicated. Two open arrows point
to the polyadenylation sites. b, Hydropathy plot of the
protein-coding region. The three open boxes represent the
regions containing the epitope sequences for monoclonal
antibodies, 4C11, 10A6 and 18A10. Hydrophobic segments
(a-i) are indicated. (from Furuichi et al., 1989)

amino acid sequences containing epitopes for the antibodies 4CII, IOA6, and I8AI0 are localized at position 679-727, 943-1,237 and 2,648-2,749 (C-terminal end), respectively. A hydropathy profile of the protein sequence shows the existence of several stretches of hydrophobic amino acids, which could represent multiple membrane-spanning sequences. Although P400 is thought to be an integral membrane protein, it lacks a definalbe signal sequence with a hydrophobic stretch at the N-terminus, which indicates that an terminal hydrophilic region is exposed on the cytoplasmic face of the membrane. The protein has 21 consensus sequences for N-linked glycosylation. Most of these are not involved in glycosylation, because digestion of P400 protein with endo-β-N-acetylglucosaminidase F causes only small changes in electrophoretic mobility. Partial digestion of P400 by Staphylococcus aureus V8 protease generated a polypeptide with a molecular weight of about 40kd, possessing both binding activity to concanavalin A and immunoreactivity with the C-terminal specific antibody I8AI0. This suggests that at least the C-terminal region is glycosylated. The large cytoplasmic domain contains two sites that may act as targets for serine phosphorylation (amino acids 1,588 and 1,755), a possibility supported by the fact that a serine residue of P400 protein is known to be phosphorylated in a cAMP-dependent manner (Furuichi et al., 1989).

Expressed InsP3 binding activity in P400-cDNA transfected NGI08-15 cell line

To examine whether the P400 cDNA clone encodes an InsP3-binding protein, we transfected neuroblastoma/glioma hybrid cell line NGI08-15 with P400 cDNA. The cell contains an endogenous protein immunoreactive to P400 monoclonal antibodies. The size of endogenous P400 protein in NGI08-15 is smaller than that of the cerebellum. In addition to the size difference, between the binding kinetics of the cerebellum and NGI08-15 there was also a difference. P400 in NGI08-15 had two specific binding components, of high- and low-affinity, with low capacity. In the transient assay, the P400 cDNA transfected cells produced a large amount of proten immunoreactive to all three of the anti-P400 monoclonel antibodies. Both the cDNA derived and endogenous P400 proteins were observed in membrane fractions, but not in soluble cytosolic fractions. In the cDNA transfected NGI08-15 cells, expression of the cerebellar-type P400 protein, derived from the cDNA, is apparently coupled with the elevation of InsP3 binding activity. The expressed protein displays high affinity (Kd, 21.7nM) and specificity for Ins(1,4,5)P3 with high capacity, as does P400 in cerebellar microsomes (Furuichi et al., 1989).

Homology to ryanodine receptor

Recently, Takeshima et al. (1989) have described the primary structure of the ryanodine receptor, which is an important molecule in mobilizing calcium from the calcium storage site, the sarcoplasmic reticulum, in the skeletal muscle during excitation-contraction coupling. Calcium mobilization is considered to be due to the ryanodine receptor channel. P400 has fragmentary sequence homology with the ryanodine receptor protein. In particular, the putative transmembrane domains h and i and the successive C-terminal

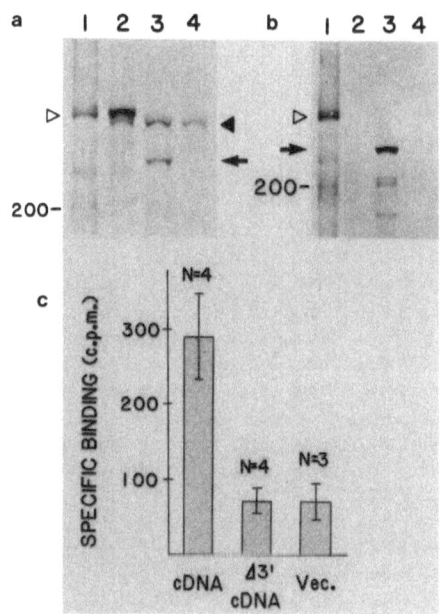

Fig.5.   Binding of [3H]InsP3 to P400 protein expressed in NG108-15
cells. a, Western blot analysis of the crude membrane
proteins using the anti-P400 monoclonal antibody, 4C11.
Lane 1, Microsomal proteins (5 μg) prepared from mouse
cerebellum; lane 2, crude membrane proteins (50 μg)
prepared from the NG108-15 transfected with P400 cDNA;
lane 3, 3'-deletion construct and lane 4, vector plasmid
DNA. Cerebellar P400 protein (△); endogenous P400 of
NG108-15 (▲); trancated P400 (→). Position is shown of a
molecular weight marker (x10⁻³). b, Western blot analysis
of soluble cytosolic proteins using 4C11. Lane 1, the
same as lane 1 of a; lane 2 to 4, soluble cytosolic
proteins (15 μg) prepared from the same cells as those in
lanes 2 to 4 of a, respectively. c, Specific binding of
[3H] InsP3 to the crude membrane proteins prepared from
NG108-15 transfected with P400 cDNA (cDNA), the 3'-dele-
tion construct (Δ3' cDNA) or vector DNA (Vec). N represent number of samples tested. The results are the means
± s.d. of N samples. (from Furuichi et al., 1989)

region in P400 have remarakable similarity to the transmembrane segments M3 and M4 and the successive C-terminal region
in the ryanodine receptor. This homology does not extend to
the binding sites for modulators (calcium, nucleotides ,
calmodulin) as proposed by Takeshima et al. (1989). The overall resemblance between these two receptors of the proposed
transmembrane topology, sharing a large N-terminal region
located on the cytoplasmic side and a short C-terminal region
spanning the membrane. These observations imply related
calcium mobilization functions.

REFERENCES

M.J. Berridge, Inositol trisphosphate and diacyl glycerol:
       two interacting second messengers, Ann. Rev. Biochem.,56:
       159-193 (1987).
F. Crepel, J.-L. Dupont, and R. Gardette, Selective absence
       of calcium spikes in Purkinje cells of staggerer mutant
       mice in cerebellar slices maintained in vitro.,
       J. Physiol.,346: 111-125 (1984).
T. Furuichi, S. Yoshikawa., A. Miyawaki, K. Wada, N. Maeda,
       and K. Mikoshiba, Primary stucture and functional expres-
       sion of the inositol 1,4,5-trisphosphate-binding protein
       P400., Nature, 342: 32-38 (1989).
N. Maeda, M. Niinobe, K. Nakahira, and K. Mikoshiba,
       Purification and characterization of P400 protein,
       a glycoprotein characteristic of Purkinje cell, from
       mouse cerebellum., J. Neurochem.,51: 1724-1730 (1988).
N. Maeda, M. Niinobe, Y. Inoue, and K. Mikoshiba, Develop-
       mental expression and intracellular location of P400
       protein characteristic of Purkinje cells in the mouse
       cerebellum., Dev. Biol.,133: 67-76 (1989).
N. Maeda, M. Niinobe, and K. Mikoshiba, A cerebellar Pur-
       kinje cell marker P400 protein is an inositol 1,4,5-tris-
       phosphate (InsP3) receptor protein. Purification and
       characterization of InsP3 receptor complex., EMBO J.,9:
       61-67 (1990).
J. Mallet, M. Huchet, R. Pougeois, and J.P. Changeux,
       Anatomical, physiological and biochemical studies on the
       cerebellum from mutant mice. III. Protein differences
       associated with weaver, staggerer and nervous mutation.,
       Brain res., 103: 291-312 (1976).
K. Mikoshiba, M. huchet, and J.P. Changeux, Biochemical
       and immunological studies on the P400 protein, a protein
       characteristic of the Purkinje cell from mouse and rat
       cerebellum., Dev. Neurosci., 2: 254-275 (1979).
K. Mikoshiba, H. Okano, and Y. Tsukada, P400 protein
       characteristic to Purkinje cells and related proteins in
       cerbella from neuropathological mutant mice: autoradio-
       graphic study by 14C-leucine and phosphorylation.,
       Dev. Neurosci.,7: 179-187 (1985).
S. Supattapone, P.F. Worley, J.M. Baraban, and S.H. Snyder,
       Solubilization, purification, and characterization of an
       inositol trisphosphate receptor., J. Biol. Chem.,263:
       1530-1534 (1988a).
S. Supattapone, S.K. Danoff, A. Theibert, S.K. Joseph, J.
       Steiner, and S.H. Snyder, Cyclic AMP-dependent phosphory
       lation of a brain inositol trisphosphate receptor

decreases its release of calcium., <u>Proc. Natl. Acad. Sci.
USA</u>,85: 8747-8750 (1988b).

H. Takeshima, S. Nishimura, T. Matsumoto, H. Ishida, K.
Kangawa, N. Minamino, H. Matsuo, M. Ueda, M. Hanaoka,
T. Hirose, and S. Numa, Primary structure and expression
from complementary DNA of skeletal muscle ryanodine
receptor., <u>Nature</u>339: 439-445 (1989).

P.F. Worley, J.M. Baraban, and S.H. Snyder, Inositol 1,4,5-
trisphosphate receptor binding: autoradiographic localization in rat brain., <u>J. Neurosci.</u>9: 339-346 (1989).

H. Yamamoto, N. Maeda, M. Niinobe, E. Miyamoto, and K.
Mikoshiba, Phosphorylation of P400 protein by cyclic
AMP-dependent protein kinase and Ca^{2+}/calmodulin-
dependent protein kinase II., <u>J. Neurochem.</u>,53: 917-923
(1989).

INOSITOL 1,4,5-TRISPHOSPHATE ACTIVATES CA^{2+} CHANNELS IN THE PLASMA

MEMBRANES OF RAT BRAIN NERVE TERMINALS

M. Satoh, H. Ueda, S. Tamura, Y. Yoshihara and N. Fukushima

Department of Pharmacology
Faculty of Pharmaceutical Sciences
Kyoto University
Sakyo-ku, Kyoto 606, Japan

INTRODUCTION

There is accumulating evidence that a wide variety of stimulation of receptors by hormones and neurotransmitters results in increased phosphoinositide turnover and mobilization of Ca^{2+} from intracellular stores (Berridge and Irvine, 1989). Such post receptor mechanisms are relevant to the stimulation of phospholipase C-mediated hydrolysis of phosphatidylinositol 4,5-bisphosphate giving rise to diacylglycerol and to inositol 1,4,5-trisphosphate (IP$_3$). Furthermore, IP$_3$ mobilizes Ca^{2+} from microsomal organelles such as rough- and smooth-endoplasmic reticulum and calsiosome in various secretory cells (Henne et al., 1987; Payne and Fein, 1987; Volpe et al., 1988; see a review by Adbel-Latif, 1986). Thus, it is likely that the IP$_3$-induced calcium mobilization from intracellular organelles is involved in hormone secretion by receptor stimulation.

On the other hand, the release of neurotransmitters predominantly occurs from the nerve terminals and is regulated by intracellular concentration of free calcium ion ([Ca^{2+}]$_i$). Although phosphatidylinositol-specific phospholipase C (PLC), PLC-I, was reported to exist in the nerve terminals (Gerfen et al., 1988) and IP$_3$ formed by this enzyme is expected to play an important role in the release of neurotransmitters, details of relevant mechanisms remain to be revealed. In this report, a novel mechanism of IP$_3$-induced calcium mobilization in the nerve terminals and an example of possible involvement of such a mechanism in receptor-mediated regulation of [Ca^{2+}]$_i$ will be described.

BACKGROUND OF THIS STUDY

Kyotorphin is a dipeptide (Tyr-Arg) originally found by Takagi et al. (1979) as an endogenous analgesic substance in the bovine brain. This dipeptide was formed from tyrosine and arginine by a specific synthetase (kyotorphin synthetase: Ueda et al, 1987) and was also possibly synthesized from precursor proteins by calpain-like proteases (Yoshihara et al., 1990) in rat brain synaptosomes, a nerve ending-rich subfraction. The levels of kyotorphin and kyotorphin synthetase were closely correlated in various regions of the rat brain and were subcellularly localized densely in synaptosomes (Ueda et al., 1980).

Kyotorphin induced a naloxone-antagonizable analgesia which may be mediated via a release of Met-enkephalin from the brain (Takagi et al., 1979; Rackham et al., 1982). Furthermore, this dipeptide enhanced the influx of $^{45}Ca^{2+}$ added extracellularly into synaptosomes without affecting membrane potentials (Ueda et al., 1986). However, the dipeptide did not elicit [^{3}H]noradrenaline, [^{3}H]GABA, [^{3}H]D-aspartate or β-endorphin (Janicki and Lipkowski, 1983). Thus, such a selective effect of kyotorphin on neurotransmitter release would suggest the presence of the specific kyotorphin-receptor. In fact, the kyotorphin receptor was recently identified using [^{3}H]kyotorphin (Ueda et al., 1989). Kyotorphin-specific binding was composed of high and low affinity components. The high affinity component of [^{3}H]kyotorphin binding disappeared in the presence of GTPγS and $MgCl_2$, a finding which suggests that the kyotorphin receptor in the high affinity binding state may be coupled to GTP-binding proteins (G-proteins).

This view was supported by evidence that kyotorphin stimulated membrane low K_m GTPase activity, an event known as a post-receptor mechanism, following the GDP-GTP exchange on the G-proteins triggered by the stimulation of coupled receptors (Gilman, 1987). Further, the kyotorphin stimulation of low K_m GTPase was abolished by pretreatment of membranes with islet activating protein (IAP), pertussis toxin, at a concentration of 10 μg/ml (Fig. 1, left-hand graph). However, when highly purified G_i or G_o was reconstituted in the IAP-pretreated membranes, the stimulatory effect of kyotorphin on low K_m GTPase was recovered by reconstitution with G_i but not with G_o (Fig. 1, right-hand graph), while the basal low K_m GTPase activity increased to a higher level than that in intact membranes, in both cases. These findings suggest that the kyotorphin receptor is functionally coupled to G_i but not to G_o. Indeed, we preliminarily demonstrated that kyotorphin inhibited the cyclic AMP formation by forskolin (100 μM) in the rat brain stem membrane preparations. In this context, a synthetic dipeptide, Leu-Arg, is regarded as a specific antagonist to kyotorphin, since Leu-Arg showed a potent displacement of [^{3}H]kyotorphin binding, yet in itself had no effect on the low K_m GTPase activity and cyclic AMP formation. But, Leu-Arg inhibited the effects of kyotorphin on them.

Moreover, we found that kyotorphin stimulated phospholipase C in crude membranes of brain homogenates and lysed synaptosomal membranes in a concentration-dependent manner, as determined by the hydrolysis of [^{3}H]-phosphatidylinositol 4,5-bisphosphate into [^{3}H]inositol phosphates, including [^{3}H]inositol 1,4,5-trisphosphate ([^{3}H]IP$_3$). This effect was much more evident in the latter preparation than in the former and was antagonized by Leu-Arg. The kyotorphin stimulation of phospholipase C was also abolished by pretreating membranes with IAP. When highly purified G_i or G_o was reconstituted in IAP-pretreated membranes, the stimulatory effect of kyotorphin on phospholipase C was recovered by reconstitution with G_i but not with G_o (Fig. 2B). These results suggest that the kyotorphin receptor is functionally coupled to stimulation of phospholipase C, through G_i.

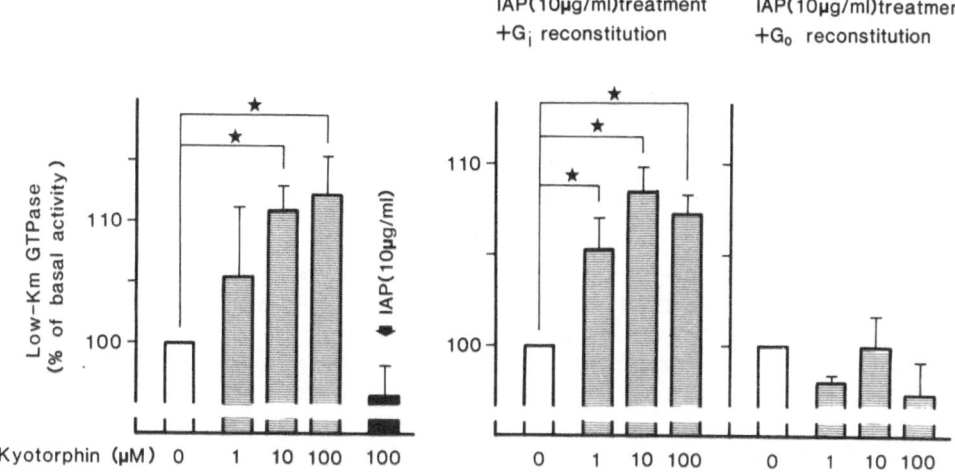

Fig. 1. (Left-hand graph): Stimulatory effect of kyotorphin
on low K_m GTPase and an inhibition of the kyotorphin
effect by IAP (islet activating protein, pertussis toxin).
(Right-hand graph): Kyotorphin stimulation of low K_m GTPase
in IAP-pretreated membranes in which G_i was reconstituted.

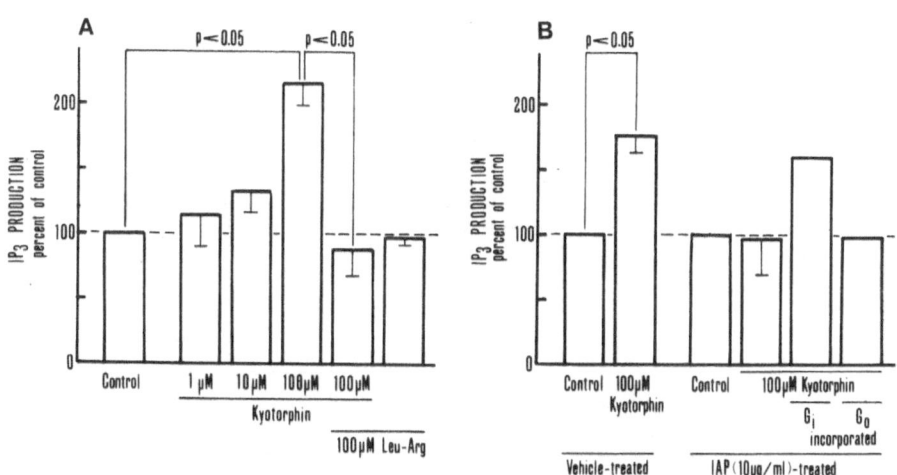

Fig. 2. (A): A concentration-dependent stimulation of IP_3 production
(phospholipase C activity) by kyotorphin and its inhibition
by Leu-Arg. (B): Kyotorphin stimulation of IP_3 production
in IAP-pretreated membranes in which G_i was reconstituted.

As described above, kyotorphin binds to its specific receptor present in the presynaptic palasma membrane, activates G_i coupled to the kyotorphin-receptor, stimulates phospholipase C activity, consequently facilitates IP_3 production, and furthermore elevates $[Ca^{2+}]_i$. To reveal the mechanism of the final step of this chain reaction, elevation of $[Ca^{2+}]_i$ by IP_3, we studied the effect of IP_3 on the release of $^{45}Ca^{2+}$ from resealed vesicles prepared mainly from synaptic plasma membranes of the brain.

RESEALED VESICLES

Methods of Preparation

Male Sprague-Dawley rats (200 - 250 g) were decapitated and the whole brains were removed and homogenized in 10 vol. of 0.32 M sucrose. The homogenates were centrifuged at 1,000 x g for 10 min and the supernatant was further centrifuged at 12,000 x g for 20 min. Resulting pellets were used for preparation of myelin, synaptosomes and mitochondria, while the supernatant was used for preparation of microsomes by further centrifugation at 100,000 x g for 60 min, according to Gray and Whittaker (1962).   Further subcellular fractions were prepared by discontinuous density gradient centrifugation of lysed synaptosomes, composed of 0.4, 0.6, 0.8, 1.0 and 1.2 M sucrose (Whittaker et al., 1964). Synaptic vesicles were obtained from the interface between 0.4 - 0.6 M sucrose, synaptic plasma membranes (SPM) from that between 0.6 - 0.8 M and 0.8 - 1.0 M sucrose and synaptic mitochondria from the pellet. Activities of Na$^+$,K$^+$-ATPase and NADPH-cytochrome c reductase were measured, according the methods of Verity (1972) and Kasper (1971), as marker-enzymes of plasma membranes and endoplasmic reticulum, respectively (Table 1).

Table 1.   Activities of marker enzymes in subfractions of the rat brain

Subfractions	Na$^+$,K$^+$-ATPase[a]	NADPH-cytochrome c reductase[b]
Microsomes	2.27	1.33
Myelins	2.22	0.83
Synaptosomes	1.65	0.54
Mitochondria	1.55	0.93
Synaptic vesicles	<0.01	0.35
Synaptic plasma membranes (SPM)	1.67	0.33
Synaptic mitochondria	0.97	0.33

[a] and [b]: Ratios of activities of Na$^+$,K$^+$-ATPase (a marker enzyme of plasma membranes) and NADPH-cytochrome c reductase (a marker enzyme of endoplasmic reticulum) in each sub-fraction to those in homogenates which were 0.112 μmol/ mg protein/min and 5.98 nmol/mg protein/min, respectively.

To prepare resealed vesicles, each subcellular preparation per rat brain was hypotonically lysed with 10 ml of 5 mM Tris-HCl buffer (pH 7.5) containing 1 mM $MgCl_2$, 0.574 mM $CaCl_2$ and 1 mM EGTA [TMC buffer] by a Potter-Elvehjem homogenizer and centrifuged at 10,000 x g for 5 min. The concentration of free calcium ion in TMC buffer was calculated to be 0.1 μM (Fabiato and Fabiato, 1979). The obtained pellets were resuspended in TMC buffer. Aliquots (10 mg of protein) were incubated in 10 ml of TMC buffer with $^{45}Ca^{2+}$ (0.5 μCi) at 37° C for 0.5 - 35 min in the absence of ATP. In some experiments, A23187 (a calcium ionophore; 5 μM) was added to the incubation medium. To prepare "once resealed vesicles", the SPM were incubated in the absence of $^{45}Ca^{2+}$ and ATP for 30 min at 37° C. Thereafter, once resealed vesicles were incubated with $^{45}Ca^{2+}$ in the presence or absence of ATP and/or calmodulin and/or A23187, for 0.5 - 35 min at 37° C. At various periods of incubation, an aliquot (100 μl) was removed and passed through a GF/C filter (Whatman), followed by three washes with 3 ml of TMC buffer. The level of $^{45}Ca^{2+}$ was determined by measuring radioactivity on the filter. An electron micrograph of resealed vesicles is shown in Fig. 3. Empty vesicles of 0.2 μm or less in diameter were obtained.

Incorporation of $^{45}Ca^{2+}$ into Resealed Vesicles

When the SPM were incubated with $^{45}Ca^{2+}$ in TMC buffer at 37° C, the radioactivity accumulated in the preparations (resealed vesicles), with the time of incubation. There was a rapid increase in $^{45}Ca^{2+}$ level within 1 min, then a slow but linear accumulation within 20 min, thereafter a plateau. An addition of A23187 (5 μM) to the incubation medium at 10 min produced a decrease in $^{45}Ca^{2+}$ with further incubation (Fig. 4). The maximal decrease in $^{45}Ca^{2+}$ level by A23187 was 43 % of the control at 20 min after the beginning of the incubation. A similar effect of A23187 was obtained at 10

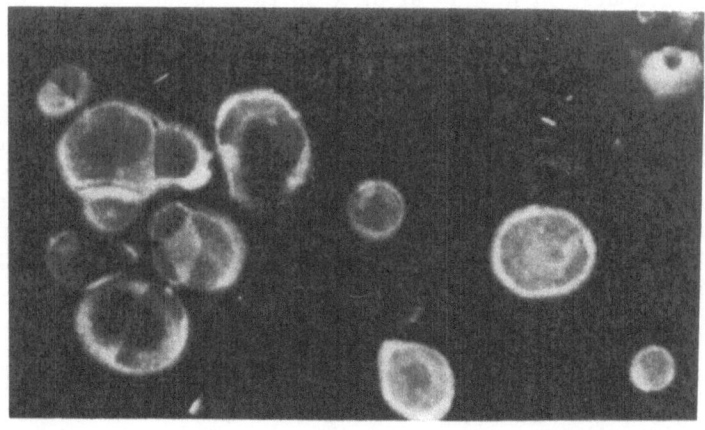

0.5 µm

Fig. 3.   An electron micrograph (negatively staining) of the resealed vesicles prepared from SPM by incubation in TMC buffer for 30 min in the absence of ATP.

μM. It is conceivable that 43 % of the control accumulation is attributed to the amount of $^{45}Ca^{2+}$ "incorporated" or "enclosed" into resealed vesicles which is supposed to be formed during incubation of SPM at 37° C. As neither ATP nor Na$^+$ was present in this preparation, the accumulation of $^{45}Ca^{2+}$ was unlikely due to uptake by means of the ATP-dependent calcium pump or Na$^+$-K$^+$ exchange mechanism.

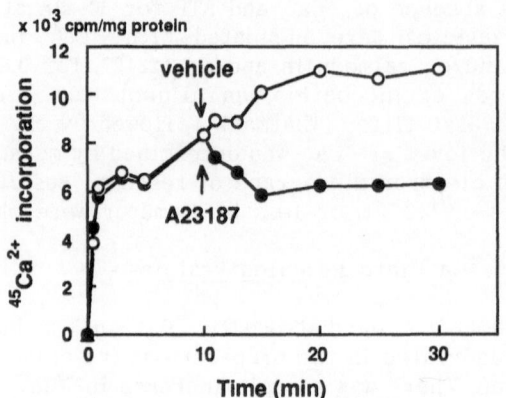

Fig. 4. Incorporation of $^{45}Ca^{2+}$ into resealed vesicles prepared from the synaptic plasma membranes. Representative profiles of the time-course.

On the other hand, when once resealed vesicles prepared from the SPM were incubated with $^{45}Ca^{2+}$ at 37° C in the absence of ATP, only a rapid increase in the level of $^{45}Ca^{2+}$ was observed within 5 min after the start of the incubation. The addition of A23187 did not produce any significant changes in $^{45}Ca^{2+}$ level (Fig. 5). As the plateau level was very similar to that in the presence of A23187 in the former preparation (Fig. 4), it is likely that the increase in $^{45}Ca^{2+}$ level in the once resealed vesicles is due to binding of $^{45}Ca^{2+}$ to the surface of the vesicles.   However, when the once resealed vesicles were incubated with $^{45}Ca^{2+}$ in the presence of ATP (1 mM), there was a further accumulation of $^{45}Ca^{2+}$, compared to that in the absence of ATP. Such a further accumulation by ATP was reduced by A23187 (Fig. 6). Thus, the ATP-activated and A23187-sensitive accumulation of $^{45}Ca^{2+}$ can be produced in the once resealed vesicles.

Furthermore, when the once resealed vesicles from the SPM were incubated with $^{45}Ca^{2+}$ in the presence of calmodulin (5 μg/ml) as well as ATP (1 mM), calmodulin augmented the ATP-activated accumulation of $^{45}Ca^{2+}$. However, calmodulin (5 μg/ml) alone did not affect the incorporation of $^{45}Ca^{2+}$ into the once resealed vesicles. In this type of experiments, the concentration of free calcium ion in the incubation medium was adjusted to 10 μM which is required for activation of Ca^{2+}-activated ATPase (calcium

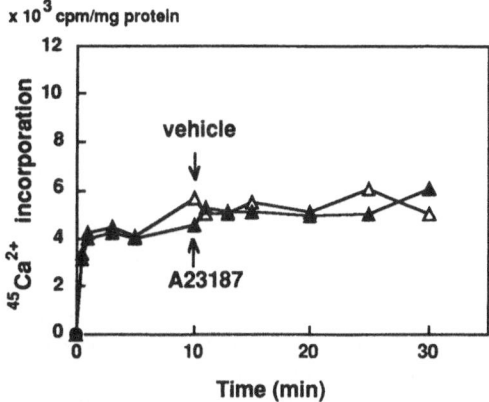

Fig. 5. Incorporation of $^{45}Ca^{2+}$ into once resealed vesicles prepared from the synaptic plasma membranes. Representative profiles of the time-course.

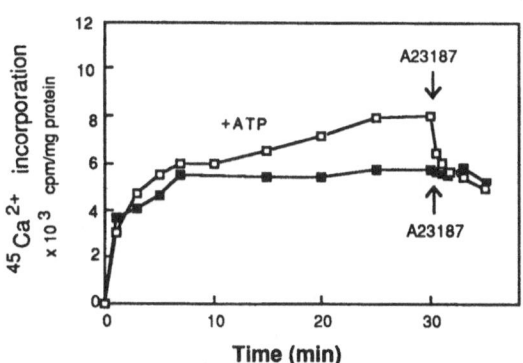

Fig. 6. The effects of ATP and A23187 on incorporation of $^{45}Ca^{2+}$ into once resealed vesicles prepared from the synaptic plasma membranes. Representative profiles of the time-course.

pump) by calmodulin. As the ATP- and calmodulin-binding sites in the ATP-dependent calcium pump are regarded to be located inside of cells from the study on the primary structure of the calcium pump (Verma et al., 1988), the existence of ATP- and calmodulin-dependent incorporation of $^{45}Ca^{2+}$ strongly indicates that at least a part of resealed vesicles prepared in this study was in an inside-out type.

We examined the effect of saponin on A23187-sensitive (ATP-activated) incorporation of $^{45}Ca^{2+}$ into resealed vesicles prepared from the SPM, because saponin is well known to perforate the plasma membranes through forming a characteristic ring-like micelle with cholesterol which is rich in the plasma membranes but sparse in endoplasmic reticulum or mitochondria (Inamitsu and Otsuki, 1984). Treatment of the SPM with saponin (5 - 30 μg/ml) markedly inhibited the A23187-sensitive incorporation in a concentration-dependent manner (Fig. 7). On the other hand, the A23187-sensitive incorporation of $^{45}Ca^{2+}$ into intact microsomes was inhibited to 50 % of control by 100 μg/ml of saponin. Such an inhibition is probably due to considerable amounts of postsynaptic plasma membranes in the microsomes, since Inamitsu and Ohtsuki (1984) reported that the ATP-dependent calcium incorporation into resealed vesicles from plasma membranes (saponin-sensitive) was 70 % or more of the total ATP-dependent incorporation into resealed microsomes. This view was supported by the data that the activity of Na$^+$,K$^+$-ATPase, a marker enzyme of plasma membranes, was relatively high in the microsomes (Table 1). These results confirmed that most of the incorporation of $^{45}Ca^{2+}$ into the resealed vesicles from the SPM was through the plasma membranes.

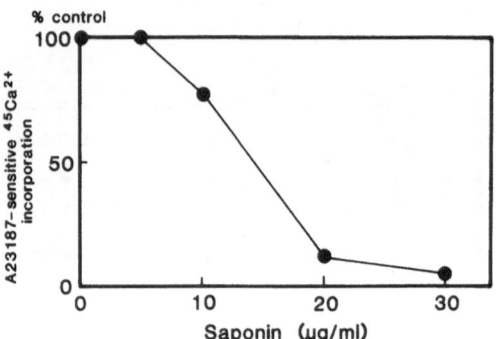

Fig. 7. Inhibitory effect of saponin on A23187-sensitive incorporation of $^{45}Ca^{2+}$ into resealed vesicles prepared from the synaptic plasma membrane.

RELEASE OF $^{45}CA^{2+}$ FROM RESEALED VESICLES

Methods of Superfusion

Newly prepared resealed vesicles (300 - 500 μg protein: incubation with $^{45}Ca^{2+}$ for 30 min at 37° C) suspended in buffer A (TMC buffer but 50 mM Tris-HCl) were placed on a GF/C filter (6 mm in diameter) fixed in a chamber in the superfusion system (Fig. 8) and were superfused with the same buffer at a flow rate of 1 ml/min. Superfusate was collected every 1 min. IP$_3$ and other agents were added to buffer A in tube B and applied to the resealed vesicles by switching the cock.

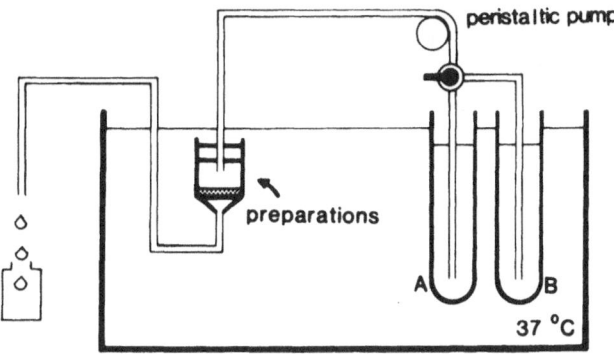

Fig. 8.   A system used for superfusion of resealed vesicles

Effect of IP$_3$ on the Release of ^{45}Ca^{2+}

The basal release of ^{45}Ca^{2+} rapidly decreased and reached a plateau 20 min after the start of superfusion. IP$_3$ was applied for 2 min from 24 min after the start of superfusion, namely the 25th and 26th sampling time. In standard experiments, the release of ^{45}Ca^{2+} was represented as the fractional release, namely the ratio (%) of the amount (cpm) of ^{45}Ca^{2+} in each sample to the total amounts (cpm) at real time, because the fractional release was fairly constant from experiment to experiment. The IP$_3$-evoked release of ^{45}Ca^{2+} (%) was then expressed as [the total of fractional releases of the 25th - 27th samples (in the presence of the agent)] - [the mean of basal fractional releases of the 22nd - 24th samples and of the 28th - 30th samples].

An addition of IP$_3$ (5 μM) to the superfusion medium increased in fractional release of ^{45}Ca^{2+} from resealed vesicles prepared by the SPM and resting levels were restored after discontinuance of the application of IP$_3$. However, a similar application of IP$_3$ did not alter the fractional release of ^{45}Ca^{2+} from intact synaptosomes preloaded with it, which must be only in right side-out type. It should be noted that kyotorphin can act on such intact synaptosomes and increase in [Ca^{2+}]$_i$. The facilitatory effect of IP$_3$ on the release of ^{45}Ca^{2+} from resealed vesicles of the SPM was concentration-dependent in a range of 0.5 - 10 μM and this effect appeared to be saturable. The double reciprocal plot showed that the apparent K$_m$ was 1.5 μM and that the maximal response was 4.16 %. Inositol 1,3,4,5-tetrakisphosphate (IP$_4$; 1 - 10μM) also increased the release of ^{45}Ca^{2+} from resealed vesicles of the SPM in smaller extent than IP$_3$ did, but inositol, inositol 4-mono phosphate and inositol 1,4-bisphosphate did not produce any significant increase in the release of ^{45}Ca^{2+}.

On the other hand, the IP$_3$-evoked release of ^{45}Ca^{2+} from resealed vesicles of the SPM was inhibited in a concentration-dependent manner by an addition of heparin, a putative antagonist to IP$_3$ (Worley et al., 1987), from 9 min after the start to the end of superfusion. The IC$_{50}$ of heparin was 4.8 μg/ml, a value in good accord with its IC$_{50}$ in IP$_3$ binding in the cerebellar membranes (Worley et al., 1987). These results strongly suggest the IP$_3$-evoked release of ^{45}Ca^{2+} was mediated through a specific IP$_3$-receptor. Furthermore, treatment of the SPM with saponin markedly inhibited the basal and IP$_3$ (5 μM)-release of ^{45}Ca^{2+}. The IC$_{50}$ of saponin was

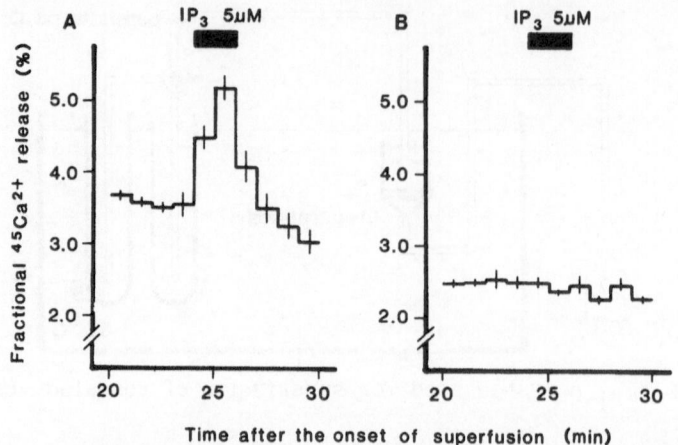

Fig. 9.   Effects of IP$_3$ on the release of ^{45}Ca^{2+} from resealed
vesicles of SPM (A) and from intact synaptosomes (B).

9 μg/ml. When the SPM was preincubated for 20 min at 37° C with trypsin
(100 μg/ml), pronase (100 μM), N-ethylmaleimide (5 mM), phospholipase A$_2$ (0.5
U/ml) or phospholipase C (0.5 U/ml), the IP$_3$-evoked release of ^{45}Ca^{2+} from
resealed vesicles of the SPM was significantly inhibited to 53 %, 28.9 %,
29.3 %, 10.7 % or 14.4 % of control, respectively. However, nitrendipine (10
μM), verapamil (10 μM), ω-conotoxin (3 mM) and ryanodine (10 μM) did not
affect the IP$_3$-evoked release of ^{45}Ca^{2+}. From the above mentioned findings,
it is conceivable that exogenously applied IP$_3$ acts on its specific
receptor exposed on the outside of an inside-out type of resealed vesicle
prepared from the synaptic plasma membranes, and activates calcium
channels in a voltage-independent fashion. In this context, the involvement
of IP$_3$ in regulation of calcium channels in the plasma membranes has been
claimed in electrophysiological experiments using lymphocytes and mast
cells (Kuno and Gardner, 1987; Penner et al., 1988). Further, Furuichi et al.
(1989) found the protein expressed in NG108-15 cells transfected with
P$_{400}$cDNA, which encodes a protein specific for Purkinje cells in the
cerebellum, showed the IP$_3$-binding activity. An immunoelectron microscopic
study using monoclonal antibody 4C11 demonstrated P$_{400}$ protein (IP$_3$
receptor) was present at the plasma membranes as well as the endoplasmic
reticulum and postsynaptic density (Maeda et al., 1989).

To examine whether the IP$_3$-evoked release of ^{45}Ca^{2+} is specific to the
SPM, the effects of IP$_3$ on the release of ^{45}Ca^{2+} were studied in various
subcellular preparations shown in Table 1. Those respective preparations
(300 - 500 μg/assay) were lysed hypotonically and preloaded with ^{45}Ca^{2+} in
TMC buffer for 30 min at 37° C and then placed on a GF/C filter in the
superfusion chamber. The extent of basal fractional release of ^{45}Ca^{2+} was
similar among various preparations. However, the IP$_3$-evoked release was
larger in the preparation from lysed synaptosomes than that from myelins,
mitochondria or microsomes. Among the preparations from further
subfractions of lysed synaptosomes, the IP$_3$-evoked release of ^{45}Ca^{2+} was
largest in the SPM preparation (Fig. 10). As the greater part of plasma
membranes of synaptosomes are in general regarded as presynaptic origin,
the IP$_3$-evoked release of ^{45}Ca^{2+} is likely more specific at presynaptic
plasma membranes, although the possibility that endoplasmic reticulum and
postsynaptic membranes are targets of IP$_3$ in neurons is not excluded.

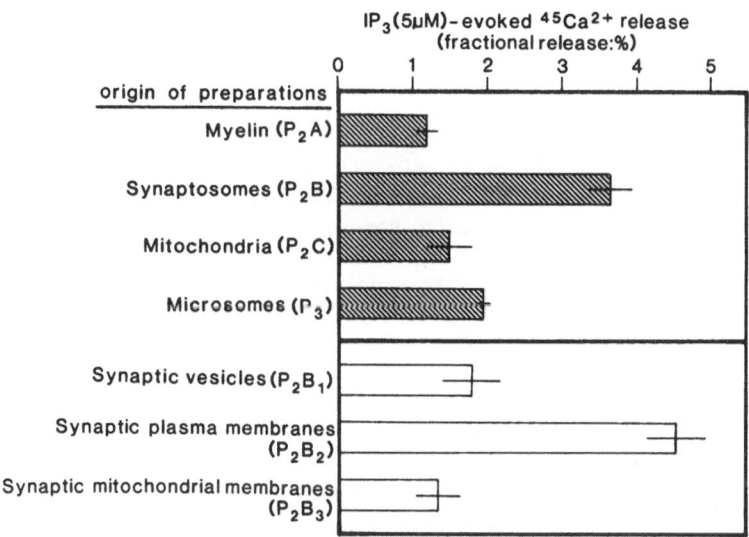

Fig. 10. IP_3-evoked release of $^{45}Ca^{2+}$ from various
preparations made of subfractions of the
rat brain

Effect of kyotorphin on the release of $^{45}Ca^{2+}$

Kyotorphin at concentrations of 1 - 100 μM evoked the release of $^{45}Ca^{2+}$ from resealed vesicles which were prepared by incubation of lysed synaptosomes of the rat brain with $^{45}Ca^{2+}$ in the absence of ATP for 30 min at 37° C, in a concentration-dependent manner, with the help of GppNHp (10 μM), an enzymatically unhydrolyzable analog of GTP. Such a kyotorphin-evoked release is considered to be produced in a right-side-out type of resealed vesicles. The kyotorphin-evoked release of $^{45}Ca^{2+}$ from the resealed vesicles was inhibited by pretreatment of lysed synaptosomal membranes with neomycin (0.03 - 0.3 mM), an inhibitor of phospholipase C (Cockcroft and Gomperts, 1985), or IAP (30 - 50 μM). The inhibition by the latter was recovered by the reconstitution of the IAP-pretreated membranes with highly purified G_i1. These results suggest that kyotorphin-evoked release of $^{45}Ca^{2+}$ is probably mediated by IP_3 generated through activation of a G-protein and phospholipase C.

CONCLUSION

From the above described findings, we speculate on a possible role of IP_3 in calcium mobilization across the plasma membranes of the rat brain nerve terminals, as shown in Fig. 11. In brief, there are specific IP_3 receptors at intracellular side of the plasma membranes of nerve terminals. IP_3 acts on its specific receptors and activates calcium channels to uptake Ca^{2+} from extracellular space, in a voltage-independent fashion. The level of IP_3 in the nerve terminals is regulated by various receptor-operated mechanisms. For example, kyotorphin acts on its specific receptors, activates G_i, and then stimulates phospholipase C to increase IP_3 formation.

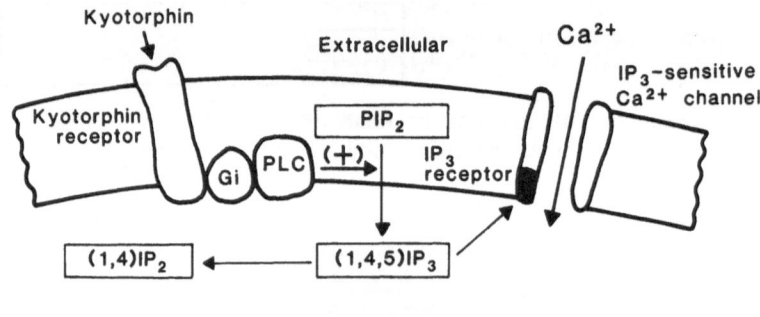

Fig. 11.   A working hypothesis on functional coupling between IP_3-sensitive Ca^{2+} channel and kyotorphin receptor in the nerve terminal

REFERENCES

Abdel-Latif, A. A., 1986, Calcium-mobilizing receptors, polyphosphoinositides, and the generation of second messengers, Pharmacol. Rev., 38:227-272.
Berridge, M. J. and Irvine, R. F., 1989, Inositol phosphates and cell signalling, Nature, 341:197-205.

Cockcroft, S. and Gomperts, B. D., 1985, Role of guanine nucleotide binding protein in the activation of polyphosphoinositide phosphodiesterase, <u>Nature</u>, 314:534-536.

Fabiato, A. and Fabiato, F., 1979, Calculator programs for computing the composition of the solutions cocntaining multiple metals and ligands used for experiments in skinned muscle cells, <u>J. Physiol. (Paris)</u>, 75:463-505.

Furuichi, T., Yoshikawa, S., Miyawaki, A., Wada, K., Maeda, N. and Mikoshiba, K., 1989, Primary structure and functional expression of the inositol 1,4,5-trisphosphate-binding protein P_{400}, <u>Nature</u>, 342:32-38.

Gerfen, C. R., Choi, W. C., Suh, P. G. and Rhee, S. G., 1988, Phospholipase C I and II brain isozymes: immunohistochemical localization in nueronal systems in rat brain, <u>Proc. Natl. Acad. Sci. USA</u>, 85:3208-3212.

Gilman, A. G., 1987, G protein: transducers of receptor generated signals, <u>Ann. Rev. Biochem.</u>, 56:615-649.

Gray, E. G. and Whittaker, V. P., 1962, The isolation of nerve endings from brain: an electronmicroscopic study of fragments derived from homogenization and centrifugation, <u>J. Anat. (Lond.)</u>, 96:79-86.

Henne, V., Piiper, A. and Soling, H.-D., 1987, Inositol 1,4,5-trisphosphate and 5'-GTP induced calcium release from different intracellular pools, <u>Febs Lett.</u>, 218:153-158.

Inamitsu, T. and Ohtsuki, I., 1984, Characterization of ATP-dependent Ca^{2+} uptake by canine brain microsomes with saponin, <u>Eur. J. Biochem.</u>, 145:115-121.

Janicki, P. K. and Lipkowski, A. W., 1983, Kyotorphin and D-kyotorphin stimulate met-enkephalin release from rat striatum in vitro, <u>Neurosci. Lett.</u>, 43:73-77.

Kasper, C. B., 1971, Biochemical distinctions between the nuclear and microsomal membranes from rat hepatocytes, <u>J. Biol. Chem.</u>, 246:577-581.

Kuno, M. and Gardner, P., 1987, Ion channels activated by inositol 1,4,5-trisphosphate in plasma membrane of human T-lymphocytes, <u>Nature</u>, 326:301-304.

Maeda, N., Niinobe, M., Inoue, Y. and Mikoshiba, K., 1989, Developmental expression and intracellular location of P_{400} protein characterstic of Purkinje cells in the mouse cerebellum, <u>Dev. Biol.</u>, 133:67-76.

Payne, R. and Fein, A., 1987, Inositol 1,4,5-trisphosphate releases calcium from specialized sites with Limulus photoreceptors, <u>J. Cell Biol.</u>, 104; 933-937.

Penner, R., Matthews, G. and Neher, E., 1988, Regulation of calcium influx by second messengers in rat mast cells, <u>Nature</u>, 334:499-504.

Racknam, A., Wood, P. L. and Hudgin, R. L., 1982, Kyotorphin (tyrosine-arginine): further evidence for indirect opiate receptor activation,<u>Life Sci.</u>, 30:1337-1342.

Takagi, H., Shiomi, H., Ueda, H. and Amano, H., 1979, A novel analgesic dipeptide from bovine brain is a possible met-enkephalin releaser, <u>Nature</u>, 282:410-412.

Ueda, H., Shiomi, H. and Takagi, H., 1980, Regional distribution of a novel analgesic dipeptide, kyotorphin (Tyr-Arg) in the rat brain and spinal cord, <u>Brain Res.</u>, 198:460-464.

Ueda, H., Yoshihara, Y., Fukushima, N., Shiomi, H., Nakamura, A. and Takagi, H., 1987, Kyotorphin (tyrosine-arginine) synthetase in rat brain synaptosomes, <u>J. Biol. Chem.</u>, 262:8165-8173.

Ueda, H., Yoshihara, Y., Misawa, H., Fukushima, N., Kakada, T., Ui, M., Takagi, H. and Satoh, M., 1989, The kyotorphin (Tyrosine-Arginine) receptor and a selective reconstitution with purified G_i, measured with GTPase and phospholipase C assays, J. Biol. Chem., 264:3732-3741.

Ueda, H., Yoshihara, Y. and Takagi, H., 1986, A putative met-enkephalin releaser, kyotorphin enhances intracellular Ca^{2+} in the synaptosomes, Biochem. Biophys. Res. Commun., 137:897-902.

Verity, M. A., 1972, Cation modulation of synaptosomal respiration, J. Neurochem., 19:1305-1317.

Verma, A. K., Filoteo, A. G., Stanford, D. R., Wieben, E. D., Penniston, J. T., Strehler, E. E., Fischer, R., Heim, R., Vogel, G., Mathews, S., Strehler-Page, M.-A., James, P., Vorherr, T., Krebs, J. and Carafoli, E., 1988, Complete primary structure of a human plasma membrane Ca^{2+} pump, J. Biol. Chem., 263:14152-14159.

Volpe, P., Krause, K.-H., Hashimoto, S., Zorzato, F., Pozzan, T., Meldolesi, J. and Lew, D. P., 1988, "Calciosome", a cytoplasmic organelle: the inositol 1,4,5-trisphosphate-sensitive Ca^{2+} store of nonmuscle cells?, Proc. Natl. Acad. Sci. USA, 85:1091-1095.

Whittaker, V. P., Michaelson, I. A. and Kirkland, R. J., 1964, The separation of synaptic vesicles from nerve-ending particles (synaptosomes), Biochem. J., 90:293-303.

Worley, P. F., Baraban, J. M., Supattapone, S., Wilson, V. S. and Snyder, S. H., 1987, Characterization of inositol trisphosphate receptor binding in brain, J. Biol. Chem., 262:12132-12136.

Yoshihara, Y., Ueda, H., Fujii, N., Shide, A., Yajima, H. and Satoh, M., 1990, Purification of a novel type of calcium-activated neutral protease from rat brain - possible involvement in production of the neuropeptide, kyotorphin, from calpastatin fragments, J. Biol. Chem. (in press).

FORMATION OF INOSITOL POLYPHOSPHATES IN CULTURED ADRENAL CHROMAFFIN CELLS

Nobuyuki Sasakawa, Toshio Nakaki and Ryuichi Kato

Department of pharmacology, Keio University School of Medicine, 35 Shinanomachi, Shinjuku-ku, Tokyo 160, Japan

INTRODUCTION

In a variety of cell types, the rapid hydrolysis of phosphoinositides to 1,2-diacylglycerol and inositol 1,4,5-trisphosphate (Ins (1,4,5) P_3) plays an important role in transmembrane signalling (Nishizuka, 1984). Diacylglycerol acts through stimulating protein kinase C, whereas Ins (1,4,5) P_3 releases calcium from intracellular storage sites (Berridge and Irvine, 1984). It has been reported that these second messengers are generated by a signal transduction process comprising three main component s: a receptor, a GTP-binding protein and phospholipase C (Berridge and Irvine, 1984). However, it still possible that changes in cytosolic calcium concentration have physiologically relevant effects on phospholipase C, and it remains controversial as to whether the activation of phospholipase C is a consequence or forerunner of the rise in intracellular calcium concentration. Furthermore, details of some of the metabolic pathways linking them have been elucidated. The metabolism of inositol pentakisphosphate ($InsP_5$) and inositol hexakisphosphate ($InsP_6$) occurs through pathways which are separate from the agonist-sensitive pathways and is not clear in intact cells. Therefore, $InsP_5$ and $InsP_6$ are generally assumed to have a "housekeeping" function rather than being acute second messengers (Berridge and Irvine, 1989).

In the present study, we have investigated the effects of various kind of agents on formation of inositol polyphosphates and cellular calcium uptake in chromaffin cells. The results demonstrate the existence of two different mechanisms of $InsP_3$ formation, i.e. calcium uptake-dependent and -independent mechanisms, in these cells. Moreover, $InsP_5$ is rapidly increased by nicotine, a nicotinic agonist, in cultured bovine adrenal chromaffin cells.

MATERIALS AND METHODS

Primary culture of bovine adrenal chromaffin cells

Chromaffin cells were isolated from fresh bovine adrenal medulla as described previously (Sasakawa et al., 1989a; Sasakawa et al., 1989b). The chromaffin cells were purified by differential plating (Sasakawa et al., 1989b) and were plated on 35 mm diameter dishes (2.5×10^6 cells/dish) in a volume of 3 ml minimum essential medium (Gibco; Grand Island, NY), supplemented with 10% fetal calf serum (Hyclone; Logan, UT). The purity of chromaffin cells was over 95%. They were cultured at 37 °C in an atmosphere of 95% air/5% CO_2. The chromaffin cells were used for experiments 5 days after plating.

Measurement of [^{3}H]InsP$_3$ accumulation

The cells were labelled with [^{3}H]inositol (10 μCi/ml) in 35 mm diameter dishes for 2 days, as described previously (Nakaki et al., 1988; Sasakawa et al., 1987; Sasakawa et al., 1989a; Sasakawa et al., 1989b). The culture medium was aspirated and was replaced by a Locke's solution, consisting of 154 mM NaCl, 5.6 mM KCl, 2 mM $CaCl_2$, 1 mM $MgCl_2$, 3.6 mM $NaHCO_3$, 5.6 mM glucose, 10 mM Hepes and 0.1% bovine serum albumin (pH 7.4). The cell sheets were washed three times with the Locke's solution. The cells were preincubated at 37 °C for 20 min in 10 mM LiCl-containing Locke's solution. Stimulants were added to the incubation mixture and the cells were incubated for various time. At the end of the incubation, the Locke's solution was aspirated and inositol polyphosphates were extracted as described previously (Nakaki et al., 1988; Sasakawa et al., 1987; Sasakawa et al., 1989a; Sasakawa et al., 1989b). Inositol polyphosphates were separated by anion-exchange chromatography according to the method of Batty et al. (Batty et al., 1985). The fraction of InsP$_3$ which was eluted with 0.8 mM-ammonium formate / 0.1 M-formic acid from the anion exchange column was collected and its radioactivity was measured by scintillation counting (Sasakawa et al., 1989a).

HPLC analysis of [^{3}H]inositol polyphosphates isomers

Isomers of inositol polyphosphates were separated by HPLC (Nakaki et al., 1988; Sasakawa et al., 1989a). In brief, the samples were subjected to HPLC with a Whatman Partisil SAX 10 column. We used a gradient delivered from two independent pumps drawing on reservoirs containing water (pump A) and 3.5 M ammonium formate(pH 3.7 with phosphoric acid at 25 °C; pump B) at a flow rate of 1.2 ml/min as follows: 0 min, 0% B; 5 min, 29% B; 10 min, 29% B; 18 min, 49% B; 27 min, 100% B; 30 min, 100% B; 55 min, 0% B. The fractions were collected at 20 s intervals. Radioactivity of each fraction was determined by scintillation counting with 10 ml of ACSII (Amersham).

Measurement of $^{45}Ca^{2+}$ uptake

Cellular $^{45}Ca^{2+}$ uptake was measured as described previously (Sasakawa et al., 1983; Sasakawa et al., 1984).

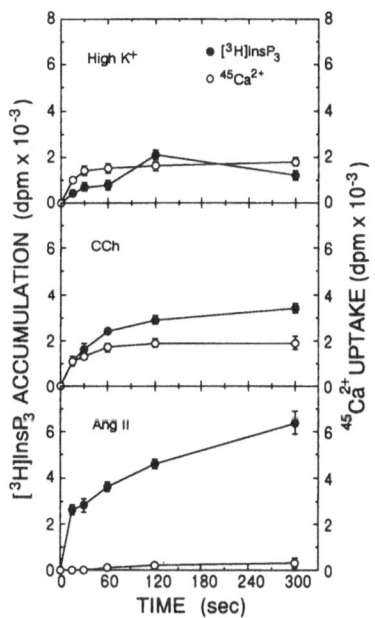

Fig. 1. Time courses of the $[^3H]InsP_3$ accumulation and $^{45}Ca^{2+}$ uptake stimulated by high K^+, carbamylcholine or AngII.
The cells were preincubated with Lock's solution containing LiCl (10 mM) and bovine serum albumin (0.1%) for 20 min. Thereafter the cells were treated with or without stimulant for the indicated time period.$[^3H]InsP_3$ accumulation and cellular $^{45}Ca^{2+}$ uptake induced by several stimulants were determined as described in Materials and Methods.The basal values were subtracted from the data. High K^+; KCl (56 mM), CCh; carbamylcholine (300 μM), AngII; angiotensin II (10 μM).

RESULTS AND DISCUSSION

Fig. 1 shows the time course of $[^3H]InsP_3$ accumulation and $^{45}Ca^{2+}$ uptake induced by several stimulants in cultured adrenal chromaffin cells. When the $[^3H]$inositol-prelabelled chromaffin cells were stimulated by high K^+, CCh or angiotensin II (Ang II), a rapid accumulation of $[^3H]InsP_3$ was observed. At the same time, high K^+ or CCh induced rapid increases in $^{45}Ca^{2+}$ uptake, whereas AngII did not induce a significant $^{45}Ca^{2+}$ uptake. High K^+ induced a significant increase in $[^3H]InsP_3$ accumulation at 30 sec. The accumulation reached the maximum level at 2 min. The high K^+ induced $^{45}Ca^{2+}$ uptake preceded $[^3H]InsP_3$ accumulation and reached maximum level at 1 min. CCh- and Ang II- induced $[^3H]InsP_3$ accumulation were more rapid in onset than the high K^+-induced accumulation. Table 1 shows the effects of nifedipine, a calcium channel antagonist, on $[^3H]InsP_3$ accumulation and $^{45}Ca^{2+}$ uptake. Nifedipine completely inhibited high K^+-induced both $[^3H]InsP_3$ accumulation and $^{45}Ca^{2+}$ uptake. Nifedipine inhibited CCh-induced $^{45}Ca^{2+}$ uptake but did not inhibit CCh-induced $[^3H]InsP_3$ accumulations effectively as high K^+-induced accumulation. Nifedipine had no effect on AngII-induced $[^3H]InsP_3$ accumulation.

As shown in Table 2, KCl induced a concentration-dependent increase in $[^3H]InsP_3$ accumulation and $^{45}Ca^{2+}$ uptake with a similar potency and exerted its maximum effects at 56 mM.

In order to examine the Ca^{2+} uptake-mediated (without receptor stimulation) $[^3H]InsP_3$ accumulation, we used several types of intracellular Ca^{2+}-increasing

Table 1. Effects of nifedine on $[^3H]InsP_3$ accumulation and $^{45}Ca^{2+}$ uptake.

	$[^3H]InsP_3$ Accumulation (dpm x 10^{-3})	$^{45}Ca^{2+}$ Uptake (dpm x 10^{-2})
None	2.1 ± 0.1	7.0 ± 0.8
Nifedipine (10 μM)	1.9 ± 0.1	6.8 ± 1.0
High K^+	3.3 ± 0.1	23.0 ± 2.0
High K^+ + Nifedipine	2.3 ± 0.2	8.2 ± 1.2
Carbamylcholine (300 μM)	4.4 ± 0.3	24.0 ± 1.4
Carbamylcholine+ Nifedipine	3.6 ± 0.2	9.0 ± 0.8
Angiotensin II (10 μM)	5.7 ± 0.5	8.0 ± 0.4
Angiotensin II+ Nifedipine	5.1 ± 0.3	7.7 ± 0.2

$[^3H]InsP_3$ accumulation and cellular $^{45}Ca^{2+}$ uptake induced by several stimulants for 2 min were determined as described in Materials and Methods. Nifedipine was added to the preincubation medium 3 min before the addition of the stimulants.

Table 2. Concentration-dependent effects of KCl on $[^3H]InsP_3$ accumulation and $^{45}Ca^{2+}$ uptake.

Concentration of KCl (mM)	$[^3H]InsP_3$ Accumulation (dpm x 10^{-3})	$^{45}Ca^{2+}$ Uptake (dpm x 10^{-2})
5.6	1.8 ± 0.2	7.0 ± 0.1
28.0	2.4 ± 0.1	12.0 ± 0.8
56.0	3.2 ± 0.1	17.0 ± 1.2
84.0	3.3 ± 0.2	19.0 ± 1.0

$[^3H]InsP_3$ accumulation and cellular $^{45}Ca^{2+}$ uptake induced by several concentration of KCl for 2 min were determined as described in Materials and Methods.

agents (Fig. 2). BAY K 8644, a calcium channel activator (Schramm et al., 1983), plus a partially depolarizing concentration of KCl (14 mM), induced $^{45}Ca^{2+}$ uptake and $[^3H]InsP_3$ accumulation. Veratridine, a sodium channel activator, was as effective as 56 mM KCl for $^{45}Ca^{2+}$ uptake and $[^3H]InsP_3$ accumulation. Ionomycin, a calcium ionophore, also induced $^{45}Ca^{2+}$ uptake and $[^3H]InsP_3$ accumulation.

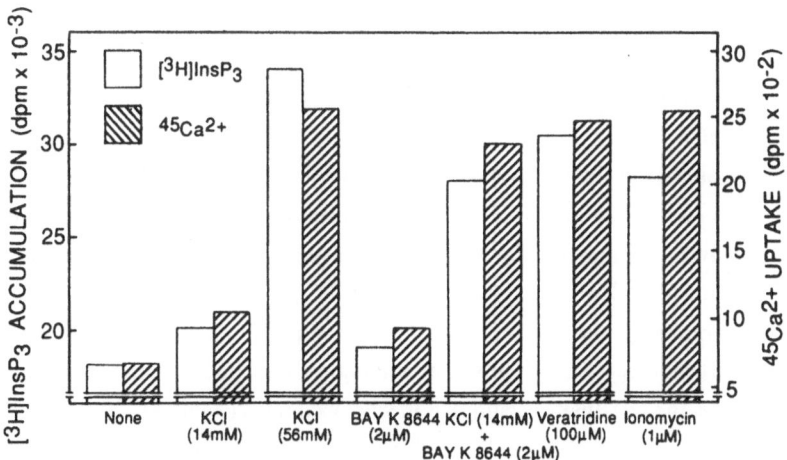

Fig. 2   Effects of several types of intracellular Ca^{2+} concentration-increasing agents on $[^3H]InsP_3$ accumulation and $^{45}Ca^{2+}$ uptake.
$[^3H]InsP_3$ accumulation and cellular $^{45}Ca^{2+}$ uptake induced by several stimulants for 2 min were determined as described in Materials and Methods.

Table 3. Effects of cholinergic drugs on $[^3H]InsP_3$ accumulation in adrenal chromaffin cells cultured for 5 days.

	$[^3H]InsP_3$ Accumulation (%)
Acetylcholine (300 μM)	100
Nicotine (10 μM)	65 ± 6
Muscarine (100 μM)	10 ± 4
Acetylcholine + Atropine (10 μM)	70 ± 5
Acetylcholine + Hexamethonium (100 μM)	4 ± 1
Acetylcholine + d-Tubocurarine (100 μM)	7 ± 3

$[^3H]InsP_3$ accumulation induced by several stimulants for 2 min were determined as described in Materials and Methods. Antagonist was added to the preincubation medium 3 min before the addition of the stimulants.
$[^3H]InsP_3$ accumulation is expressed as a percentage, where 0% represents the basal level and 100% the maximal level attained by 300 μM acetylcholine. The basal level and maximal level were 1300 ± 500 and 3100 ± 920 dpm , respectively.

These results suggest the existence of the Ca^{2+} uptake-triggered mechanism of $InsP_3$ accumulation represented by high K^+ stimulation, and also the Ca^{2+} uptake-independent mechanism of $InsP_3$ accumulation represented by Ang II stimulation in cultured adrenal chromaffin cells. Since ionomycin induced $[^3H]InsP_3$ accumulation, the depolarization of the cell membrane itself does not lead to the stimulation of $InsP_3$ accumulation. It is highly unlikely that the high K^+-induced $[^3H]InsP_3$ accumulation is due to some substances which are secondarily released from chromaffin cells by high K^+ stimulation, because 1000-fold dilution of incubation medium does not affect the $[^3H]InsP_3$ accumulation induced by high K^+ (Sasakawa et al., 1987).

The physiological significance of this Ca^{2+} uptake-triggered mechanism of $InsP_3$ accumulation is unclear but it is possible that phospholipase C with low sensitivity of Ca^{2+}, which is activated by massive Ca^{2+} entry, is involved in the amplification system of receptor stimulation-mediated $InsP_3$ formation in these cells.

To determine whether the activation of muscarinic or nicotinic receptors was responsible for the observed increase in $[^3H]InsP_3$ accumulation, the effects of specific agonist and antagonists on the $[^3H]InsP_3$ accumulation were examined (Table 3). Nicotine showed a greater effect than muscarine on $[^3H]InsP_3$ accumulation. The nicotinic antagonist hexamethonium and d-tubocurarine showed greater inhibitory effects than the muscarinic antagonist atropine on $[^3H]InsP_3$ accumulation. These results suggest that nicotinic receptors mainly mediate the $InsP_3$ accumulation in cultured adrenal chromaffin cells. Our previous results showed that cholinergic receptors linked to the $InsP_3$ accumulation functionally shift from muscarinic (< 2 days) to nicotinic (> 4 days) during primary culture of adrenal chromaffin cells (Nakaki et al., 1988)

The effects of 12-O-tetradecanoyl phorbol-13-acetate (TPA), a protein kinase C activator (Blumberg, 1981), on [^{3}H]InsP$_3$ accumulation induced by high K$^+$, nicotine and Ang II were examined. As shown in Table 4, TPA (100 nM) partially inhibited nicotine- and Ang II-induced [^{3}H]InsP$_3$ accumulation but failed to inhibit the high K$^+$-induced [^{3}H]InsP$_3$ accumulation. It has been reported that TPA inhibits the receptor-mediated InsP$_3$ accumulation in several types of cells. At least three mechanisms are reportedly responsible for the inhibition of InsP$_3$ accumulation by TPA: (1) phosphorylation of receptor (Leeb-Lundberg et al., 1985), (2) phosphorylation of G-protein (Orellana et al., 1987), and (3) stimulation of Ins(1,4,5)P$_3$ degradation (Connolly et al., 1986). In cultured adrenal chromaffin cells, the inhibitory effects of TPA on nicotine- and Ang II-induced [^{3}H]InsP$_3$ accumulation seem to be due to the inhibition of InsP$_3$ formation at the site of either the receptor or G-protein, because the high K$^+$-induced [^{3}H]InsP$_3$ accumulation was not affected by TPA pretreatment.

Islet-activating protein (IAP), the active component of pertussis toxin, catalyses ADP-ribosylation of the GTP-binding α subunit of Gi and prevents the effects of inhibitory hormones on adenylate cyclase (Katada and Ui, 1982). It has been shown that a GTP-binding protein, which is different from Gi and is also ADP-ribosylated by IAP, participates in the control of cellular events such as the regulation of the linkage between receptors and phospholipase C (Brandt et al., 1985). Botulinum toxin D type (BTD) also ADP-ribosylates a 21 KDa protein which is derived from the membrane fraction of bovine adrenal gland (Ohashi and Narumiya, 1987). Therefore, we examined the effects of IAP and BTD on [^{3}H]InsP$_3$ accumulation induced by high K$^+$, nicotine and Ang II. As shown in Fig. 3, IAP and BTD partially inhibited nicotine- and Ang II-induced [^{3}H]InsP$_3$ accumulation but failed to inhibit the

Table 4. Effects of TPA on [^{3}H]InsP$_3$ accumulation in cultured adrenal chromaffin cells.

	[^{3}H]InsP$_3$ accumulation (dpm x 10^{-3})	
	TPA (-)	TPA (+)
None	2.5 ± 0.3	2.3 ± 0.1
High K$^+$	4.4 ± 0.2	4.2 ± 0.3
Nicotine (10μM)	6.2 ± 0.3	4.5 ± 0.2
Angiotensin II (10μM)	7.3 ± 0.8	5.5 ± 0.2

[^{3}H]InsP$_3$ accumulation induced by several stimulants for 2 min was determined as described in Materials and Methods. TPA (100 nM) was added to the preincubation medium 10 min before the addition of the stimulants.

high K$^+$-induced [^{3}H]InsP$_3$ accumulation. These results suggest that GTP-binding proteins may contribute to the receptor-mediated [^{3}H]InsP$_3$ accumulation .

These results suggest that the Ca^{2+} uptake-triggered mechanism of InsP$_3$ formation represented by high K$^+$ differ from Ca^{2+} uptake-independent mechanisms of InsP$_3$ formation represented by Ang II.   CCh induced increases in both ^{45}Ca^{2+} uptake and [^{3}H]InsP$_3$ accumulation. Nifedipine inhibited the CCh-induced ^{45}Ca^{2+} uptake similar to that of the inhibition of high K$^+$-induced ^{45}Ca^{2+} uptake whereas nifedipine failed to inhibit CCh-induced [^{3}H]InsP$_3$ accumulation (Fig.1. & Table 1). Furthermore, nicotine-induced [^{3}H]InsP$_3$ accumulation was partially inhibited by TPA, IAP and BTD pretreatment (Table 4 & Fig.3). These findings suggest the mechanism of nicotinic receptor-mediated [^{3}H]InsP$_3$ accumulation has an intermediate property between those of high K$^+$ and Ang II.

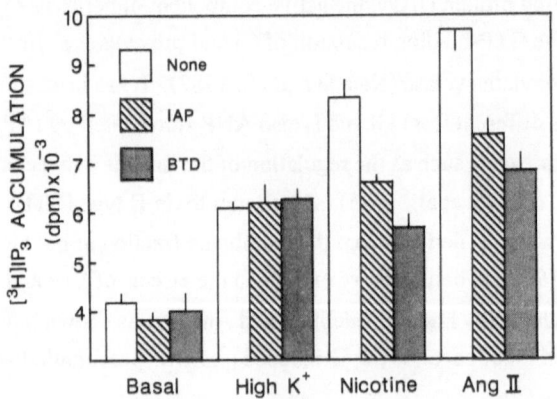

Fig. 3.   Effects of IAP and BTD on [^{3}H]InsP$_3$ accumulation in cultured adrenal
chromaffin cells.
The cells were pretreated with IAP (200 ng/ml) or BTD (10 mg/ml) for 20
hr. [^{3}H]InsP$_3$ accumulation induced by several stimulants for 2 min was
determined as described in Materials and Methods. IAP; islet-activating
protein (pertussis toxin), BTD; botulinum toxin type D.

Ins (1,4,5)P$_3$ has been well established as an intracellular messenger, whose function is to release calcium from intracellular store site (Berridge and Irvine, 1984). In contrast, the evidence that other inositol polyphosphate isomers might act as second messengers is so far controversial. Although Ins(1,3,4,5)P$_4$ has been proposed to participate in cellular calcium homeostasis (Hill et al., 1988; Higashida and Brown, 1986), the function and the biosynthesis pathway of highly phosphorylated inositol such as InsP$_5$ and InsP$_6$ remain to be explored. Vallejo has shown that exogenously added InsP$_5$ and InsP$_6$ regulate blood pressure via the central nervous system (Vallejo et al., 1987). This rises the possibility that the biosynthesis of these inositol polyphosphates is regulated by stimulation by hormones and neurotransmitters. Thus, we have examined the effects of nicotinic stimulation on the formation of several kinds of inositol polyphosphate isomers in cultured adrenal chromaffin cells.

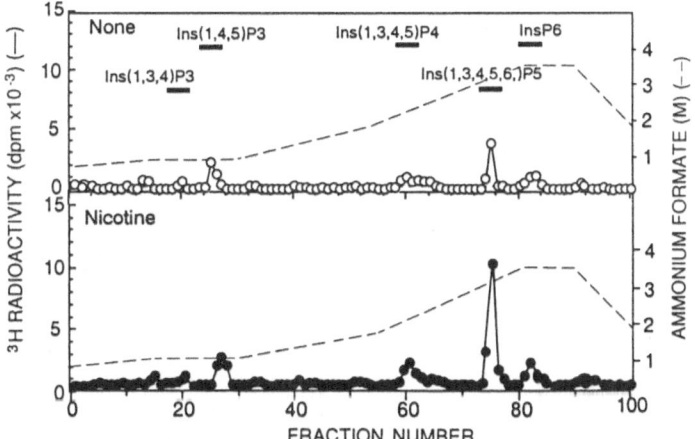

Fig. 4. Elution profile of [^{3}H]inositol polyphosphates produced in intact adrenal chromaffin cells and separated by high performance anion exchange chromatography.
Upper panel: unstimulated cells. Lower panel: cells stimulated by 10 μM nicotine. The cells were labelled with 50 μCi/ml myo-[2-^{3}H]Inositol in 35 mm diameter dishes for 2 days. Nicotine solution was added to the incubation mixture and the cells were incubated for 15 s in the presence of 10mM LiCl. At the end of the incubation, the Locke's solution was aspirated and inositol polyphosphates were extracted as described in Materials and Methods.

Fig. 4 shows a typical elution profile of the separation of the inositol polyphosphate isomers from the extract of [^{3}H]inositol-prelabeled chromaffin cells. [^{3}H]Ins(1,4,5)P$_3$ and [^{3}H]Ins(1,3,4)P$_3$ were eluted at 1.0 M ammonium formate under the conditions we used, consistent with the results of Jackson et al (Jackson et al., 1987), who used a similar gradient. The peak at 2 M ammonium formate was considered as [^{3}H]InsP$_4$, since the standard [^{3}H]Ins(1,3,4,5)P$_4$ was eluted at the same retention time and Jackson et al. also observed

$[^3H]InsP_4$ in the identical position (Jackson et al., 1987). The peak at 3 M ammonium formate was considered as $[^3H]InsP_5$, since the retention time was identical with that of a standard $[^3H]Ins(1,3,4,5,6)P_5$. The more polar $[^3H]$inositol-containing material may be $[^3H]InsP_6$, since the standard $[^3H]InsP_6$ eluted at the same position. Stimulation with nicotine increased $InsP_3$ in the chromaffin cells (Fig. 4), consistent with previous reports (Nakaki et al., 1988; Sasakawa et al., 1987; Sasakawa et al., 1989a; Eberhard and Holz, 1987). Nicotine induced a large and rapid increase in $InsP_5$, which peaked at 15 s after stimulation. This suggests the presence of active degradative pathways as well as the stimuli-sensitive productive ones. $InsP_4$ and $InsP_6$ were also increased with a slower time course and a lesser magnitude than $InsP_5$. The results of a representative experiment among three separate ones is shown in Table 5.

Table 5. Time course of $[^3H]InsP_3$, $[^3H]InsP_4$, $[^3H]InsP_5$ and $[^3H]InsP_6$ accumulation stimulated by nicotine in adrenal chromaffin cells.

	$[^3H]$Inositol Polyphosphate Accumulation (dpm x 10^{-3})			
Time (sec)	0	5	15	120
None	3.5	3.7	4.5	3.9
$Ins(1,3,4)P_3$				
Nicotine	-	4.2	4.9	10.0
None	4.8	5.0	4.3	5.1
$Ins(1,4,5)P_3$				
Nicotine	-	9.2	10.5	3.9
None	2.8	2.2	2.4	3.2
$InsP_4$				
Nicotine	-	2.5	4.3	7.6
None	8.0	7.8	7.9	7.6
$InsP_5$				
Nicotine	-	12.2	16.0	9.4
None	2.8	3.3	2.7	3.1
$InsP_6$				
Nicotine	-	4.0	5.7	7.2

The chromaffin cells were labelled with $[^3H]$inositol as described in Materials and Methods. The peak heights of each inositol polyphosphate were presented. The concentrations of nicotine was 10 μM. The experiments were repeated twice and gave similar results.

In chromaffin cells the agonist-stimulated increase in Ins(1,4,5)P$_3$ is small.(Nakaki et al., 1988; Sasakawa et al., 1989a; Sasakawa et al., 1989b; Sasakawa et al., 1988). Most of the agonist-induced InsP$_3$ fraction consists of Ins(1,3,4)P$_3$. In fact, InsP$_5$ is the most pronounced inositol polyphosphate identified so far in the chromaffin cells. Furthermore, the time course was comparable to Ins(1,4,5)P$_3$. In many other systems, receptor stimulation does not increase unequivocally InsP$_5$. A current hypothesis proposes that InsP$_5$ is formed by phosphorylation of InsP$_4$ (Stephens et al., 1988a; Stephens et al., 1988b). Alternatively, it is possible that InsP$_5$ is generated independently of InsP$_4$ formation, although more rigorous research is necessary for establishing this possibility. Although inositol pentakisphosphate-containing phospholipid has not been shown, it is possible that such phospholipids exist in mammalian plasma membranes.

SUMMARY

Formation of inositol polyphosphates has been characterized in cultured bovine adrenal chromaffin cells in terms of calcium dependency and isomers of inositol polyphosphates. There are two distinct pathways of generation of InsP$_3$. Stimulants such as high K$^+$ induce InsP$_3$ accumulation by a calcium uptake-dependent mechanism. Stimulants such as Ang II induce InsP$_3$ accumulation by a calcium uptake-independent mechanism. Both mechanisms are involved in nicotinic stimulation. These results suggest that calcium entry as well as receptor-mediated mechanisms play a significant role in phosphoinositides hydrolysis through phospholipase C in adrenal chromaffin cells. Nicotinic receptor stimulation induces a rapid and transient increase in Ins(1,4,5)P$_3$ accumulation followed by a slower accumulation of Ins(1,3,4)P$_3$. Moreover, nicotine induces a large and rapid increase in Ins(1,3,4,5,6)P$_5$ accumulation with an extent and time course similar to Ins(1,4,5)P$_3$, which peaks at 15 sec after stimulation. Nicotine also induced Ins(1,3,4,5)P$_4$ and InsP$_6$ accumulation with a slower time course and a lesser magnitude than Ins(1,3,4,5,6)P$_5$. These results indicate that adrenal chromaffin cells possess fine regulation of inositol polyphosphates metabolism and that inositol polyphosphates are involved with the control of cellular function in these cells.

REFERENCES

Batty, I. R., Nahorski, S. R., and Irvine, R. F., 1985, Rapid formation of inositol 1,3,4,5-tetrakisphosphate following muscarinic receptor stimulation of rat cerebral cortical slices, Biochem. J., 232:211-215.

Berridge, M. J., and Irvine, R. F., 1984, Inositol trisphosphate, a novel second messenger in cellular signal transduction, Nature, 312:315-321.

Berridge, M. J., and Irvine, R. F., 1989, Inositol phosphates and cell signalling, Nature, 341:197-205.

Blumberg, P. M., 1981, In vitro studies on the mode of action of phorbol esters, potent tumor promoters, part 2, CRC Crit. Rev. Toxicol., 9:199-234.

Brandt, S. J., Dougherty, R. W., Lapetina, E. G., and Niedel, J. E., 1985, Pertussis toxin inhibits chemotactic peptide-stimulated generation of inositol phosphates and lysosomal enzyme secretion in human leukemic (HL-60) cells, Proc. Natl. Acad. Sci. USA, 82:3277-3280.

Connolly, T. M., Lawing Jr., W. J., and Majerus, P. W., 1986, Protein kinase C phosphorylates human platelet inositol trisphosphate 5'-phosphomonoesterase, increasing the phosphatase activity, Cell, 46:951-958.

Eberhard, D. A., and Holz, R. W., 1987, Cholinergic stimulation of inositol phosphate formation in bovine adrenal chromaffin cells: Distinct nicotinic and muscarinic mechanisms, J. Neurochem., 49:1634-1643.

Higashida, H., and Brown, D. A., 1986, Membrane current responses to intracellular injections of inositol 1,3,4,5-tetrakisphosphate and inositol 1,3,4-trisphosphate in NG108-15 hybrid cells, FEBS lett., 208:283-286.

Hill, T. D., Dean, N. M., and Boynton, A. L., 1988, Inositol 1,3,4,5-tetrakisphosphate induces Ca^{2+} sequestration in rat liver cells, Science, 242:1176-1177.

Jackson, T. R., Hallam, T. J., Downes, C. P., and Hanley, M. R., 1987, Receptor coupled events in bradykinin action: rapid production of inositol phosphates and regulation of cytosolic free Ca^{2+} in a neural cell line, EMBO J., 6:49-54.

Katada, T., and Ui, M., 1982, Direct modification of the membrane adenylate cyclase system by islet-activating protein due to ADP-ribosylation of a membrane protein, Proc. Natl. Acad. Sci. USA, 79:3129-3133.

Leeb-Lundberg, L. M. F., Cotecchia, S., Lomasney, J. W., Debernaardis, J. F., Lefkowitz, R. J., and Caron, M. G., 1985, Phorbol esters promote α_1-adrenergic receptor phosphorylation and receptor uncoupling from inositol phospholipid metabolism, Proc. Natl. Acad. Sci. USA, 82:5651-5655.

Nakaki, T., Sasakawa, N., Yamamoto, S., and Kato, R., 1988, Functional shift from muscarinic to nicotinic cholinergic receptors involved in inositol trisphosphate and cyclic GMP accumulation during the primary culture of adrenal chromaffin cells, Biochem. J., 251:397-403.

Nishizuka, Y., 1984, Turnover of inositol phospholipids and signal transduction, Science, 225:1365-1370.

Ohashi, Y., and Narumiya, S., 1987, ADP-ribosylation of a Mr 21000 membrane protein by type D botulinum toxin, J. Biol. Chem., 262:1430-1433.

Orellana, S., Solski, P. A., and Brown, J. H., 1987, Guanosine 5'-O-(thiotriphosphate)-dependent inositol trisphosphate formation in membranes is inhibited by phorbol ester and protein kinase C, J. Biol. Chem., 262:1638-1643.

Sasakawa, N., Kumakura, K., Yamamoto, S., and Kato, R., 1983, Effects of W-7 on catecholamine release and $^{45}Ca^{2+}$ uptake in cultured adrenal chromaffin cells, Life Sci., 33:2017-2024.

Sasakawa, N., Yamamoto, S., Ishii, K., and Kato, R., 1984, Inhibition of calcium uptake and catecholamine release by 8-(N,N-diethylamino)-octyl-3,4,5-trimethoxybenzoate hydrochloride (TMB-8) in cultured adrenal chromaffin cells, Biochem. Pharmacol., 33:4063-4067.

Sasakawa, N., Nakaki, T., Yamamoto, S., and Kato, R., 1987, Inositol trisphosphate accumulation by high K^+ stimulation in cultured adrenal chromaffin cells, FEBS lett., 223:413-416.

Sasakawa, N., Nakaki, T., Yamamoto, S., and Kato, R., 1989a, Calcium uptake-dependent and -independent mechanisms of inositol trisphosphate formation in adrenal chromaffin cells: comparative studies with high K^+, carbamylcholine and angiotensin II, Cell. Signal., 1:75-84.

Sasakawa, N., Nakaki, T., Yamamoto, S., and Kato, R., 1989b, Stimulation by ATP of inositol trisphosphate accumulation and calcium mobilization in cultured adrenal chromaffin cells, J. Neurochem., 52:441-447.

Sasakawa, N., Yamamoto, S., Nakaki, T., and Kato, R., 1988, Effects of islet-activating protein on the catecholamine release, Ca^{2+} mobilization and inositol trisphosphate formation in cultured adrenal chromaffin cells, Biochem. Pharmacol., 37:2485-2487.

Schramm, M., Thomas, G., Towart, R., and Franckowiak, G., 1983, Novel dihydropyridines with positive inotropic action of Ca^{2+} channels, Nature, 303:535-537.

Stephens, L. R., Hawkins, P. T., Barker, C. J., and Downes, C. P., 1988a, Synthesis of *myo*-inositol 1,3,4,5,6-pentakisphosphate from inositol phosphates generated by receptor activation, Biochem. J., 253:721-733.

Stephens, L. R., Hawkins, P. T., Morris, A. J., and Downes, C. P., 1988b, L-*myo*-inositol 1,4,5,6-tetrakisphosphate 3-hydroxyl kinase, Biochem. J., 249:283-292.

Vallejo, M., Jackson, T., Lightman, S., and Hanley, M. R., 1987, Occurrence and extracellular actions of inositol pentakis- and hexakisphosphate in mammalian brain, Nature, 330:656-658.

EFFECTS OF OXYGEN DEPLETION ON PHOSPHOINOSITIDE BREAKDOWN IN RAT BRAIN

SLICES

Haruaki Ninomiya[1], Takashi Taniguchi[2] and
Motohatsu Fujiwara[1]

[1]Department of Pharmacology, Faculty of Medicine
Kyoto University, Kyoto 606 and [2]Department of Neurobiology
Kyoto Pharmaceutical University, Kyoto 607, Japan

INTRODUCTION

Electrophysiological studies indicate that hypoxia exerts dual effects on neuronal excitability. The first effect of hypoxia is supression
of excitability during hypoxia period which would lead to the impared
synaptic transmission during hypoxia (Gorman, 1966; Dolivo, 1974; Hansen
et al., 1982), and the second effect is an enhancement of the excitability
after hypoxic period (Shiff and Balestrino, 1985; Shiff and Somjen, 1984,
1985, 1987). Biochemical studies have clearly established that hypoxia
disrupts the biosynthesis of various neurotransmitters, including acetylcholine (Gibson and Duffy, 1981; Gibson et al., 1981) or catecholamines
(Davis and Carlsson, 1973; Miwa et al., 1986). The first effect of hypoxia, viz., supression of excitability, may be readily explained by decreased neurotransmitter synthesis in the presynaptic terminals, but the
cellular and molecular mechanisms underlying the posthypoxic hyperexcitability remain unclear. PI breakdown is an ubiquitous intracellular signal
transduction system in the brain, and our initial motivation to start this
study was to examine whether any alterations in PI metabolism would play a
role in changes in neuronal excitability caused by hypoxia.

MATERIALS AND METHODS

Myo-2-[3H] inositol (14.0 Ci/mmol) and [3H] prazosin (80.9 Ci/mmol)
were purchased from New England Nuclear (Boston, MA, U.S.A.). All other
chemicals were were obtained commercially.

Assay of PI turnover and in vitro anoxia

Assay of PI turnover was done according to the method of Brown et
al. (1984) with a modification detailed below; in vitro anoxia was induced
as described by Bosley et al. (1983). Adult male Wistar rats were used in
all experiments.

Tissue preparation and phospholipid labeling

Rats were decapitated and the whole brain was rapidly removed. Brain
slices of 1 mm thickness were made manually using a tissue chopper and
four regions of the brain (cerebral cortex, striatum, hippocampus and

cerebellum) were dissected out on an ice-cold plate. Cross-chopped slices (350 μM x 350 μM x 1 mm) of the tissue preparations were cut with a MacIlwain tissue chopper and dispersed in Krebs buffer (120 mM NaCl;4.7 mM KCl;1.3 mM $CaCl_2$;1.2 mM KH_2PO_4;1.2 mM $MgSO_4$;25 mM $NaHCO_3$;11.7 mM glucose) equilibrated to pH 7.4 with 95% O_2/5% CO_2 and incubated at 37 ^oC for 45 min with gentle shaking and an intermediate change of buffer. Slices were then incubated at 37 ^oC for one hour in Krebs buffer containing 0.3 μM [³H] inositol to label the phospholipid pool.

PI breakdown

Following phospholipid labeling with 0.3 μM [³H] inositol under oxygenated conditions for 60 min, the slices were washed with an excess volume of fresh buffer containing 5 mM myo-inositol to remove free [³H] inositol. 50 μl of packed slices were then incubated at 37 ^oC in 240 μl Krebs buffer containing 10 mM LiCl (final concentration; 8 mM), previously equilibrated with 95% O_2/5% CO_2 for controls and with 95% N_2/5% CO_2 for anoxic incubations. Following the addition of 10 μl of agonist solution (or buffer for determinating basal breakdown), the reaction was allowed to continue for various lengths of time under a continuous gas flow of the appropriate composition. The samples were reoxygenated by bubbling the incubation medium with 95% O_2/5% CO_2 for 10 min. Reaction was halted by adding 0.94 ml of chloroform/methanol (1:2 v/v) and 0.31 ml of chloroform, then 0.31 ml of water was added to separate the phases. After centrifugation at 1,000 g for 10 min, 0.75 ml of the upper aqueous phase was diluted to 3 ml with water and added to 0.5 ml of slurry (50% w/w) of Dowex-1 resin. After four washes with 3 ml of 5 mM myo-inositol, the ³H-inositol phosphates (IPs) fraction was eluted with 1 ml of 0.1 M formic acid/1.0 M ammonium formate. A portion (0.8 ml) of this eluate was added to 10 ml of scintillation fluid and counted for radioactivity.

After evaporation of the organic phase, proteins were dissolved in 0.5 M NaOH and measured according to the method of Lowry et al. (1951).

Exposure to in vivo hypoxia

We used the hypoxic chamber described previously (Ninomiya et al., 1988). In brief, a gas mixture of 8% O_2/92% N_2 was continuously passed through the chamber at a flow rate of 3 l/min. A CO_2 scrubber containing soda lime was used to eliminate CO_2. The concentration of O_2 in the chamber was monitored with an oxygen measuring device. The hypoxic rats were kept in this chamber and the control rats were kept in a chamber filled with room air.

Following exposure to either hypoxia or room air for various lengths of time, the rats were decapitated and assay of PI turnover in brain slices was done as described above, using the 95% O_2/5% CO_2 to gas the incubation medium.

Radioligand Binding assay

Tissue samples were homogenized in 50 mM sodium-potassium phosphate buffer (pH 7.4) with a Polytron homogenizer (PT10) at a setting of 8 for 10 sec. The homogenate was centrifuged twice at 48,000 g for 20 min. The final pellet was resuspended in buffer to obtain a crude membrane fraction.

[³H] Prazosin binding was carried out by incubation of membrane preparations (30-50 μg of protein) with [³H] prazosin adjusted to a final volume of 250 μl with buffer, in glass tubes at 37 ^oC for 60 min. The reaction was terminated by the addition of 3 ml of ice-cold buffer and then filtered under reduced pressure through Whatman GF/B glass-fiber filters. Filters were washed two times with 3 ml ice-cold buffer, dried, and then counted for radioactivity in 6 ml of scintillation fluid. Specific binding was calculated as the difference between binding in the absence and presence of 100 μM phentolamine.

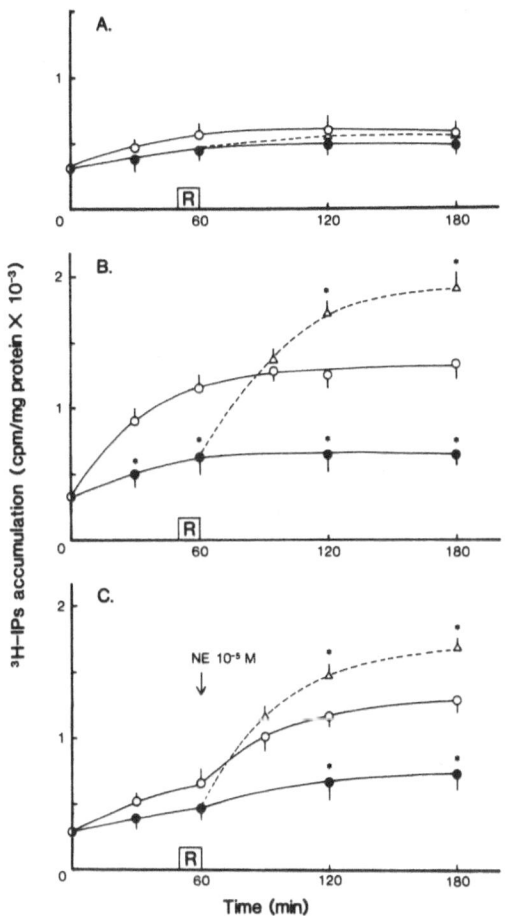

Fig. 1 Effect of anoxia and oxygenation on NE-stimulated accumulation of ^{3}H-IPs in cortical slices. After labeling
phospholipid pool with 0.3 μM [^{3}H] inositol under oxygenated conditions for 60 min, slices were washed with fresh
buffer containing 5 mM myo-inositol, and then incubated
in the absence (A) or presence (B) of 10^{-5} M NE under
oxygenated or anoxic conditions. All of the oxygenated
samples and some of the anoxic samples were oxygenated by
bubbling the incubation medium with 95% O_2/5% CO_2 for 10
min at the time indicated (R in the figure). In some
experiments (C), incubation was begun in the absence of
NE, and 10^{-5} M NE was added immediately after oxygenation. At various times, the reaction was stopped by
adding chloroform-methanol (1:2 v/v) and ^{3}H-IPs accumulation was determined as described in "Materials and Methods".   ○ : oxygenated,   ● : anoxia,   △ : oxygenated
following anoxia. Values shown are means ± SEM of at
least four determinations. * : p < 0.01 ; significantly
different from values for oxygenated incubations. (cited
from Ninomiya et al., 1989 with permission)

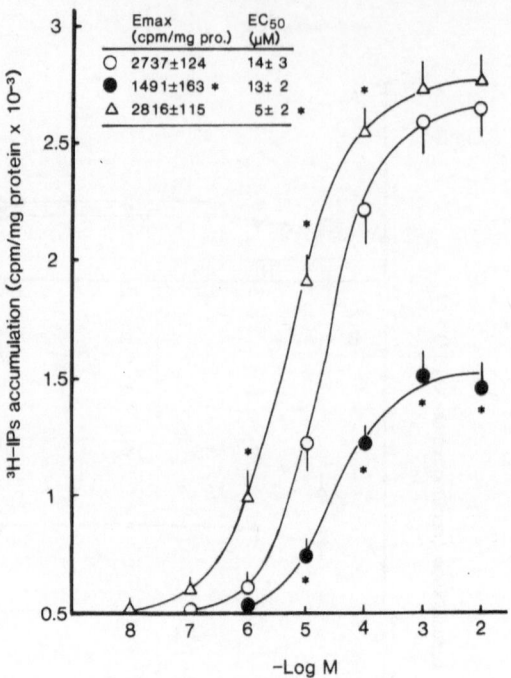

	Emax (cpm/mg pro.)	EC50 (µM)
O	2737±124	14± 3
●	1491±163 *	13± 2
△	2816±115	5± 2 *

Fig. 2 Dose-response curves for NE-stimulated accumulation of ^{3}H-IPs. After lipid labeling and washing as described in legend for Fig. 1, cortical slices were incubated with increasing concentrations of NE for 120 min under oxygenation (O), for 120 min under anoxia (●), or for another 60 min following 50 min-anoxia and 10-min oxygenation (△). Values shown are means ± SEM of at least four determinations. * : p < 0.01 ; significantly different from values for oxygenated incubations. (cited from Ninomiya et al., 1989 with permission)

Data analysis

All values are expressed as means ± SEM of n experiments. Student's t test or ANOVA was used for statistical analysis.

RESULTS AND DISCUSSION

In the first series of experiments, we examined the effects of in vitro anoxia on PI breakdown in rat brain slices stimulated by norepine-phrine (NE), carbachol (Carb) or glutamic acid (Glu), using cortical or hippocampal slices. After labeling the phospholipid for 60 min in the presence of oxygen, the slices were washed with fresh buffer containig 5 mM myo-inositol to remove free [^{3}H]inositol and then incubated in the presence of 8 mM LiCl, either in the presence or absence of agonists.

Fig. 1-A shows the time-dependent change in the basal accumulation of ^{3}H-IPs in cortical slices (in the absence of agonists) under oxygenated or anoxic conditions. There was a slight (but not statistically significant) decrease in the basal accumulation in the absence of O_2 but subseqent uoxygenation restored accumulation to control levels.

Fig. 1-B shows the time-dependent change in the accumulation of ^{3}H-IPs stimulated by 10^{-5} M NE. NE-stimulated accumulation was significantly reduced by anoxia. The decrease caused by anoxia was, however, reversed to

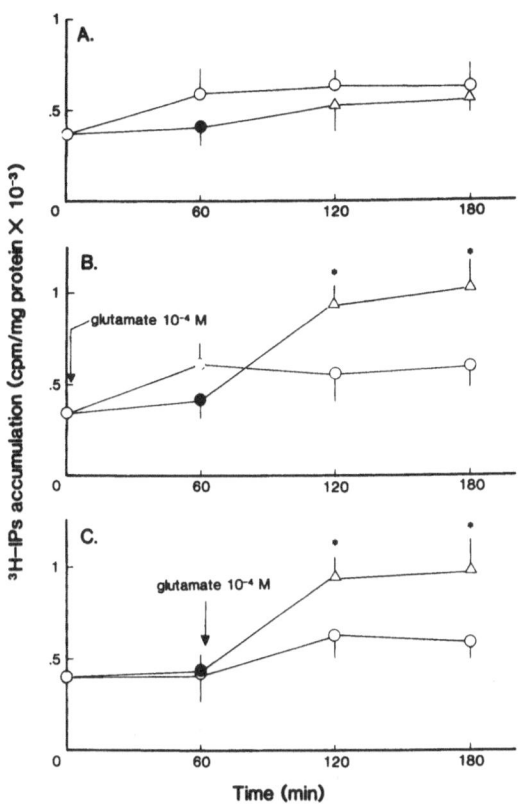

Fig. 3 Effect of anoxia and oxygenation on Glu-stimulated
accumulation of ^{3}H-IPs in hippocampal slices. After
phospholipids labeling, slices were washed and then incubated in the absence (A) or presence (B) of 10^{-4} M Glu
under oxygenated or anoxic conditions. At 55 min, all of
the samples were oxygenated by bubbling the incubation
medium with 95% O_2/5% CO_2 for 5 min and the reaction was
allowed to continue under oxygenated conditions. In some
experiments (C), incubation was begun in the absence of
Glu, and 10^{-4} M Glu was added immediately after oxygenation. In any case, the reaction was stopped by adding
chloroform-methanol (1:2 v/v) and ^{3}H-IPs accumulation was
determined as described in "Materials and Methods."
○ : oxygenated, ● : anoxia, △ : oxygenated
following anoxia. Values shown are means $\pm$ SEM of three
determinations each done in triplicate. * : $p < 0.01$;
significantly different from values for oxygenated incubations.

an increase by 10-min of oxygenation followed by a 60-min incubation under
oxygenated condition and the accumulation overshot the corresponding
control, oxygenated level. Such overshooting in NE-stimulated accumulation
induced by oxygenation following anoxia was also observed when NE was
added just after oxygenation (Fig. 1-C). Dose-response curves for NE (Fig.
2) showed that the decreased accumulation under anoxia was due to a decrease in Emax value, while the overshooting caused by oxygenation following anoxia was due to a decrease in EC_{50} value. Thus, in vitro anoxia

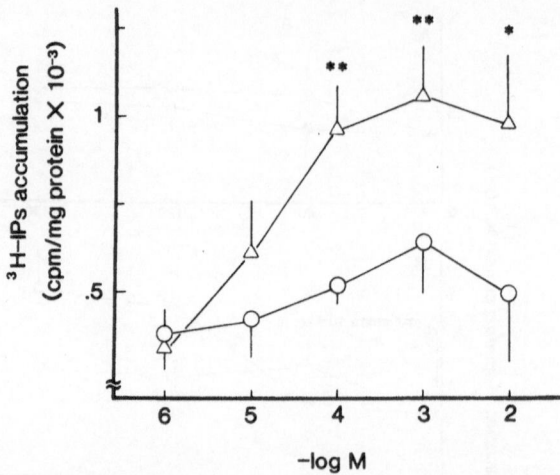

Fig. 4 Dose-response curves for Glu-stimulated accumulation of
^{3}H-IPs in hippocampal slices. Hippocampal slices were
incubated with increasing concentrations of Glu for 60
min under oxygenation (O) or anoxia (Δ), and all
samples were oxygenated for another 60 min. Values shown
are means ± SEM of three determinations each done in
triplicate. ** : p < 0.01, * : p < 0.05 ; significantly
different from values for oxygenated incubations.

exerted dual effects on PI breakdown stimulated by NE. The first effect is
the supression of PI breakdown under anoxia, and the second effct is the
enhancement of the breakdown, that is, enhanced accumulation of ^{3}H-IPs
after anoxia. We arbitarily named this second phenomenum as postanoxic
overshooting of PI breakdown.

    The second effect of in vitro anoxia on PI breakdown, the enhancement
of ^{3}H-IPs accumulation following an exposure to in vitro anoxia, was also
observed with Glu (Fig. 3, 4). Glutamate, at 10^{-4} M, caused a little
increase in ^{3}H-IPs accumulation in hippocampal slices and there was no
difference between control and anoxic incubations for 60 min. Sixty-min
oxygenation following 60-min anoxic incubations, however, resulted in a
significant increase in the accumulation of ^{3}H-IPs compared with control
60-min oxygenation following 60-min oxygenated incubations. This postanoxic induction of the effect of glutamate was also observed when the slices
were exposed to anoxia for 60 min in the absence of glutamate and then
simulated by 10^{-5} M glutamate for 60 min under oxygenated conditions.
Thus, the presence of exogenous glutamate during the anoxic period was not
necessary to induce the effect.

    The enhancement of ^{3}H-IPs accumulation following an exposure to in
vitro anoxia, was not observed with Carb. Fig. 5-A shows the time-dependent change in the accumulation of ^{3}H-IPs stimulated by Carb in cortical
slices. Carb-stimulated accumulation was significantly reduced by anoxia
and was completely restored to control levels by oxygenation. The same
results were obtained when Carb was added just after oxygenation (Fig. 5-
B) instead of before the anoxic incubation. Dose-response curves for Carb
(Fig. 6) indicated that the decreased level of accumulation at low O_2
tension was due to a decrease in Emax value, and the stimulatory effect of
carbachol was fully restored by oxygenation.

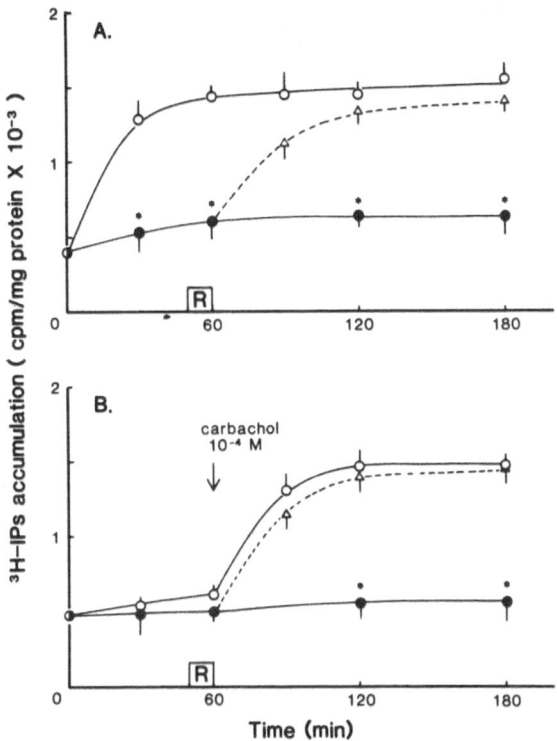

Fig. 5 Effect of anoxia and oxygenation on Carb-stimulated
accumulation of ^{3}H-IPs in cortical slices. Experimental
procedures are similar to those described for Fig. 1,
with the exception that the accumulation of ^{3}H-IPs was
stimulated with Carb. (A) Incubation was started in the
presence of 10^{-4} M Carb. (B) Incubation was started in
the absence of Carb, and 10^{-4} M Carb was added immediately after oxygenation. O : oxygenated, ● : anoxia,
△ : oxygenated following anoxia. Values shown are
means ± SEM of at least three determinations. * : p <
0.01 ; significantly different from values for oxygenated
incubations. (cited from Ninomiya et al., 1989 with
permission)

As for NE and Carb, we examined the effect of in vitro anoxia in
slices from four regions of the rat brain, cortex, striatum, hippocampus
and cerebellum. Shown in Fig. 7 is the accumulation of ^{3}H-IPs stimulated
by 10 μM NE or 100 μM Carb. The accumulation of ^{3}H-IPs either stimulated
by NE or by Carb, was significantly decreased under in vitro anoxia in
cortical slices and this effect was also seen in striatal and hippocampal
lices. Oxygenation following anoxia resulted in overshooting of the accumulation stimulated by NE in cortical slices. This postanoxic enhancemant
of the effect was also observed in striatal and hippocampal slices but not
with Carb. An exception was cerebellar slices in which neither NE nor Carb
had much effect on PI breakdown and either in vitro anoxia or oxygenation
following anoxia caused no significant change.

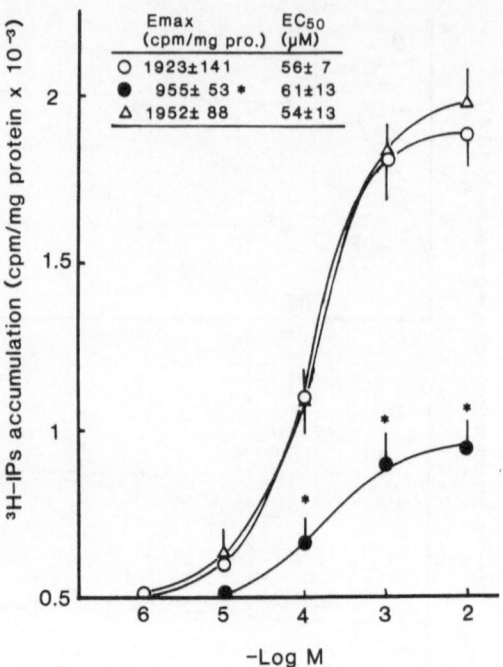

Fig. 6 Dose-response curves for Carb-stimulated accumulation of
^{3}H-IPs. Cortical slices were incubated with increasing
concentrations of Ccarb for 120 min under oxygenation
(O), for 120 min under anoxia (●), or for another 60
min following 50-min anoxia and 10-min oxygenation
(Δ). Values shown are means ± SEM of at least three
determinations. * : p < 0.01 ; significantly different
from values for oxygenated incubations. (cited from
Ninomiya et al., 1989 with permission)

In the next series of experiments, we tried to examine whether the
effects of oxygen depletion which we had observed in vitro could be also
observed in vivo. Fig. 8 shows the time-dependent change in the accumulation of ^{3}H-IPs stimulated by NE or Carb in slices from rats exposed to in
vivo hypoxia (8% O_2/92% N_2). NE-stimulated accumulation was significantly
increased following a 6-hour exposure to in vivo hypoxia and remained high
for at least 24 hours. By contrast, there was no significant change in
Carb-stimulated accumulation. Dose-response curves for NE (Fig. 9) showed
that the increased accumulation following in vivo hypoxia was due to a
decrease in EC_{50} value. We have preliminary results that the enhancement
of PI breakdown in brain slices following an exposure to in vivo hypoxia,
observed with NE, was also observed with Glu.

The data presented above can be summarized as Table 1. Both NE and
Carb are effective stimulants on PI breakdown in brain slices under normal
conditions, and lose their effects during in vitro anoxia. Glu is a weak
stimulant under control conditions and its effect does not change during
in vitro anoxia. Oxygenation following in vitro anoxia induces an apparent
enhancement of PI breakdown to NE and also a sensitivity to Glu. The same
kind of induction was also observed by an exposure to in vivo hypoxia, but
was not observed with Carb.

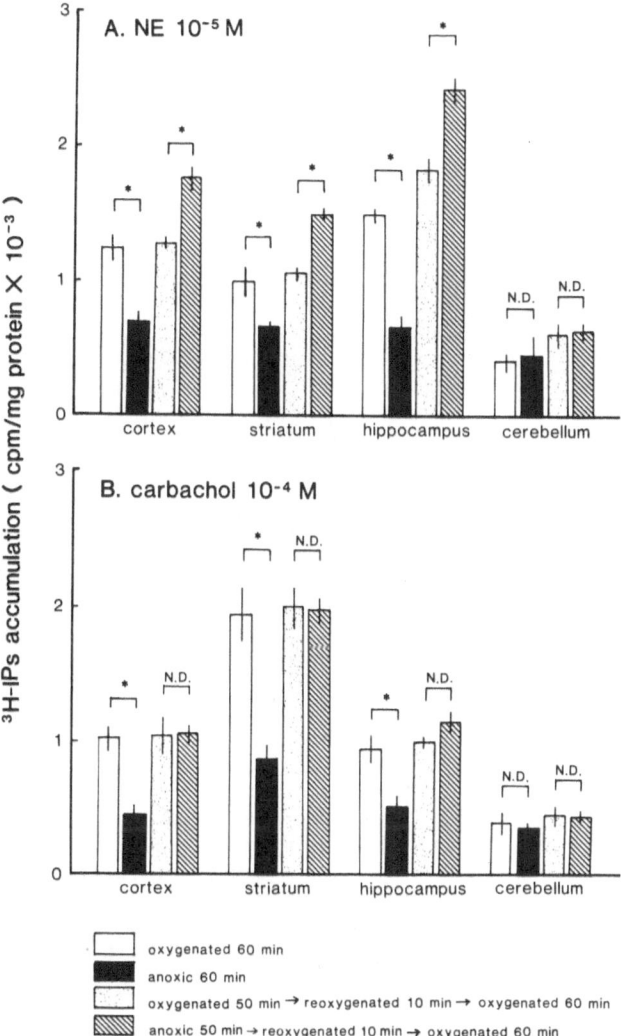

Fig. 7 Effect of anoxia and oxygenation on NE- and Carb-stimulated accumulation of ^{3}H-IPs in slices from discrete regions of brain. Preincubation and lipid labeling were done as described in Fig. 1. Slices were then incubated in the presence of 10^{-5} M NE (A) or 10^{-4} M Carb (B) for 60 min under oxygenation or anoxia, or for another 60 min following reoxygenation for 10 min.   N.D. : not significantly different ; * : p < 0.01 ; significantly different from values for oxygenated incubations.

133

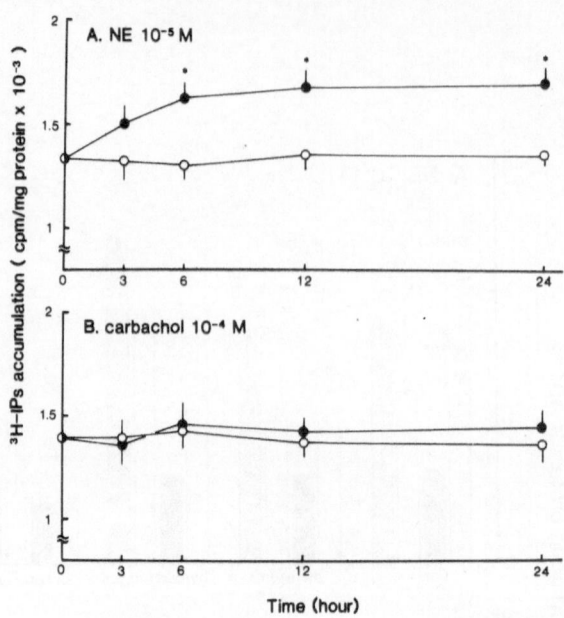

Fig. 8 Time course of the effect of in vivo hypoxia on NE- and
Carb-stimulated in vitro accumulation of ^{3}H-IPs in cortical slices. Following exposure to in vivo hypoxia (8% O_2,
92% N_2) (●) or room air (○) for various lengths of
time, rats were decapitated and cortical slices were
prepared as described under "Materials and Methods".
After labeling the phospholipid pool, slices were washed
and then incubated in the presence of 10^{-5} M NE (A) or
10^{-4} M Carb (B) under oxygenated condition for 60 min.
The reaction was stopped by adding chloroform-methanol
(1:2 v/v) and ^{3}H-IPs accumulation was determined as
described in "Materials and Methods". Basal accumulation
of ^{3}H-IPs in these experiments was between 500-600 cpm/mg
protein and was not altered by exposure to hypoxia.
Values shown are means ± SEM of at least three determinations each done in triplicate. ∗ : $p < 0.01$; significantly different from values for room air-exposure.
(cited from Ninomiya et al., 1989 with permission)

    While we had been working on this subject, two reports were published,
dealing with the ischemia/hypoxia model. Chen et al. (1988) demonstrated
that combined ischemia and hypoxia enhanced PI breakdown stimulated by
quisqualate in the immature rat brain, and suggested that glutamate receptor-coupled PI metabolism was involved in ischemic/hypoxic brain injury.
Seren et al. (1989) examined the effects of transient ischemia induced by
four vessel occulsion on PI breakdown in rat brain slices. They found that
a transient ischemic insult (30 min's occulsion of cerebral arteries),
with a time lag of 24 hours, resulted in an enhanced PI breakdown in rat
cortical and hipocampal slices stimulated by glutamate receptor agonists
or by NE. They also reported that this kind of enhancement was not observed with Carb. The findings of these two groups on in vivo ischemia
models were very similar in nature to what we had observed using in vitro
or in vivo hypoxia model.

    Two questions should be addressed regarding to the alterations in PI
breakdown in brain slices caused by hypoxia or ischemia.

    The first question; what are the physiological meanings of these
changes?

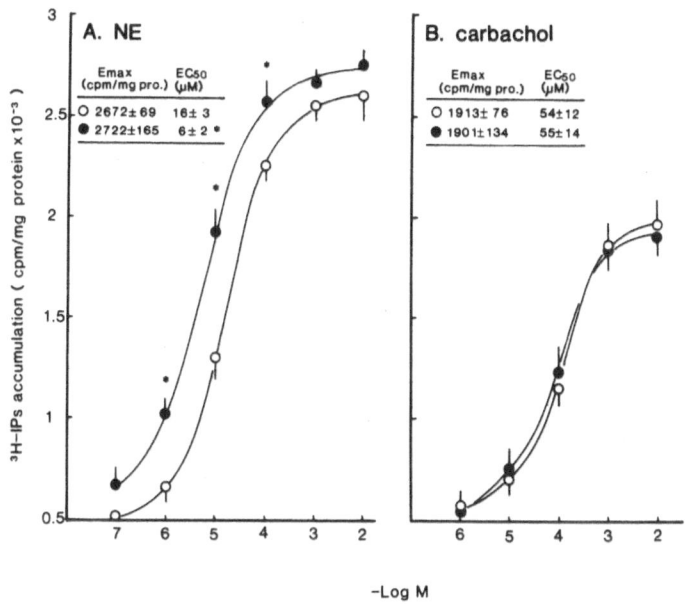

Fig. 9 Dose-response curves for NE- and Carb-stimulated in
vitro accumulation of ^{3}H-IPs in cortical slices following
a 6-hour exposure to in vivo hypoxia (8% O_2/92% N_2).
Cortical slices were prepared from rats exposed to hypoxia (●) or room air (Δ) for 6 hours. After lipid
labeling and washing, slices were incubated with increasing concentrations of NE (A) or Carb (B). Values shown
are means ± SEM of at least three determinations each
done in triplicate. * : p < 0.01 ; significantly different from values for room air-exposure. (cited from Ninomiya et al., 1989 with permission)

Table 1. Summary of the effects of oxygen depletion on PI
breakdown in rat brain slices

	^{3}H-IPs accumulation stimulated by		
	Norepinephrine	Glutamate	Carbachol
During in vitro anoxia	↓	→	↓
After in vitro anoxia	↑	↑	→
After in vio hypoxia	↑	↑	→

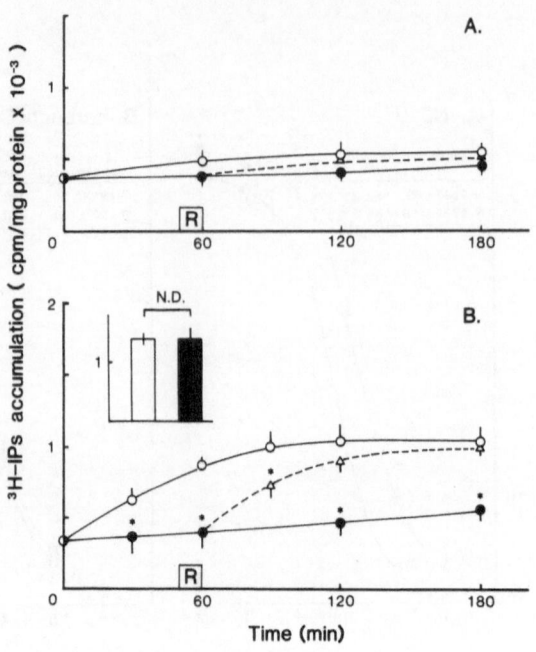

Fig. 10 Effect of anoxia and oxygenation on the NE-stimulated accumulation of ^{3}H-IPs in cortical slices in the absence of added CaCl$_2$. Experimental procedures are similar to those described for Fig. 1 with the exception that stimulation was performed in Krebs buffer containing no CaCl$_2$. (A) Basal accumulation in the absence of NE. (B) 10^{-5}M NE was present throughout incubation. O : oxygenated, ● : anoxia, Δ : oxygenated following anoxia. Values shown are means ± SEM of at least three determinations. * : p < 0.01 ; significantly different from values for oxygenated incubations. inset 1.3 mM CaCl$_2$ was added at 60 min immediately following reoxygenation and the incubation was continued for another 60 min. white bar ; reoxygenated following oxygenation, dark bar ; reoxygenated following anoxia. N.D. ; no significant difference between the two conditions. (cited from Ninomiya et al., 1989 with permission)

   As stated in INTRODUCTION, electrophysiological studies have shown that oxygen depletion exerts dual effects on the excitability of neurons, the first effect is supression of excitability during hypoxic period which would presumably leads to the impared synaptic transmission during hypoxia, and the second effect is the postsynaptic hyperexcitability. Besides, biochemical studies have shown that ischemic or hypoxic damage on brain slices causes an enhanced Ca^{2+} influx or an enhanced release of free fatty acids. PI breakdown is an intracellular signal transduction system closely related to both Ca^{2+} and lipid metabolism and it is quite possible that the effect of hypoxia is related to all of these events under hypoxia or ischemia. We expect, that, in the future, we or some other group will reveal quantitative relationships between these events under this kind of pathological conditions.
   The second question; what are the biochemical mechanisms underlying the effects of oxygen depletion on PI metabolism? Some results from our work which had been conducted trying to answer this question are presented below.

Table 2. Effect of anoxia and oxygenation on incorporation of [^{3}H] inositol into phospholipids in cortical slices

	PI	PIP	PIP$_2$
Before incubation	91. 5 $\pm$ 1. 0	7. 4 $\pm$ 1. 3	1. 0 $\pm$ 0. 1
Oxygenation	89. 8 $\pm$ 0. 4	8. 0 $\pm$ 0. 4	2. 2 $\pm$ 0. 3 [a]
Oxygenation + reoxygenation	87. 8 $\pm$ 1. 9	9. 9 $\pm$ 1. 4	2. 4 $\pm$ 0. 6 [a]
Anoxia	92. 1 $\pm$ 1. 0	7. 0 $\pm$ 1. 0	0. 9 $\pm$ 0. 1
Anoxia + oxygenaiton	89. 0 $\pm$ 2. 1	8. 4 $\pm$ 2. 2	2. 6 $\pm$ 0. 5 [a]

After the phospholipid pool was labeled under oxygenated conditons for 60 min, cortical slices were washed and then incubated under oxygenated or anoxic conditions. At 50 min, samples were oxygenated and were allowed to incubate for another 60 min under oxygenated conditons. The reaction was stopped just after washing, after a 50-min oxygenated incubation, after reoxygenation following oxygenated incubation, after a 50-min period of anoxia, or after oxygenation following anoxia. Phospholipids were extracted and analysed by thin layer chromatography according to the methods described by Jolles et al. (1981). Data shown are percentage radioactivity (mean $\pm$ SEM values from at least three determinations) of individual phospholipid from the organic layer. Neither anoxia nor oxygenation caused any change in the total incorporation values. (cited from Ninomiya et al., 1989 with permission)

Table 3. [^{3}H] Prazosin binding parameters in crude homogenate of cortical slices exposed to anoxia and oxygenation.

		Saturation isotherms		Inhibition by NE	
		Kd	Bmax	Ki	nH
(a)	oxygenated	209 $\pm$ 9	175 $\pm$ 7	6. 2 $\pm$ 0. 5	0. 88 $\pm$ 0. 02
(b)	anoxia	212 $\pm$ 5	170 $\pm$ 11	5. 2 $\pm$ 0. 6	0. 84 $\pm$ 0. 02
(c)	oxygenated after anoxia	211 $\pm$ 8	170 $\pm$ 15	5. 5 $\pm$ 0. 3	0. 83 $\pm$ 0. 03

Cortical slices were homogenized following either 60-min incubation under oxygenation (a), 60-min incubation under anoxia (b), or 60-min incubation under oxygenation following 50-min anoxia and 10-min oxygenaiton (c). Kd and Bmax values were obtained from Scatchard analysis of saturation isotherms using [^{3}H] prazosin in concentrations from 50 to 1000 pM. Kd values are given as pM, and Bmax values are given as fmol/mg protein. Ki (given as μM) and nH value was obtained from inhibition data. Values shown are means $\pm$ SEM of at least five determinations. (cited from Ninomiya et al., 1989 with permission)

Fig. 10 shows the effect of Ca^{2+} removal on the anoxia-induced changes in NE-stimulated accumulation of ^{3}H-IPs. In the absence of added extracellular Ca^{2+}, anoxia still caused a significant reduction in NE-stimulated accumulation, but the postanoxic overshooting was not observed. Nor did the addition of 1.3 mM $CaCl_2$ just after reoxygenation have any significant effect on the accumulation in reoxygenated samples (Fig. 11-B inset).

Table 2 shows the effects of in vitro anoxia and subsequent oxygenation on the incorporation of $[^3H]$ inositol into phosphoinositides in cortical slices evaluated by thin layer chromatography. The incubation of slices for one or two hours under normal oxygenated conditions resulted in a doubling of the label in phosphatidylinositol-di-phosphate (PIP_2), and although anoxic incubations supressed this increase in PIP_2, oxygenation following anoxia restored the labelling to the control level. Therefore, the postanoxic enhancement of PI breakdown induced by oxygenation following anoxia in vitro, seems not due to an increased label in PIP_2.

Table 3 shows $[^3H]$ prazosin binding parameters in crude homogenates of cortical slices before and after exposure to in vitro anoxia. None of the parameters, Kd and Bmax values obtainted from saturation isotherms, or Ki and Hill coefficient values obtained from displacement experiments with NE, showed any significant changes.

In summary, we have shown an enhancement of PI breakdown in rat brain slices stimulated by some agonists following an exposure to in vitro or in vivo hypoxia. This effect was dependent upon the presence of added extracellular Ca^{2+} during an exposure to hypoxia and seemed neither due to changes in phospholipid labeling nor in receptors. In our view, the most likely explanation for these changes is that the coupling mechanisms between some receptors, namely alpha1-adrenergic receptors or a subtype of glutamate receptors, is converted to "an activated form" following an exposure to hypoxia, and it remains to be clarified what the exact nature of the activation is.

Finally, brain slices are too complicated for us to evaluate the individual steps of PI breakdown and now we are planing to examine the effects of oxygen depletion on PI metabolism in a simpler system, such as culture cell lines, using a more direct assay procedures of PI breakdown, maybe HPLC. We hope this experimental approach will provide us more detailed information on the biochemical mechanisms of the effects of oxygen depletion on PI metabolism.

REFERENCES

Bosley T. M., Woodhams P. L., Gordon R. D., and Balázs R. (1983) Effects of anoxia on the stimulated release of amino acid neurotransmitters in the cerebellum in vitro. J. Neurochem., 40, 189-201.

Brown E. D., Kendall D. A., and Nahorski S. R. (1984) Inositol phospholipid hydrolysis in rat cerebral cortical slices: I. Receptor characterization. J. Neurochem., 42, 1379-1387.

Chen C. -K., Silverstein F. S., Fisher S. K., Statman D., and Johnston M. V. (1988) Perinatal hypoxic-ischemic brain injury enhances quisqualic acid-stimulated phosphoinositide turnover. J. Neurochem., 51, 353-359.

Davis J. N. and Carlsson A. (1973) The effect of hypoxia on monoamine synthesis, levels and metabolism in rat brain. J. Neurochem., 21, 783-790.

Dolivo M. (1974) Metabolism of mammalian sympathetic ganglia. Fed. Proc., 33, 1043-1048.

Gibson G. E. and Duffy T. E. (1981) Impaired synthesis of acetylcholine by mild hypoxic hypoxia or nitrous oxide. J. Neurochem., 36, 28-33.

Gibson G. E., Peterson, C. and Sansone J. (1981) Decreases in amino acid and acetylcholine metabolism during hypoxia. J. Neurochem., 37, 192-201.

Gorman A. L. F. (1966) Differential patterns of activation of the pyramidal system elicited by surface anodal and cathodal cortical stimulation. J. Neurophysiol., 29, 547-564.

Hansen A. J., Haunsgaard J. and Jahnsen H. (1982) Anoxia increases potassium conductance in hippocampal nerve cells, Acta Physiol. Scand., 115, 301-310.

Jolles J., Zwiers H., Dekker A., Wirtz K. W. A. and Gispen W. H. (1981) Corticotropin-(1-24)-tetracosapeptide affects protein phosphorylation and polyphosphoinositide metabolism in bat brain. Biochem. J., 194, 283-291.

Lowry O. H., Rosebrough N. J., Farr A. L., and Randall R. J. (1951) Protein measurement with the Folin phenol reagent. J. Biol. Chem., 193, 265-275.

Miwa S., Fujiwara M., Inoue M. and Fujiwara M. (1986) Effects of hypoxia on the activities of noradrenergic and dopaminergic neurons in the rat brain. J. Neurochem., 47, 63-69.

Ninomiya H., Taniguchi T., Fujiwara M. and Kameyama M. (1988) Increased binding of [^{3}H]muscimol and [^{3}H]flunitrazepam in the rat brain under hypoxia. J. Neurochem., 51, 1111-1117.

Ninomiya H., Taniguchi T., Fujiwara M. and Kameyama M. (1989) Effects phosphoinositide turnover in rat brain slices. J. Neurochem., 53, 183-190.

Shiff S. J. and Balestrino M. (1985) Possible mechanism of posthypoxic hyperexcitability in hippocampal slices. Soc. Neurosci. Abstr. 11, 374.

Schiff S. J. and Somjen G. G. (1984) Hypoxia and synaptic function. Clin. Neuropharmacol. 7 (suppl 1), 498-499.

Schiff S. J. and Somjen G. G. (1985) Hyperexcitability following moderate hypoxia in hippocampal tissue slices. Brain Res., 337, 337-340.

Schiff S. J. and Somjen G. G. (1987) The effect of graded hypoxia on the hippocampal slice: A in vitro model of the ischemic penumbra. Stroke, 18, 30-37.

Seren M. S., Aldinio C., Zanoni R., Leon A. and Nicoletti F. (1989) Stimulation of inositol phospholipid hydrolysis by excitatory amino acids is enhanced in brain slices from vulnerable regions after transient global ischemia. J. Neurochem., 53, 1700-1705.

EXPRESSION OF G-PROTEIN α-SUBUNIT cDNA

Toshihide Nukada

Department of Medical Chemistry
Kyoto University Faculty of Medicine
Kyoto 606, and Department of Biochemistry
Institute of Brain Reseach, Faculty of
Medicine, University of Tokyo, Tokyo 113
Japan

INTRODUCTION

A family of membrane-associated G-proteins are essential tor receptor-effector coupling (Gilman, 1987; Neer and Clapham, 1988; Ross, 1989). G-proteins are composed of three subunits termed α, β and γ. The α-subunit contains a binding site for guanine nucleotides and possesses the GTPase activity. It is believed that the GTP-bound α-subunit is an active form for the effector system and the functional life time is determined by the GTPase activity. Molecular cloning of G-protein α-subunit (Gα) has revealed the existence of nine or more Gα genes and a high degree of amino acid sequence homology. However, it is not clear whether each Gα transduces a distinct signal or not.

TRANSDUCIN α-SUBUNIT (α_t)

Transducin is a major component of the disks of the retinal rod outer segment. The availability of antibodies to $G_t\alpha$ and partial amino acid sequence permitted the cloning of cDNAs corresponding to α_t (Tanabe et al., 1985; Yatsunami and Khorana, 1985; Medynski et al., 1985). Lochrie et al. (1985) have reported the cloning of another cDNA encoding the transducin α-subunit of cone photorceptor outer segments: two sequences differ in approximately 20% of the encoded amino acid residues. Thus there exist two isoforms of G_t: one that activates the photosensitive cyclic GMP-specific phosphodiesterase of rod outer segments ($G_{t\ rod}$) and one that plays the analogous role in cones ($G_{t\ cone}$). However the precise function of $G_{t\ cone}$ has not been determind.

α-SUBUNIT OF ADENYLATE CYCLASE-STIMULATING G-PROTEIN (α_S)

G_S α-subunit, first defined functionally by its ability to activate adenylate cyclase, is found on purification to be a mixture of two polypeptides with different molecular weights of 52,000 and 45,000 (Northup et at., 1980). The cDNAs corresponding to the 52-kd (Robishaw et al., 1986a; Nukada et al., 1986a; Itoh et al., 1986) and 45-kd form (Robishaw et al., 1986b) of α_S have been cloned. In addition two other forms of α_S have been cloned (Bray et al., 1986). These four mRNAs are presumed to arise from a single gene for α_S by alternative splicing.

Recently α_S has been shown to directly regulate cardiac calcium channels (Yatani et al., 1987). Three of the four splice variants are shown to stimulate both adenylate cyclase and calcium channels (Mattera et al., 1989).

α-SUBUNIT OF ADENYLATE CYCLASE-INHIBITING G-PROTEIN (α_i)

The G-protein, G_i, which inhibits adenylate cyclase, has been shown to have three subtypes of the α-subunit by molecular cloning [α_{i1} (Nukada et al., 1986b), α_{i2} (Itoh et al., 1986) and α_{i3} (Didsbury and Snyderman, 1987)]. They share about 90% amino-acid sequence homology and so are difficult to separate from one another. The 41-kd polypeptides corresponding to α_{i1} are purified from brain and shown to effectively inhibit the forskolin-stimulated adenylate cyclase activity, as compared with the 40- and 39-kd polypeptides corresponding to α_{i2} and α_o, respectively (Katada et al., 1987). On the other hand, pretreatment of platelet membranes with C-terminal antibodies reactive with α_{i2}, but not with antibodies to α_{i3}, block α_2-adrenergic inhibition of adenylate cyclase (Simonds et al., 1989). This discrepancy may be accounted for partly by the contamination of other subtype of G-protein α-subunit during purification steps or the cross-reactivity of antibodies with subtypes of α_i.

For the other functions of G_i, α_{i3} is firstly defined as the stimulatory G-protein of receptor-regulated K^+ channels (Codina et al., 1988). But it has been reported later that all three subtypes of α_i are active on opening the K^+ channels (Yatani et al., 1988).

α-SUBUNIT OF G_O (α_O)

The cDNA for α_O has been cloned by Itoh et al. (1986) and Van Meurs et al. (1987). Several lines of evidence available concerning the function of α_O are accumulated: α_O mediates receptor-induced inhibition of Ca^{2+} currents (Hescheler et al., 1987; Harris-Warrick et al., 1988; MacFadzean et al., 1989); α_O activates the opening of brain K^+ channels (VanDongen et al., 1988); α_O evokes a Cl^- current by mobilizing Ca^{2+} from intracellular inositol 1,4,5-trisphosphate-sensitive stores in *Xenopus* oocyte (Moriarty et al., 1990). Recently, the existence of four subtypes of α_O or α_O-like G-protein α-subunits has been demonstrated by Kobayashi et al. (1989). So this will lead to confusion over which of the subtypes (or α_O-like proteins) is involved in the functions.

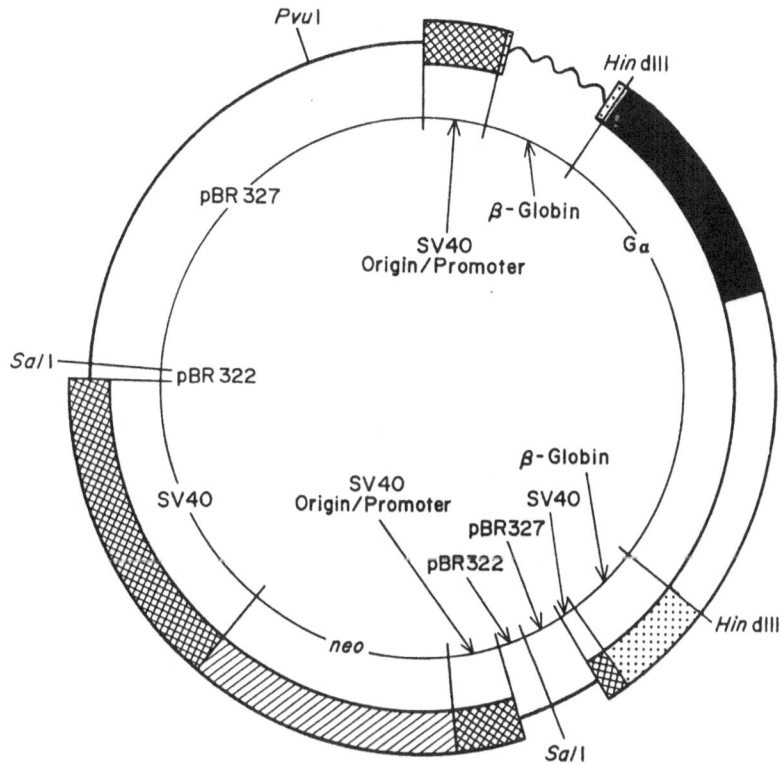

Fig. 1. Schematic representation of the recombinant plasmid used for expression of the G-protein α-subunit. The origins of the segments are shown on the inner circle. The protein-coding region of the G-protein α-subunit cDNA is indicated by a closed box, and the 5'- and 3'- noncoding regions of the G-protein α-subunit cDNA by open boxes. The *neo* DNA segment is represented by a hatched box. The SV40 sequences are shown by cross-hatched boxes; the orientation of the two sets of the early gene promoter is such that transcription occurs clockwise. The exonic and 3'-flanking sequences of the rabbit β-globin gene are indicated by stippled boxes and its intronic sequence by a wavy line. The sequences of the plasmids pBR327 and pBR322 are represented by lines. Only the relevant restriction endonuclease sites are shown.

Table 1. Effect of isoproterenol on the cellular content of cyclic AMP

Cell line		Cyclic AMP (pmol/10^7 cells)	
		Control	+isoproterenol
	wild-type S49	71 ± 43	4979 ± 2193
	cyc⁻ S49	3 ± 1	6 ± 1
Vector	Transformant		
pKGSαN	CGSα1	54 ± 31	1541 ± 885
	CGSα4	22 ± 12	1475 ± 719
	CGSα7	26 ± 10	1305 ± 424
pSV2-neo	CN1, CN2 and CN3	15 ± 3	15 ± 3

Cells (~10^6) were incubated at 37°C for 10 min in 0.5 ml of Dulbecco's modified Eagle's medium containing 0.2 mM 3-isobutyl-1-methylxanthine in the presence and absence of 10 μM isoproterenol. Cellular cyclic AMP was then determined using a radioimmunoassay kit (Yamasa shoyu) accoding to the procedure described by the vendor. Data are given as means ± SD from three experiments, except that the values obtained with the three different pSV2-neo-transformed clones (one experiment each) are averaged. Reproduced from Nukada et al., 1987.

OTHER G-PROTEIN α-SUBUNIT

Two research groups have isolated the cDNA for a G-protein α-subunit ($\alpha_{x(z)}$) that lacks an apparent ADP-ribosylation site for pertussis toxin (Fong et al., 1988; Matsuoka et al., 1988). The function of $\alpha_{x(z)}$ has not been defined. The primary structure of another G-protein , an olfactory neuron specific G-protein α-subunit ($G_{olf}\alpha$), has been deduced from the cDNA sequence (Jones and Reed, 1989). $G_{olf}\alpha$ shares 88% amino acid identity with α_S and has a capacity to stimulate adenylate cyclase.

EXPRESSION OF Gα cDNA

To clarify the function of each Gα, two recombinant DNA techniques are mainly used to eliminate the contamination by other subtypes: synthesis of a recombinant Gα protein in *Eschericia coli* (Graziano et al., 1987; VanDongen et al., 1988; Yatani et al., 1988; Mattera et al., 1989) and expression of Gα cDNA in mammalian cultured cells (Nukada et al., 1987; Sullivan et al, 1987). The former technique was applied to α_O and subtypes of α_S and α_i. However the potency of the recombinant Gα is much lower than that of native Gα (Graziano et al., 1987; VanDongen et al., 1988, Yatani et al., 1988; Mattera et al., 1989) and the selective roles of each subtypes have not yet been observed (Yatani et al., 1988; Mattera et al., 1989). This may be partly due to a failure of the recombinant α-subunit to undergo a post-translational modification (Buss et al., 1987).

Functional expression of cloned α_S cDNA (Nukada et al., 1987) has been shown using an expression vector carrying the SV40 early gene promoter and the neomycin-resistance marker gene (Fig.1). The expression vector for α_S cDNA (pKGsαN) was introduced by electroporation (Potter et al., 1984) to the cyc⁻ variant (Bourne et al., 1975) of S49 murine lymphoma cells, wihich is defective in α_S (Northup et al., 1983; Harris et al., 1985). Neomycin-resistant clones that expressed RNA species (with sizes expected from the expression constructs) encoding the α_S were selected by RNA blot hybridization analysis. Stimulation of β-adrenergic receptors by isoproterenol increased the cellular content of cyclic AMP in the α_S-transformed clones (clones CGSα1, CGSα4 and CGSα7) but not in non-transfected cyc⁻ cells and pSV2-neo-transformed clones: the plasmid pSV2-neo (Southern and Berg, 1982) contained no α_S cDNA (Table 1). Sullivan et al. (1987) have also reported the functional expression of α_S using a retroviral vector expressed in cyc⁻ cells. This Gα cDNA expression system may reveal the physiological roles of each Gα since the exogenous Gα derived from the expression of a Gα cDNA may be subjected to the same post-translational modification as the endogeneous Gα proteins in cultured cells. For this, it is particularly fascinating to express α_i and α_O cDNAs and to test the cellular changes in the α_i- and α_O-transformed clones.

REFERENCES

Bourne, H. R., Coffino, P. and Tomkins, G. M. (1975). Selection of a variant lymphoma cell deficient in adenylate cyclase. Science 187, 750-752.

Bray, P., Carter A., Simons, C., Puckett, C., Kamholz, J., Spiegel, A. and Nirenberg, M. (1986). Human cDNA clones for four species of Gα_s signal transduction protein. Proc. Natl. Acad. Sci. USA 83, 8893-8897.

Buss, J. E., Mumby, S. M., Casey, P. J., Gilman, A. G. and Sefton, B. M. (1987). Myristoylated α subunits of guanine nucleotide-binding regulatory proteins. Proc. Natl. Acad. Sci. USA 84, 7493-7497.

Codina, J., Olate, J., Abramowitz, J., Mattera R., Cook, R. G. and Birnbaumer, L. (1988). α_i-3 cDNA endodes the α subunit of G_K, the stimulatory G protein of receptor-regulated K^+ channels. J. Biol. Chem. 263, 6746-6750.

Didsbury, J. R. and Snyderman, R. (1987). Molecular cloning of a new human G protein: evidence for two $G_{i\alpha}$-like protein families. FEBS Lett. 219, 259-263.

Fong, H. K. W., Yoshimoto, K. K., Eversole-Cire, P. and Simon, M. I. (1988). Identification of a GTP-binding protein α subunit that lacks an apparent ADP-ribosylation site for pertussis toxin. Proc. Natl. Acad. Sci. USA 85, 3066-3070.

Gilman, A. G. (1987). G proteins: transducers of receptor-generated signals. Annu. Rev. Bichem. 56, 615-649.

Graziano, M. P., Casey, P. J. and Gilman A. G. (1987). Expression of cDNAs for G proteins in *Escherichia coli*: two forms of $G_{s}\alpha$ stimulate adenylate cyclase. J. Biol. Chem. 262, 11375-11381.

Harris, B. A., Robishaw, J. D., Mumby, S. M. and Gilman, A. G. (1985). Molecular cloning of complementary DNA for the alpha subunit of the G protein that stimulates adenylate cyclase. Science 229, 1274-1277.

Harris-Warrick, R. M., Hammond, C., Paupardin-Tritsch, D., Homburger, V., Rouot, B., Bockaert, J. and Gerschenfeld, H. M. (1988). An α_{40} subunit of a GTP-binding protein immunologically related to G_O mediates a dopamine-induced decrease of Ca^{2+} current in snail neurons. Neuron 1, 27-32.

Hescheler, J., Rosenthal, W., Trautwein, W. and Schultz, G. (1987). The GTP-binding protein, G_O, regulates neuronal calcium channels. Nature 325, 445-447.

Itoh, H., Kozasa, T., Nagata, S., Nakamura, S., Katada, T., Ui, M., Iwai, S., Ohtsuka, E., Kawasaki, H., Suzuki, K.,and Kaziro, Y. (1986). Molecular cloning and sequence determination of cDNAs for α subunits of the guanine nucleotide-binding proteins G_S, G_i, and G_O from rat brain. Proc. Natl. Acad. Sci. USA 83, 3776-3780.

Jones, D. T. and Reed, R. R. (1989). G_{olf}: an olfactory neuron specific-G protein involved in odorant signal transduction. Science 244, 790-795.

Katada, T., Oinuma, M., Kusakabe, K. and Ui, M. (1987). A new GTP-binding protein in brain tissues serving as the specific substrate of islet-activating protein, pertussis toxin. FEBS Lett. 213, 353-358.

Kobayashi, I., Shibasaki, H., Takahashi, K., Kikkawa, S., Ui, M. and Katada, T. (1989). Purification of GTP-binding proteins from bovine brain membranes: identification of heterogeneity of the α-subunit of G_O proteins. FEBS Lett. 257, 177-180.

Lochrie, M. A., Hurley, J. B. and Simon, M. I. (1985). Sequence of the alpha subunit of photoreceptor G protein: homologies between transducin, *ras*, and elongation factors. Science 228, 96-99.

Matuoka, M., Itoh, H., Kozasa, T. and Kaziro, Y. (1988).
Sequence analysis of cDNA and genomic DNA for a
putative pertussis toxin-insensitive guanine
nucleotide-binding regulatory protein α subunit. Proc.
Natl. Acad. Sci. USA 85, 5384-5388.

Mattera, R., Graziano, M. P., Yatani, A., Zhou, Z.,
Graf, R., Codina, J., Birnbaumer, L., Gilman, A. G. and
Brown, A. M. (1989). Splice variants of the α subunit
of the G protein G_S activate both adenylyl cyclase and
calcium channels. Science 243, 804-807.

McFadzean, I., Mullaney, I., Brown, D. A. and Milligan, G.
(1989). Antibodies to the GTP binding protein, G_O,
antagonize noradrenaline-induced calcium current
inhibition in NG108-15 hybrid cells. Neuron 3, 177-182.

Medynski, D. C., Sullivan, K., Smith, D., Van Dop, C.,
Chang, F.-H., Fung, B. K.-K., Seeburg, P. H. and
Bourne, H. R. (1985). Amino acid sequence of the α
subunit of transducin deduced from the cDNA sequence.
Proc. Natl. Acad. Sci. USA 82, 4311-4315.

Moriarty, T. M., Padrell, E., Carty, D. J., Omri, G.,
Landau, E. M. and Iyengar, R. (1990). G_O protein as
signal transducer in the pertussis toxin-sensitive
phosphatidylinositol pathway. Nature 343, 79-82.

Neer, E. J, and Clapham, D. E. (1988). Roles of G protein
subunits in transmembrane signalling. Nature 333,
129-134.

Northup, J. K., Sternweis, P. C., Smigel, M. D.,
Schleifer, L. S., Ross, E. M. and Gilman, A. G. (1980).
Purification of the regulatory component of adenylate
cyclase. Proc. Natl. Acad. Sci. USA 77, 6516-6520.

Northup, J. K., Smigel, M. D., Sternweis, P. C. and
Gilman, A. G. (1983). The subunits of the stimulatory
regulatory component of adenylate cyclase. J. Biol.
Chem. 258, 11369-11376.

Nukada, T., Tanabe, T., Takahashi, H., Noda, M., Hirose, T.,
Inayama, S. and Numa, S. (1986). Primary structure of
the α-subunit of bovine adenylate cyclase-stimulating
G-protein deduced from the cDNA sequence. FEBS Lett.
195, 220-224.

Nukada, T., Tanabe, T., Takahashi, H., Noda, M., Haga, K.,
Haga, T., Ichiyama, A., Kangawa, K., Hiranaga, M.,
Matsuo, H. and Numa, S. (1986). Primary structure of
the α-subunit of bovine adenylate cyclase-inhibiting
G-protein deduced from the cDNA sequence. FEBS Lett.
197. 305-310.

Nukada, T., Mishina, M. and Numa, S. (1987). Functional
expression of cloned cDNA encoding the α-subunit of
adenylate cyclase-stimulating G-protein. FEBS Lett.
211, 5-9.

Potter, H., Weir, L. and Leder, P. (1984). Enhancer-dependent expression of human κ immunoglobulin genes introduced into mouse pre-B lymphocytes by electroporation. Proc. Natl. Acad. Sci. USA 81, 7161-7165.

Robishaw, J. D., Russell, D. W., Harris, B. A., Smigel, M. D. and Gilman, A. G. (1986). Deduced primary structure of the α subunit of the GTP-binding stimulatory protein of adenylate cyclase. Proc. Natl. Acad. Sci. USA 83, 1251-1255.

Robishaw, J. D., Smigel, M. D. and Gilman, A. G. (1986). Molecular basis for two forms of the G protein that stimulates adenylate cyclase. J. Biol. Chem. 261, 9587-9590.

Ross, E. M. (1989). Signal sorting and amplification through G protein-coupled receptors. Neuron 3, 141-152.

Simonds, W. F., Goldsmith, P. K., Codina, J., Unson, C. G. and Spiegel, A. M. (1989). G_{i2} mediates α_2-adrenergic inhibition of adenylyl cyclase in platelet membranes: *in situ* identification with G_α C-terminal antibodies. Proc. Natl. Acad. Sci. USA 86, 7809-7813.

Southern, P. J. and Berg, P. (1982). Transformation of mammalian cells to antibiotic resistance with a bacterial gene under control of the SV40 early region promoter. J. Mol. Appl. Genet. 1, 327-341.

Sullivan, K. A., Miller, R. T., Masters, S. B., Beiderman, B., Heideman, W. and Bourne, H. R. (1987). Identification of receptor contact site involved in receptor-G protein coupling. Nature 330, 758-760.

Tanabe, T., Nukada, T., Nishikawa, Y., Sugimoto, K., Suzuki, H., Takahashi, H., Noda, M., Haga, T., Ichiyama, A., Kangawa, K., Minamino, N., Matsuo, H. and Numa, S. (1985). Primary structure of the α-subunit of transducin and its relationship to *ras* proteins. Nature 315, 242-245.

VanDongen, A. M. J., Codina, J., Olate, J., Mattera, R., Joho, R., Birnbaumer, L. and Brown, A. M. (1988). Newly identified brain potassium channels gated by the guanine nucleotide binding protein Go. Science 242, 1433-1437.

Van Meurs, K. P., Angus, C. W., Lavu, S., Kung, H.-F., Czarnecki, S. K., Moss, J. and Vaughan, M. (1987). Deduced amino acid sequence of bovine retinal $G_{o}\alpha$: similarities to other guanine nucleotide-binding proteins. Proc. Natl. Acad. Sci. USA 84, 3107-3111.

Yatani, A., Codina, J., Imoto, Y., Reeves, J. P., Birnbaumer, L. and Brown, A. M. (1987). A G protein directly regulates mammalian cardiac calcium channels. Science 238, 1288-1292.

Yatani, A., Mattera, R., Codina, J., Graf, R., Okabe, K., Padrell, E., Iyengar, R., Brown, A. M. and Birnbaumer, L. (1988). The G protein-gated atrial K$^+$ channel is stimulated by three distinct $G_i\alpha$-subunits. Nature 336, 680-682.

Yatsunami, K. and Khorana, H. G. (1985). GTPase of bovine rod outer segments: the amino acid sequence of the α subunit as derived from the cDNA sequence. Proc. Natl. Acad. Sci. USA 82, 4316-4320.

SEROTONIN RECEPTOR SUBTYPES IN BRAIN: LIGAND BINDING PROPERTIES AND COUPLING WITH G PROTEINS

Yasuyuki Nomura, Yoshihisa Kitamura, Michihisa Tohda, Shin-ichi Imai, Toshiaki Katada[*] and Michio Ui[+]

Department of Pharmacology, Faculty of Pharmaceutical Sciences, Hokkaido University, Sapporo 060, [*]Department of Life Science, Faculty of Science, Tokyo Institute of Technology, Yokohama 227, and [+]Department of Physiological Chemistry, Faculty of Pharmaceutical Sciences, Tokyo University, Tokyo 113, Japan

INTRODUCTION

Serotonin (5-hydroxytryptamine; 5-HT) receptors are classified as 5-HT_1, 5-HT_2 and 5-HT_3, and evidence is emerging to suggest heterogeneity within 5-HT_1 receptor category, e.g., 5-HT_{1A}, 5-HT_{1B}, 5-HT_{1C} and 5-HT_{1D} receptors (Peroutka, 1988). Several GTP-binding proteins (G proteins) in mammalian brain were identified, e.g., $\alpha_{52/45}$, α_{41}, α_{40}, α_{39} (α-subunits of G_s, G_i1, G_i2 and G_o) (Katada et al., 1986a; Katada et al., 1987; Itoh et al., 1988; Casey and Gliman, 1988) and several G proteins with low molecular weight (20 ~ 30 kDa), including 24-kDa G protein (24 K-G) (Katada and Ui, 1988), ADP-ribosylation factor (ARF) (Kahn and Gliman, 1986), a substrate of botulinum toxin (G_b, rho product) (Narumiya et al., 1988) and other small molecular G proteins (smg) (Takai et al., 1989). N-ethylmaleimide (NEM) has been used as a useful probe to alkylate sulfhydryl residues in receptors and G proteins involved in their coupling (Katada et al., 1986b; Kitamura and Nomura, 1987; Nomura et al., 1988). To classify 5-HT receptor subtypes in the CNS from binding characteristics and the aspect of coupling properties of these receptors with G proteins, we here examined the effects of GTPγS, NEM and several 5-HT receptor ligands on specific binding of [^{3}H]8-hydroxy-2-(di-n-propylamino)tetralin (8-OH-DPAT) (5-HT_{1A}), [^{125}I]iodocyanopindolol (ICYP) (5-HT_{1B}), [^{3}H]mesulergine (5-HT_{1C}), [^{3}H]4-bromo-2,5-dimethoxyphenyliso-propylamine (DOB) (5-HT_2) and [^{3}H]ketanserin (5-HT_2) to crude synaptic membranes of rat brain.

Table I Radioligands used in binding experiments for 5-HT receptor subtypes and distribution in the CNS.

5-HT_{1A}	5-HT_{1B}	5-HT_{1C}	5-HT_{1D}	5-HT_2	5-HT_3

A. Radioligand:

5-HT_{1A}	5-HT_{1B}	5-HT_{1C}	5-HT_{1D}	5-HT_2	5-HT_3
[^{3}H]5-HT	[^{3}H]5-HT	[^{3}H]5-HT	[^{3}H]5-HT	[^{3}H]ketanserin*	[^{3}H]GR 65630
[^{3}H]8-OH-DPAT*	[^{125}I]ICYP*	[^{3}H]mesulergine*		[^{3}H]DOB*	[^{3}H]ICS 205-930
[^{3}H]ipsapirone		[^{3}H]mianserin		[^{3}H]spiperone	[^{3}H]quipazine
[^{3}H]WB 4101		[^{125}I]LSD		[^{3}H]mianserin	
[^{3}H]buspirone				[^{125}I]LSD	

B. Distribution:

5-HT_{1A}	5-HT_{1B}	5-HT_{1C}	5-HT_{1D}	5-HT_2	5-HT_3
Raphe nuclei	(only rat and mouse)	Choroid plexus	Basal ganglia	Layer IV cortex	Entorhinal cortex
Hippocampus	Substantia nigra				Vagus nerve

*, Radioligands used in this study.

MATERIALS AND METHODS

Purification of G Protein from Rat Brain and Treatment of Rat Brain Membranes with NEM or Pertussis Toxin

Membrane-bound $G_i1\alpha$, $G_o\alpha$ and 24 K-G were purified from about 30 rat brains by the method of Katada et al. (1986, 1987, 1988).

The treatment with NEM or pertussis toxin (PTX) was performed according to Kitamura and Nomura (1987). In the NEM-treatment, brain membranes (about 2 mg protein/ml) were incubated with various concentrations (1 µM ~ 1 mM) of NEM at 0 °C for 30 min and 5 mM dithiothreitol (DTT) was then added. In the PTX-treatment, the membranes were incubated with preactivated PTX (0.1 ~ 10 µg) in the presence of 5 mM NAD and 10 mM thymidine at 30 °C for 60 min. After these incubations, the membranes were then washed three times and stored at -80 °C.

5-HT Receptor Binding Assay and Data Analysis

The reaction mixture (100 µl) was composed of 50 mM Tris-HCl (pH 7.4), treated membranes (100 ~ 200 µg protein) and each radiolabeled ligand ([^{3}H]8-OH-DPAT, [^{125}I]ICYP, [^{3}H]mesulergine, [^{3}H]DOB and [^{3}H]ketanserin) with or without GTPγS. After the reaction mixture was incubated at 30 °C for 40 min, it was filtered under the reduced pressure through a Whatman GF/C filter and washed three times with 2 ml of ice-cold buffer. Specific binding was defined as the radioactivity bound after subtraction of the nonspecific binding (in the presence of 10 µM 5-HT for 5-HT_{1A} and 5-HT_{1B} sites, 10 µM mianserin for 5-HT_{1C} site and 10 µM methysergide for 5-HT$_2$ site) from the total binding. The displacement curves by serotonergic ligands were computer-analyzed with one- or two-site model using the modified LIGAND (Munson and Rodbard, 1980) by a nonlinear least-squares curve-fitting procedure, SIMPLEX method (Yamaoka et al., 1981). The values of the dissociation constants of agonists (K_D) and those of antagonists (K_i) of the serotonergic ligands were calculated using the Cheng and Prusoff equation (1973). Statistical differences between control and test values were analyzed by Student's t-test.

RESULTS

Characterization of 5-HT$_{1A}$, 5-HT$_{1B}$, 5-HT$_{1C}$ and 5-HT$_2$ Receptor Bindings in Rat Brain Membranes

To examine the binding characteristics in 5-HT receptor subtypes, we used specific radioligands for each receptor (Table I): e.g. [^{3}H]8-OH-DPAT (for 5-HT$_{1A}$ receptor), [^{125}I]ICYP (for 5-HT$_{1B}$ receptor), [^{3}H]mesulergine (for 5-HT$_{1C}$ receptor) and [^{3}H]ketanserin (for 5-HT$_2$ receptor). The binding of [^{125}I]ICYP to 5-HT$_{1B}$ receptors was already established that the assay was done in the presence of 30 µM isoproterenol to exclude effectively any interference from the binding of β-adrenoceptors (Hoyer et al., 1985; Offord et al., 1988). Since [^{3}H]mesulergine can bind to both 5-HT$_{1C}$ and 5-HT$_2$ recognition sites, assay of [^{3}H]mesulergine binding to 5-HT$_{1C}$ receptor has been performed to use porcine choroid plexus which do not possess 5-HT$_2$ receptors (Pazos et al., 1984). In contrast, spiperone-competition curves of [^{3}H]mesulergine binding to rat brain membranes showed biphasic behavior, a major component (77 ± 3 % of specific binding, concentration which inhibits 50 % of control binding (IC$_{50}$) = 17 ± 8 nM) representing the binding to 5-HT$_2$ receptors and a minor component (23 ± 3 %, IC$_{50}$ = 20,000 ± 9,000 nM) representing the binding to 5-HT$_{1C}$ receptors. The inhibitory concentration of spiperone to 5-HT$_2$ binding sites was an order of magnitude lower than that for 5-HT$_{1C}$ sites. Therefore, we carried out the assay of [^{3}H]mesulergine binding to 5-HT$_{1C}$ receptors in rat brain membranes in the presence of 0.1 µM spiperone to exclude the binding to 5-HT$_2$ receptors. The -log [K_D] (pK$_D$) and -log [K_i] (pK$_i$) values obtained from the present assay system showed a significant correlation (correlation coefficient (r) = 0.918, p < 0.001) to pK$_D$ and pK$_i$ values estimated from the binding to pig choroid plexus membranes (Hoyer, 1988).

We estimated K_D and K_i values of each drug for each radioligand binding: drugs examined were 5-HT, methysergide, 8-OH-DPAT, TFMPP, DOI, spiperone, propranolol, mianserin, yohimbine, ketanserin, ritanserin, and ICS 205-930. The correlation in the K_i values of these drugs were evaluated between two categories of 5-HT receptors (Imai et al, 1989). Highly significant correlations in affinities were found among (i) 5-HT$_{1A}$, 5-HT$_{1B}$ and 5-HT$_{1D}$ sites one another; (ii) 5-HT$_{1C}$ and 5-HT$_2$ sites. There was a reciprocal correlation between 5-HT$_{1A}$ and 5-HT$_2$ sites (Table II).

Table II Correlation coefficients between 5-HT receptor subtypes.

:	5-HT_{1B}	5-HT_{1D}	5-HT_{1A}	5-HT_2	5-HT_{1C}
5-HT_{1B} :	—				
5-HT_{1D} :	0.970^b	—			
5-HT_{1A} :	0.856	0.906^a	—		
5-HT_2 :	-0.158	-0.516	-0.944^b	—	
5-HT_{1C} :	0.165	-0.108	-0.388	0.754^b	—

Significance: $^a p < 0.05$, $^b p < 0.01$.

Effects of GTPγS and NEM in the Coupling of G Proteins with 5-HT$_{1A}$ Receptor and 5-HT$_{1B}$ Receptors

In the brain membranes, the binding of [^{3}H]8-OH-DPAT, a specific 5-HT$_{1A}$ agonist, was reduced to the level in the presence of GTPγS by NEM, an irreversible sulfhydryl alkylating agent. In addition, PTX-treatment also decreased the binding in a concentration dependent manner and 100 µg/ml PTX reduced to the level in the presence of GTPγS but cholera toxin did not (Kitamura et al., 1988).

In the presence of 30 µM isoproterenol, [^{125}I]ICYP, a 5-HT$_{1B}$ antagonist, specifically labeled 5-HT$_{1B}$ receptors. GTPγS did not shift antagonist competition (including propranolol, mianserin, ICS 205-930) for the binding. In contrast, GTPγS clearly affected the competition of several agonists (the magnitude order of 5-HT > TFMPP > methysergide > DOI > 8-OH-DPAT) for [^{125}I]ICYP-labeled 5-HT$_{1B}$ receptors and shifted the competition curve to the right. The apparent agonist affinity, based on K_D, was declined in the presence of GTPγS. In addition, NEM partially induced rightward-shift of 5-HT competition for [^{125}I]ICYP binding. GTPγS, however, significantly affected both agonist competition curves in the control and NEM-treated membranes. The two-site model analysis, showed that the amount of high affinity site is not changed by NEM-treatment (Imai et al., 1988).

Effects of GTPγS and NEM in the Coupling of G Proteins with 5-HT$_2$ and 5-HT$_{1C}$ Receptors

[^{3}H]DOB has been described as an agonist with high affinity to 5-HT$_2$ receptor (Lyon et al., 1987). [^{3}H]DOB binding was increased by NEM-treatment in a concentration-dependent manner and the maximum effect was at 0.1 mM NEM. In the presence of 20 µM GTPγS, the binding was decreased and was not changed by NEM (Imai et al., 1988). Data obtained from the saturation experiments on specific [^{3}H]ketanserin (5-HT$_2$ antagonist) binding were analyzed by a linear regression least-squares method. Both values of K_D and B_{max} in [³H]ketanserin binding were not changed by 20 μM GTPγS and 1 mM NEM. However, GTPγS and NEM clearly affected the agonist competition for [³H]ketanserin-labeled 5-HT₂ receptors. In the absence of GTPγS, 5-HT competed for [³H]ketanserin binding with "shallow slopes," i.e., the Hill coefficient (nH) was significantly lower than unity. NEM at 0.1 mM induced the interconversion to high affinity states from low affinity states. In the presence of GTPγS, 5-HT competition curves for [³H]ketanserin binding to control and NEM-treated membranes were "steep", showing nH close to unity. In the presence of GTPγS, however, the K_D value in NEM-treated membranes was significantly ($p < 0.05$) decreased versus that in control membranes (Imai et al., 1988).

In the presence of 0.1 μM spiperone, [³H]mesulergine could specifically bind to 5-HT₁C receptors in brain membranes. The characteristics of [³H]mesulergine binding were very similar to those of [³H]ketanserin binding. In addition, NEM caused a leftward-shift of a 5-HT competition curve for [³H]mesulergine binding, in the presence and absence of GTPγS, similar to those of [³H]ketanserin binding. GTPγS, however, shifted to the right the 5-HT competition curve for [³H]mesulergine binding to control and NEM-treated membranes (Imai et al., 1989).

Coupling Activity of 5-HT₁A Receptors with $G_i1α$, $G_oα$ and 24 K-G in Brain Membranes

5-HT₁A agonists such as 5-HT, 8-OH-DPAT, and buspirone inhibited forskolin-stimulated adenylate cyclase activity in rat hippocampal membranes. NEM and PTX attenuated the inhibitory activity of these drugs (Okada et al., 1989). We tried the reconstitution experiment using purified G_i1, G_o and 24 K-G from rat brain into NEM (0.3 mM)-pretreated membranes (Kitamura et al., 1988). $G_i1α$ ($α_{41}$) recovered NEM-reduced [³H]8-OH-DPAT binding more than $G_oα$ ($α_{39}$), but 24 K-G did not affect this binding. In the presence of GTPγS, $G_i1α$, $G_oα$ and 24 K-G did not affect the binding (Fig. 1).

DISCUSSION

Heterogeneity of 5-HT receptors has been studied but there are still many arguments. We here tried to rearrange the subtypes using rat brain membranes. Examining the ligand binding to rat brain membranes, highly significant correlations in affinities were found among 5-HT₁A, 5-HT₁B and 5-HT₁D receptors. The significant correlation was also the case between 5-HT₁C and 5-HT₂ receptors. It is suggested that 5-HT₁A, 5-HT₁B and 5-HT₁D receptors belong to one category and that 5-HT₂ and 5-HT₁C receptors belong to another category. From a comparison of each coefficient, both 5-HT₁D and 5-HT₁B types seem to branch from 5-HT₁A type, and 5-HT₁C type from 5-HT₂ type. 5-HT₂ type, however, is negatively correlated with 5-HT₁A type. 5-HT receptors, initially, were classified

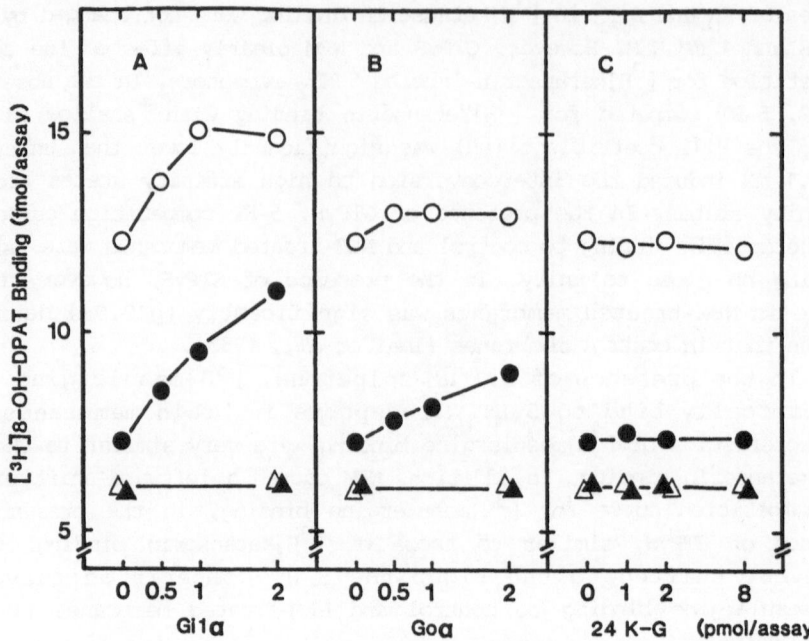

Fig. 1. Effects of $G_i1\alpha$, $G_o\alpha$ and 24 K-G on [^{3}H]8-OH-DPAT binding in NEM-treated membranes. Rat brain membranes were treated with vehicle ($\bigcirc$, $\triangle$) or 0.3 mM NEM ($\bullet$, $\blacktriangle$) as described in **MATERIALS AND METHODS.** The treated membranes were incubated with the various concentrations of $G_i1\alpha$ **(A)**, $G_o\alpha$ **(B)** or 24 K-G **(C)**, for 30 min at 0 °C, and binding assay was carried out with [^{3}H]8-OH-DPAT (0.2 nM) in the presence ($\triangle$, $\blacktriangle$) or absence ($\bigcirc$, $\bullet$) of 20 µM GTPγS at 30 °C for 40 min. $G_i1\alpha$, $G_o\alpha$ and 24 K-G were purified from rat brain.

into two: the one is 5-HT_1 receptor as the site labeled with high affinity by [^{3}H]5-HT and various 5-HT receptor agonists, and the other is 5-HT_2 receptor as that labeled by [^{3}H]spiperone and various 5-HT receptor antagonists (Peroutka and Snyder, 1979) (Table III). 5-HT_1 receptor are then further classified into 5-HT_{1A} receptors with high affinity for spiperone and 5-HT_{1B} receptors with low affinity for spiperone (Pedigo et al., 1981). Noting that 5-HT_{1B} type exists in rat and mouse brain but not in guinea-pig, bovine, or human brain (Peroutka, 1988), the present evidence is interesting from an ontogenetic aspect of 5-HT receptor subtypes. Thus, it seems that 5-HT_{1A} receptor and 5-HT_2 receptor are the basic classification of 5-HT receptors not including the rather unique 5-HT_3 receptor (Table III).

Activation of 5-HT_{1A} (Okada et al., 1989), 5-HT_{1B} (Bouhelal et al., 1988) and 5-HT_{1D} (Hoyer and Schoeffter, 1988) receptors inhibits the adenylate cyclase activity through PTX-sensitive G proteins, and the 5-HT_2 (Conn and Sanders-Bush, 1985) and 5-HT_{1C} receptors (Conn et al, 1986;

Table III **5-HT receptor classification from properties of binding and coupling to effectors and G proteins**

Binding property	Effector coupling	G protein coupling[*]
A. $5\text{-}HT_1$:	**A.** Enzyme	
high affinity site	i) Adenylate cyclase	
for agonists	(inhibition)	**A.** G_i coupling
$5\text{-}HT_{1A}$ ——————→	$5\text{-}HT_{1A}$ ——————→	$5\text{-}HT_{1A}$ (G-21)
$5\text{-}HT_{1B}$ ——————→	$5\text{-}HT_{1B}$ ------------→	?
$5\text{-}HT_{1C}$	$5\text{-}HT_{1D}$ ------------→	?
$5\text{-}HT_{1D}$		
	ii) Phospholipase C	
B. $5\text{-}HT_2$:	(stimulation)	**B.** G_p coupling
high affinity site		
for antagonists	$5\text{-}HT_2$ ——————→	$5\text{-}HT_{2A}$
$5\text{-}HT_2$ ——————→	$5\text{-}HT_{1C}$ ——————→	$5\text{-}HT_{2B}$
C. $5\text{-}HT_3$:	**B.** Ion channel	
"M" receptor	Ligand-gated ion channel	
$5\text{-}HT_3$ ——————→	$5\text{-}HT_3$	

*, speculated from primary structure (in Fig. 2)
G_i, inhibitory G protein of adenylate cyclase system
G_p, stimulatory G protein of phospholipase C

Nomura et al., 1987) activate phospholipase C through PTX-sensitive and/or PTX-insensitive G proteins. $5\text{-}HT_3$ receptors seem to activate ion channels inducing membrane depolarization in neuroblastoma cell lines (Watling, 1988) and directly gate Ca^{2+} channels in NG 108-15 (M. Tohda, and Y. Nomura, in preparation). The $5\text{-}HT_3$ receptor may couple to ion channels or may be a channel itself (Table III).

NEM induced uncoupling of $5\text{-}HT_{1A}$ receptors from G proteins as well as PTX (Kitamura et al., 1988). $5\text{-}HT_{1A}$ receptors couple with G_i1 rather than G_o but not 24 K-G in brain membranes (Fig. 1). In addition, $5\text{-}HT_{1B}$, $5\text{-}HT_{1C}$ and $5\text{-}HT_2$ receptors also couple with G proteins. Binding experiments using an agonist in $5\text{-}HT_{1B}$ receptors indicate partial uncoupling from G proteins at 0.3 mM NEM, whose concentration completely uncouples $5\text{-}HT_{1A}$ receptors from G proteins. GTPγS (20 μM) did not completely cause uncoupling of $5\text{-}HT_{1B}$ receptors from G proteins, and the pretreatment with PTX (30 μg/ml) did not affect the coupling of $5\text{-}HT_{1B}$ receptors with G proteins. Stratford et al. (1988) reported that a high concentration (1 mM) of NEM causes the uncoupling of $5\text{-}HT_{1B}$ receptors from G proteins and that agonist binding is recovered by G_i/G_o mixture as well as $5\text{-}HT_{1A}$ and $5\text{-}HT_{1D}$ receptors. Therefore, $5\text{-}HT_{1B}$ receptors so

A.

EXTRACELLULAR

		N-terminus	E-1	E-2	E-3
α_2	(human):	37	10	15	5
5-HT$_{1A}$	(human):	35	10	13	7
5-HT$_2$	(rat):	50	11	13	10
5-HT$_{1C}$	(rat):	52	11	14	10
α_1	(hamster):	49	10	11	7
β_1	(human):	62	10	18	7
β_2	(human):	37	10	18	7

EXTRACELLULAR

H₂N YY   I II III IV V VI VII   COOH

INTRACELLULAR

	C-1	C-2	C-3	C-Terminus
α_2 :	13	18	147	20
5-HT$_{1A}$:	13	17	128	18
5-HT$_2$:	13	18	66	87
5-HT$_{1C}$:	13	18	76	86
α_1 :	13	18	68	163
β_1 :	13	18	77	97
β_2 :	13	18	51	84

B.

Membrane
Spanning No.

I	5-HT$_{1A}$(human):	QVITSLLL-GTLIF-CAVLGNACVVAA
	5-HT$_2$(rat):	EKNWSALLTTVVII-LTIAGNILVIMA
	5-HT$_{1C}$(rat):	VQNWPALS-IVVIIIMTIGGNILVIMA
II	5-HT$_{1A}$:	LIGSLAVTDLMVSVLVLPMAALYQV
	5-HT$_2$:	FLMSLAIADMLLGFLVMPVSMLTIL
	5-HT$_{1C}$:	FLMSLAIADMLVGILVMPLSLLAIL
III	5-HT$_{1A}$:	CDLFIALDVLCCTSSILHLCAIALDR
	5-HT$_2$:	CAIWIYLDVLFSTASIMHLCAISLDR
	5-HT$_{1C}$:	CPVWISLDVLFSTASIMHLCAISLDR
IV	5-HT$_{1A}$:	PRALISLTWLIGPLISIP-PMLGWRTP
	5-HT$_2$:	AFLKIIAVWTISVGISMPIPVFGLQDD
	5-HT$_{1C}$:	AIMKIAIVWAISIGVSVPIPVIGLRDE
V	5-HT$_{1A}$:	DHGYTIYSTFGAFYIPLLLMLVLYGR
	5-HT$_2$:	DNFVLIGS-FVAFFIPLTIMVITYFL
	5-HT$_{1C}$:	PNFVLIGS-FVAFFIPLTIMVITYFL
VI	5-HT$_{1A}$:	TLGTIIMGTFILCWLPFFIVALVLPFC
	5-HT$_2$:	VLGIVFFLFVVMWCPFFITNIMAVIC
	5-HT$_{1C}$:	VLGIVFFVFLIMWCPFFITNILSVLC
VII	5-HT$_{1A}$:	TLLGAIINWLGYSNSLLNPVIYAYFN
	5-HT$_2$:	ALLNVFV-WIGYLSSAVNPLVYTLFN
	5-HT$_{1C}$:	KLLNVFV-WIGYVCSGINPLVYTLFN

Fig. 2. Model for the transmembrane structure of the G protein
← **receptor superfamily. A,** The number of amino acid residues of N-terminus, 6 loops and C-terminus in each receptor: human $5-HT_{1A}$ (G-21) (Kobilka et al., 1987b), rat $5-HT_{1C}$ (Julius et al., 1988), $5-HT_2$ (Pritchett et al., 1988) receptor, hamster α_1- (Cotecchia et al., 1988), human α_2-(Kobilka et al., 1987a), β_1- and β_2-adrenoceptors (Cotecchia et al., 1988). Similar length of amino acid residue is indicated by box. **B,** Amino aci. sequences for the putative transmembrane spanning domains of $5-HT_{1A}$, $5-HT_2$ and $5-HT_{1C}$ receptor clones. Matched amino acids are indicated by boxes, and * indicate the position of only one cysteine residue, existing in $5-HT_2$ and $5-HT_{1C}$ receptor but not in $5-HT_{1A}$ receptor.

tightly couple with G proteins that the coupling may show low sensitivity to NEM and PTX (Imai et al., 1988).

GTPγS caused a decrease in [^{3}H]DOB binding that was inversely increased by NEM. NEM does not affect GTPγS-caused uncoupling of $5-HT_{1C}$ and $5-HT_2$ receptors from G proteins (Imai et al., 1989). It has been reported that NEM has biphasic influences on muscarinic receptors coupling to G proteins in porcine caudate nucleus: at a low concentration the compound causes uncoupling of muscarinic receptors from G proteins by alkylating G proteins and at a high concentration it causes increase in the affinity for agonist to muscarinic receptors by alkylating receptors (Nukada et al., 1983). In addition, alkylated muscarinic receptors can couple with G proteins and the coupling causes further increase in the affinity for agonist (T. Haga, personal communication). It has also been reported that muscarinic receptors purified from porcine brain are composed of high and low affinity sites for agonist, that reduction of an intramolecular disulfide bond(s) by DTT decrease the binding affinities of both agonist and antagonist to the receptors, and that the interconversion from low affinity to high affinity sites by 5,5'-dithiobis(2-nitrobenzoic acid) can be induced by changes in the redox state of sulfhydryl groups in the receptor molecules (Berstein et al., 1988). Thus, sulfhydryl groups in receptor molecule seem to play the important roles in agonist and antagonist binding.

The primary structures of $5-HT_{1A}$ (Kobilka et al., 1987b; Fargin et al., 1988), $5-HT_{1C}$ (Julius et al., 1988) and $5-HT_2$ (Pritchett et al., 1988) receptors have been cloned by three different laboratories and discussed (Hartig, 1989). These reports describe that the predicted amino acid sequence of the $5-HT_{1C}$ receptor is 51 % identical to that of the $5-HT_2$ receptor. However, the amino acid sequence of the $5-HT_{1A}$ receptor is 35 % identical to both $5-HT_{1C}$ and $5-HT_2$ receptors (Pritchett et al., 1988). In addition, site-directed mutagenesis studies (Dixon et al., 1987) and a recent study constructing chimeric adrenoceptors (Kobilka et al., 1988) have demonstrated the importance of the third cytoplasmic loop (C-3 in Fig. 2A) for the coupling of the receptors to G proteins and the membrane spanning regions for ligand binding (Lefkowitz and Caron, 1988). The third cytoplasmic loop and C-terminus of 5-HT$_{1A}$ receptor are very similar length to those of α_2- (Kobilka et al., 1987a) and D$_2$-receptors (Bunzow et al., 1988), causing inhibition of adenylate cyclase (Fig. 2A), while those of 5-HT$_{1C}$ and 5-HT$_2$ receptors are similar to α_1-receptors (Cotecchia et al., 1988), activating phospholipase C, and/or β-receptors activating adenylate cyclase, rather than α_2-receptors (Table III). In comparing the amino acid sequence among 5-HT$_{1A}$, 5-HT$_{1C}$ and 5-HT$_2$ receptors, a cysteine residue (Cys) in the sixth of the membrane spanning region is the same only in 5-HT$_2$ (Cys^{315}) and 5-HT$_{1C}$ receptors (Cys^{327}), but not in the 5-HT$_{1A}$ receptor (Leu^{359}) (Fig. 2B). NEM is known to irreversibly alkylate cysteine residues. Therefore, this cysteine residue in 5-HT$_{1C}$ and 5-HT$_2$ receptors may be alkylated by NEM resulting in the change of agonist-binding affinity to each receptor.

CONCLUSION

The subtypes of 5-HT receptors have been classified and characterized in binding and coupling properties with effectors (Table III). Amino acid sequences of 5-HT$_{1A}$, 5-HT$_{1C}$ and 5-HT$_2$ were recently determined: 5-HT$_{1A}$ receptor clone (G-21) was screened using human β_2-adrenoceptor cDNA (Kobilka et al., 1987b), the cloning of 5-HT$_{1C}$ receptor was done using the electrophysiological assay in Xenopus oocytes injected with choroid plexus RNA (Julius et al., 1988) and the 5-HT$_2$ receptor clone was screened using oligonucleotide probe of the 5-HT$_{1C}$ receptor (Pritchett et al., 1988). Coupling with G proteins, receptors are classified as followings: (i) 5-HT$_{1A}$ receptor in G$_i$-coupling family, (ii) 5-HT$_2$ and 5-HT$_{1C}$ receptor in same G$_p$-coupling family (Table III). From these close structural and functional relationships between 5-HT$_2$ and 5-HT$_{1C}$ receptors, Prichett et al. (1988) proposed that 5-HT$_2$ and 5-HT$_{1C}$ should change into 5-HT$_{2A}$ and 5-HT$_{2B}$, respectively (Table III). We expect cloning and sequencing of 5-HT$_{1B}$ and 5-HT$_{1D}$ subtypes in the near future.

ACKNOWLEDGMENT

We thank Dr. R. W. Olsen for his advice of the manuscript. This study was in part supported by Grant-in-Aid from the Ministry of Education, Science and Culture, Japan.

REFERENCES

Berstein, G., Haga, K., Haga, T., and Ichiyama, A., 1988, Agonist and antagonist binding of muscarinic acetylcholine receptors purified from porcine brain: interconversion of high- and low- affinity sites by sulfhydryl reagents, J. Neurochem. 50:1687-1694.

Bouhelal, R., Smounya, L., and Bockaert, J., 1988, 5-HT$_{1B}$ receptors are negatively coupled with adenylate cyclase in rat substantia nigra, Eur. J. Pharmacol., 151:189-196.

Bunzow, J. R., Van Tol, H. H. M., Grandy, D. K., Albert, P.,Salon, J., Christie, M., Machida, C. A., Neve, K. M., and Civelli O.,, 1988, Cloning and expression of a rat D$_2$ dopamine receptor cDNA, Nature 336:783-787.

Casey, P. J., and Gilman, A. G., 1988, G protein involvement in receptor-effector coupling, J. Biol. Chem., 263:2577-2580.

Cheng, Y. C., and Prusoff, W. H., 1973, Relationship between the inhibition constant (K$_I$) and the concentration of inhibitor which causes 50 per cent inhibition (I$_{50}$) of an enzymatic reaction, Biochem. Pharmacol., 22:3099-3108.

Conn, P. J., and Sanders-Bush, E., 1985, Serotonin-stimulated phospho-inositide turnover; mediation by the S$_2$ binding site in rat cerebral cortex but not in subcortical regions, J. Pharmacol. Exp. Ther., 234:195-203.

Conn, P. J., Sanders-Bush, E., Hoffman, B. J., and P. R. Hartig, P. R., 1986, A unique serotonin receptor in choroid plexus is linked to phosphatidylinositol turnover, Proc. Natl. Acad. Sci. USA 83:4086-4088.

Cotecchia, S., Schwinn, D. A., Randall, R. R., Lefkowitz, R. J., Caron, M. G., and Kobilka, B. K., 1988, Molecular cloning and expression of the cDNA for the hamster α_1-adrenergic receptor, Proc. Natl. Acad. Sci. USA, 85:7159-7163.

Dixon, R. A. F., Sigal, I. S., Candelore, M. R., Register, R. B., Scattergood, W., Rands, E., and Strader, C. D., 1987, Structural features required for ligand binding to the β-adrenergic receptor, EMBO J., 6:3269-3275.

Fargin, A., Raymond, J. R., Lohse, M. J., Kobilka, B. K., Caron, M. G., and Lefkowitz, R. J., 1988, The genomic clone G-21 which resembles a β-adrenergic receptor sequence encodes the 5-HT$_{1A}$ receptor, Nature, 335:358-360.

Hartig, P. R., 1989, Molecular biology of 5-HT receptors, Trends Pharmacol. Sci., 10:64-69.

Hoyer, D., Engel G., and Kalkman, H. O., 1985, Characterization of the 5-HT$_{1B}$ recognition site in rat brain: binding studies with (-)-[^{125}I]iodocyanopindolol, Eur. J. Pharmacol., 118:1-12.

Hoyer, D., 1988, Molecular pharmacology and biology of 5-HT$_{1C}$ receptors, Trends Pharmacol. Sci. 9:89-94.

Hoyer, D., and Schoeffter, P., 1988, 5-HT$_{1D}$ receptor-mediated inhibition of forskolin-stimulated adenylate cyclase activity in calf substantia nigra, Eur. J. Pharmacol., 147:145-147.

Imai, S., Kitamura, Y., and Nomura, Y., 1988, Effects of N-ethylmaleimide on the coupling of 5-HT receptor subtypes with GTP-binding protein in rat brain, Bulletin of Japanese Neurochemical Society, 27:132-133 (in Japanese).

Imai, S., Kitamura, Y., and Nomura, Y., 1989, Classification of 5-hydroxytryptamine receptor subtypes from binding characteristics in rat brain membranes, Bulletin of Japanese Neurochemical Society, **28**:184-185 (in Japanese).

Itoh, H., Katada, T., Ui, M., Kawasaki, H., Suzuki, K., and Kaziro, Y., 1988, Identification of three pertussis toxin substrates (41, 40 and 39 kDa proteins) in mammalian brain, FEBS Lett., **230**:85-89.

Julius, D., MacDermott, A. B., Axel, R., and Jessell, T. M., 1988, Molecular characterization of a functional cDNA encoding the serotonin 1c receptor, Science, **241**:558-564.

Kahn, R. A., and Gilman, A. G., 1986, The protein cofactor necessary for ADP-ribosylation of G_s by cholera toxin is itself a GTP binding protein, J. Biol. Chem., **261**:7906-7911.

Katada, T., Oinuma, M., and Ui, M., 1986a, Two guanine nucleotide-binding proteins in rat brain serving as the specific substrate of islet-activating protein, pertussis toxin: interaction of the α-subunits with βγ-subunits in development of their biological activities, J. Biol. Chem., **261**:8182-8191.

Katada, T., Kurose, H., Oinuma, M., Hoshino, S., Shinoda, M., Amanuma, S., and Ui, M., 1986b, Role of GTP-binding proteins in coupling of receptors and adenylate cyclase, in: "Gunma Symposia on Endocrinology, Vol. **23**", VNU Science Press BV, Tokyo, p.45-67.

Katada, T., Oinuma, M., Kusakabe, K., and Ui, M., 1987, A new GTP-binding protein in brain tissues serving as the specific substrate of islet-activating protein, pertussis toxin, FEBS Lett., **213**:353-358.

Katada, T., and Ui, M., 1988, Unique properties of a new GTP-binding protein with a molecular mass of 24,000 daltons purified from porcine brain membranes, in: "Cold Spring Harbor Symposia on Quantitative Biology, Vol. **53**; Molecular Biology of Signal Transduction", Cold Spring Harbor Laboratory, New York, p.255-261.

Kitamura, Y., and Nomura, Y., 1987, Uncoupling of rat cerebral cortical α_2-adrenoceptors from GTP-binding proteins by N-ethylmaleimide, J. Neurochem., **49**:1894-1901.

Kitamura, Y., Imai, S., and Nomura, Y., 1988, Coupling of 5-HT_{1A} receptor with GTP-binding protein in rat brain membranes, Japan. J. Pharmacol., **46**:251P.

Kobilka, B. K., Matsui, H., Kobilka, T. S., Yang-Feng, T. L., Francke, U., Caron, M. G., Lefkowitz, R. J., and Regan, J. W., 1987a, Cloning, sequencing, and expression of the gene coding for the human platelet α_2-adrenergic receptor, Science, **238**:650-656.

Kobilka, B. K., Frielle, T., Collins, S., Yang-Feng, T., Kobilka, T. S., Francke, U., Lefkowitz, R. J., and Caron, M. G., 1987b, An intronless gene encoding a potential member of the family of receptors coupled to guanine nucleotide regulatory proteins, Nature, **329**:75-79.

Kobilka, B. K., Kobilka, T. S., Daniel, K., Regan, J. W., Caron, M. G., and Lefkowitz, R. J., 1988, Chimeric α_2-, β_2-adrenergic receptors:

delineation of domains involved in effector coupling and ligand binding specificity, Science, **240**:1310-1316.

Lefkowitz, R. J., and Caron, M. G., 1988, Adrenergic receptors: models for the study of receptors coupled to guanine nucleotide regulatory proteins, J. Biol. Chem., **263**:4993-4996.

Lyon, R. A., Davis, K. H., and Titeler M., 1987, ^{3}H-DOB (4-bromo-2,5-dimethoxyphenylisopropylamine) labels a guanyl nucleotide-sensitive state of cortical 5-HT$_2$ receptors, Mol. Pharmacol., **31**:194-199.

Munson, P. J., and Rodbard D., 1980, LIGAND: a versatile computerized approach for characterization of ligand-binding systems, Anal. Biochem., **107**:220-239.

Narumiya, S., Sekine, A., and Fujiwara, M., 1988, Substrate for botulinum ADP-ribosyltransferase, G$_b$, has an amino acid sequence homologous to a putative rho gene product, J. Biol. Chem. **263**:17255-17257.

Nomura, Y., Kaneko, S., Kato, K., Yamagishi, S., and Sugiyama, H., 1987, Inositol phosphate formation and chloride current responses induced by acetylcholine and serotonin through GTP-binding proteins in Xenopus oocyte after injection of rat brain messenger RNA, Mol. Brain Res., **2**:113-123.

Nomura Y., Kitamura, Y., and Kawata, K., 1988, Function and mechanism of the interaction of GTP-binding proteins with α_2-adrenoceptors in the brain, in: "Neurotransmitters and Signal Transduction," Plenum, New York, p.301-311.

Nukada, T., Haga, T., and Ichiyama, A., 1983, Muscarinic receptors in porcine caudate nucleus: II. different effects of N-ethylmaleimide on ^{3}H]cis-methyldioxolane binding to heat-labile (guanyl nucleotide-sensitive) sites and Heat-stable (guanyl nucleotide-insensitive) sites, Mol. Pharmacol. **24**:374-379.

Offord, S. J., Ordway, G. A., and Frazer, A., 1988, Application of ^{125}I]-iodocyanopindolol to measure 5-hydroxytryptamine$_{1B}$ receptors in the brain of the rat, J. Pharmacol. Exp. Ther., **244**:144-153.

Okada, F., Tokumitsu, Y., and Nomura, Y., 1989, Pertussis toxin attenuates 5-hydroxytryptamine$_{1A}$ receptor-mediated inhibition of forskolin-stimulated adenylate cyclase activity in rat hippocampal membranes, J. Neurochem., **52**:1566-1569.

Pazos, A., Hoyer D., and Palacios, J. M., 1984, The binding of serotonergic ligands to the porcine choroid plexus: characterization of a new type of serotonin recognition site, Eur. J. Pharmacol., **106**:539-546.

Pedigo, N. W., Yamamura, H. I., and Nelson, D. L., 1981, Discrimination of multiple ^{3}H]5-hydroxytryptamine binding sites by the neuroleptic spiperone in rat brain, J. Neurochem., **36**:220-226.

Peroutka, S. J., and Snyder, S. H., 1979, Multiple serotonin receptors: differential binding of ^{3}H]5-hydroxytryptamine, ^{3}H]lysergic acid diethylamide and ^{3}H]spiroperidol, Mol. Pharmacol., **16**:687-699.

Peroutka, S. J., 1988, 5-Hydroxytryptamine receptor subtypes: molecular, biochemical and physiological characterization, _Trends Neurosci._, **11**:496–500.

Pritchett, D. B., Bach, A. W., Wozny, M., Taleb, O., Toso, R. D., Shih, J. C., and Seeburg, P. H., 1988, Structure and functional expression of cloned rat serotonin 5-HT-2 receptor, _EMBO J._, **7**:4135–4140.

Stratford, C. A., Tan, G. L., Hamblin, M. W., and Ciaranello, R. D., 1988, Differential inactivation and G protein reconstitution of subtypes of [^{3}H]5-hydroxytryptamine binding sites in brain, _Mol. Pharmacol._, **34**:527–536.

Takai, Y., Kikuchi, A., Yamashita, T., Yamamoto, K., Kawata, M., and Hoshijima, M., 1989, Small molecular weight GTP-binding proteins from bovine brain membranes: purification, characterization and possible functions, _in_: "Physiology and Pharmacology of Transmembrane Signalling," Elsevier, Amsterdam, p.77–86.

Watling, K. J., 1988, Radioligand binding studies identify 5-HT$_3$ recognition sites in neuroblastoma cell lines and mammalian CNS, _Trends Pharmacol. Sci._, **9**:227–229.

Yamaoka, K., Tanigawara, Y., Nakagawa, T., and Uno, T., 1981, A pharmacokinetic analysis program (MULTI) for microcomputer, _J. Pharm. Dyn._, **4**:879–885.

A CEREBRAL ENDOGENOUS FACTOR REGULATES THE ACTIVITY OF THE SEROTONERGIC RECEPTORS MODULATING THE NEURONAL RELEASE OF ACETYLCHOLINE

Gilles Fillion, Pascal Barone, Isabelle Cloëz, Marie-Paule Fillion, Catherine Harel, Olivier Massot, Jean-Claude Rousselle, and Emilie Zifa

Unité de Pharmacologie Neuroimmunoendocrinienne Institut Pasteur, 28 rue du Dr Roux, F75724 Paris Cedex 15 (France)

INTRODUCTION

The 5-hydroxytryptamine (5-HT) receptors involved in the function of the serotonergic system in the brain of mammalians are numerous and their knowledge is still not complete. A first classification was produced by Bradley et al. (1986) based on the existence of 3 classes of receptors $5-HT_1$, $5-HT_2$ and $5-HT_3$. The $5-HT_1$ class corresponded to the existence of 3 subtypes of sites called $5-HT_{1A}$, $5-HT_{1B}$ and $5-HT_{1C}$ and 3 subtypes were also proposed for the $5-HT_3$ class. Recent results allowed to simplify the classification of these receptors on the basis of their transduction system. It is, indeed, possible to distinguish serotonergic receptors coupled directly to an ionic channel; they correspond to the $5-HT_3$ receptor class (Richardson and Buchheit, 1988). The existence of subtypes of $5-HT_3$ receptors proposed on the basis of pharmacological heterogeneities is not yet fully established (Richardson and Engel, 1986). The second main class of receptor consists of those coupled to G proteins; they are $5-HT_1$ and $5-HT_2$ sites. Moreover, a $5-HT_4$ receptor type has been proposed (Dumuis et al., 1988) and may correspond to a G protein related receptor, however, the corresponding recognition sites for $5-HT_4$ have not yet been identified. $5-HT_1$ receptors are subdivided into 4 subtypes $5-HT_{1A}$, $5-HT_{1B}$, $5-HT_{1C}$ and $5-HT_{1D}$ (Hoyer et al., 1987) and an additional $5-HT_{1E}$ has been proposed by Leonhardt et al. (1989).

These various subtypes are either related to the activation or the inhibition of the adenylate cyclase activity or to the stimulation of the metabolism of inositol phosphates (see Peroutka, 1988).

In the recent years, we proposed the existence of $5-HT_{1D}$ receptors on the basis of the observation of pharmacological heterogeneities of the 5-HT$_1$ binding leading to the characterisation of a 5-HT$_{1\text{non A, non B, non C}}$ site (Fillion et al., 1988; Fayolle et al., 1988). The pharmacology of this site closely corresponded (correlation = 0.91) to that of the effect of 5-HT on the adenylate cyclase activity strongly suggesting the existence of a functional receptor (5-HT$_{1D}$). Similar results were obtained by Hoyer and Schoefter (1988). In the present work, the physiological role of this receptor was studied and we payed attention to the regulation mechanisms affecting this receptor during physiopathological processes. Finally, the activities of various drugs on this receptor were examined leading to specific interactions which could be considered for developing new ways of therapeutic actions.

MATERIAL AND METHODS

Binding assays of [^{3}H]-5HT to 5-HT$_{1D}$ sites in guinea-pig brain sections

Guinea-pig brains were frozen immediately after sacrifice and frozen-sections were obtained using a Cryostat-Microtome Reichert; these 20 μm coronal sections were placed on glass slides. They were incubated for one hour at 22°C in the presence of 2 nM [^{3}H]5-HT (total binding) and of non radioactive 8-OH-DPAT (0.1 μM) and mesulergine (1 μM) to mask the binding of the amine to 5-HT$_{1A}$ and 5-HT$_{1C}$ sites. The presence of pindolol, inhibiting the binding to 5-HT$_{1B}$ receptors had no effect since the latter sites are absent in guinea-pig brain.

Non-specific binding was determined in the presence of an excess of non-radioactive 5-HT (10 μM). After 5 minutes washes (3 times) at 0° C, the labelled sections were apposed to Amersham Hyperfilm and exposed for 2 to 3 months in the dark at 4°C. The films were then processed and the obtained autoradiograms analysed using a Biocom RAG 200 image analyser.

Synaptosomal release assays

These experiments were performed using two types of protocols.

a) Superfusion experiments

Cerebral tissue (hippocampus or cortex as specified in the text) were dissected out from brain of rat or guinea-pig as indicated. The tissue was homogenized, using a Potter-Elvehjhem (10 cycles, 300 rpm) in 15 vol of sucrose solution (0.32 M) at 4°C buffered with 5 mM Tris-HCl containing PMSF (100 μM), aprotinine (5 U/L) and EGTA (2 mM). The homogenate was centrifuged (1000 g for 5 min at 4°C) and washed twice after resuspension in the same buffered medium. Then, the suspension was centrifuged at 17500 g for 20 min at 4°C. The obtained pellet represented the crude synaptosomal fraction. Synaptosomes were resuspended in Krebs-Hepes buffer containing NaCl 118 mM, KCl 4.8 mM, Mg SO$_4$ 1.2 mM Hepes 17 mM, glucose 11 mM and ascorbate 0.01 % and adjusted to pH 7.4 and oxygenized in 95 % O^2, 5 % CO$_2$ at 37°C. [^{3}H]choline (100 nM) was added to the suspension for 15 min at 37°C to be taken up. The excess of radioactivity in the medium was discarded by two successive centrifugations. Aliquots of the suspension containing the [^{3}H]choline were distributed in perfusion chambers (300 µl) and superfused for 20 min with gassed nutritive medium. Then, the effluent was collected per unit of time and at the indicated times, stimulations of tritium release were obtained by K$^+$ pulses-under the indicated conditions. The effects of drugs were determined by measuring successively the K$^+$ induced release of tritium in the presence or the absence of the drug. The release or radioactivity corresponding to the first and the second K$^+$ induced tritium efflux were compared in measuring the P$_2$/P$_1$ ratio, P$_1$ and P$_2$ representing the excess of efflux of tritium over the basal release upon the first and second evoked release respectively.

b) Release in tubes

In this series of experiments, the release of tritium from the same synaptosomal material was determined using the technique previously described by Minnema et al. (1986). Aliquots (500 µl) of labelled synaptosomal suspension were incubated for 3 min at 37°C in polypropylene tubes in the absence (basal release) or the presence of KCl (10 mM) (total release) in a Krebs-Hepes buffer containing CaCl$_2$ (1.3 mM). Incubation was stopped by rapid centrifugation in a Mikroliter Centrifuge (total duration 2 min) and the supernatant was collected and its radioactivity determined by scintillation spectrometry. Drug to be tested were added in the tube (10 µl/tube) before the synaptosomal fraction.

Results

I. Distribution of 5-HT$_{1D}$ receptors in guinea-pig brain

The distribution of 5-HT$_{1D}$ receptors in guinea-pig brain was studied in serial sections. The results obtained showed that 5-HT$_{1D}$ were widely and heterogeneously distributed within the brain. The areas exhibiting the highest density were the dentate gyrus and the substantia nigra (> 3000 fmoles/mg prot.) a marked density in globus pallidus, superior colliculi and CA1 of hippocampus and lateralis dorsalis septal nucleus (> 2000 fmoles/mg prot.) and a lower density in dorsal subiculum, CA4 of hippocampus, periaqueductal grey, lateralis intermediate septum and putamen (> 1000 fmoles/mg prot.). The cortex was also clearly labelled (> 900 fmoles/mg prot.) whereas no significant binding was observed in thalamus, central nuclei or white matter (corpus-callosum).

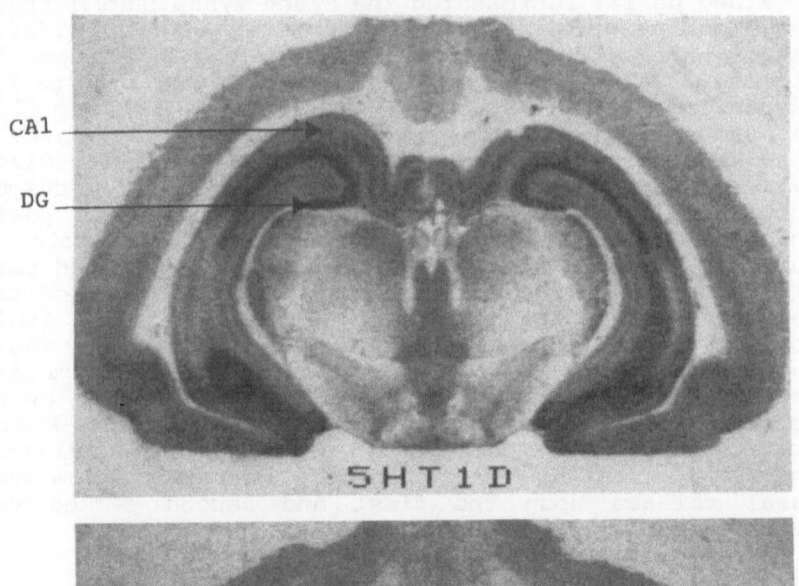

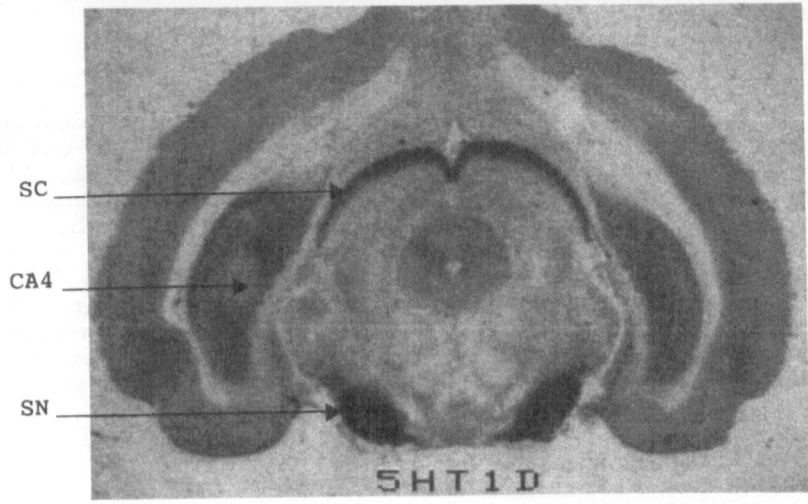

Fig. 1   Autoradiograms of the distribution of ^{3}H-5HT binding
         sites in coronal section of a guinea-pig brain
         representing the 5-HT$_{1D}$ receptors.
         The binding is obtained in the presence of 8-OH-
         DPAT (0.1 μM) and mesulergine (1 μM).
         The non-specific binding (not shown) observed in the
         presence of 10 μM 5-HT represents a negligible
         binding.
         The upper section shows the 5-HT$_{1D}$ binding at the
         level of hippocampus (CA1), dentate gyrus (DG), the
         lower section shows this binding in superior col-
         liculus (SC), substantia nigra (SN) and the posterior
         part of the hippocampus (CA4).

These results were in agreement with the hypothesis of the existence of neuromodulating function mediated by these receptors present in large brain areas.

II.   Cellular function of 5-HT$_{1D}$ receptor

The release of [^{3}H]acetylcholine evoked by K$^+$ pulses was measured in hippocampus synaptosomal preparations using superfusion techniques and release measured in tubes. It was possible to show that in rat or in guinea-pig brain, 5-HT itself or TFMPP, a 5-HT agonist were able to alter the release of [^{3}H]acetylcholine. This effect was dose dependent and the ED$_{50}$ values were similar (close to 50 µM) in the two species; the maximal inhibiting effect of TFMPP on the release was 80 % in rat and 70 % in guinea-pig. The assays performed to measure the effects of drugs corresponded to inhibiting effects close to 40 % of the maximum. Interestingly it was shown that the pharmacological properties of the serotonergic inhibiting effect in rat and in guinea-pig were different: the phenomenon corresponded to the activity of a 5-HT$_{1D}$ receptor subtype in the guinea-pig hippocampus and surprisingly it was mediated by a 5-HT$_{1B}$ in rat. Indeed, propranolol which inhibits 5-HT$_{1B}$ receptors, antagonized the inhibiting effect in rat and was inefficient in guinea-pig. In agreement with this observation, a 5-HT$_{1B}$ agonist (RU 24969) mimicked the effect of TFMPP in rat brain and was inefficient in guinea-pig. Dihydroergotamine and methiotepin which are non-selective serotonergic antagonists markedly blocked the effect of TFMPP whereas selective antagonists for 5-HT$_3$ (MDL 72222), 5-HT$_{1C}$ (mesulergine) and 5-HT$_2$ (spiperone) receptors were totally inefficient. Moreover, in guinea-pig, the 5-HT$_{1D}$ agonist CGS 12066 was more efficient than TFMPP and mCPP at the same concentration, whereas 8-OH-DPAT, a 5-HT$_{1A}$ agonist, and quipazine, a 5-HT$_3$ agonist, were devoid of any effect.

These results strongly suggest that serotonergic receptors located on cholinergic terminals modulate the evoked release of acetylcholine. These receptors are so-called heterologous presynaptic receptors and appear to correspond to 5-HT$_{1B}$ in rat and mouse and 5-HT$_{1D}$ in guinea-pig; according to preliminary assays they are also 5-HT$_{1D}$ in monkey.

Therefore, it seems that the same cellular function is mediated by two different receptors in these two groups of species. These two ways of regulation were presumably determined during evolution possibly originating from common ancestor genes. The function of these receptors is quite important since it corresponds to the capacity of the serotonergic system to modulate the activity of the cholinergic system. It is not known, at the present time, whether the modulating effect of 5-HT on the activity of other neurotransmitter systems involves also heterologous presynaptic receptors and whether they belong to the 5-HT$_{1D}$ - 5-HT$_{1B}$ subtype.

III. Regulation mechanisms

It was of interest to examine whether this system capable of modulating the cholinergic system was regulated during a physiopathological event.

The experimental model used in these assays was the isolation of mice which consisted in housing mice one per cage for a week. They received food and water ad libitum and except for isolation were submitted to identical conditions as the control mice kept 10 per cage. After a week isolation, mice were sacrificed and hippocampal tissues examined for the capacity of TFMPP to inhibit the K^+ evoked ACh release. In 6 different series of animals, it was observed that the isolation of mice did not affect the choline uptake, nor the evoked release of [^{3}H]ACh. However, the antagonizing effect of TFMPP on the evoked release was markedly and consistently reduced (control inhibition was 24.7 ± 2.2 % whereas in isolated mice it was 13.8 ± 2.2 % -p < 0.01). Interestingly the uptake, the evoked release of [^{3}H]5-HT and the inhibiting effect of TFMPP on the evoked release of[^{3}H]5-HT involving the autoreceptors was not modified in the same animals. These results indicate that the serotonergic heterologous presynaptic receptors capable of modulating the function of the cholinergic system are themselves regulated by stress. This regulation is selective since homologous presynaptic receptors which presumably obey the same pharmacology are not

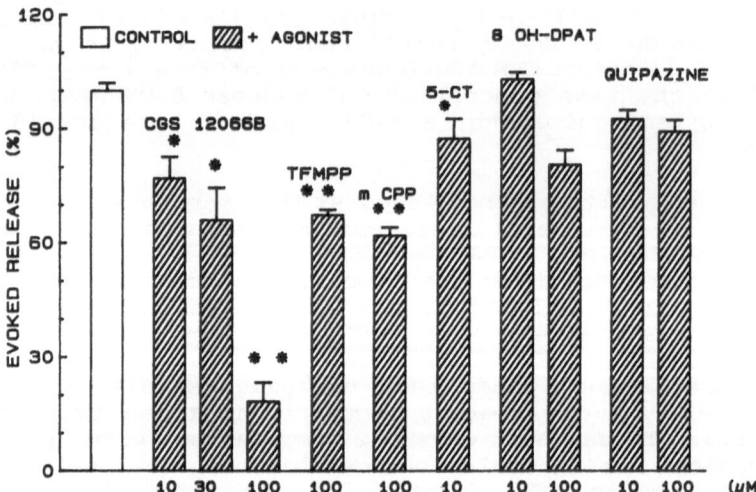

Fig. 2   Effect of serotonergic agonists on the ^{3}H ACh release evoked by K^+ in guinea-pig brain synaptosomes.
Drugs at the indicated concentrations were added together with K^+.
Each bar represents the mean ± SEM of 3-4 independent assays performed in triplicates.
Statistical analysis using Student t-test corresponded to the following:
   *   = < 0.05
   ** = < 0.01
for the results in the presence of the drug compared to the control values.

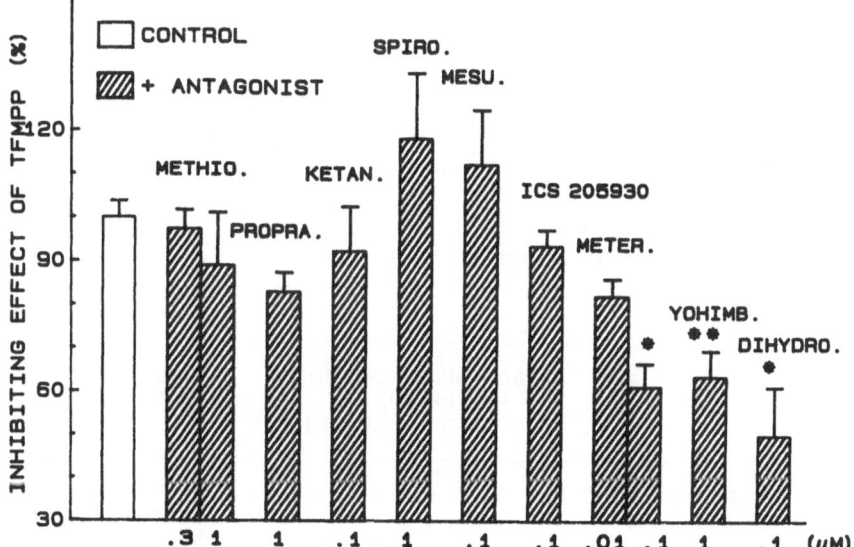

Fig. 3　Effect of serotonergic antagonists on the inhibition
induced by TFMPP of the ³H ACh evoked release.
TFMPP (100 μM) and K⁺ are added together and compared
to the effect of K⁺ alone.
The drugs are added to the medium in the presence of
TFMPP (Methio.= methiotepine; propra.= propranolol;
ketan.= ketanserin; spiro.= spiroperidol; mesu.=
mesulergine; meter.= metergoline; yohumb.= yohumbine,
dihydro.= dihydroergotamine).
Each bar represents the mean ± SEM of 3 to 7 independent assays performed in triplicates.
Statistical analysis using Student t-test corresponded in the following:
$$* = p < 0.05$$
$$** = p < 0.01$$
for the results in the presence of the drug compared
in the presence of TFMPP alone.

regulated in parallel. This mechanism may be of importance to explain how the brain is capable of regulating its own function.

IV. Pharmacological interactions

The serotonergic receptors 5-HT_{1D} in guinea-pig and 5-HT_{1B} in rat thus appeared heterologous presynaptic receptors modulating the activity of other neurotransmitters. It was interesting to examine the pharmacological interactions affecting this cellular function. Among drugs which have been studied, a series of antidepressants, inhibiting the uptake of 5-HT or norepinephrine (NA) have been tested. It was interesting to note that most of the studied drugs showing antidepressive properties were also able to antagonize the inhibitory effect of TFMPP on the evoked release of ACh. Table 1 illustrates their activities.

Table 1 Effect of drugs on the K^+ evoked release of ACh and on the inhibitory effect of TFMPP on this ACh release.

	EFFECT ON K^+-EVOKED ACH RELEASE (% OF CONTROL RELEASE)	EFFECT ON TFMPP INHIBITION (% OF TFMPP EFFECT)	
CHLORIMIPRAMINE (100nM)	98.4 ± 2.5 (3)	55.6 ± 11.0 (7)	***
IMIPRAMINE (50nM)	96.2 ± 1.7 (3)	52.1 ± 20.1 (3)	*
FLUVOXAMINE (50nM)	108.0 ± 4.9 (4)	55.3 ± 4.3 (3)	*
FLUOXETINE (100nM)	88.0 ± 4.2 (4)	69.7 ± 4.6 (3)	*
CITALOPRAM (10nM)	98.0 ± 3.4 (3)	59.2 ± 1.3 (3)	**
MINAPRINE (10nM)	99.7 ± 2.2 (3)	51.9 ± 5.8 (15)	***
COCAINE (50nM)	93.3 ± 2.5 (3)	96.6 ± 2.5 (3)	
- (100nM)	nd	126.1 ± 21.9 (3)	
CHLORPROMAZINE (50nM)	107.6 ± 5.0 (5)	89.2 ± 9.2 (3)	
- (100nM)	nd	92.5 ± 0.9 (3)	
CLOBAZAM (50nM)	105.8 ± 2.7 (3)	89.8 ± 12.0 (3)	
- (100nM)	nd	107.5 ± 9.2 (3)	

* $p<0.05$, ** $p<0.01$, *** $p<0.001$ by paired two-tailed Student's t test
nd: non determined

Particular characteristics of these interactions have to be pointed out: i.e. the effects of these drugs are usually already observed at low concentrations (in the 10 nM range), they are dose-dependent and appear selective for a series of antidepressants since clobazam, chlorpromazine and cocaine are devoid of any effect. Special attention should be payed to minaprine since its antidepressant properties are not yet fully established. It possesses a strong antagonizing property of the observed serotonergic effect (already active at 1 nM) but is devoid of any activity as an amine uptake inhibitor. This observation strongly suggests that the activity of antidepressants in this mechanism is presumably independent of their capacity to inhibit the amine uptake.

V. Peptidergic fraction interacting with 5-HT$_{1D}$ receptor

Antidepressants interact with 5-HT$_{1D}$ receptor through a mechanism which is likely distinct from a simple competitive effect. This is suggested by the fact that at low concentrations these drugs do not displace competitively the amine. Moreover, specific sites have been found for [^{3}H]minaprine (Fillion et al. 1988; Usuki-Ito et al., this symposium). These sites might be involved in the complex interaction of minaprine and possibly antidepressants with the heterologous presynaptic serotonergic receptor. On the basis of these observations, the hypothesis of the existence of a cerebral endogenous ligand acting at this site has been suggested. The purification of a ligand corresponding to that property has been undertaken. It includes various extraction procedures from cerebral tissues dissected out from horse brain followed by several separation steps involving HPLC. This procedure allowed us to obtain a fraction corresponding to likely peptidergic compounds having an apparent molecular weight of 3 to 4 kD as determined by gel filtration (TSK 40 S) and an amino acid content rich in ASP, GLU, GLY, THR, ALA. This fraction is not yet fully purified to homogeneity.

The biological activity of this fraction is interesting since it is able to interact with the binding of [^{3}H]5-HT to its 5-HT$_1$ receptors in a dose-dependent pattern. Moreover, it is able to antagonize the effect of TFMPP in inhibiting the [^{3}H]ACh release from synaptosomal preparation in guinea-pig and therefore its effect corresponds to an increase of the [^{3}H]ACh evoked release.

Assays are currently in progress to purify to homogeneity the compound(s) present in the fraction possessing the biological activity.

CONCLUSION

It is now well known from the initial work of Dahlström and Fuxe (1965) that serotonin cell bodies are for their vast majority located in the nucleus raphe particularly the raphe pallidus, obscurus and dorsalis and especially the raphe magnus. This very centralized location of the cell bodies suggests that information may be delivered at the same time in various brain areas. Moreover, many data (Steinbush and Nieuwenhuys, 1981; Ruda et al., 1982; Sano et al., 1982) have been reported showing that the unmyelinated serotonergic fibers are divided into two groups: fibers which are long and straight with few varicosities and have been considered as "tract fibers" (Sano et al., 1982) and fibers forming a very dense network, especially in globus pallidus and substantia nigra forming very numerous varicosities. These fibers were called "ring-like" fibers and considered by Pasik and Pasik (1982) to correspond to a neuromodulator system since most of these varicosities were not forming synapses. A similar hypothesis was previously proposed by Beaudet and Descarries (1976) showing that in the rat brain cortex 80 % of the varicosities were devoid of any synaptic profiles. Therefore, the structural organization of this part of the serotonergic cerebral system appears to favor a neuromodulating function in addition to a more "classical" neurotransmitter function.

It was shown previously that 5-HT was able to modulate the evoked release of acetylcholine (Vizi 1981; Maura and Raiteri 1984,1986; Gillet et al., 1985) and we have shown that 5-HT$_{1B}$ in rat and 5-HT$_{1D}$ in guinea-pig and presumably in monkey and man, are responsible for this effect. The "neuroendocrine-like" modulating function of the serotonergic system proposed by Pasik and Pasik (1982) involves non-synapsing serotonergic varicosities releasing 5-HT. Since the concentration of 5-HT diffusing from the varicosities in the three directions of space will rapidly and markedly decrease, the micromolar affinity of the synaptic receptors generally observed in "classical" neurotransmission would not allow a satisfactory function; on the contrary, 5-HT$_1$ subtypes (5-HT$_{1B}$ and 5-HT$_{1D}$) which have nanomolar affinities for 5-HT are well-fitted with the neuroendocrine mechanism.

The fact that a particular subtype of 5-HT receptor is involved in the neuromodulation, suggests that specific pharmacological agents will selectively act at this level. This specificity allows to reduce the interaction of these agents at other sites and is beneficial for therapeutic purposes. This can be illustrated by an exemple. The degeneration of the cholinergic system is likely involved, among others, in the pathology of Alzheimer's disease, therefore, presumably, it would be of therapeutic value to facilitate the cholinergic function. The specific blockade of 5-HT$_{1D}$ receptors inhibiting the ACh release would likely increase the cholinergic function and thus restore partially cognitive functions without having the numerous secondary effects of therapeutic treatments which are based on the use of cholinergic agonists. It is also particularly interesting to observe that an endogenous fraction extracted from brain cerebral tissue acts at the 5-HT$_{1D}$ receptor. Although the results are not yet complete enough, they suggest that an additional step of regulation exists and may correspond to an other level of specificity.

The heterologous presynaptic receptor for 5-HT is also interesting in the domain of mood and depression. It has been shown that 5-HT could presumably play an important role in depression (Coppen et Swade 1988), therefore, it is not surprising that various 5-HT receptors are involved in the pathology of depression. Accordingly, 5-HT$_{1A}$ and 5-HT$_2$ have been proposed for this role respectively (see rev. by Grahame-Smith, 1988). However, the fact that antidepressants interact with 5-HT$_{1D}$ suggest that the latter receptors may also be involved. Their implication as modulators of other neurotransmitter functions represent an important property which might be crucial in the mechanisms inducing the pathology or in those corresponding to the therapeutic activity of antidepressants. However, since depression is a complex pathology, other aspects often observed together with depression have to be considered i.e. impulsivity, inhibition or anxiety. The serotonergic system, through its neuromodulating capacity may play an important role in these psychiatric disorders. More work is needed to shed light on these aspects.

REFERENCES

Beaudet, A., and Descarries, L., 1976, Quantitative data on serotonin nerve terminals in adult rat neocortex, Brain Res., 111:301-309.

Bradley, P.B., Engel, G., Feniuk, W., Fozard, J.R., Humphrey, P.P.A., Middlemiss, D.N., Mylecharane, E.J., Richardson, B.P., and Saxena, P.R., 1986, Proposals for the classification and nomenclature of functional receptors for 5-hydroxytryptamine, Neuropharmacol., 25:563-576.

Coppen, A., and Swade, C., 1987, 5-HT and depression: the present position, in: "New concept in depression," M. Briley & G. Fillion, eds, Vol. 2, Pierre Fabre Monograph series, Publ., pp 120-136.

Dahlström, A., and Fuxe, K., 1965, Evidence for the existence of monoamine-containing neurons in the central nervous system. I. Demonstration of monoamines in cell bodies of brain neurons, Acta Physiol. Scand., 62(suppl. 232):1-55.

Dumuis, A., Sebben, M., and Bockaert, J., 1988a, Pharmacology of 5-hydroxytryptamine$_{1A}$ receptors which inhibit cAMP production in hippocampal and cortical neurons in primary culture, Mol. Pharmacol., 33:178-186.

Fayolle, C., Fillion, M.-P., Barone, P., Oudar, P., and Rousselle J.-C., Fillion, G., 1988, 5-hydroxytryptamine stimulates two distinct adenylate cyclase activities in rat brain, the high affinity activation is related to a 5-HT$_1$ subtype different from 5-HT$_{1A}$, 5-HT$_{1B}$, and 5-HT$_{1C}$, Fundam. Clin. Pharmacol., 2:195-214.

Fillion, G., Dufois, S., Fayolle, C., Fillion, M.-P., Oudar, P., Barone, P., 1988, Molecular interactions of antidepressants with the serotonergic receptor 5-HT$_{1D}$ coupled to a high-affinity adenylate cyclase activity: importance of the site labelled by ^{3}H-minaprine, in: "New concept in depression," M. Briley & G. Fillion, eds, Vol. 2, Pierre Fabre Monograph series, Publ., 42-59.

Gillet, G., Ammor, S., and Fillion, G., 1985, Serotonin inhibits acetylcholine release from rat striatum slices: evidence for a presynaptic receptor mediated effect. J. Neurochem., 45:1687-1691.

Grahame-Smith, D.G., 1988, Neuropharmacological adaptive effects in the actions of antidepressant drugs, ECT and lithium, in: "New concept in depression," M. Briley & G. Fillion, eds, Vol. 2, Pierre Fabre Monograph series, Publ., pp. 1-14.

Hoyer, D., Waeber, C., and Palacios, J.M., 1987, Identification of a 5-HT$_1$ recognition site in pig and human brain different from 5-HT$_{1A}$, 5-HT$_{1B}$, and 5-HT$_{1C}$, Br. J. Pharmacol., 92:561P.

Hoyer, D., and Schoeffter, P., 1988, 5-HT$_{1D}$ receptor-mediated inhibition of forskoline-stimulated adenylate cyclase activity in calf substantia nigra, Eur. J. Pharmacol., 147:145-147.

Leonhardt, S., Herrick-Davis, K., and Titeler, M., 1989, Detection of a novel serotonin receptor subtype (5-HT$_{1E}$) in human brain: interaction with a GTP-binding protein, J. Neurochem., 53:465-471.

Maura, G., and Raiteri, M., 1984, Functionnal evidence that chronic drugs induce adaptative changes of central autoreceptors regulating serotonin release, Eur. J. Pharmacol., 97:309-313.

Maura, G., and Raiteri, M., 1986, Cholinergic terminals in rat hippocampus possess 5-HT$_{1B}$ receptors mediating inhibition of acetylcholine release, Eur. J. Pharmacol., 129:333-337.

Minnema, D.J., Greenland, R.D., and Michaelson, I.A., 1986, Effect of in vitro inorganic lead on dopamine release from superfused rat striatal synaptosomes, Tox. Appl. Pharmacol., 84:400-411.

Pasik, T., and Pasik, P., 1982, Serotonergic afferents in the monkey neostriatum, Acta Biol. Acad. Sci. Hung., 33:277-288.

Peroutka, S.J., 1988, Functional correlates of central 5-HT binding sites, in: "Neuronal serotonin," N.N. Osborne & M. Hamon, ed., J. Wiley & Sons Ltd, Publ, Chichester, pp. 423-448.

Richardson, B.P. and Buchheit, K.-H., 1988, The pharmacology, distribution and function of 5-HT$_3$ receptors, in: "Neuronal serotonin," N.N. Osborne & M. Hamon, ed., J. Wiley & Sons Ltd, Publ, Chichester, 465-506.

Richardson, B.P., and Engel, G., 1986, The pharmacology and function of 5-HT$_3$ receptors, Trends Neurosci., 9:424-428.

Ruda, M.A., Coffieds, J., and Steinbush, H.W.M., 1982, Immunocytochemical analysis of serotonergic axons in laminae I and II of the lumbar spinal cord of the cat, J. Neurosci., 2:1660-1671.

Sano, Y., Takeuchi, Y., Kimura, H., Goto, M., Kawata, M., Kojima, T., Matsuura, T., Ueda, S., and Yamada, Y., 1982, Immunohistochemical studies on the processes of serotonin neurons and their ramification in the central nervous system, with regard to the possibility of the existence of Golgi's rete nervosa diffusa, Histol. Jap., 45:305-316.

Steinbush, H.W.M., and Nieuwenhuys, R., 1981, Localization of serotonin-like immunoreactivity in the central nervous system and pituitary of the rat, with special reference to the innervation of the hypothalamus, in: "Serotonin," Current aspects of Neurochemistry and functions. Adv. Exp. Med. Biol. Vol. 133, B. Haber, S. Gabay, M.R. Issidorides and S.G.A. Alivisatos eds, pp. 7-36, Plenum, New York.

Usuki-Ito, C., Muramatsu, M., Okuyama, S., and Otomo, S., 1990, Characteristics and distribution of specific ^{3}H-minaprine binding to rat brain slices, This Symposium.

Vizi, E.S., Harsing, L.G., and Zsilla, G., 1981, Evidence of the modulatory role of serotonin in acetylcholine release from striatal interneurons, Brain res., 212:89-99.

SEROTONIN RECEPTOR HETEROGENEITY AND THE ROLE OF POTASSIUM CHANNELS IN

NEURONAL EXCITABILITY

J.S. Kelly, P. Larkman, N.J. Penington, D.G. Rainnie, H.
McAllister-Williams and J. Hodgkiss

Department of Pharmacology, Edinburgh University, 1 George
Square, Edinburgh, EH8 9JZ

SUMMARY

Intracellular recordings in vitro from a variety of central neuronal
types have shown both inhibition and excitation to be modulatory
consequences of serotonin (5-HT) receptor activation. These responses
can be seen in isolation or in some cases (e.g. hippocampal pyramidal
cells) as a complex biphasic combination of hyperpolarisation followed by
depolarisation, suggesting overall control of neuronal excitability may
be dependent on the interaction between activation of more than one postsynaptic receptor and/or mechanism. Our studies have confirmed the 5-HT
evoked depolarisation of rat facial motorneurones (FM's) and the
hyperpolarisation seen in presumed serotonergic neurones of the dorsal
raphe nucleus (DRN) to be the result of opposite effects on K^+ ion
permeability. Suppression of a resting K^+ conductance leads to
depolarisation while activation leads to hyperpolarisation. The same
mechanisms appear to be responsible for the 5-HT evoked responses in
hippocampal pyramidal cells but in addition there is also a suppression
of a Ca^{++} dependent K^+ conductance responsible for the long spike after
hyperpolarisation (AHP). Data from the hippocampus and DRN indicate the
5-HT induced hyperpolarisation to be sensitive to Pertussis Toxin (PTX)
and irreversibly mimicked by GTPγS, a non-hydrolysable analogue of GTP,
suggesting the involvement of a G protein in K^+ channel activation. The
mechanism of K^+ channel closure is less clear as it is unaffected by PTX
or activation of adenylate cyclase, however there is indirect evidence
that the phosphoinositide pathway may be involved from the cloned $5-HT_{1C}$
receptor which also closes a K^+ channel in cell lines. The results show
that hyperpolarisation evoked by 5-HT in the hippocampus and DRN to be
mimicked and blocked by $5-HT_{1A}$ agonists and antagonists. However, the
depolarisations in the hippocampus and FM's are mediated by site-
dependent receptors with profiles which do not fit into the current 5-HT
receptor subtype classification.

INTRODUCTION

The heterogeneity of 5-HT actions is best illustrated by the several
reports that emphasise the biphasic actions of 5-HT on the excitability
of hippocampal neurones (Andrade and Nicoll, 1987b; Colino and Halliwell,
1987; Baskys et al, 1989). Inhibition, originally described as a

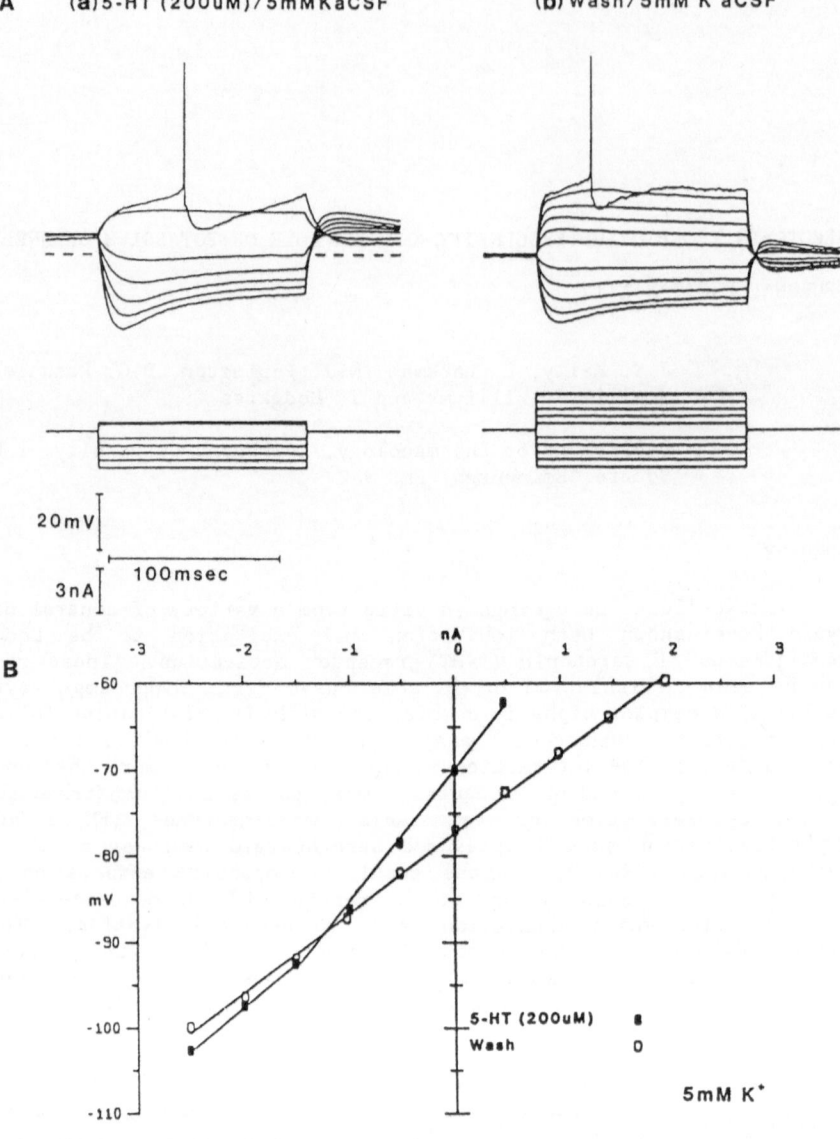

A (a) 5-HT (200uM)/5mMK⁺aCSF (b) Wash/5mM K⁺aCSF

20mV
3nA
100msec

nA

B

5-HT (200uM) ●
Wash ○

5mM K⁺

Figure 1

The action of 5-HT (200 μM) in aCSF containing 5 mM K⁺
A) Electrotonic potentials (upper traces) evoked by current
pulses of varying amplitude (lower traces) taken during (a) and
after (b) 5-HT superfusion. 5-HT evoked a 6 mV depolarisation
from a resting potential of -76 mV. Control peak R_m was 8.2 MΩ
and increased to 15.3 MΩ in the presence of 5-HT. Recovery to
-77 mV was associated with a fall in peak R_m to 9.3 MΩ. (Spike
amplitude is attenuated on the voltage records).
B) Peak deflection current-voltage plots were linear up to the
point of intersection giving a reversal potential of -89 mV for
the 5-HT evoked depolarisation. Note the rectification of the
5-HT I/V relation after the point of intersection.

suppression of spontaneous or evoked firing of action potentials (Segal, 1976), has been shown to be mediated by a mechanism common to 5-HT evoked inhibition in other cell types. Intracellular recordings from neurones of the (DRN) and the lateral septal nucleus (LSN) as well as the hippocampus have shown the primary effect of a 5-HT mediated inhibition to be a hyperpolarisation associated with a decrease in neuronal input resistance (R_m) and a reduction in the membrane time constant, τ (Rainnie, 1988; Joels et al, 1988; Segal, 1980). The reversal potential of this action was dependent on the external potassium ion concentration ($[K^+]_o$) and not affected by chloride loading indicating that activation of a potassium conductance mediated hyperpolarisation. Despite some differences, between cell types, in the intrinsic activity of specific agonists used to define the receptor subtype mediating this hyperpolarisation, pharmacological studies by and large implicate the involvement of a 5-HT_{1A} receptor (Andrade and Nicoll 1987a,b; Rainnie, 1988; Joels et al, 1987) (See Table 1).

The slow depolarisation seen to follow the hyperpolarisation or even to occur in isolation in some hippocampal pyramidal cells is associated with a small increase in input resistance. The difficulty in obtaining this response in isolation has led us to examine a similar depolarising action of 5-HT on FM's in the rat brainstem. In vivo, iontophoretic application of 5-HT and noradrenaline (NA) evoke a sub-threshold depolarisation which facilitates excitability (VanderMaelen and Aghajanian, 1980). Depolarisation of FM's by 5-HT and NA was examined in greater detail using an in vitro brainstem slice preparation.

RESULTS

Addition of 5-HT (200 μM) to the superfusing artificial cerebro-spinal fluid (aCSF) evoked a slow monophasic depolarisation which was sustained as long as 5-HT was present in the aCSF (Figure 1A). The membrane voltage records (Fig. 1A upper traces) obtained by injecting current pulses of varying amplitude (Fig. 1A lower traces) indicate that 5-HT evoked a depolarisation of 6 mV from a resting potential of -76 mV. This was accompanied by an increase in neuronal input resistance as indicated by the increased slope of the current voltage relationship in the presence of 5-HT (Fig. 1B). When measured at the peak of the voltage deflection, before the onset of time dependent inward rectification, the input resistance showed an 85% increase during the exposure to 5-HT from 8.2 to 15.3 MΩ. Concurrently, the time constant of membrane charging, τ, increased from 2.4 to 4 ms. Thus, increased excitability can be attributed to an interaction of all three factors, a depolarisation bringing the membrane potential closer to the action potential threshold, an increase in R_m which enhances the size of synaptic signals and a lengthening of τ increasing the probability of temporal summation.

The peak deflection current-voltage relationships shown in Fig. 1B are linear up to the point of intersection giving a reversal potential for the 5-HT evoked depolarization of -89 mV. The preservation of the depolarisation while using 3M KCl filled microelectrodes, which would be expected to reverse the chloride gradient across the cell membrane, suggests that the event is mediated by a change in permeability to potassium ions. The predicted value for the reversal potential of an event mediated by a potassium conductance according to the equation:

$$K^+ = 61.5 \log [K^+]_o/[K^+]_i \qquad\qquad \text{Equation 1}$$

when $[K^+]_o$ equals 5 mM and $[K^+]_i$ = 150 mM, is -91 mV, which is close to the experimentally determined value of $E_{5\text{-HT}}$ (Figure 1).

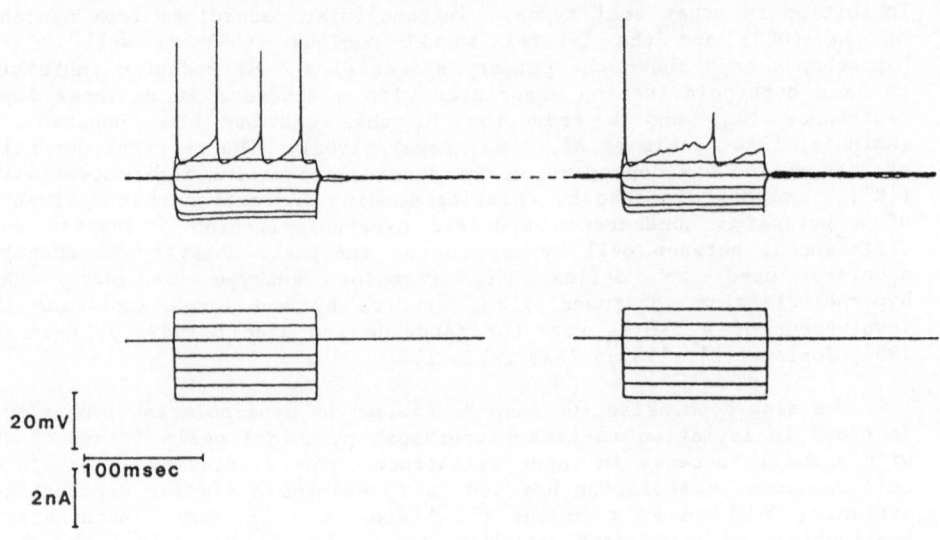

A (a) 5-HT(200uM)/15mM K⁺aCSF   (b) Wash/15mM K⁺aCSF

20mV
100msec
2nA

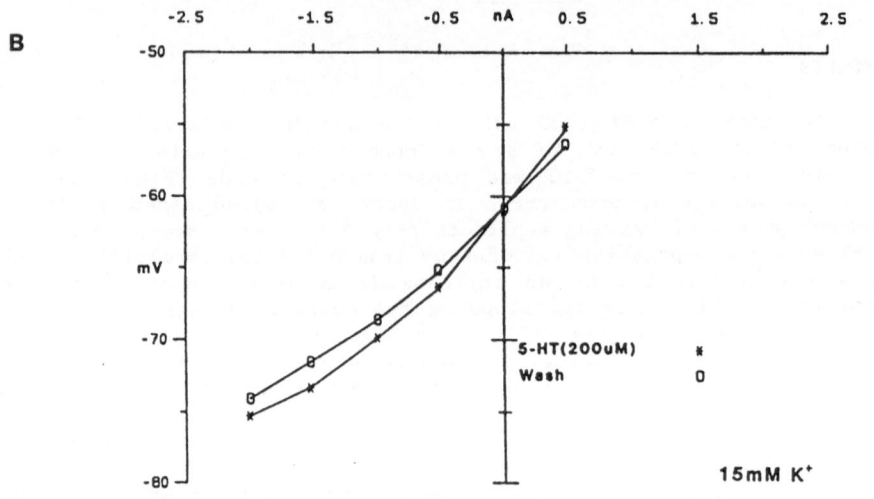

B

| -2.5 | -1.5 | -0.5 | nA | 0.5 | 1.5 | 2.5 |

-50
-60
mV
-70

5-HT(200uM) ✳
Wash □

15mM K⁺

-80

Figure 2

The action of 5-HT (200 μM) in aCSF containing 15 mM K⁺
A) The facial motoneurone had a resting potential of -61 mV
which was unchanged when 5-HT (200 μM) was added to the
superfusing aCSF. Increased excitability in response to a +1nA
current step indicated an underlying increase in R_m in the
presence of 5-HT. (Spike amplitude is attenuated on the voltage
records).
B) Current-voltage relationships were linear over only a small
range around -61 mV. R_m increased from 8.7 to 11 MΩ in the
presence of 5-HT. Thus in this case the reversal potential for
the 5-HT evoked depolarisation and the resting potential of -61
mV coincide.

The dependence of the 5-HT evoked depolarisation on $[K^+]_o$ was further investigated. Figure 2 shows a FM resting at -61 mV after changing to superfusion with 15 mM from 5 mM K^+-containing aCSF. Superfusion of 5-HT (200 μM) failed to evoke any change in membrane potential however a clear increase in input resistance was observed as indicated by the increased excitability evoked by depolarising current pulses (Fig 2A). The peak deflection current-voltage plot (Fig. 2B) was linear only in a narrow range around the resting potential but showed a 27% increase in R_m from 8.7 to 11 MΩ in the presence of 5-HT. The implication is that for this FM in 15 mM K^+ the reversal potential for the 5-HT evoked depolarisation is actually the same as the resting potential, which is in agreement with the predicted value from equation 1.

Figure 3 shows the effects of varying $[K^+]_o$ on the reversal potentials of both 5-HT and NA evoked depolarisations obtained from several current clamp experiments. The experimentally determined reversal potentials are plotted against values predicted from equation 1. In 10 and 15 mM $[K^+]_o$ the experimentally determined values for both 5-HT (closed squares) and NAd (open spaces) correlate closely with predicted values. When tested in 5 mM $[K^+]_o$ the reversal potentials for the 5-HT depolarisation vary widely around the predicted value while the values for NA evoked depolarization are more positive than both the values predicted by equation 1 and and the values determined for 5-HT. When tested on the same cells it was observed that for an equivalent sized depolarisation NA consistently evoked a larger increase in R_m, the resulting effect being that the reversal potential for the NA depolarisation was always more positive than that for 5-HT.

The inward currents evoked by 5-HT and NA at different membrane potentials from the same FM were studied under voltage-clamp conditions (Fig.4). The currents were obtained by subtraction of current-voltage curves obtained by evoking step voltage commands of varying amplitude from a holding potential of -67 mV, in the presence and absence of each transmitter. The currents represent instantaneous currents taken before the onset of time dependent inward rectification though the reversal levels determined from the steady-state current levels were identical (not shown). The reversal potentials for the two transmitters taken as the potential at which no net current flows is significantly different being -117 mV for 5-HT and -100 mV for NA. It is also apparent that under these conditions, the reversal potential for the 5-HT effect is considerably more negative than the predicted value. When voltage clamped or manually clamped E_{NA} is close to the value predicted by equation 1 while E_{5-HT} is more negative.

The reason for the difference in reversal potentials for the individual transmitters is unclear. However, the data shows that both events are dependent on the external potassium ion concentration and that the actions of NA can be attributed solely to this based on the close correlation between experimental and predicted reversal potentials. The reasons for the more negative reversal potentials of 5-HT evoked depolarisations can only be speculated on. A differential localisation of 5-HT and NA receptors on the somatic-dendritic tree, the former being more distally situated may lead to attenuation of 5-HT-evoked but not the NA-evoked responses when recording intra-somatically (Johnston and Brown, 1983). Little anatomical evidence exists for this proposal. A second alternative explanation could be that 5-HT evokes not only a conductance change for potassium but also for one or more other species of ion. A dendritic conductance increase to Na^+ and/or Ca^{++} in addition to a

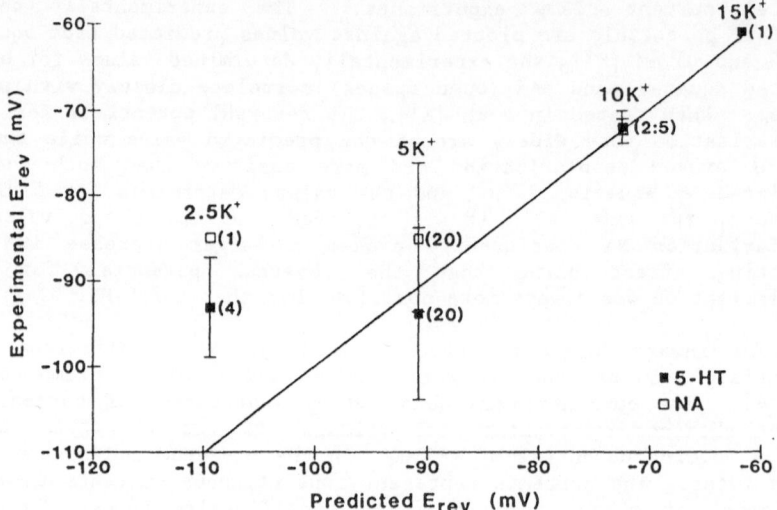

Effects of $[K^+]_o$ on 5-HT and NA Reversal Potentials

Figure 3

Experimentally determined changes in reversal potential for the actions of 5-HT (closed squares) and NA (open squares) on facial motoneurones plotted against values predicted from the Nernst equation (equation 1, see text) when superfusing with aCSF containing different concentrations of K^+ indicated above each group of points. The line represents the relationship predicted by equation 1. (Numbers in parentheses equals η for each point).

decrease in a K^+ conductance has been proposed to account for over-estimation of the reversal potential of carbachol induced depolarisation of hippocampal pyramidal cells (Benson et al, 1988).

In the hippocampus the 5-HT evoked depolarisation is blocked by extracellular Ba^{++} and intracellular Cs^+ and is maintained in the presence of a reversed Cl^- gradient thus inferring a decrease in a resting potassium conductance (Andrade and Nicoll, 1987; Colino and Halliwell, 1987). The 5-HT evoked inward current is comprised of two components one strongly voltage sensitive at potentials above -55 mV and the other voltage insensitive. The voltage sensitivity is claimed to be due to a suppression of the M-current though the exact contribution of this to the inward current is contested.

Recently a more detailed study of 5-HT evoked depolarisation of central neurones has been reported. Neurones of the nucleus accumbens in vitro are depolarised by 5-HT through a reduction of a Ba^{++} sensitive inwardly rectifying potassium conductance. This contributes to the resting potential throughout the physiological range and may in part be modulated by Ca^{++} (North and Uchimura, 1989). The E_{5-HT} was dependent on $[K^+]_o$ and the depolarisation could be seen to readily reverse polarity when the neurone was held at potentials below E_{K^+}.

5-HT has also exhibited two other distinct mechanisms by which neurones can be depolarised, both associated with an increase in conductance. Neurones of the rat nucleus prepositus hypoglossi (PrH) and the guinea-pig and cat medial and dorsal lateral geniculate nucleus (LGN d/m) are depolarised by a 5-HT evoked augmentation of the time and voltage dependent inward rectifier, I_h, which is active at the resting potential in these neurones. I_h is a mixed Na^+/K^+ conductance with a reversal potential positive to the normal resting potential of these neurones. The effects of 5-HT are slow on onset and do not desensitise (Bobker and Williams, 1989; Pape and McCormick, 1989). In contrast, iontophoretic or pressure ejected, but not superfused, 5-HT evokes a transient, rapidly desensitising depolarisation of guinea-pig sub-mucous plexus neurones (Surprenant and Crist, 1988; Derkach et al, 1989). The response is similar to that evoked by acetylcholine, acting through a nicotinic receptor, in that it is due to increased cation conductance (mainly Na^+) and can be blocked by d-tubocurarine, but can be distinguished by selective blockade with 5-HT$_3$ receptor antagonists.

While the conformity of the pharmacology of the 5-HT$_{1A}$ receptor mediating an increase in K^+ conductance is rarely challenged, the possibility that a single receptor type is linked to a depolarisation mediated by a reduction in K^+ conductance is less apparent. Our work on FM's has shown that the 5-HT$_{1A}$ ligands 8-OH-DPAT and dipropyl-5-CT neither mimic nor block 5-HT evoked depolarisation thereby differentiating the receptor mediating this response from the 5-HT$_{1A}$ receptor seen to mediate inhibition in other cell types. The 5-HT$_3$ receptor agonist, 2-CH$_3$-5HT and antagonist ICS 205 930 were likewise ineffective. Methysergide, an LSD analogue, shown to be a peripheral D receptor antagonist has been shown to be effective against both 5-HT$_1$ and 5-HT$_2$ receptors in the CNS. We found that superfusion of methysergide effectively antagonised 5-HT-evoked depolarisation of FM's without affecting the NA evoked response. Additionally LY-53857 a methysergide analogue with greater 5-HT$_2$/5-HT$_{1C}$ receptor selectivity was also an effective antagonist. Figure 5 illustrates the reduction in the 5-HT (150 μM) evoked depolarisation and increase in R_m when tested in the presence of LY-53857 (50 μM). An attempt to further isolate the mediating receptor using the 5-HT$_2$ receptor antagonists ketanserin and spiperone was less successful. Spiperone at high micromolar

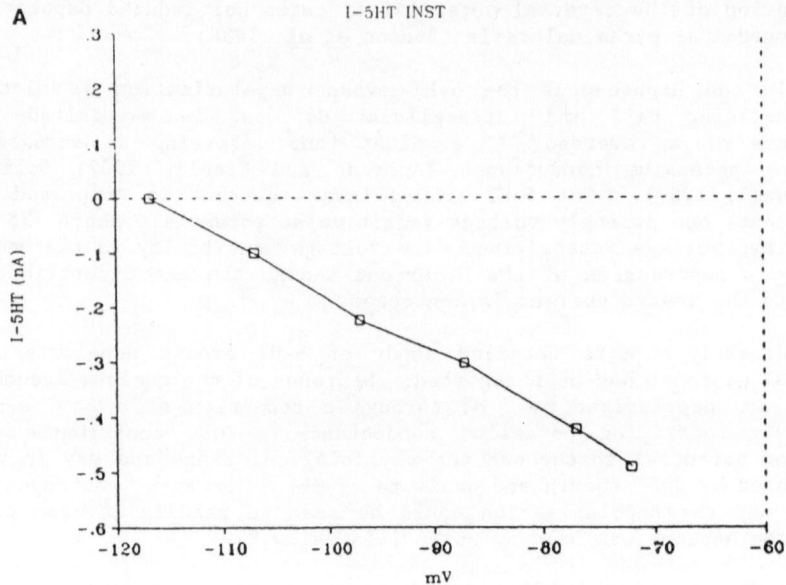

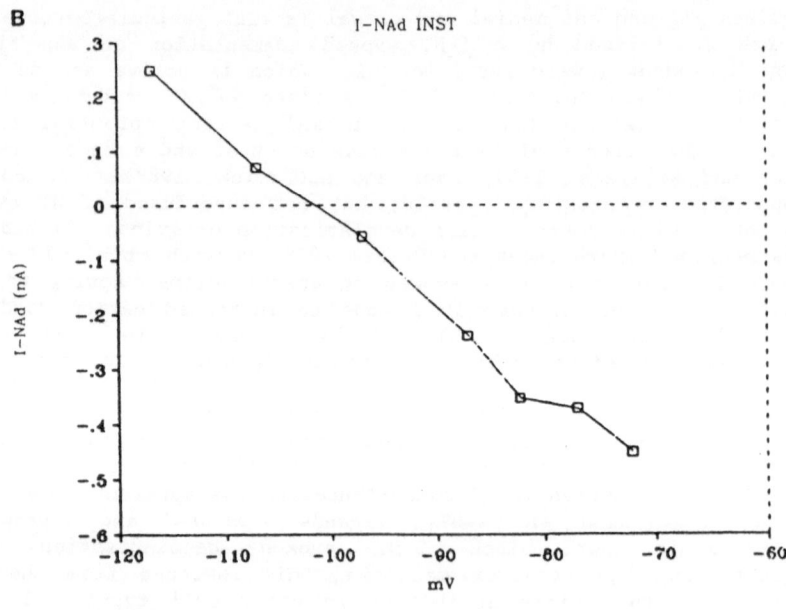

Figure 4

Instantaneous 5-HT (A) and NA (B) induced inward currents
recorded under voltage clamp at different holding potentials in
the same facial motoneurone. The neurone was stepped from the
resting potential (-67 mV) to different holding potentials for
250 ms in the absence and presence of 5-HT (200 μM) or NA (50
μM). I_{5-HT} and I_{NA} were obtained by subtraction of step
currents evoked under control conditions from those obtained at
the plateau of the transmitter evoked response and measured
before the onset of the time dependent inward rectifier.

concentrations superfused for extended periods failed to antagonise 5-HT evoked depolarisation while ketanserin under the same conditions reduced but was unable to fully block 5-HT responses in the manner achieved using methysergide and LY-53857. This data is summarised in Table 1 where the pharmacology of other 5-HT evoked effects are also detailed.

As with 5-HT evoked depolarisation of FM's, spiperone was an ineffective antagonist of the hippocampal depolarisation despite blocking the hyperpolarisation. Unlike FM's, methysergide and ketanserin were both unable to block hippocampal depolarisation. Debate exists over the effectiveness of ICS 205 930 as an antagonist (Andrade and Nicoll, 1987b; Colino and Halliwell, 1987). The 5-HT$_{1B}$ antagonist RU 24969 has been claimed to be an antagonist while a separate study saw no effects of the agonist TFMPP. In the nucleus accumbens a more definitive pharmacology is claimed in which antagonism occurs with ketanserin, mianserin and low concentrations of spiperone, while 5-HT$_1$ receptor ligands are all inactive. Thus in the nucleus accumbens a 5-HT$_2$ receptor mediates 5-HT closure of potassium channels, in the FMN the distinction is less clear and a 5-HT$_{1C}$ or 5-HT$_2$ receptor may be responsible while the identity of the hippocampal receptor remains obscure. 5-HT mediated augmentation of I$_h$ in PrH neurones could be antagonised with high concentrations of spiperone but not ketanserin, mianserin or ICS 205930 while 5-CT was an agonist. In the LGNd methysergide was an antagonist while 8-OH-DPAT, TFMPP, ipsapirone and 2-CH$_3$-5HT were all without effect. Both sets of data suggesting an as yet undefined 5-HT$_1$-like receptor.

DISCUSSION

Although to date no attempt has been made to identify the signal transduction mechanisms involved in 5-HT evoked depolarisation of FM's it is worth commenting that this is as yet a poorly investigated aspect of this action of 5-HT. While a pertussis-toxin (PTX) sensitive G protein links 5-HT$_{1A}$ receptors, probably directly, to potassium channels mediating hyperpolarisation (Andrade et al, 1986; Innis et al, 1988; Williams et al, 1988) the second messenger pathway involved in K$^+$ channel closure is purely speculative. Clearly depolarisation in the hippocampus is not via a PTX sensitive G protein though one insensitive to PTX has not been ruled out. Substance P (SP) induced closure of a TEA sensitive inwardly rectifying K$^+$ channel in cultured nucleus basalis neurones is insensitive to PTX but on the addition of GTPγS, a non-hydrolysable activator of G proteins, to the patch pipette the SP effect becomes irreversible (Nakajima et al, 1988). While intracellular GTPγS irreversibly activates the 5-HT sensitive K$^+$ conductance in hippocampal neurones, the actions of this ligand and the GTP antagonist GTPβS on PTX treated cells have not been examined.

Other evidence suggests 5-HT$_2$ and 5-HT$_{1C}$ receptors may be G protein linked. Molecular biological studies have shown these receptors to possess the same properties as other G protein linked receptors. They are single sub-unit proteins of about 450 amino acids with seven hydrophobic transmembrane domains and a large intracellular loop between domain V and VI which has been proposed to be the site of G protein interaction (Hartig, 1989). Binding studies have shown that [^{3}H] DOB, a 5-HT$_2$ agonist, binding to cerebral cortex shows GTP sensitivity. This binding site may be either an agonist substate of the 5-HT$_2$ receptor or a separate sub-type (Wood et al 1989; Pierce and Peroutka 1989; Lyon et al, 1987).

Injection of mRNA coding for 5-HT$_2$ or 5-HT$_{1C}$ receptors into xenopus oocytes has been shown to lead to expression of these receptors on the

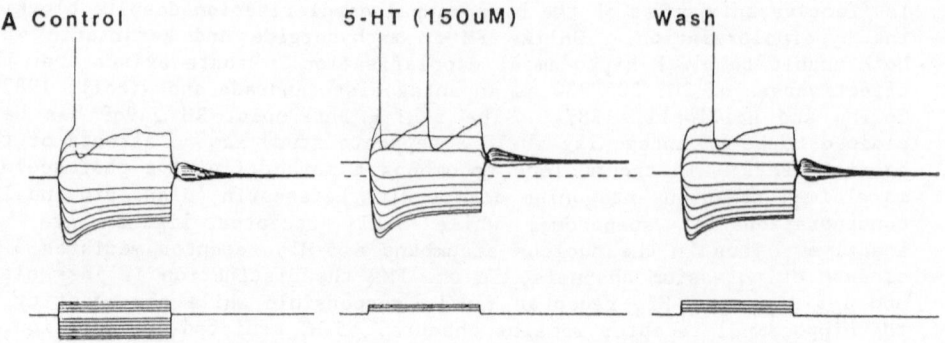

A Control   5-HT (150uM)   Wash

LY-53857 (50uM)   LY-53857/5-HT (150uM)

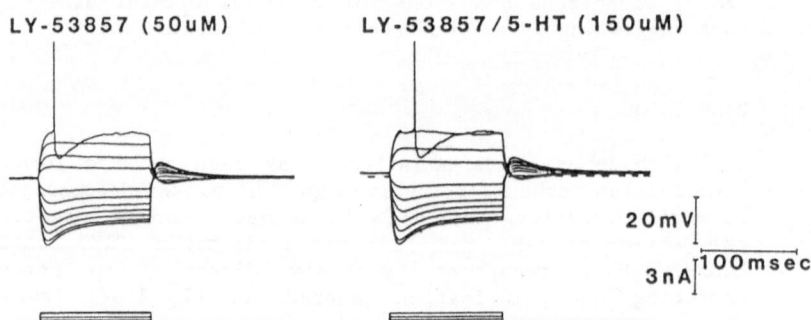

20mV
3nA⌐100msec

Figure 5

Antagonism of the 5-HT evoked depolarisation by LY-53857 (50 μM).

A) The FM was resting at -72 mV and peak R_m over its linear range was 16.3 MΩ. 5-HT (150 μM) evoked a reversible 5 mV depolarisation associated with an increase in peak R_m to 20 MΩ. LY-53857 (50 μM) was superfused for 20 minutes during which time the membrane potential drifted to -70 mV though peak R_m remained constant at 16.7 MΩ. Co-application of 5-HT (150 μM) with LY-53857 evoked a depolarization reduced in amplitude to 2mV and an associated increase in R_m to 17.8 MΩ. The control hyperpolarising current steps also apply to the other electrotonic records.

C) The peak deflection current-voltage plots in the presence and absence of 5-HT (150 μM) show LY-53857 (50 μM) to attenuate the response to 5-HT.

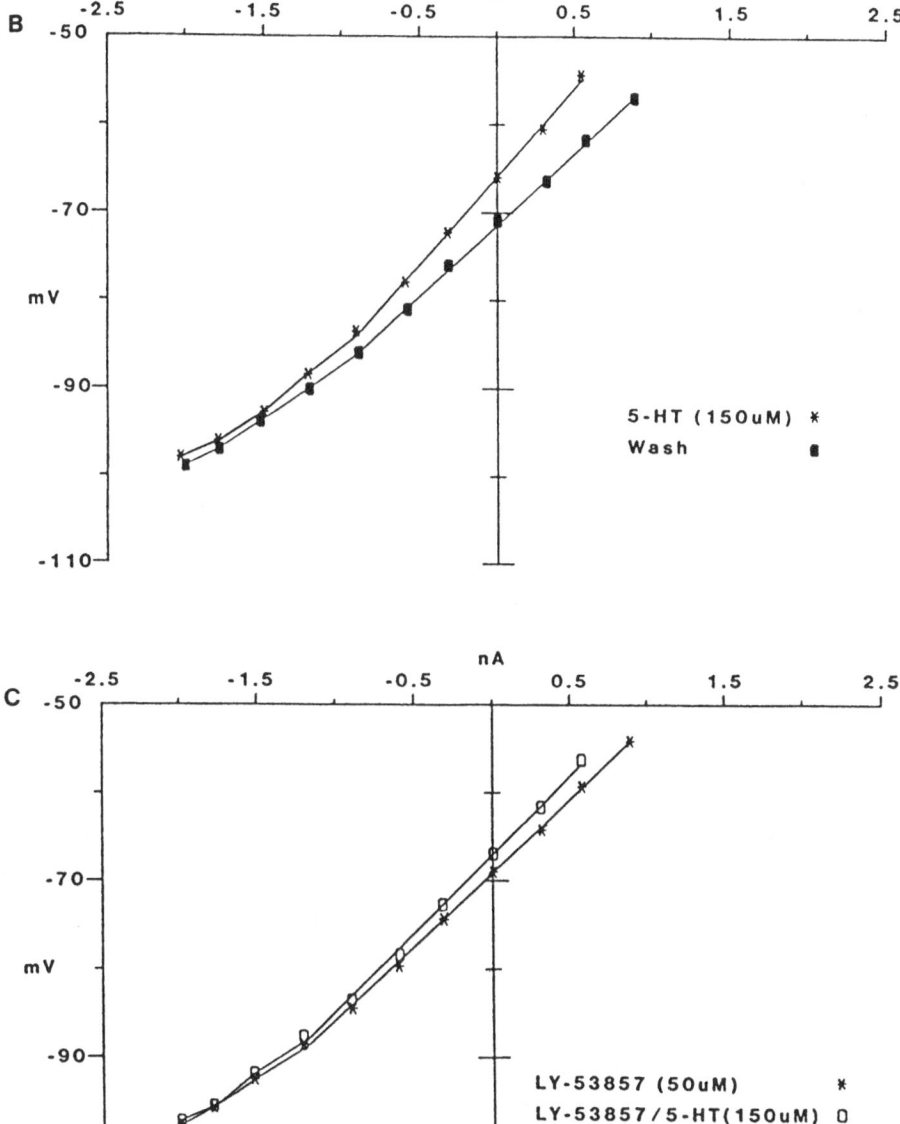

Table 1. Classification of 5-HT mediated depolarizations and hyperpolarizations by mechanism, brain region and receptor sub-type using specific ligands.

NE = No Effect
+ = Agonist
- = Antagonist
+/- = Partial Agonist

Mechanism Brain Region References Sub-type	LIGAND	DEPOLARIZATION Decrease in gK+ FMN [1]	N.Acc. [2]	Hippo [3]	Increase in I_h PrH/LGNd [4]	HYPERPOLARIZATION Increase in gK+ DRN [5]	LSN [6]	Hippo [3]
5-HT$_{1a}$	8-OH-DPAT	NE	NE	NE	NE	+	+	+/-
	Buspirone			NE	NE	+	+	+/-
	Ipsapirone			NE			+/-	+/-
	Gepirone			NE				+/-
	Dipr-5-CT	NE				+		
5-HT$_{1b}$	RU24969			-			NE	NE
	TFMPP			NE	NE		NE	NE
5-HT$_1$	5-CT		NE	NE	+	+		+
5-HT$_{1c}$	mCPP		NE	NE				
5-HT$_{1a/2}$	Spiperone	NE	-	NE	-	-		-
5-HT$_{1c/2}$	LY 53857	-		-				
5-HT$_{1/2}$	Methysergide	-		NE	-	NE		+?
	Methiothepin	NE		NE	NE			-
	Mianserin		-	NE	NE			-
	Cyproheptadine			NE				-
5-HT$_2$	Ketanserin	-?	-	NE	NE			NE
5-HT$_3$	ICS-205 930	NE	NE	NE/-				NE
	GR38032							
	2-CH3-5-HT	NE			NE			NE

[1] Kelly et al, 1988
Larkman and Kelly, 1988

[2] North and Uchimura, 1989

[3] Colino and Halliwell, 1987
Andrade and Nicoll, 1987a,b
Beck, 1988

[4] Bobker and Williams, 1989
Pape and McCormick, 1989

[5] Rainnie, 1988
Williams et al, 1988

[6] Joels et al, 1987

cell surface. Activation of these receptors leads to the opening of a Ca^{++} dependent Cl^- channel via a stimulation of membrane bound phosphatidylinosital-4,5,-biphosphate (PIP_2) hydrolysis releasing the second messengers IP_3 and diacylglycerol (DAG) (Lubbert et al 1987; Julius et al 1988; Pritchett et al 1988). The involvement of a G protein has been indicated by the partial effectiveness of PTX in blocking this effect as well as the mimicking effects of GTPγS (Dascal et al, 1986). Caution should be exercised when translating these actions in the oocyte to those in the neurone, however biochemical evidence does support the involvement of 5-HT_2 and 5-HT_{1C} receptors as modulators of the PIP_2 pathway in the central nervous system (Kendall and Nahorski 1985; Conn and Sanders-Bush, 1986).

In nucleus accumbens neurones superfusion of the protein kinase C activator phorbol-12,13-diacetate evokes a depolarisation similar to that evoked by 5-HT, the effects of which it can occlude (North and Uchimura, In press). Thus in this instance at least, there is some evidence that 5-HT_2 receptors may be linked not directly to a potassium channel by a G protein but indirectly through hydrolysis of PIP_2 and subsequent protein kinase C activation by DAG. In the hippocampus intracellular injection of IP_3 leads to a slow blockade of M-current and development of an inward current, however outside the activation range of I_m it has no effects (Dutar and Nicoll, 1988). Unlike, the nucleus accumbens neurones, superfusion of the protein kinase C activator phorbol-12,13-dibutyrate on hippocampal neurones did not evoke a depolarisation however it did completely suppress I_{AHP}, the current underlying the spike afterhyperpolarisation, an action also evoked by 5-HT and carbachol (Madison et al, 1987; Dutar and Nicoll, 1988; Colino and Halliwell, 1987).

The depolarising action of 5-HT mediated by augmenting I_h in PrH and LGNd neurones appears to be mediated through activation of adenylate cyclase as superfusion of forskolin and 8-bromo-cAMP both mimic the effects of 5-HT receptor activation. The involvement of a G protein in this pathway has not been investigated (Bobker and Williams, 1989; Pape and McCormick, 1989).

In conclusion it is clear that the different actions of 5-HT on central neurones reflect a heterogeneity of receptor subtypes as well as involving different signal transduction mechanisms leading to a final effect on membrane conductances.

ACKNOWLEDGEMENTS

We are grateful to the SERC, Glaxo and The Wellcome Trust for support. LY53857 was a gift from The Lilly Research Laboratories, Ketanserin and Spiperone from Janssen and Methiothepin from Glaxo.

REFERENCES

Andrade, R., Malenka, R. C., and Nicoll, R. A., 1986, A G protein couples serotonin and GABA$_B$ receptors to the same channels in hippocampus, <u>Science</u>, 234:1261-1264.
Andrade, R., and Nicoll, R. A., 1987a, Novel anxiolytics descriminate between postsynaptic serotonin receptors mediating different physiological responses on single neurones of the rat hippocampus, <u>Naunyn-Schmeideberg's Archives of Pharmacology</u>, 336:5-10.

Andrade, R., and Nicoll, R. A., 1987b, Pharmacologically distinct actions of serotonin on single pyramidal neurones of the rat hippocampus recorded in vitro, Journal of Physiology, 394:99-124.

Baskys, A., Niesen, C. E., Davies, M. F., and Carlen, P. L., 1989, Modulatory actions of serotonin on ionic conductances of hippocampal dentate granule cells, Neuroscience, 29:443-451.

Beck, S. G., 1989, 5-Carboxyamidotryptamine mimics only the 5-HT elicited hyperpolarisation of hippocampal pyramidal cells via 5-HT$_{1A}$ receptor, Neuroscience Letters, 99:101-106.

Benson, D. M., Blitzer, R. D., and Landau, E. M., 1988, An analysis of the depolarization produced in guinea-pig hippocampus by cholinergic receptor stimulation, Journal of Physiology, 404:479-496.

Bobker, D. H., and Williams, J. T., 1989, Serotonin augments the cationic current Ih in central neurones, Neuron 2:1535-1540.

Colino, A., and Halliwell, J. V., 1987, Differential modulation of three separate K-conductances in hippocampal CA1 neurones by serotonin, Nature, 328:73-77.

Conn, P. J., and Sanders-Bush, E., 1986, Agonist-induced phosphoinosotide hydrolysis in choroid plexus, Journal of Neurochemistry, 47:1754-1760.

Dascal, N., Ifune, C., Hopkins, R., Snutch, T. P., Lubbert, H., Davidson, N., Simon, M. I., and Lester, H. A., 1986, Involvement of a GTP-binding protein in mediation of serotonin and acetylcholine responses in Xenopus oocytes injected with rat brain messenger RNA, Molecular Brain Research, 1:201-209.

Derkach, V., Surprenant, A., and North, R. A., 1989, 5-HT3 receptors are membrane ion channels, Nature, 339:706-709.

Dutar, P., and Nicoll, R. A., 1988, Classification of muscarinic responses in hippocampus in terms of receptor subtypes and second messenger systems:Electrophysiological studies in vitro, Journal of Neuroscience, 8:4214-4224.

Hartig, P. R., 1989, Molecular biology of 5-HT receptors, Trends in Pharmacological Sciences, 10:64-69.

Innis, R. B., Nestler, E. J., and Aghajanian, G. K., 1988, Evidence for G protein mediation of serotonin- and GABA$_B$- induced hyperpolarisation of dorsal raphe neurones, Brain Research, 459:27-36.

Joels, M., and Gallagher, J.P., 1988, Actions of serotonin recorded intracellularly in rat dorsal lateral septal neurons, Synapse, 2:45-53.

Joels, M., Shinnick-Gallagher, P., and Gallagher, J.P., 1987, Effect of serotonin and serotonin-analogues on passive membrane properties of lateral septal neurons in vitro, Brain Research, 417:99-107.

Johnston, D., and Brown, T. H., 1983, Interpretation of voltage-clamp measurements in hippocampal neurons, J. Neurophysiol., 50:464-486.

Julius, D., MacDermott, A. B., Axel, R., and Jessell, T. M., 1988, Molecular characterization of a functional cDNA encoding the serotonin 1c receptor, Science, 241:558-564.

Kelly, J. S., Larkman, P. M., and Rainnie, D. G., 1988, Hyperpolarizing and depolarizing responses to 5-hydroxytryptamine in the rat brain stem in vitro, British Journal of Pharmacology, 95:500P (Abstract).

Kendall, D. A., and Nahorski, S. R., 1985, 5-Hydroxytryptamine-stimulated inositol phospholipid hydrolysis in rat cerebral cortex slices:pharmacological characterization and effects of antidepressants, Journal of Pharmacology and Experimental Therapeutics, 233:473-479.

Larkman, P. M., and Kelly, J. S., 1988, The effects of serotonin (5-HT) and antagonists on rat facial motoneurones in the in vitro brainstem slice, Journal of Neuroscience Methods, 24 No.2:199 (Abstract).

Lubbert, H., Snutch, T. P., Dascal, N., Lester, H. A., and Davidson, N., 1987, Rat brain 5-HT1C receptors are encoded by a 5-6 kbase mRNA size class and are functionally expressed in injected Xenopus oocytes, Journal of Neuroscience, 7:1159-1165.

Lyon, R. A., Davis, K. H., and Titeler, M., 1987, [3H]-DOB (4-bromo-2,5-dimethoxyphenylisopropylamine) labels a guanyl nucleotide sensitive state of cortical 5-HT$_2$ receptors, Molecular Pharmacology, 31:194-199.

Madison, D. V., Lancaster, B., and Nicoll, R. A., 1987, Voltage clamp analysis of cholinergic action in the hippocampus, Journal of Neuroscience, 7:733-741.

Nakajima, Y., Nakajima, S., and Inoue, M., 1988, Pertussis toxin-insensitive G protein mediates substance P-induced inhibition of potassium channels in brain neurones, Proceedings of the National Academy of Sciences.USA., 85:3643-3647.

North, R. A., and Uchimura, N., 1989, Muscarine acts at M1 receptors to reduce potassium conductance in rat nucleus accumbens neurones, Journal of Physiology, (In press).

North, R. A., and Uchimura, N., 1989, 5-Hydroxytryptamine acts at 5-HT$_2$ receptors to decrease potassium conductance in rat nucleus accumbens neurones, Journal of Physiology, 417:1-12.

Pape, H-C., and McCormick, D. A., 1989, Noradrenaline and serotonin selectively modulate thalamic burst firing by enhancing a hyperpolarization-activated cation current, Nature, 340:715-718.

Pierce, P. A., and Peroutka, S. J., 1989, Evidence for distinct 5-hydroxytryptamine 2 binding site subtypes in cortical membrane preparations, Journal of Neurochemistry, 52:656-658.

Pritchett, D. B., Bach, A. W. J., Wozny, M., Taleb, O., Dal Toso, R., Shih, J. C., and Seeburg, P. H., 1988, Structure and functional expression of cloned rat serotonin 5HT$_2$ receptor, The EMBO Journal, 7:4135-4140.

Rainnie, D.G. The biophysical and pharmacological properties of presumptive serotonergic neurones recorded intracellularly from the dorsal raphe nucleus in the in vitro slice preparation, University of Edinburgh.:PhD Thesis, 1988.

Segal, M., 1976, 5-HT antagonists in rat hippocampus, Brain Research, 103:161-166.

Segal, M., 1980, The action of serotonin in the rat hippocampal slice preparation, Journal of Physiology, 303:423-439.

Surprenant, A., and Crist, J., 1988, Electrophysiological characterisation of functionally distinct 5-hydroxytryptamine receptors on guinea-pig submucous plexus, Neuroscience, 24:283-295.

VanderMaelen, C. P., and Aghajanian, G. K., 1980, Intracellular studies showing modulation of facial motoneurone excitability by serotonin, Nature, 287:346-347.

Williams, J. T., Colmers, W. F., and Pan, Z. Z., 1988, Voltage- and Ligand-Activated Inwardly Rectifying Currents in Dorsal Raphe Neurons in vitro, Journal of Neuroscience, 8:3499-3506.

Wood, M. D., Alexander, B. S., and Reilly, Y., 1989, Modulation by GTP of the inhibition of [3H]-DOB binding, British Journal of Pharmacology, 97 Supp.:412P (Abstract).

MODULATORY ACTIONS OF NOREPINEPHRINE ON NEURAL CIRCUITS

D.J. Woodward, H.C. Moises*, B.D. Waterhouse**, H.H. Yeh***
and J.E. Cheun***

University of Texas Southwestern Medical Center at Dallas
*University of Michigan, **Hahnemann University
***University of Rochester

Studies on the actions of the NE fiber system in the CNS have
traditionally been conducted along two primary dimensions. At the systems
level a major issue has been to resolve the consequences of NE release on
circuit functions in noradren-ergically innervated areas. At the cellular
level, the major question is how noradrenergic receptor activation leads
to subsequent membrane actions and how such effects relate to events seen
at the synaptic level. Our experience from past investigations is that
there is a critical synergy between these two levels of analysis, and that
progress at one level may generate insights critical for pursuing issues
at the other.

Early investigations of the physiology of the noradrenergic receptor
actions were facilitated by the initial demonstration of the diffuse
distribution of NE-containing fibers originating primarily from the
brainstem nucleus, locus coeruleus (LC) (Ungerstedt, 1971). These studies
of the actions of NE for the most part defined NE as a putative
neurotransmitter based mainly in terms of its ability to excite or inhibit
cellular activity in different regions of the brain. NE applied by
microiontophoresis or released synaptically via stimulation of the LC most
often caused an inhibition of target cells in cerebral cortex (Stone,
1973; Olpe et al., 1980), cerebellum (Hoffer et al., 1973), hippocampus
(Segal and Bloom, 1973), thalamus (Phillis et al., 1967; Nakai and
Takaori, 1974), and hypothalamus (Miyahara and Oomura, 1982). Appropriate
agonists and antagonists provided further evidence for the involvement of
a specific noradrenergic receptor in mediating these
electrophysiologically observed actions. Such early evidence argued
primarily for an "inhibitory effect" of NE.

Considerable difficulties exist beyond this point in formulating an answer
to the question "What does NE do in the CNS?". Unlike the heart, where
a faster, more forceful beat is a result of noradrenergically-mediated
actions, there is no obvious corresponding function of the brain which is
regulated by NE. The "inhibition" attributed to NE could be a function
of the iontophoretic dose levels or stimulation parameters employed, or
a unique interaction with the source of the background spike generator of
the cells assayed. NE is known to activate a membrane-bound beta
adrenergic complex which elevates adenyl cyclase activity and thereby
raises cyclic AMP levels within the cell. This second messenger has long
been identified as having the potential to initiate many intracellular
events through subsequent phosphorylation reactions (Siggins et al.,
1971). The phenomenon of inhibition of cell firing may reliably indicate
a successful beta receptor activation, but may only hint at possible
physiological effects on target circuit activity and behavioral
consequences. Determination of precisely what parameter of neuronal
network function NE is acting upon remains a central problem.

Neuroreceptor Mechanisms in Brain, Edited by S. Kito *et al.*
Plenum Press, New York, 1991

NE as a Neuromodulator: The question of what is being regulated by
activation of noradrenergic receptors has motivated our investigation of
the possible modulatory actions of NE. Early study of membrane actions
of NE indicated that a conductance increase was not a direct consequence
of receptor activation (Hoffer, 1973) and it was not clear then what the
functional consequences of spike suppression were. A search was therefore
initiated for alternate mechanisms and additional actions of NE. One
initial striking effect found was the ability of NE to enhance the
inhibitory action of GABA. This effect appeared at dose levels where no
direct suppression of firing occurred, or as a lingering effect after NE
iontophoresis was terminated (Fig. 1). Modulation became defined
experimentally as the capacity of a substance to influence the
responsiveness of a neuron indirectly through effects on other synaptic
systems which exert direct excitatory or inhibitory actions.

To further develop this concept of neuromodulatory action (Woodward et
al., 1979), an extended series of studies on interactions between NE and
other modes of synaptic activation was conducted. These experiments
employed microionto-phoresis, extracellular recording and stimulation of
cerebellar Purkinje cells via total application of a variety of excitatory
and inhibitory putative transmitters and by stimulation of afferent
synaptic inputs (Moises et al., 1979, 1980, 1981, 1983; Waterhouse et
al., 1980a, b, 1981, 1982; Yeh et al., 1981; Yeh and Woodward, 1983 a, b).
In general, facilitating modulatory-type enhancement effects were found
on both inhibitory and excitatory synaptic systems. With respect to
inhibition, NE was found to enhance selectively GABA or muscimol induced
inhibition but not that induced by taurine or glycine. Furthermore,
iontophoretically applied NE was found to enhance inhibitory responses of
Purkinje cells to local stimulation of parallel fibers on the surface of
the cerebellar cortex. This observation is consistent with the
anticipated enhancement of physiologically released GABA from basket and
stellate interneurons onto Purkinje cells. With respect to excitation,
NE also enhanced the ability of climbing fiber synaptic inputs to generate
bursts of spikes and induced complex shifts in the shape of the
extracellularly recorded complex potentials generated by climbing fibers.
Responses to parallel fiber stimulation were also enhanced. Overall,
these influences on synaptically evoked responses were characterized as
an enhancement of excitatory input, although the interpretation was
somewhat obscured by the mixture of effects possibly secondary to the
depressions of background firing.

Stimulation of the locus coeruleus to evoke synaptic release of NE
provided a critical test of the modulation concept. Careful examination
of this paradigm revealed the need to position the stimulation electrode
within the confines of the nucleus locus coeruleus to avoid direct
stimulation of the efferents and afferents within the superior cerebellar
peduncle since a rapid stimulation of mossy fiber afferents could evoke
a long lasting rebound fatigue or depression of Purkinje cell activity.
The most useful experiments were conducted under conditions in which very
low levels of stimulation intensity were employed so that no transient
synaptic activation of Purkinje cells or other direct effects on
spontaneous firing occured (Moises and Woodward, 1980). These conditions,
were found to enhance GABA-induced inhibitions for a period 10 to 60
seconds after the LC stimulation (10/sec for 10 sec) had ceased. This
result paralleled the earlier description of long lasting inhibitions of
spontaneous firing (Hoffer et al., 1979; Siggins et al., 1971) induced by
similar stimulus trains. These long-lasting effects were attributed to
activation of a second messenger system which outlasted the duration of
the stimulation by tens of seconds. The activa-tion of beta-receptor
linked adenyl cyclase and the intracellular generation of cyclic AMP, has
been the presumed mechanism underlying these electrophysiological
phenomena.

Pairing of LC stimulation with synaptic activation of cerebellar cortical
circuits revealed a wide assortment of modulatory phenomena (Moises et
al., 1983). Stimulation of the LC by short trains of pulses (3-6)
revealed an enhancement of basket/stellate cell-mediated inhibition of
Purkinje cells starting at conditioning-test intervals of 80 msec.
Climbing fiber and parallel fiber excitatory responses were also enhanced.
Likewise, mossy fiber-granule cell-parallel fiber activation of molecular
layer interneurons (presumed stellate or basket cells) showed an absolute
enhancement of spiking action preceded by LC conditioning stimulation.

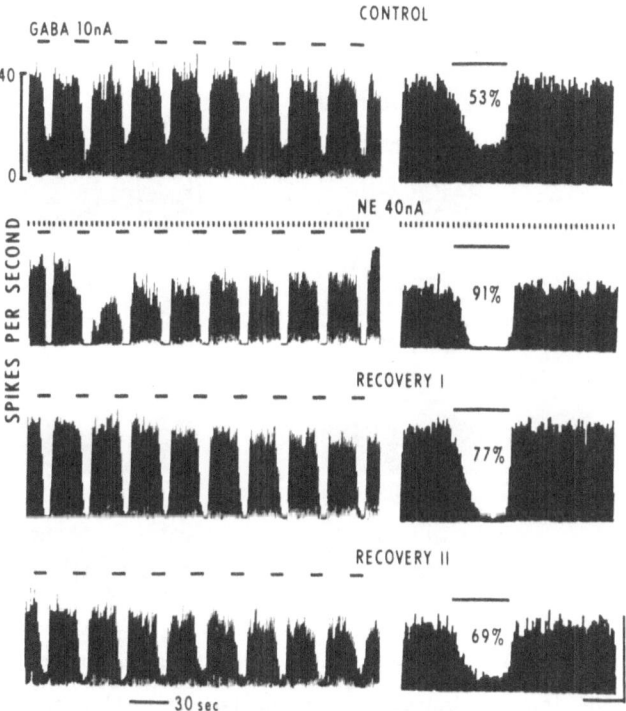

Figure 1. Effect of NE on inhibition of Purkinje cell spontaneous activity induced by iontophoretic pulses of GABA. Continuous rate meter records (left) and corresponding drug response histograms (right) show augmentation of GABA inhibition during NE iontophoresis. Drug currents in nano-amperes are shown immediately after the abbreviation for the drug. Inhibition during the GABA pulse was quantitated by comparing rate during application of GABA to firing rate between applications. Note than an enhancement of the GABA response was produced prior to the direct depressant effect of NE, and that GABA inhibition is still augmented in both rate meter and histogram records even after spontaneous discharge has returned to control levels following cessation of NE iontophoresis (R1). Horizontal lines above all records indicate the period and duration of drug ejection. Numbers below the bar on all histogram records indicate the percent inhibition of activity. Histogram calibrations: horizontal, 5 sec; vertical, 25 counts per address. From Woodward et al., 1979.

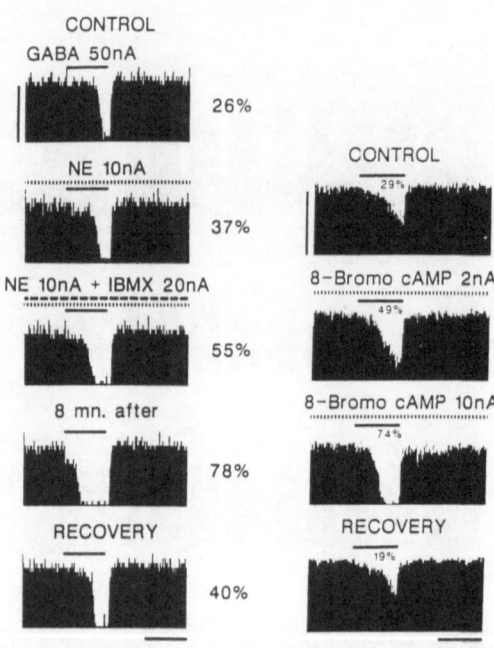

Figure 2. (Left) Phosphodiesterase inhibition enhances noradrenergic potentiation of GABA in vitro. Histogram records illustrate the effects of NE and NE + IBMX on GABA-induced inhibition of cerebellar cell discharge. Each histogram sums unit activity during 5 GABA applications. Initially, continuous microiontophoresis of NE 10 nA (broken bar) augmented the depressant responses to iontophoretic pulses of GABA 50 nA (solid bars) from 26 to 37% inhibition of spontaneous firing. Local application of the phosphodiesterase inhibitor IBMX 20 nA (dotted line) further increases the GABA response to 55% suppression of background discharge. Immediately following cessation of NE and IBMX administration, the response was further potentiated before finally returning toward the control level of response. Calibration: horizontal, 20 s (left), 10 s (right); vertical, 50 spikes/s.
(Right) Dose-response relationship of cAMP-induced augmentation of GABA inhibitory responses, in vitro. Histogram records computed from segments of continuous recordings show responses of an in vitro cerebellar neuron to pulsatile iontophoretic application of GABA (25 nA), before, during and after microiontophoretic administration of 8 BcAMP at 2 and 10 nA ejection currents. Note that the magnitude facilitating effect of BcAMP on GABA depressant responses is dependent on the ejection current of cAMP analog. Calibration: horizontal, 10 s; vertical, 80 spikes/s.

<u>Pharmacologic Studies of NE Modulation</u>: Two lines of evidence argued that the NE-induced enhancement of GABA inhibition is mediated primarily by a beta-one noradrenergic receptor subtype. X-irradiation could be applied to neonatal rat cerebellum to reduce the size of the cerebellum by 80% due to loss of granule cells, glial cells, vascular and other structures. The beta-two receptors were reduced in proportion to the tissue loss, but the absolute number of beta-one receptors, as well as the number of Purkinje cells, was unchanged (Minneman et al., 1981). This indicated that beta-one receptors were likely to reside on Purkinje cells. In studies of normal cerebellum, Yeh and Woodward (1983a) showed that the noradrenergic enhancement of GABA-induced P-cell inhibition in rat cerebellum could be evoked by the beta-one specific agonist practolol but was not blocked by the beta-two antagonist zinterol. Waterhouse et al. (1981, 1982) showed that both isoproterenol and NE induced augmentation of GABA responses in rat cerebellum and cerebral cortex. The alpha adrenergic agonist, phenylephrine, did not enhance GABA actions but did enhance excitatory responses of somato-sensory cortical units to tactile stimulation of the forepaw or iontophoretic application of Ach. Furthermore, it is worth noting that over the course of many different studies, NE, isoproterenol, NE, phenylephrine and dopamine were all observed to suppress spontaneous firing, yet enhancement of GABA appeared to be specific to beta-one receptor activitation.

Objections have often been raised that the observed modulatory actions might depend on complex, multiple interactions reflecting both pre- or postsynaptic within local circuits. For this reason, the modulatory effects were examined in a variety of preparations. Yeh and Woodward (1983b,c) demonstrated that NE caused modulatory enhancement of GABA applied iontophoretically or when released by synaptic activation via parallel fiber volleys in Purkinje cells during the first ten postnatal days in rat. At these ages, when inhibitory synaptic development is either absent or extremely immature, the firing rates are slower. The modulation also exists in the X-irradiated rat where synaptic densities and cell types are greatly reduced. These studies provide evidence that the effect is most likely postsynaptic. The halothane anesthetic is also not critical since modulation of GABA responses by NE has been observed in in vitro slice preparations of cerebellum (Sessler et al., 1989). Finally, preliminary results of West and Woodward (unpublished) in awake, locomoting rats show that iontophoretically applied NE causes a robust enhancement of GABA in the context of normal physiological activation of Purkinje neurons. These many permutations of related studies were necessary to fully explore conditions which might have been essential for the modulatory phenomenon.

<u>Studies of Cellular Mechanisms</u>: Further work has demonstrated the involvement of second messengers in inducing noradrenergic modulatory effects of the type described above. Tissue slices from rat cerebellum were studied in vitro with iontophoretic techniques similar to those employed in vivo (Sessler et al., 1989). Application of 8-bromo-3',5' cyclic AMP (BcAMP), a drug presumed to penetrate into neurons, was found to increase the inhibitory effectiveness of GABA (Fig. 2, from Sessler et al., 1989). This result provided evidence that the cyclic AMP elevation generated by elevated adenyl cyclase could be effective in inducing the membrane changes. The phosphodiesterase inhibitor, 3-isobutyl-1-methyl xanthine (IBMX) exhibited little or no effect when introduced alone but caused a substantial effect when administered in combination with NE. Forskolin, a direct activator of adenyl cyclase, also amplified GABA inhibitory effects. These results provide further evidence that noradrenergic enhancement of GABA inhibition may result from a sequence of events starting with beta-adrenergic receptor activation, subsequent adenyl cyclase stimulation and increased intracellular production of cAMP. The prolongation of the NE effect suggests that a behavioral reaction may be involved in the phenomenon of GABA receptor effectiveness.

One hypothesis is that a component of the GABA/benzodiazepine receptor may be phosphorylated by a receptor-associated protein kinase (Sweetnam et al., 1987) resulting in an enhancement of ion channel function. An alternative possibility is that an as yet unidentified phosphorylation substrate conditions the membrane to regulate GABA receptor efficacy.

Further advances can be anticipated from whole cell patch clamp studies. Recent studies (Cheun and Yeh, unpublished) have shown that single Purkinje cells can be successfully isolated from 7-14 day old rat cerebellum (Fig. 3A) and whole cell clamps successfully achieved. Robust sodium and potassium currents corresponding to action potential transients are readily detected (Fig. 3B). Initial results of some whole cell clamp studies shown in Fig. 4 revealed that transient whole cell and chloride currents could be evoked by pressure ejection of GABA from pipets placed near the cell. Application of NE through an adjacent pipet revealed an enhancement of the whole cell and GABA receptor associated chloride currents. This clearly was the lowest threshold interaction since NE at these levels by itself evoked no detectable current changes.

These isolated single cell voltage clamp findings deal with outstanding questions as to whether some of the findings based on extracellular unit recording and local iontophoresis may result from complex interactions with presynaptic afferent systems or local interneurons. Not all reduced preparations readily reveal these interactions. For example, (Cheun and Yeh, unpublished), Purkinje cells grown in tissue culture appear to lose this modulatory property (Cheun and Yeh, unpublished). Likewise, loss of GABA modulation has been found after chronic treatment with flurezapam and with desmethylimipramine in the adult rat (Yeh and Woodward, 1983d). Moreover, the effect appears not to be evident in rat hippocampal slice where a suppression of a calcium activated potassium current appears to be the most significant observable effect associated with beta adrenergic receptor activation (Madison and Nicoll, 1986). Overall, we can expect whole cell patch techniques to play a major role in future studies aimed at identifying the molecular membrane conditions responsible for maintaining this phenomenon of GABA facilitation and for study of other phenomena which lead to modulatory enhancement of neuronal responsiveness by NE.

Circuit Gating Actions of NE - Studies in Cerebellum: A primary limitation in the study of cellular electrophysiological actions is that the phenomena observed in reduced preparations may not be interpretable in terms of neural circuit and behavioral functions. This consideration has motivated studies of the interaction of iontophoretically applied NE with neuronal responses evoked by presentation of relevant physiological signals. For these studies in the rat, we employed a variety of visual stimulation paradigms to evoke complex responses in visually responsive cerebrocortical and cerebellar neurons.

Drug interactions with physiological signals need to be viewed quite differently from previous types of studies. Most tests employing electrical stimulation pulses necessarily utilize synchronously activated synaptic input, whereas physiological signals usually involve temporally dispersed signals. Neural structures such as the cerebral cortex and cerebellum, possess internal feed forward and feedback excitatory and inhibitory pathways which provide numerous targets for modulatory interaction. For example, when transient signals arrive via mossy fibers to the cerebellum, the Purkinje cells are initially excited, then inhibited by basket and stellate cells. Gradual tonic signals evoked overlapping signals so that the Purkinje cells may be excited, do nothing, or be inhibited, depending on the strength of connections. Also, if Golgi cells evoke strong feedback inhibition on granule cells, a disfacilitation of both Purkinje cells and interneurons may result. The time course of synaptic delay and durations of synapses in relation to the time course of the input are not known in cerebellum. As a consequence, it is quite impossible to predict the resulting action of a modulator such as NE solely on the basis of the known circuitry and sign of synaptic action, since so much depends on the steady state strength and phasic characteristics of the interacting synaptic systems.

A long term aim of this laboratory has been to develop a clear understanding of the normal anatomy and physiology of the cerebellar system upon which to base physiological studies of the noradrenergic modulatory actions. A series of anatomical and physiological studies (Burne et al., 1978; Burne and Woodward, 1983) led to the discovery that the paraflocculus was the primary receiving area for cerebro-ponto-cerebellar pathways from secondary visual and auditory cortical areas. Combined retrograde and anterograde transport studies showed overlapping terminations in the pairs of fibers from cortex onto pontine neurons which project to the parafloccular lobule. Electrical stimulation of cortex in areas surrounding primary visual regions revealed that most (90%) of Purkinje neurons in paraflocculus could be driven by pathways originating from the predicted cortical regions.

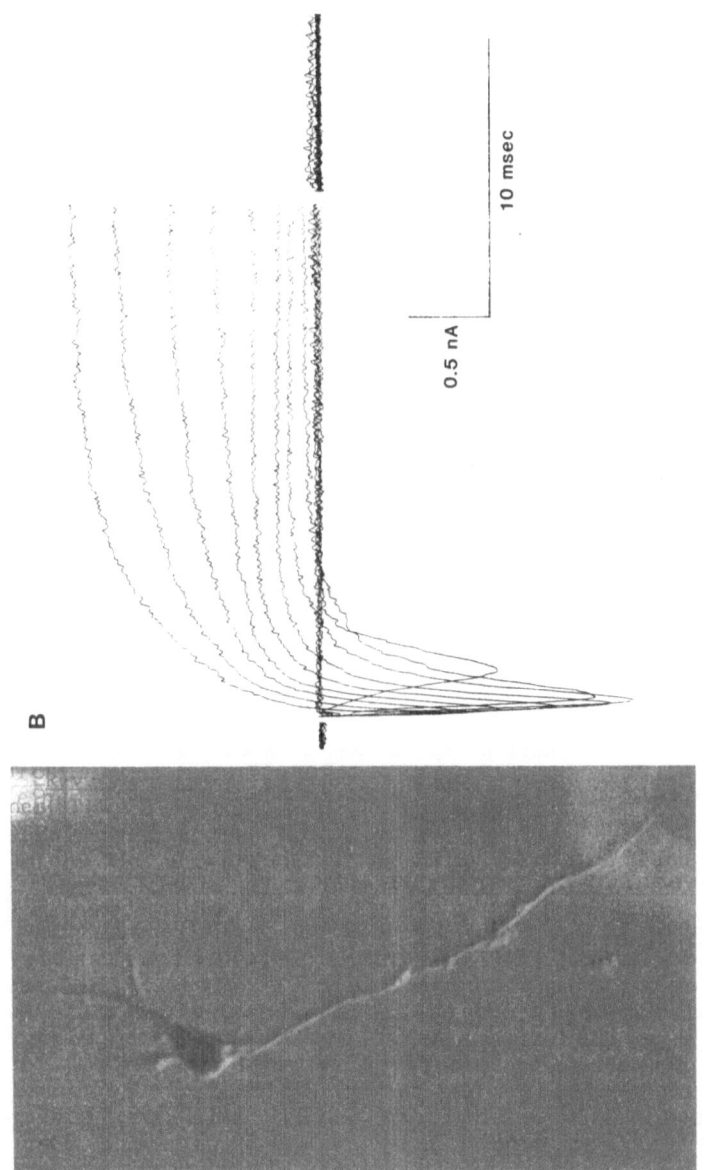

Figure 3 A. Photomicrograph of a Purkinje cell dissociated from postnatal day-7 rat cerebellum seen under Hoffman Modulation Optics. Immunoreactivity to PEP-19, a marker for Purkinje cells in the rat cerebellum, aided in its identification following dissociation. **B:** Whole-cell current profile of an identified Purkinje cell. Bath solution (mM): 140 NaCl, 2 CaCl$_2$, 1 MgCl$_2$, 10 HEPES, 10 glucose. Pipet solution (mM): 120 KCl, 1 CaCl$_2$, 2 MgCl$_2$, 5 EGTA, 10 HEPES, 3 ATP, 0.1 cAMP, 0.1 leupeptin. Capacitative transients were blanked.

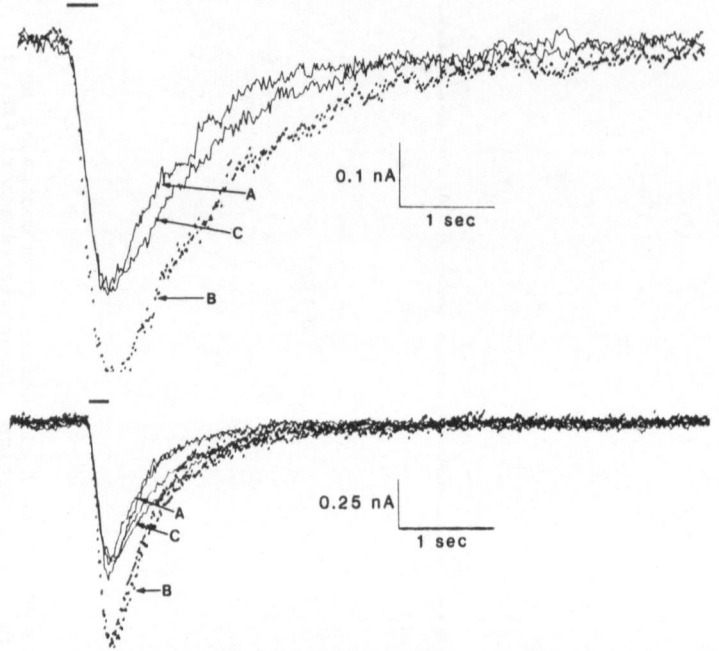

Figure 4 Top: Augmentation of GABA-activated whole-cell current by norepinephrine. In this example, pressure pulses (1 psi, 150 ms) of GABA (10 μM) were applied every 30 sec onto a freshly-dissociated Purkinje cell. The computer-generated current response records were superimposed and GABA response amplitudes were compared before (**A**), during (**B**; dotted trace) and after (**C**) application of norepinephrine (500 μM). GABA responses recovered to control levels approximately 2 min after cessation of norepinephrine aplication. Norepinephrine, by itself, did not induce obvious changes in membrane conductance (not shown). The bath and pipet solutions used were identical to those described in Figure 3. V_H = -65 mV; horizontal bar above current records indicates the period of GABA application. **Bottom:** Potentiation of GABA-activated chloride current (ICl) by norepinephrine. Macroscopic I_{cl} induced by pressure application (150 msec duration every 30 sec) of 10 μM GABA were compared before (**A**), during (**B**; dotted trace) and after (**C**) application of 500 μM norepinephrine. Two consecutive traces are illustrated for **A, B,** and **C**. For examination of I_{cl} in isolation, the bath solution contained (mM) 115 choline chloride, 20 TEA-Cl, 5 CsCl, 10 $MgSO_4$, 5 HEPES, 10 glucose and the pipet solution contained (mM) 115 CsCl, 25 TEA-Cl, 2.5 EGTA, 5 HEPES. V_H = -65 mV; horizontal bar above current records indicates the period of GABA application.

A set of studies were conducted which involved iontophoresing NE onto Purkinje cells while providing visual stimulation in the form of vertical bar-like dot patterns moved across the visual field (Moises et al., 1990). Our simplest expectation was that NE would cause a lowering of background firing, accompanied by an increase or relative preservation of excitatory input, i.e., an enhanced signal to noise. An alternate hypothesis was that selective alterations of synaptic gain would result in unique changes in the mode of neural function.

The results from visual stimulation without NE were surprising in that many cells showed little influence of the visual input (Moises et al., 1990) while others demonstrated weak to modest influences. Indeed, the primary difficulty in studying visual responses in paraflocculus was in the lack of stimulus-evoked activity in an area which clearly received a substantial visual pontocerebellar projection. It was not clear, a priori, whether the cortex failed to emit signals to pons or whether the cerebellum failed to respond. The most remarkable effect of NE was found in instances in which neurons were unresponsive to control presentations of novel visual stimuli, and when no change in baseline firing rate was observed. During the period of NE iontophoresis, these Purkinje neurons exhibited marked phasic inhibitions as the visual pattern moved across the visual field (Fig. 5), and then lost their responsiveness after cessation of NE iontophoresis. Numerous other patterns of visual response changes were found in paraflocculus. Doses of NE sufficient to cause inhibition of spontaneous firing in some instances increased the absolute excitatory component of visual evoked responses or preserved it relative to suppression of background discharge. Some instances also revealed an enhancement or decrement of directionally selective responses.

Such results are impossible to account for in terms of a simple direct inhibitory noradrenergic action, since no suppression of spontaneous discharge occurs. Multiple actions of NE could be involved, but in some cases, as in Figure 6, the simple enhance-ment of an inhibition above threshold could account for the major effects that are observed. It is clear from these results that potentially powerful signals, not simply weak subthreshold responses, are emitted from the cerebral cortex and reach the Purkinje cells; however, the circuit appears set in an unresponsive mode until NE allows responses to be expressed.

Gating Actions in Visual Cortex: NE was also administered iontophoretically to neurons in the visual cortex of halothane anesthetized rats during visual stimulation (Waterhouse et al., 1990). The visually responsive cells in rat appeared to correspond to "simple" or "complex" primary cortical neurons as defined elsewhere for cat and monkey; however, the distinctions could not be defined too clearly because of sometimes ambiguous neuronal properties in rat visual cortex. A further clarification of the nature of the variability in properties became apparent when iontophoresis of NE or serotonin (5-HT) was carried out during visual stimulation.

As shown in Fig. 7 (from Waterhouse et al., 1990), NE applied to a visual cortical neuron in rat, suppressed the background firing while leaving the excitatory component intact. Results from some cells revealed that as NE doses increased, there often occurred an absolute increase in excitatory signal followed, at higher concentrations, by some suppression of both signal and background. No change in receptive field boundaries could be observed, but a substantial increase in signal to noise is the major outcome at high NE doses. Administration of 5-HT onto these visual cortical circuits under similar stimulus conditions resulted in a progressive depression of excitatory responses with little change in background firing. Similar reciprocal effects between NE and 5-HT have been found previously in somatosensory cortex by Waterhouse et al. (1987). Further studies on visual cortical neurons which exhibited little or no response to visual input under control recording conditions revealed that hidden synaptic signals did exist which could generate substantial excitatory signals if modulated by NE (Fig. 8). Other neurons have shown emergent inhibitions after NE where no responses existed previously (Fig. 8).

It appears that the modulatory actions of NE and 5-HT have the capacity to regulate the responsiveness of visual cortical circuitry over a wide range. If this represents a normal physiological process, one would expect the tonic level of action of these agents to exert a significant

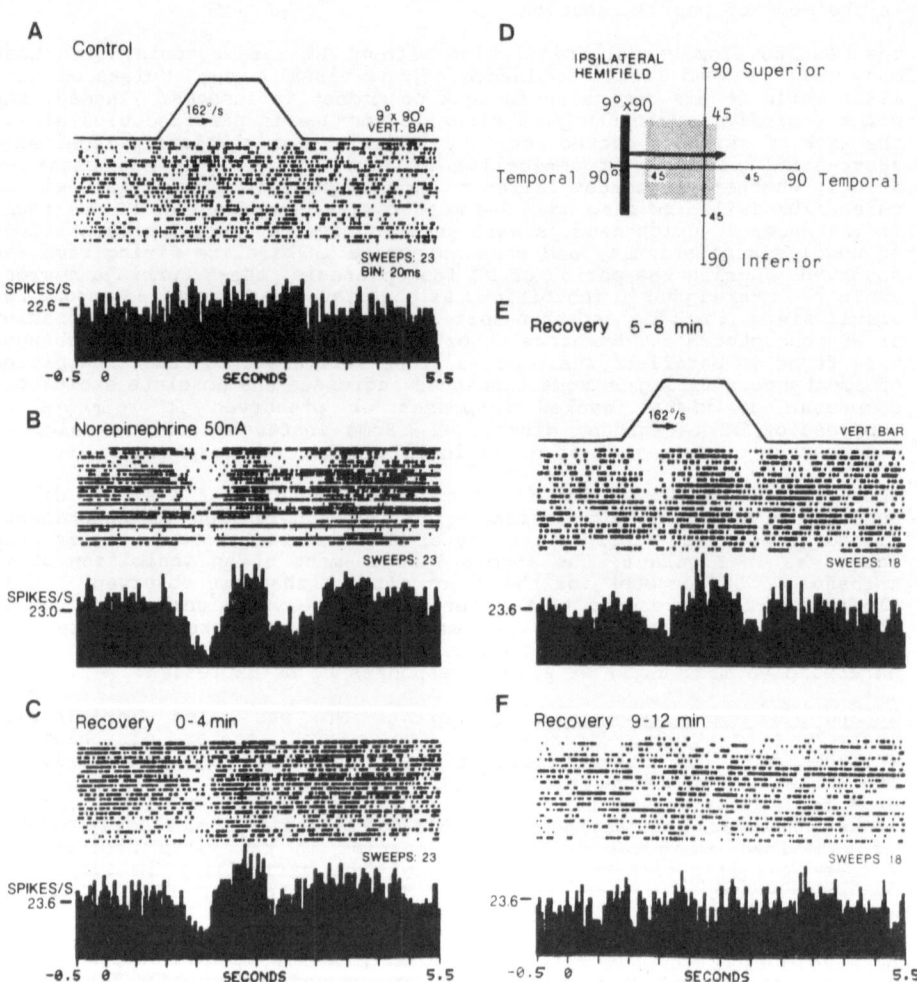

Figure 5. Expression of an inhibitory response of a neuron to visual stimulation by NE iontophoresis. Prior to administration of NE, this Purkinje cell showed no appreciable respone to presentations of a vertically oriented light bar stimulus (9° X 90°) moving from left to right (upslope of trapezoidal waveform) and back through its receptive field. (B) During iontophoresis of NE 50 nA, the cell responded in an inhibitory manner to presentations of the visual stimulus with a preference for movement in the forward (left-to-right, indicated by small arrow) direction. Note that the expression of the inhibitory response by NE was accompanied by an elevation in background discharge of the neuron. (C,E) Note also that the cell continued to show inhibitory responses to visual stimulation for upwards of 8 min following the termination of NE administration. (D) The movement trajectory of the light bar stimulus (long arrow) through the inhibitory zone of the cell's receptive field (shading).

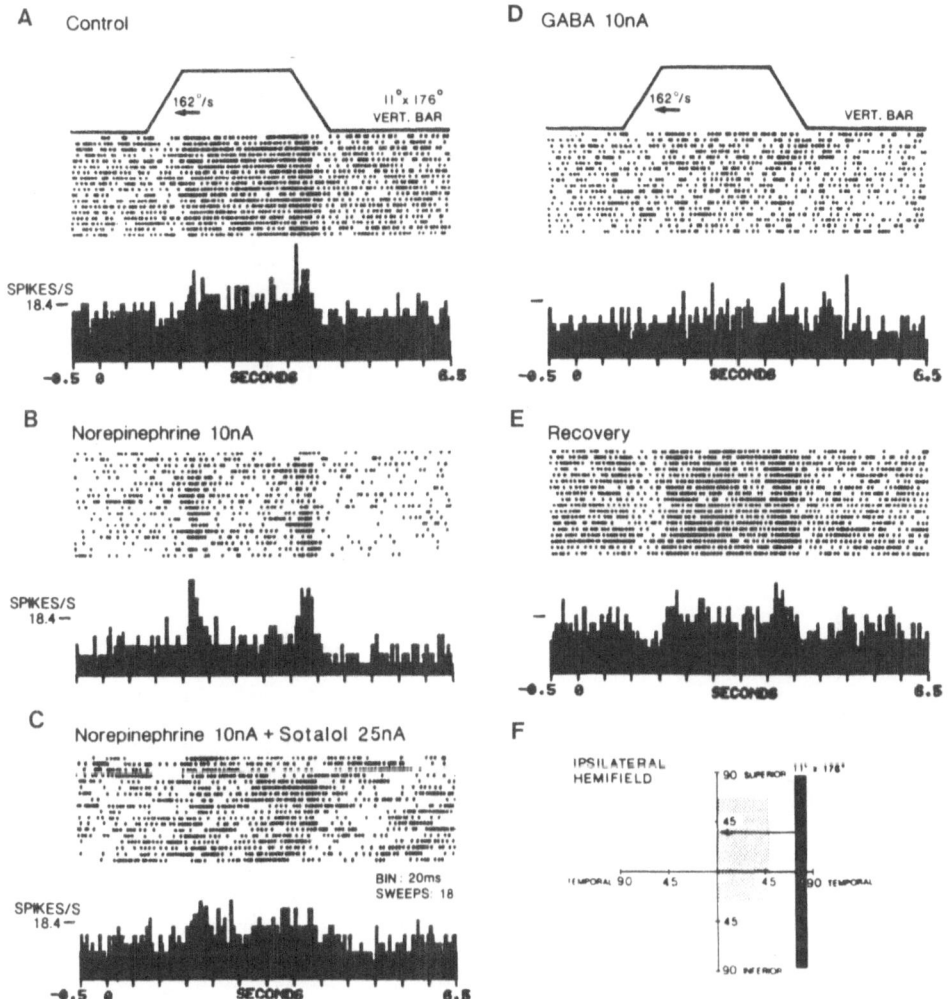

Figure 6. Comparison of the effects of iontophoresis of NE and GABA on visually evoked and spontaneous simple spike discharge of a Purkinje cell. (A) This neuron was stimulated by movement of a vertical bar of light (11° X 176°) from right to left (upslope of trapezoidal waveform) and in the reverse (preferred) direction through its receptive field. (B) During iontophoresis of NE 10 nA, spontaneous discharge was markedly reduced, whereas activity evoked by stimulus movement in the preferred direction (left-to-right) was only slightly decreased, resulting in the enhancement of the ratio of signal to noise. Note that NE had the additional effect of enhancing the cell's excitatory response to movement of the stimulus in the non-preferred direction (from right to left). (C) This enhancement of the visual responses by NE was blocked when the specific beta-adrenergic antagonist sotalol 25 nA was applied concurrently to the cell by iontophoresis. Note that sotalol antagonized both the depression in spontaneous activity by NE and the enhancement in the response to right-to left movement of the visual stimulus. Drug iontophoresis was suspended fo r20 min following the generation of this histogram record to allow for complete recovery to the control levels of visually evoked and spontaneous activity before further testing with GABA was initiated. (D) In contrast to the effects observed during NE, iontophoresis of GABA 10 nA produced a smaller suppression in background firing and yet virtually eliminated the responses to visual stimulation. Recovery of the visually evoked responses to near control levels was observed following the cessation of NE (C) and GABA iontophoresis (E). (F) The receptive field location of the excitatory zone (shading) and forward movement trajectory (arrow) of the light bar stimulus.

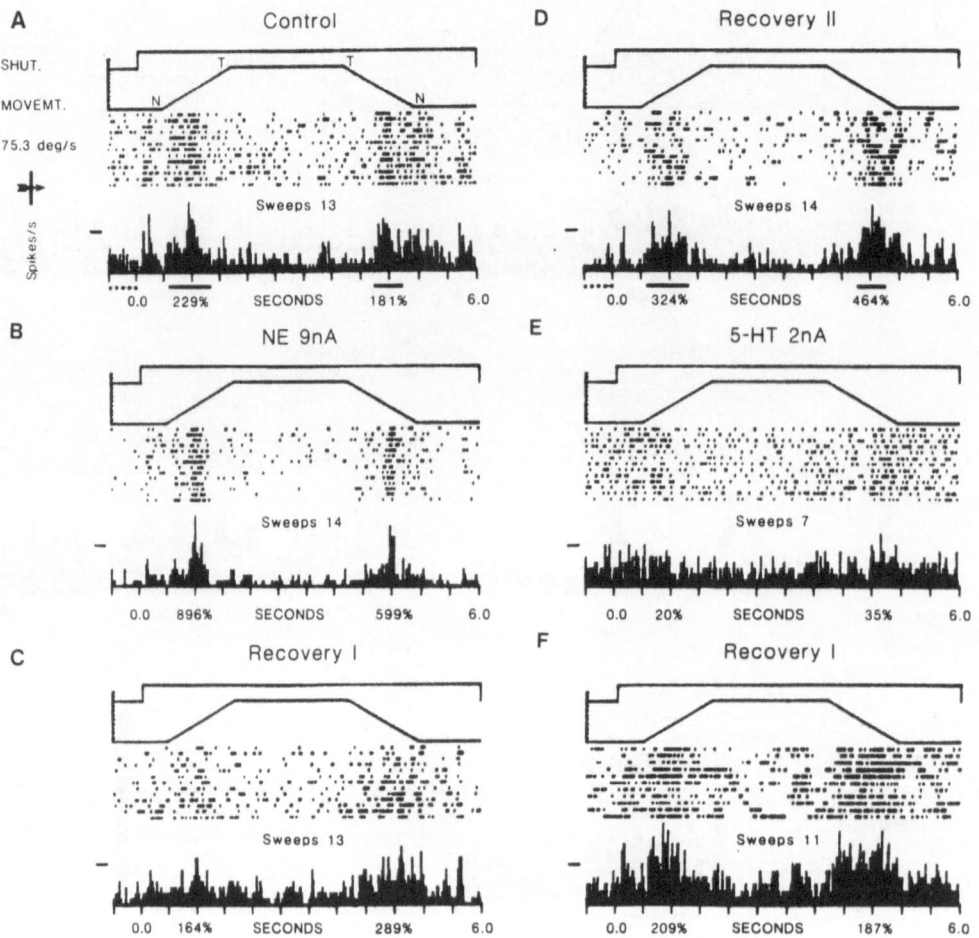

Figure 7. Effects of NE and 5-HT on the visually-evoked response of a type II cell. (Presence of orientation preference, absence of inhibitory sidebands, see Waterhouse et al., 1990). Raster and peri-event histogram records illustrate the response of a type II neuron to a moving (75.3 deg./s), vertically-oriented bar of light before, during and after microiontophoretic application of NE or 5-HT. (A) The initial response of this cell was an excitatory discharge as the stimulus moved across the contralateral half of the visual field in the nasal (N) to temporal (T) direction and a similar period of increased activity as the stimulus traveled in the reverse direction (T to N). These excitatory responses (solid bars) were expressed as a percentage increase in firing rate over spontaneous activity (dotted line). (B) Local application of NE (9 nA) suppressed spontaneous discharge to a greater extent than activity during the period of visually-evoked excitations, thus yielding a net enhancement of the stimulus bound excitatory signals. (C,D) Recovery to the control pattern of response was gradually observed after cessation of NE iontophoresis. (E) Microiontophoretic administration of 5-HT (2 nA) suppressed both excitatory response peaks, while slightly elevating spontaneous firing rate; thus randomizing the cell's activity with respect to movement of the visual stimulus. (F) Recovery to the control level of response was again noted following termination of the 5-HT ejection current. Histogram bin width = 15 ms.

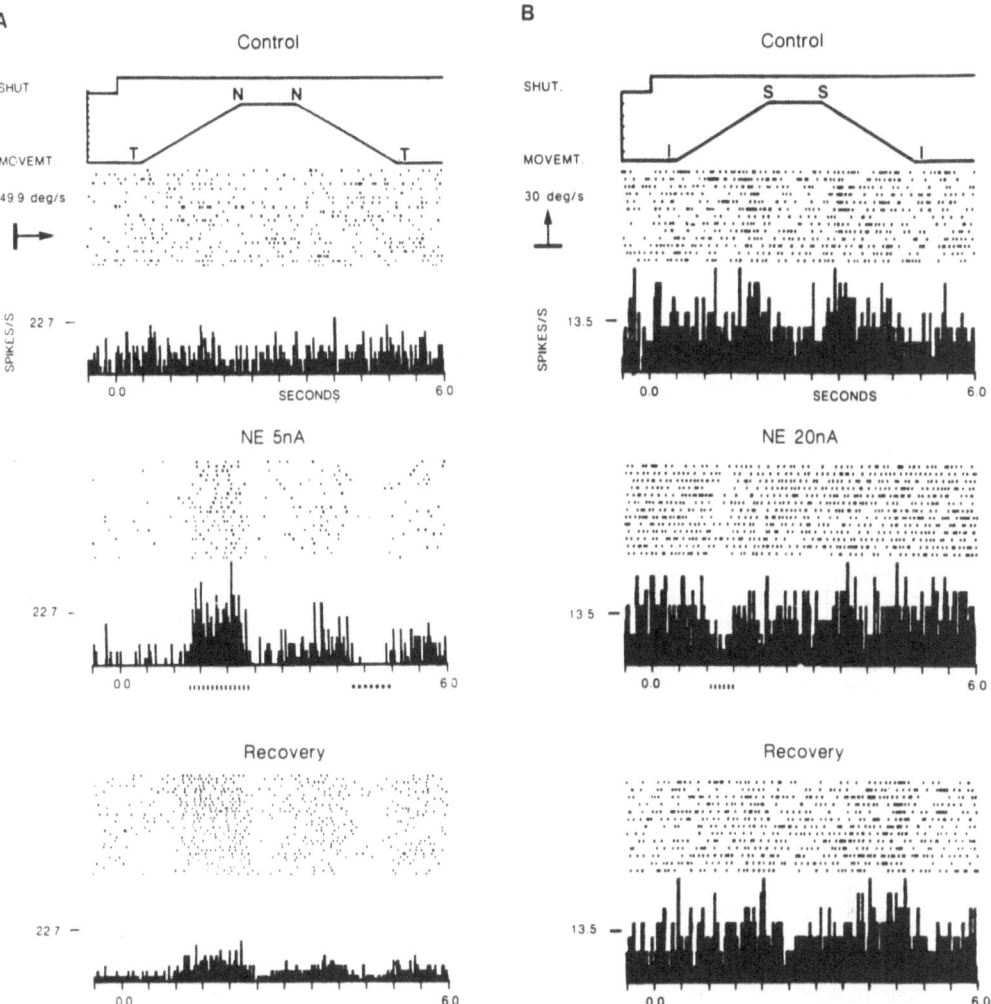

Figure 8. NE-induced potentiation of previously subthreshold responses. (A) Raster and peri-event histogram records show the spike activity of an area 17 cortical cell during presentation of vertical bar of light moving (49.9 deg./s) across the visual field in the temporal (T) to nasal (N) and then N to T direction. In the control condition (top), there was no discernible response to the moving bar of light; however, with NE (5 nA) administration (middle) an obvious stimulus-evoked excitation (broken bar beneath histogram) and a stimulus-bound inhibition (dotted line beneath histogram) became evident. After cessation of NE administration (bottom), the cell gradually lost its sensitivity to the visual stimulus. Control and NE histograms sum unit activity during 22 trials, recovery histogram = 31 trials; bin width = 15 ms. (B) Records from a second visual cortical neuron illustrate a similar enhancement of an initially subthreshold inhibitory input. In the control condition (top) no measurable response was observed as a horizontal bar of light was moved (30 deg./s) across the visual field from an inferior (I) to a superior (S) position and then back again (S to I). During 20 nA ejection of NE (middle) a weak, but consistent inhibitory trough (broken bar beneath histogram) appeared as the visual stimulus was moved from I to S. This depressant response was again absent during the period of recovery (bottom) from NE application. Each histogram sums unit activity during 13 trials; bin width = 15 ms.

influence on the responsiveness of cortex to exogenous and endogenous signals.

Overall, the visual cortex and cerebellum exhibit similarities in that regulation by NE of threshold gating of signals can take place with a minimal alteration in the background activity of recorded cells. In visual cortex, the predominant control response of cells to presentation of appropriate visual stimuli is a phasic excitatory component with inhibitory side bands. In the case of a pre-existing well-defined response, the action of NE was to sharpen the transition between excitatory and inhibitory phases of discharge. The normal operating mode in cerebellar paraflocculus under the conditions tested is quite different from cerebral cortex in that strong signals arrive without influencing the Purkinje cells. One postulate is that a potent gate is normally tuned to reject signals, which becomes unbalanced by increases in inhibitory gain induced by NE.

Summary: A spectrum of studies has been conducted on a single aspect of NE function in which, through a beta-one receptor activation, NE appears to mediate a degree of physiological control over the gain of GABA mediated inhibition. It is significant that this single effect has been observed in numerous interrelated preparations ranging from single isolated Purkinje cells from young rats to adult Purkinje cells in awake locomoting rats. With respect to the functional conse-quences of these effects, our best current speculation as to "what NE does" is that NE acts to regulate the strength of these tuned gating mechanisms in both cerebral and cerebellar cortices.

There are numerous unanswered questions raised by the past work. One pressing issue is -- when and for what reason in normal function does the modulation take place? When does NE release normally occur (is it phasic or tonic), and which of the demonstrated actions appears and for how long in relation to period of receptor activation? Does NE release cause the circuit to "react" to conditions which need "improved neurocomputation" or does NE stabilize the circuit to react predictably in the face of stress? Finally, what is the molecular sequence of events between receptor activation and an alteration of GABA receptor channel opening? What additional molecular control mechanisms exist and how can the diverse inhibitory and modulatory phenomena be reconciled, both short and long term? Issues are defined which need to be clarified at all levels of the current skeleton of basic understanding. Our prediction is that pursuit of these issues will benefit from an exchange of insight gained from investigations at all levels.

Supported by AFOSR-87-0138, NIHNS-18081 to BDW; NS-01340, NS-24830 to HY; DA02338, and an award from the Biological Humanics Foundation to DJW.

REFERENCES

Burne, R.A., and Woodward, D.J., 1983, Visual cortical projections to the paraflocculus in the rat. an electrophysiologic study. Exp. Brain Res., 49:55-67.

Burne, R.A., Mihailoff, G.A., and Woodward, D.J., 1978, Visual corticopontine input to the paraflocculus: a combined autoradiographic and horseradish peroxidase study. Brain Res., 143:139-146.

Hoffer, B.J., Siggins, G.R., Oliver, A.P., and Bloom, F.E., 1973, Activation of the pathway from locus coeruleus to rat cerebellar Purkinje neurons: pharmacological evidence of noradrenergic central inhibition. J. Pharmacol. Exp. Ther., 184:553-569.

Madison, D.V., and Nicoll, R.A., 1986, Cyclic adenosine 3',5'-monophosphate mediates β-receptor actions of noradrenaline in rat hippocampal pyramidal cells. J. Physiol. (Lond.), 372:245-259.

Miyahara, S., and Oomura, Y., 1982, Inhibitory action of the ventral noradrenergic bundle on the lateral hypothalamic neurons through alp noradrenergic mechanisms in the rat. Brain Res., 234:459-463.

Moises, H.C., Waterhouse, B.D., and Woodward, D.J., 1983, Locus coeruleus stimulations potentiates local inhibitory processes in rat cerebellum. Brain Res. Bull., 10:795-804.

Moises, H.C., Waterhouse, B.D., and Woodward, D.J., 1981, Locus coeruleus stimulations potentiates Purkinje cell responses to afferent input: the climbing fiber system. Brain Res., 222:43-64.

Moises, H.C., and Woodward, D.J., 1980, Potentiation of GABA inhibitory action in cerebellum by locus coeruleus stimulation. Brain Res., 182:327-344.

Moises, H.C., Woodward, D.J., Hoffer, B.J., and Freedman, R., 1979, Interactions of norepinephrine with Purkinje cell responses to putative amino acid neuro-transmitters applied by microiontophoresis. Exp. Neurol., 64:489-515.

Moises, H.C., Burne, R.A., and Woodward, D.J., Modification of the visual response properties of cerebellar neurons by norepinephrine. Brain Res. (In review).

Nakai, Y., and Takaori, S., 1974, Influence of norepinephrine-containing neurons derived from the locus coeruleus on lateral geniculate neuronal activities ofcats. Brain Res., 71:47-60.

Olpe, H., Glatt, A., Lazlo, J., and Schellenberg, A., 1980, Some electrophysiological and pharmacological properties of the cortical, noradrenergic projection of the locus coeruleus in the rat. Brain Res., 186:9-19.

Phillis, H.W., Tebecis, A.K., and York, D.H., 1967, The inhibitory action of monoamines on lateral geniculate neurones. J. Physiol., 190:563-581.

Segal, M., and Bloom, F.E., 1974, The action of norepinephrine in the rat hippocampus. I., Iontophoretic studies. Brain Res., 107:513-525.

Sessler, F.M., Mouradian, R.D., Cheng, J.-T., Yeh, H.H., Liu, W., and Waterhouse, B.D., 1989, Noradrenergic potentiation of cerebellar Purkinje cell responses to GABA: evidence for mediation through the β-adrenoceptor-coupled cyclic AMP system. Brain Res., 499:27-38.

Siggins, G.R., Oliver, A.P., Hoffer, B.J., and Bloom, F.E., 1971, Cyclic adenosine monophosphate and norepinephrine: effects on transmembrane properties of cerebellar Purkinje cells. Science, 171:192-194.

Sweetnam, P.M., Tallman, J.F., Gallager, D.W., and Nestler, E.J., 1987, Phosphorylation of a subunit of the GABA/benzodiazepine receptor by a receptor-associated protein kinase. Soc. Neurosci. Abstr., 13:964.

Waterhouse, B.D., Sessler, F.M., Cheng, J.-T., Woodward, D.J., Azizi, S.A., and Moises, H.C., 1988, New evidence for a gating action of norepinephrine in central neuronal circuits of mammalian brain. Brain Res. Bull., 21:425-432.

Waterhouse, B.D., Sessler, F.M., Cheng, J.-T., and Yeh, H.H., 1987, Noradrenergic potentiation of cerebellar Purkinje cell responses to GABA: evidence for mediation by an intracellular second messenger system. Soc. Neurosci. Abstr., 13:1347.

Waterhouse, B.D., Moises, H.C., Yeh, H.H., and Woodward, D.J., 1982, Norepinephrine enhancement of inhibitory synaptic mechanisms in cerebellum and cerebral cortex: mediation by beta adrenergic receptors. J. Pharmacol. Exp. Ther., 221:495-506.

Waterhouse, B.D., Moises, H.C., and Woodward, D.J., 1981, Alpha receptor mediated facilitation of somatosensory cortical neuronal responses to excitatory synaptic inputs and iontophoretically applied acetylcholine. Neuropharmacology, 20:907-920.

Waterhouse, B.D., Moises, H.C., and Woodward, D.J., 1980, Noradrenergic modulation of somatosensory cortical neuronal responses to iontophoretically applied putative neurotransmitters. Exp. Neurol., 69:30-49.

Waterhouse, B.D., and Woodward, D.J., 1980, Interaction of norepinephrine with cerebro-cortical activity evoked by stimulation of somatosensory afferent pathways. Exp. Neurol., 67:22-34.

Waterhouse, B.D., Azizi, S.A., Burne, R.A., and Woodward, D.J., Modulation of rat cortical area 17 neuronal responses to moving visual stimuli during norepinephrine and serotonin microiontophoresis.

Woodward, D.J., Waterhouse, B.D., Hoffer, B.J., and Freedman, R., 1979, Modulatory actions of norepinephrine in the central nervous system. Fed. Proc., 38:2109-2116.

Yeh, H.H., and Woodward, D.J., 1983a, Beta-I adrenergic receptors mediate noradrenergic facilitation of Purkinje cell responses to gamma-aminobutyric acid in cerebellum of rat. Neuropharmacology, 22:629-639.

Yeh, H.H., and Woodward, D.J., 1983b, Noradrenergic action in the developing rat cerebellum: interaction between norepinephrine and γ-aminobutyric acid applied microiontophoretically to immature Purkinje cells. Develop. Brain Res., 10:49-62.

Yeh, H.H., and Woodward, D.J., 1983c, Alterations in beta adrenergic physiological response characteristics after long-term treatment with desmethyl imipramine: interaction between norepinephrine and γ-aminobutyric acid in rat cerebellum. J. Pharmacol. Exp. Ther., 226:126-134.

Yeh, H.H., and Woodward, D.J., 1983d, Noradrenergic action in the developing rat cerebellum: interaction between norepinephrine and synaptically-evoked responses of immature Purkinje cells. Develop. Brain Res., 11:207-218.

Yeh, H.H., Moises, H.C., Waterhouse, B.D., and Woodward, D.J., 1981, Modulatory interactions between norepinephrine and taurine, beta-alanine, gamma-aminobutyric acid and muscimol, applied iontophoretically to cerebellar Purkinje cells. Neuropharmacology, 20:549-560.

GENETIC ANALYSIS OF THE β-ADRENERGIC RECEPTOR

Catherine D. Strader and Richard A.F. Dixon

Departments of Biochemistry
and Molecular Biology
Merck Sharp & Dohme Research Laboratories
Rahway, NJ and West Point, PA

INTRODUCTION

Perhaps the largest class of hormone and neurotransmitter receptors is those whose signal transduction pathways involve the activation of guanine nucleotide binding regulatory proteins (G-proteins). Hormones which activate such receptors are prevalent in both the central nervous system and the periphery, and include glycoproteins (lutropin, thyrotropin), peptides (neurokinins, angiotensin), and small molecules (retinal, biogenic amines). The signal tranduction pathway common to these systems is initiated by the binding of the ligand to the cell-surface receptor. The agonist-bound receptor interacts with a G-protein, forming a high-affinity ternary hormone-receptor-G-protein complex, and catalyzing the exchange of GDP for GTP in the nucleotide binding site of the G-protein (Gilman, 1987). This nucleotide exchange reaction activates the G-protein and destabilizes the ternary complex. The activated G-protein then interacts with effector systems, leading to the modulation of intracellular second messenger levels. Effector systems which are activated by G-protein coupled pathways include adenylyl cyclase, guanylyl cyclase, phospholipases A and C, phosphodiesterases, Ca^{++} and K^+ channels, and ion co-transport systems.

Several of the G-protein coupled receptors have been cloned in recent years, revealing structural similarities among these proteins which presumably reflect their common mechanism of action (Strader, et al, 1989a). The structural motif shared by all known receptors of this class consists of seven discrete stretches of 20-25 hydrophobic amino acids, presumed to form transmembrane α-helices, separating eight hydrophilic domains of varying lengths. The majority of the primary sequence conservation among these receptors lies within the putative transmembrane domain, with the loop regions being more divergent. Based on electron diffraction data for rhodopsin, which activates the cGMP phosphodiesterase via coupling to the G-protein transducin, a model has been proposed for the transmembrane disposition of G-protein coupled receptors (Findlay and Pappin, 1986). This proposed structure is shown in Figure 1 for the β-adrenergic receptor (βAR), which couples through the G-protein G_s to activate adenylyl cyclase. Because the N-terminus of the receptor contains two sites of N-linked glycosylation (Rands, et al, 1990), this region is proposed to be exposed extracellularly. The postulated transmembrane orientation of the seven hydrophobic domains would then dictate the alternating intracellular and extracellular exposure of the remaining hydrophilic loops, leaving the C-terminus of the protein exposed to the cytoplasm. Because the C-terminal and proposed third intracellular loop regions of the G-protein coupled receptors are abundant in Ser and Thr residues, their proposed intracellular orientation is consistent with evidence that they may serve as substrates for intracellular kinases. Recent immunological evidence is consistent with the transmembrane orientation of the βAR which is depicted in Figure 1 (Aoki, et al, 1989; Wang, et al, 1989).

LIGAND BINDING DOMAIN

Mutagenesis of the βAR gene has been used to identify amino acid residues which contribute to the ligand binding site of that receptor. To determine which domains of the protein are involved in the binding of adrenergic ligands, a systematic deletion analysis was undertaken, in which the nucleotides encoding each of the hydrophilic and hydrophobic regions of the protein were independently deleted from the βAR gene (Dixon, et al, 1987a,b). The resulting mutant receptors were then expressed in mammalian cells, where they were characterized biochemically and pharmacologically. It was determined that most of the hydrophilic loop regions were not required for ligand binding to the βAR. In contrast, deletion of the putative transmembrane helices, either singly or in pairs, resulted in imcorrectly folded receptor protein, which could not be analyzed pharmacologically. Thus, deletion mutagenesis demonstrated that the ligand binding domain of the βAR must reside somewhere within the hydrophobic core of the protein, and that this transmembrane core region is critical for maintaining the tertiary structure of the receptor (Dixon, et al, 1987a,b). Similar results have been determined for rhodopsin by biophysical techniques: the binding of retinal to the opsin protein involves the formation of a Schiff base between the aldelyde moiety of retinal and the amine side chain of Lys[296] in the seventh transmembrane helix of opsin (Findlay and Pappin, 1986).

In order to identify specific molecular interactions within the transmembrane domain of the βAR which are involved in the binding of adrenergic agonists and antagonists, specific amino acid replacements within this region were made by site-directed mutagenesis of the βAR gene. Adrenergic ligands are catecholamines, and the free prinary or secondary amine group of the ligand has been established to be critical for the binding interaction to occur. To identify specific amino acids which might interact directly with the amine group of the ligand,

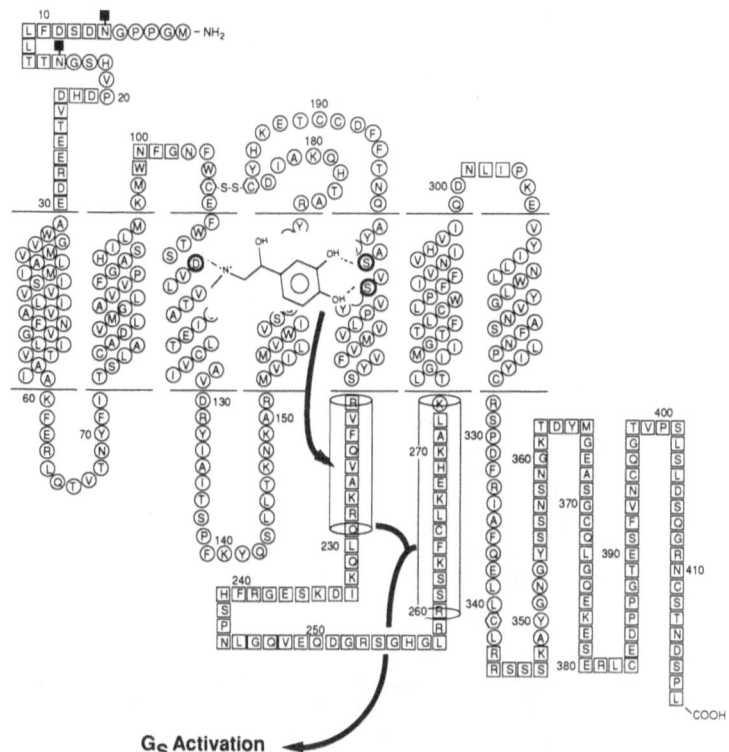

G$_S$ Activation

Fig. 1. Model for the structure of the βAR. The predicted
transmembrane topology of the receptor is shown, with
the extracellular region at the top. The amino acid
residues are designated by the single letter code.
Residues which couple be deleted without affecting
ligand binding are shown in squares. The sites of N-
linked glycosylation are designated with solid squares.
The agonist epinephrine is shown in the hydrophobic
binding pocket of the receptor. The postulated ionic
interaction with Asp[113] and hydrophobic interactions
with Ser[204] and Ser[207] are shown. The cylinders
outline the regions of the third intracellular loop
which are required for coupling to G$_S$. The solid
arrows represent the conformational changes in the
binding and G-protein coupling domains which lead to
the interaction of the receptor with G$_S$.

mutations were introduced into the βAR to substitute neutral
residues for acidic amino acids within the hydrophobic core
of the receptor (Strader, et al, 1987a; 1988). Substitution
of Asp[113] in the 3rd transmembrane helix of the βAR with a
Glu residue reduced the affinity of the receptor for
catecholamines by 2 orders of magnitude, whereas the
substitution of an Asn residue at this position resulted in
a 10,000-fold decrease in affinity. The affinity of the βAR
for antagonists was similarly reduced by these amino acid
substitutions. These data indicate a requirement for the
acidic side chain of Asp[113] for the interaction of the βAR
with agonist and antagonist ligands, and are consistent with

the formation of an ion pair between this carboxylate side chain and the protonated amine group of the ligand. In contrast, substitution of Glu[107] with an Ala residue had no effect on the affinity of the βAR for agonists or antagonists. Replacement of Asp[79] in transmembrnae helix 2, which is highly conserved among all G-protein coupled receptors, with an Ala residue, decreased the affinity and efficacy of agonists for the receptor but did not affect antagonist binding, suggesting a role for this residue in mediating receptor activation (Strader, et al, 1987a).

A similar combination of receptor mutagenesis and chemical modification of the ligands has suggested a role for two Ser residues in transmembrane helix 5 in binding the catechol hydroxyls on the aromatic ring of catecholamine agonists (Strader, et al, 1989b). Substitution of either Ser[204] or Ser[207] with an Ala residue decreased the affinity and efficacy of catecholamine agonists at the βAR. An examination of the additivity of the effects of the removal of either of the Ser residues from the receptor with the effects of substitution of either of the catechol hydroxyl groups on the ligand has suggested that the catechol ring of the agonist is anchored to the receptor via two hydrogen bonds. One bond appears to link the para hydroxyl group of the ligand with the hydroxyl side chain of Ser[207], with a second bond linking the hydroxyl side chain of Ser[204] to the meta hydroxyl group of the ligand. The model shown in Figure 1 would predict these two Ser residues to be located approximately one helical turn apart on the same face of transmembrane helix 5 in the receptor. This conformation would allow the simultaneous binding of both of the catechol hydroxyl groups of the ligand to the receptor, consistent with the results of pharmacophore mapping experiments.

Further mutagenesis experiments have revealed residues within other transmembrane helices to also be involved in ligand binding to the βAR. Molecular replacement of Phe[290] in transmembrane helix 6 has suggested a role for this residue in the binding of agonists to the receptor, perhaps through aromatic interactions with the catechol ring of the ligand (Dixon, et al, 1988). Replacement of Asn[318] or Ser[319] in the seventh transmembrane helix of the receptor has suggested the involvement of this hydrophobic domain in agonist binding, as well (Strader, et al, 1987a; Dixon, et al, 1988). The construction of hybrid β_1/β_2 adrenergic receptors suggests that the determinant(s) for the subtype specificity of catecholamine agonists also lies within the transmembrane domain of the βAR, involving residues in transmembrane helices 4 and 5 (Frielle, et al, 1988; Dixon, et al, 1989). However, smaller molecular replacements in this region have failed to identify a specific amino acid side chain(s) responsible for the subtype selectivity of the receptor, suggesting that this distinction arises from more subtle conformational effects (Dixon, et al, 1989). Since the subtype specificity of endogenous catecholamines involves the distinction between a hydrogen and a methyl substituent on the amine moiety of the ligand, these results suggest the presence of very specific hydrophobic interactions between the ligand and the receptor within the transmembrane domain of the protein. Similar mutagenesis experiments are currently underway to identify other specific molecular interactions involved in the binding of agonists and antagonists to the βAR.

Examination of the sequences of other G-protein coupled receptors reveals patterns of conservation of residues within the transmembrane binding domains of these proteins. An Asp residue is present at a position analogous to that of Asp^{113} in the third hydrophobic domain of the βAR in all known receptors of this class which bind protonated amines as ligands, but not in receptors whose ligands are not amines (Strader, et al, 1988). Similarly, the presence of Ser residues at positions analogous to Ser^{204} and Ser^{207} is predictive of receptors which bind catechol ligands (Strader, et al, 1989b). These data, together with the evidence for the involvement of these residues in the binding of ligands to the βAR, suggest that all G-protein coupled receptors share a common binding domain, in which various functional groups on the ligand interact with the amino acid side chains at key positions within the hydrophobic transmembrane domain. The identity of the residues at these key positions would then determine the ligand binding specificity of the receptor. According to this hypothesis, knowledge of most of the sites of interaction between several receptors and their ligands should allow the prediction of the binding specificity of a newly cloned receptor from analysis of the distribution of amino acid residues in its binding site. The development of such a model will be useful for the rational design of pharmacological agents which act therapeutically at specific G-protein coupled receptors.

G-PROTEIN COUPLING DOMAIN

The coupling of the βAR to G-proteins was also investigated by site-directed mutagenesis of the receptor. Deletion mutagenesis has implicated the intracellular loops of the receptor in its coupling to G_s (Strader, et al, 1987b; O'Dowd, et al, 1988; Cheung, et al, 1989). In particular, deletion of a large portion of the third intracellular loop resulted in an absence of G-protein coupling, as assessed by the absence of GTP-sensitive agonist binding to the receptor and the inability of the mutant receptor to activate adenylyl cyclase (Strader, et al, 1987b). Further analysis of this region of the receptor, using smaller deletions, revealed that the middle portion of the loop could be deleted without affecting the ability of the receptor to couple to G_s. In contrast, deletion of the C-terminal 12 amino acids from this loop severely attenuated G-protein coupling, and this function was completely abolished by the deletion of the N-terminal 8 amino acids from the third intracellular loop of the receptor (Cheung, et al, 1989). These data demonstrate a requirement for these regions of the receptor for its interaction with G_s, and suggest that these regions may be directly involved in the coupling reaction. Further evidence for the involvement of residues within the the third intracellular loop in the interactions between receptors and G-proteins arises from analysis of hybrid M_1/M_2 muscarinic acetylcholine receptors. These two subtypes of the muscarinic receptor couple to different G-proteins. In these studies, the third intracellular loop of each receptor subtype was substituted for the other (Kubo, et al, 1988). The specificity of G-protein coupling of the hybrid receptors correlated with the source of this loop, suggesting that the determinants for G-protein specificity lie within this region of the receptor. Activation of G-proteins by synthetic peptides, as well as proteolysis studies, have also pointed to a role for the third intracellular loops of the βAR and rhodopsin in G-protein coupling (Hamm, et al, 1988; Palm, et al, 1989; Findlay and Pappin, 1986). There is also some evidence for the involvement of other intracellular loops of the βAR and rhodopsin from mutagenesis and peptide activation studies. However, the extent to which these regions contribute to the activation of G-proteins remains to be determined.

The regions at the N- and C-termini of the third intracellular loop of the βAR that are required for G-protein coupling would be predicted by secondary structure prediction algorithms to form amphipathic α-helices (Cheung, et al, 1989). The predicted secondary structure of these regions is conserved among all G-protein coupled receptors, although this region encompasses one of the most divergent primary sequences of the entire protein. Thus, it seems likely that the secondary structural characteristics of this region are important in mediating the coupling of receptors to G-proteins. This hypothesis is attractive in light of evidence that the wasp venom peptide mastoparan, which forms an amphipathic α-helix in solution, can activate the G-protein G_o, suggesting that this structural motif may be an important component in G-protein activation (Higashijima, et al, 1988). These regions at the N- and C-termini of the third intracellular loop may be considered as amphipathic extensions of the fifth and sixth transmembrane helices of the receptor into the cytoplasm. This model suggests a mechanism by which ligand binding to the transmembrane portion of the receptor can mediate G-protein coupling (Strader, et al, 1989b). This hypothesis is outlined schematically in Figure 1. The binding of adrenergic agonists to the βAR involves specific interactions of the catechol ring substituents with the side chains of Ser residues in transmembrane helix 5. Additional agonist-specific aromatic interactions also probably exist with residues in helix 6. The interaction of the ligand with this portion of the receptor would be expected to cause conformational changes in helices 5 and 6 as the receptor-ligand complex approaches its lowest energy conformation. These conformational changes in the transmembrane helices could be transmitted to the regions at the bottom of the helices, causing alterations in the conformation of these amphipathic domains, and leading to G-protein activation. Antagonists, which do not interact with the Ser residues at positions 204 and 207 in helix 5, would not cause these conformational changes in the receptor and, therefore, would not be expected to cause G-protein activation. The model shown in Figure 1 is based on genetic and biochemical analysis of the βAR protein, and merely represents a working hypothesis for the interactions between the receptor, ligand, and G-protein. The definition of the physical chemical interactions among these molecules and an understanding of the nature of the conformational changes which promote these interactions will await direct biophysical analysis of the ternary complex.

DESENSITIZATION

Prolonged exposure to adrenergic agonists of cells expressing the βAR results in a refractoriness of the responsiveness of the receptor to further agonist stimulation (Sibley, et al, 1987). This phenomenon, termed desensitization, results from a combination of intracellular events which are only now beginning to be defined at the molecular level. The desensitized receptor is uncoupled from G-proteins, a phenomenon which develops within a time period of minutes after agonist is added to the system. During this process, the βAR is rapidly sequestered away from the cell surface to an as yet undefined intracellular compartment. Longer exposure of the cells to agonist results in down-regulation, whereby the receptor is lost from the cell, presumably by degradation. These two mechanisms for reducing the number of receptors which are available at the surface of the cell to bind ligands have been observed for a number of G-protein coupled receptors. However, some receptor systems, most notably rhodopsin, do not undergo this form of desensitization. A second phenomenon which is temporally associated with desensitization is an observed increase in receptor phosphorylation. The βAR is a substrate for several intracellular kinases, including cAMP-dependent protein kinase (PKA) (Clark, et al, 1988), protein kinase C (PKC), and a β-agonist specific kinase (βARK) (Benovic, et al, 1986). Other G-protein coupled receptors have also been observed to be phosphorylated by these kinases (Kwantra, et al, 1989), and an agonist-specific rhodopsin kinase has also been identified (Palczewski, et al, 1988). The relationships of these various sequestration and phosphorylation events to the functional uncoupling of the receptor from its G-protein-effector system, and the molecular basis for these phenomena, are currently the subjects of investigation, using both biochemical and genetic approaches.

Current models for the relationship between phosphorylation and desensitization were initially developed for rhodopsin. Upon exposure to light, rhodopsin is phosphorylated by rhodopsin kinase to a high stoichiometry on Ser and Thr residues near the C-terminus. The phosphorylated rhodopsin then is able to bind to the protein arrestin, and no longer activates transducin. The sequence homology which has been noted between the C-terminal regions of arrestin and transducin has led to the suggestion of competitive inhibition by arrestin of the interaction between phosphorylated rhodopsin and transducin as a mechanism for desensitization (Wilden, et al, 1986).

Other G-protein coupled receptors also contain Ser/Thr rich domains which might serve as substrates for kinases, either at the C-terminal tail or within the long third intracellular loop. For the βAR, the C-terminal tail is rich in Ser and Thr residues, implicating this region as the site of phosphorylation by βARK. Removal of this region of the βAR by deletion mutagenesis, or specific molecular substitution of Ser residues within this C-terminal domain with Ala residues, resulted in a mutant receptor which was no longer phosphorylated by βARK (Bouvier, et al, 1988). These mutations did not affect the final extent of receptor sequestration, nor did they prevent the uncoupling of the receptor from G_s (Strader, et al, 1987c; Kobilka, et al, 1987). However, the initial rates of both the uncoupling and the sequestration processes were slowed by the mutations (Bouvier, et al, 1988; Cheung, et al, 1989). Thus, removal of the region of the βAR which is phosphorylated by βARK decreases the rate of receptor desensitization, without affecting the final level of sequestration or uncoupling. These data are consistent with a role for phosphorylation by βARK in modulating the rate of the desensitization process. This observation, coupled with the observation that βARK is prevalent in the brain and not in peripheral tissue, has led to the suggestion that phosphorylation by βARK may be physiologically important in mediating rapid desensitization at synapses, where epinephrine is serving as a neurotransmitter and rapid desensitization is critical (Cheung, et al, 1989). In the periphery, where desensitization is a relatively slow process and βARK levels appear to be lower, other mechanisms may be more significant.

Phosphorylation of the βAR by other kinases also appears to play a role in mediating various forms of desensitization. Phosphorylation of the βAR by PKA has been proposed as a mechanism for heterologous desensitization, in which several G_s-coupled receptor systems are desensitized as a result of the stimulation of one type of receptor (Clark, et al, 1988). Activation of G_s, leading to stimulation of adenylyl cyclase, would activate PKA as a part of the signal transduction cascade. Evidence for the involvement of PKA in this process arises from the observation that the process does not occur in cells lacking PKA and from the ability to mimic this desensitization response in a cell-free system by the addition of PKA (Clark, et al, 1988; Kunkel, et al, 1989). Mutagenesis of the βAR has implicated Ser^{262} in the third intracellular loop of the receptor as the site of phosphorylation by PKA, since removal of this Ser residue prevented this heterologous densensitization process (Clark, et al, 1989). Deletion of Ser^{262} from the βAR had no effect on its ability to couple to G_s. However, it is interesting to note that this Ser residue is located in the third intracellular loop of the βAR, adjacent to the domain at the C-terminus of the loop which appears to be involved in G-protein coupling (Figure 1). It seems likely, therefore, that phosphorylation of Ser^{262} by PKA could perturb the conformation of this region in such a way as to interfere with the ability of the receptor to couple to G_s, resulting in desensitization of the receptor.

Deletion analysis of the intracellular domains of the βAR has demonstrated that the majority of the third intracellular loop and the entire C-terminal tail of the receptor are not required for agonist-mediated sequestration to occur (Strader, et al, 1987c). In contrast, the mutant βAR from which the N-terminal region of the third intracellular loop has been deleted, which does not couple to G_s, fails to undergo agonist-mediated receptor sequestration (Cheung, et al, 1989). Replacement of these amino acid residues with the analogous region of the M_1-muscarinic receptor does not restore the ability of the receptor to couple to G-proteins (Cheung, et al, 1990). However, this molecular replacement did restore the sequestration response of the receptor to agonists. Thus, a region of the βAR which is required for coupling to G-proteins is also involved in the sequestration response that accompanies desensitization. A possible explanation for the overlapping functions of this region would be as a recognition site for an arrestin-like protein involved in desensitization of the βAR. The existence of such a protein in the βAR system has been inferred from the similarities in the actions of βARK and rhodopsin kinase, although direct evidence for this pathway is currently lacking (Benovic, et al, 1987). By analogy with the rhodopsin-transducin system, such an arrestin-like protein would be expected to share homology with $G_{s\alpha}$, suggesting the likelihood of overlapping binding sites for the two proteins on the βAR. Based on the mutagenesis studies, the region at the N-terminus of the third intracellular loop of the receptor emerges as a likely candidate for such a recognition site.

Taken together, these data demonstrate that distinct structural domains of the βAR are required for the various forms of receptor desensitization. Thus, residues within the C-terminal tail of the protein are involved in a βARK-mediated process which appears to cause rapid, homologous desensitization. PKA-mediated heterologous desensitization appears to arise from the phosphorylation of Ser^{262} in the third intracellular loop. Residues near the N-terminus of this third loop are involved in receptor sequestration, via a mechanism which may involve an interaction with another protein. Under physiological conditions, desensitization probably involves contributions from all three of these mechanisms, as well as other pathways whose mechanisms are less well understood. A combination of genetic and physiological approaches should extend our understanding of the biochemical basis for these interrelated pathways.

REFERENCES

Aoki, C., Zemcik, B.A., Strader, C.D., and Pickel, V.M., 1990, Cytoplasmic loop of β-adrenergic receptors: synaptic and intracellular localization and relation to catecholaminergic neurons in the nuclei of the solitary tracts. Brain Res., in press.

Benovic, J.L., Strasser, R.H., Benovic, J.L., Daniel, K., and Lefkowitz, R.J., 1986, Beta-adrenergic receptor kinase: identification of a novel protein kinase that phosphorylates the agonist-occupied form of the receptor. Proc. Natl. Acad. Sci. (USA), 83:2797-2801.

Benovic, J.L., Kuhn, H., Weyand, I., Codina, J., Caron, M.G., and Lefkowitz, R.J., 1987, Functional desensitization of the isolated β-adrenergic receptor by the β-adrenergic receptor kinase: potential role of an analog of the retinal protein arrestin (48-kDa protein)., Proc. Natl. Acad. Sci. (USA), 84:8879-8882.

Bouvier, M.W., Hausdorff, A, DeBlasi, A., O-Dowd, B.F., Kobilka, B.K., Caron, M.G., and Lefkowitz, R.J., 1988, Removal of phosphorylation sites from the β-adrenergic receptor delays the onset of agonist-promoted desensitization. Nature, 333:370-373.

Cheung, A.H., Sigal, I.S., Dixon, R.A.F., and Strader, C.D., 1989, Agonist-promoted sequestration of the β_2-adrenergic receptor requires regions involved in functional coupling with Gs. Mol. Pharm., 34:132-138.

Cheung, A.H., Dixon, R.A.F., Hill, W.S., Sigal, I.S., and Strader, C.D., 1990, Separation of the structural requirements for agonist-promoted activation and sequestration of the β-adrenergic receptor. Mol. Pharm., in press.

Clark, R.B., Kunkel, M.W., Friedman, J., Goka, T.J., and Johnson, J.A, 1988, Activation of cAMP-dependent protein kinase is required for heterologous desensitization of adenylyl cyclase in S49 wild-type lymphoma cells. Proc. Natl. Acad. Sci. (USA), 85:1442-1446.

Clark, R.B., Friedman, J, Dixon, R.A.F., and Strader, C.D., 1989, Identification of a specific site required for rapid heterologous desensitization of the β-adrenergic receptor by cAMP-dependent protein kinase., Mol. Pharm., 36:343-348.

Dixon, R.A.F., Sigal, I.S., Rands, E., Register, R.B., Candelore, M.R., Blake, A.D., and Strader, C.D., 1987a, Ligand binding to the β-adrenergic receptor involves its rhodopsin-like core. Nature, 326:73-77.

Dixon, R.A.F., Sigal, I.S., Candelore, M.R., Register, R.B., Scattergood, W., Rands, E., and Strader, C.D., 1987b, Structural features required for ligand binding to the β-adrenergic receptor. EMBO J., 6:3269-3275.

Dixon, R.A.F., Sigal, I.S., and Strader, C.D., 1988, Structure-function analysis of the β-adrenergic receptor. Cold Spring Harbor Symp. Quant. Biol., 53:487-489.

Dixon, R.A.F., Hill, W.S., Candelore, M.R., Rands, E., Diehl, R.E., Marshall, M.S., Sigal, I.S. and Strader, C.D. (1989) Genetic analysis of the molecular basis for β-adrenergic receptor subtype specificity. Proteins: Structure, Function, and Genetics, 6:267-274.

Findlay, J.B.C. and Pappin, D.J.C., 1986, The opsin family of proteins. Biochem. J., 238:625-642.

Frielle, T., Daniel, K.W., Caron, M.G., and Lefkowitz, R.J., 1988, Structural basis of β-adrenergic receptor subtype specificity studied with chimeric β1/β2-adrenergic receptors. Proc. Natl. Acad. Sci. (USA), 85:9494-9498.

Gilman, A.G., 1987, G-proteins: transducers of receptor-generated signals. Ann. Rev. Biochem., 56:615-624.

Hamm, H.E., Deretic, D., Arendt, A, Hargrave, P.A., Koenig, B., and Hofmann, K.P., 1988, Site of G protein binding to rhodopsin mapped with synthetic peptides to the α subunit. Science, 241:832-835.

Higashijima, T., Uzu, S., Nakajima, T., and Ross, E.M., 1988, Mastoparan, a peptide from wasp venom, mimics receptors by activating GTP-binding regulatory proteins (G-proteins). J. Biol. Chem., 262:6491-6494.

Kobilka, B.K., MacGregor, C., Daniel, K., Kobilka, T.S., Caron, M.G., and Lefkowitz, R.J., 1987, Functional activity and regulation of human β2-adrenergic receptors expressed in Xenopus oocytes. J. Biol. Chem., 262:15796-15802.

Kubo, T., Bujo, H., Akiba, I., Nakai, J., Mishina, M., and Numa, S., 1988, Location of a region of the muscarinic acetylcholine receptor involved in selective effector coupling. FEBS Lett., 241:119-125.

Kunkel, M.W., Friedman, J., Shenolikar, S., and Clark, R.B., 1989, Cell-free heterologous desensitization of adenylyl cyclase in S49 lymphoma cell membranes mediated by cAMP-dependent protein kinase. FASEB J., 3:2067-2074.

Kwatra, M.M., Benovic, J.L, Caron, M.G., Lefkowitz, R.J., and Hosey, M.M., 1989, Phosphorylation of chick heart muscarinic cholinergic receptors by the β-adrenergic receptor kinase., Biochemistry, 28:4543-4547.

O'Dowd, B.F., Hnatowich, M., Regan, J.W., Leader, W.M., Caron, M.G., and Lefkowitz, R.J., 1988, Site-directed mutagenesis of the cytoplasmic domains of the human β_2-adrenergic receptor. J. Biol. Chem., 263:15985-15992.

Palczewski, K., McDowell, J.H., and Hargrave, P.A., 1988, Rhodopsin kinase: substrate specificity and factors that influence activity. Biochemistry, 27:2306-2313.

Palm, D., Munch, G., Dees, C. and Hekman, M., 1989, Mapping of β-adrenoceptor coupling domains to G_s-protein by site-specific synthetic peptides. FEBS Lett., 254:89-93.

Rands, E., Candelore, M.R., Cheung, A.H., Strader, C.D. and Dixon, R.A.F., 1990, Mutational analysis of β-adrenergic receptor glycosylation: N-linked sugar addition is required for receptor transport, but not function. J. Biol. Chem., (In press).

Sibley, D.R. and Lefkowitz, R.J., 1987, Molecular mechanisms of receptor desensitization using the β-adrenergic receptor-coupled adenylate cyclase system as a model. Nature, 317:124-129.

Strader, C.D., Sigal, I.S., Register, R.B., Candelore, M.R., Rands, E., and Dixon, R.A.F., 1987a, Identification of residues required for ligand binding to the β-adrenergic receptor. Proc. Natl. Acad. Sci., 84:4384-4388.

Strader, C.D., Dixon, R.A.F., Cheung, A.H., Candelore, M.R., Blake, A.D., and Sigal, I.S., 1987b Mutations that uncouple the β-adrenergic receptor from G_s and increase agonist affinity. J. Biol. Chem., 262:16439-16443.

Strader, C.D., Sigal, I.S., Blake, A.D., Cheung, A.H., Register, R.B., Rands, E., Candelore, M.R., and Dixon, R.A.F., 1987c, The carboxyl terminus of the hamster β-adrenergic receptor expressed in mouse L-cells is not required for receptor sequestration. Cell, 49:855-863.

Strader, C.D., Sigal, I.S., Candelore, M.R., Rands, E., Hill, W.S., and Dixon, R.A.F., 1988, Conserved aspartic acid residues 79 and 113 of the β-adrenergic receptor have different roles in receptor function. J. Biol Chem., 263:10267-10271.

Strader, C.D., Sigal, I.S., and Dixon, R.A.F., 1989a, Structural basis of β-adrenergic receptor function. FASEB J., 3:1825-1832.

Strader, C.D., Candelore, M.R., Hill, W.S., Sigal, I.S., and Dixon, R.A.F., 1989b, Identification of two serine residues involved in agonist activation of the β-adrenergic receptor. J. Biol. Chem., 264:13572-13578.

Wang, H.-Y., Lipfert, L., Malbon, C.C. and Bahouth, S., 1989, Site-directed anti-peptide antibodies define the topography of the β-adrenergic receptor. J. Biol. Chem., 264:14424-14431.

Wilden, G.J., Katial, A., Craft, C., and Shinohara, T., 1986, Sequence analysis of bovine retinal S-antigen. FEBS Letters, 196:23-28.

REVERSAL OF NIGROSTRIATAL-LESION-INDUCED RECEPTOR ALTERATIONS

BY GRAFTING OF FETAL MESENCEPHALIC DOPAMINERGIC NEURONS

James K. Wamsley[1], Ted M. Dawson[2], Valina L. Dawson[2], Mary A. Hunt[1], Lisa J. Fisher[3] and Fred H. Gage[3]

[1]Neuropsychiatric Research Institute
700 First Avenue South, Fargo, ND
[2]Department of Neurology, Hospital of the University of Pennsylvania, 3400 Spruce Street, Philadelphia, PA
[3]Department of Neurosciences, M-024, UCSD, La Jolla, CA

INTRODUCTION

The striatum (caudate nucleus and putamen; as opposed to the corpus striatum which includes the caudate nucleus and the lentiform nucleus comprised of the putamen and globus pallidus) and its connections have been studied for some time due to their importance in the coordination and integration of normal motor function, as well as disorders of movement (see Albin et al., 1989). These areas are thought to be the potential sites of action of many drugs which cause disturbances in movement (antipsychotics, amphetamines, etc.). Certain disorders such as Parkinson's disease, Tourette's syndrome, tardive dyskinesia, and Huntington's disease appear to characteristically involve receptor alterations in the caudate and putamen (Waddington and O'Boyle, 1989; Seeman et al., 1987; Hess et al., 1987; Barnett, 1986). Several important neurotransmitter systems relevant to these disorders overlap in the caudate nucleus and putamen (indistinguishable as separate entities in the rat and collectively referred to as the caudate-putamen or CP). These include the dopaminergic system, originating from the substantia nigra pars compacta (SNC) and the ventral tegmental region (Bjorklund and Lindvall, 1984), and the cholinergic system comprising, among others, a system of interneurons within the stroma of the nucleus itself (Graybiel and Ragsdale, 1983). The dopamine (DA) containing neurons projecting from the SNC terminate on cholinergic interneurons (Kubota et al., 1987b) as well as on projection neurons which are thought to contain gamma-aminobutyric acid (GABA) (Kubota et al., 1987a), substance P (Beckstead, 1987) or Leu-enkephalin (see Albin et al., 1989). The cholinergic interneurons make synaptic contacts with the GABAergic projection neurons as well as presumably other neuronal types (Izzo and Bolam, 1988). Dopamine released onto cholinergic interneurons is thought to be inhibitory in nature (Lehman and Langer, 1983), whereas the acetylcholine released onto GABAergic output neurons is hypothesized to be stimulatory (Limberger et al., 1986). Important interactions between dopamine and acetylcholine (Lloyd, 1978), as well as GABA (Scheel-Kruger, 1986) in the "extrapyramidal" system have been known for several years. The acetylcholine interneuron seems to be pivotal in this interaction (Lehman and Langer, 1983), [although the neostriatum in reality is much more complicated and diverse (Graybiel and Ragsdale, 1983)]. For example, in Parkinson's disease (due to the degeneration of dopaminergic neurons) there is a predominance of cholinergic tone. As such, muscarinic receptor antagonists can ameliorate some of the symptoms of Parkinson's disease. The mainstay of therapy, dopamine receptor agonists or l-dopa, also diminish cholinergic tone via an interaction with dopamine receptors on these neurons which inhibits cholinergic release, and results in decreased output from the GABAergic projection neurons.

Subtypes of receptors have been described in both the dopaminergic and cholinergic muscarinic neuronal systems (see Kebabian and Calne, 1979; Creese et al., 1983; Watson et al., 1986; Bonner, 1989; Sokoloff et al., 1990). Some of these receptor subtypes have been localized to the striatum (Wamsley et al., 1984; 1989; Boyson et al., 1986; Dawson, T. et al., 1986a,b; 1988; Dawson, V. et al., 1988; Regenold et al., 1987; Wang et al., 1989) and they seem to be involved to varying degrees in functional aspects of CP neurotransmission (Potter et al., 1984; Barnett et al., 1986; Dawson et al., 1986b; Watson et al., 1986; Waddington and O'Boyle, 1989; Albin et al., 1989). The cell populations which maintain these individual receptor subtypes on their surfaces have been studied by combining receptor autoradiography (see Dawson et al., 1986b) with various lesioning techniques (Joyce and Marshall, 1985; Savasta et al., 1986, 1988; Dawson, V. et al., 1988; 1990; Filloux et al., 1988a,b). Such information has increased our understanding of neurotransmitter interactions within the caudate nucleus. One type of lesion, namely the injection of the catecholamine neurotoxin 6-hydroxydopamine (6-OHDA) into the medial forebrain bundle (MFB) to destroy the dopaminergic projection from the pars compacta, has been used to produce an animal model of Parkinson's disease. Animals so treated vigorously circle in the direction ipsilateral to the lesion, in response to an injection of amphetamine (Ungerstedt, 1971). This circling behavior can be used as a test for the effectiveness and completeness of the 6-OHDA lesion (Ungerstedt and Arbuthnott, 1970; Marshall and Ungerstedt, 1977).

It is possible to graft fetal mesencephalic dopaminergic neurons into the 6-OHDA lesioned caudate-putamen such that these cells make synaptic contacts and release dopamine (Schmidt et al., 1983; Freund et al., 1985). Furthermore, following grafting, some behaviors of these animals appear to return to "normal" (Bjorklund and Stenevi, 1984; Bjorklund et al., 1987) suggesting that these new contacts are functional.

We have previously investigated the results of 6-OHDA lesions on DA receptor alterations caused by the loss of DA to the CP (Filloux et al., 1987a,b; 1988a; Gehlert et al., 1987). These results can be contrasted to those obtained by removal of cell bodies (in general) from the CP by injection of ibotenic acid (IA), or receptor alterations induced by lesion of the cholinergic interneurons (Filloux et al., 1988a; Dawson, V. et al., 1988). We have also investigated the effects of these and other lesions on the cholinergic system and muscarinic receptor subtypes (Dawson, T. et al., 1991b; Dawson, V. et al., 1990). The interaction of these two neurotransmitter systems within the caudate and the receptor changes induced by lesion of the ascending DA neurons are discussed below. Behavioral modification by grafting neurons into the 6-OHDA denervated CP could indicate the restoration of "normal" receptor activity in the brain of the animal receiving the transplant. The results of such experiments to examine receptor alterations and the consequences of the grafts are described in this review.

METHODOLOGY

The methods described below briefly summarize the type of experiments performed in each case. More specific information can be obtained from the articles referenced.

Lesions

Lesion of the dopaminergic projection from the pars compacta of the SN to the CP was accomplished by the stereotaxic injection of 6-OHDA (8 μg) into the MFB (Filloux et al., 1988a; Dawson, T. et al., 1991a). The neurotoxin was injected (unilaterally) in solution (4 μg/μl containing 0.2 μg/ml of ascorbic acid). The effectiveness of the lesion was tested by challenging each animal with an injection of amphetamine (5 mg/kg) 7-10 days after the lesion. Only experimental animals which displayed vigorous turning (7 turns per minute over 60 minutes in a direction ipsilateral to the lesion) were used in the transplant experiments (Dawson, T. et al., 1991a).

Transplants

Cell suspensions of fetal (13-15 day gestation) mesencephalon (containing dopaminergic cells from the substantia nigra) were microinjected into two different sites within the CP (Dawson, T. et al., 1991a). These cells were injected into lesioned animals which received the 6-OHDA lesion to fibers in the MFB 5 to 10 weeks prior. The response of the animals to a subcutaneous injection of amphetamine was tested every two months for a period of 10-12 months following grafting.

Autoradiography

Following perfusion, the brain of each animal was dissected and frozen rapidly. Cryostat sections, 10 μm in thickness, were obtained and mounted onto "subbed" slides. These slide-mounted tissue sections were labeled with the appropriate radioactive ligand by immersion into a buffered solution containing the ligand and various other constituents. After labeling, the sections were dried and opposed to tritium sensitive film (Ultrofilm; LKB, Gaithersburg, MD or Hyperfilm; Amersham, Arlington Heights, IL). Commercially available standards (Amersham Microscales) were included with the exposure of each film for subsequent quantitation of the autoradiographic grain densities by computerized microdensitometry (see Kuhar et al., 1986).

Receptor Labeling

Labeling of D_1 dopamine receptors was accomplished using [^{3}H]SCH23390 (Dawson et al., 1985; 1986a). Slide-mounted tissue sections were incubated for 30 minutes, at room temperature, in Coplin jars containing 1.0 nM [^{3}H]SCH23390 (Dupont-NEN, Boston, MA) in 50 mM Tris-HCl buffer (pH 7.4) containing 120 mM NaCl, 5 mM KCl, 2 mM $CaCl_2$, and 1 mM $MgCl_2$. Other sections were incubated under identical conditions with the addition of 10 μM fluphenazine or piflutixol to the incubation medium to allow determination of nonspecific binding. The sections were then given two 5 minute washes in ice-cold buffer without added radioactivity.

Dopamine D_2 receptors were labeled by giving slide-mounted tissue sections a 20 minute incubation, at room temperature, in 0.17 M Tris-HCl buffer (pH 7.6) containing 120 mM NaCl, 5 mM KCl, 2 mM $CaCl_2$, 0.001% ascorbic acid and a 20 nM concentration of [^{3}H]sulpiride (Gehlert and

Wamsley, 1984; 1985). Adjacent sections were incubated in the additional presence of 1 μM haloperidol to displace the specific binding of [^{3}H]sulpiride (Dupont-NEN). All sections were given four sequential 1 minute rinses in fresh buffer without added radioactivity.

Muscarinic receptors were labeled by incubating slide-mounted tissue sections in the presence of a 1 nM concentration of [^{3}H]QNB (Dupont-NEN) in Kreb's phosphate buffer (pH 7.4). The incubations lasted 60 minutes at room temperature. Some sections were incubated in a solution containing a 100 nM concentration of pirenzepine (PZ). This concentration was chosen to displace the majority of [^{3}H]QNB binding from the M_1 sites and to indicate the distribution of M_2 receptor binding. Other sections were incubated in the presence of 1 μM atropine to define nonspecific binding. All of the sections were given two 5 minute rinses in fresh ice-cold buffer. The M_1 receptor subtype was labeled directly by incubating tissue sections for 60 minutes, at room temperature, in 20 nM [^{3}H]PZ (Dupont-NEN) in Kreb's phosphate buffer (pH 7.4). These slides were then given a 5 minute rinse in ice-cold buffer before drying (Wamsley et al., 1981; 1984; Yamamura et al., 1983).

The presence of DA containing terminals within the CP was monitored by labeling DA uptake sites with [^{3}H]BTCP (Filloux et al., 1989). This was accomplished by incubating slide-mounted tissue sections for 1 hour at 4°C in 50 mM Tris-HCl buffer (pH 7.0) containing 120 mM NaCl and 10 nM [^{3}H]BTCP (Research Products International, Mt. Prospect, IL). Nonspecific binding was determined by adding 1 μM GBR12909 to the incubation medium. The sections were rinsed in fresh buffer without added radioactivity for 4 periods of 5 minute duration.

Likewise, the presence of cholinergic neurons was ascertained and quantitated by labeling the sodium-dependent, high-affinity choline uptake sites on the cholinergic terminals (Vickroy et al., 1985). This was accomplished by incubating sections for 30 minutes, at room temperature, in phosphate-buffered saline containing 120 mM NaCl, 4.8 mM KCl, 1.2 mM $MgSO_4$, 1.3 mM $CaCl_2$, 20.3 mM NaH_2PO_4, 10 mM D-glucose (pH 7.4) and 2.5 nM [^{3}H]hemicholinium-3 (Dupont-NEN). Nonspecific binding was ascertained by adding a 1 mM concentration of unlabeled hemicholinium-3 to the incubation media of adjacent sections. These tissue sections were given two 1-minute rinses in fresh, ice-cold buffer before drying.

Quantitation

The autoradiograms generated in the experiments described above were quantitated by computerized microdensitometry (see Kuhar et al., 1986). This involved the inclusion of Amersham microscales on each film for standardization of the autoradiographic grain densities generated by the underlying receptor bound radioactivity. Optical densities were measured in reference to the standards using a DADS 560 (Leitz; FRG) microphotometry system (a Leitz Orthoplan Microscope coupled with an MPV compact photometer interfaced with an IBM PS/2 computer system), a DUMAS system (Drexel's XENIX-based image analysis system; Philadelphia, PA) or an MCID imaging system (Microcomputer Imaging Device; Imaging Research Inc., St. Catharines, Ontario).

LOCATION OF RECEPTORS

Results of other experiments indicate that the receptors present in the CP and some of its connections are associated with separate neuronal populations (Table I). Scatchard analysis of saturation data (incubating sections in various concentrations of the radioligand) indicate that all of the receptor changes described below, are the result of actual changes in receptor number (B_{MAX}), not affinity (K_D).

Dopamine receptor subtypes exist independently on different neuronal cell types within the striatal system (Hall et al., 1983; Filloux et al., 1988a). Dopamine D_2 receptors exist postsynaptically on the cell bodies and/or the dendritic arborizations of DA neurons in the SNC. Lesion of the SN with 6-OHDA eliminates D_2 receptors in the ipsilateral SNC (Filloux et al., 1987a). These receptors serve as autoreceptors, since they reside on the neuron containing the neurotransmitter which activates them. Likewise, D_2 receptors are thought to exist on the terminals of the dopaminergic neurons of the SNC which project through the MFB and terminate within the CP. Autoradiographic evidence for the latter localization is weak, however, these receptors are thought to exist on the basis of various physiological studies (see Filloux et al., 1988a).

Other D_2 receptor populations exist in the CP itself. Some of these receptors appear to be presynaptic on cortico-striate fibers (see Filloux et al., 1988a,b). These cells, which are thought to contain excitatory amino acids (glutamate), are apparently regulated by the presence of the dopamine terminal. The latter localization is controversial (see Filloux et al., 1988b); however, the recent description of messenger RNA coding the synthesis of D_2 receptors in cells in the cortex (Weiner and Brann, 1989) would apparently support the many other observations which indicate the presence of presynaptic D_2 heteroreceptors on these corticostriate fibers.

Lesion of the cholinergic interneurons with AF64A indicates a D_2 receptor loss with no change in D_1 receptor binding (Dawson, V. et al., 1988). Thus, the dopaminergic cell type present on the cholinergic interneurons appears to be of the D_2 type (Scatton, 1982; Seiler and

Table I

Receptor Subtypes Associated With Striatal Components*

DA neurons
  D_2 receptors on cell bodies in SNC
  D_2 receptors on terminals in CP

ACh interneurons of the CP
  D_2 receptors present
  M_2 receptors present

GABAergic projection neurons
  M_1 and D_1 receptors present on cell bodies in the CP
  D_1 receptors presynaptic on terminals in SNR

---

*Abbreviations: D_1 and D_2, dopamine receptor subtypes; M_1 and M_2, muscarinic cholinergic receptor subtypes; CP, caudate-putamen; SNR, substantia nigra pars reticulata; SNC, substantia nigra pars compacta

Markstein, 1984; Fage and Scatton, 1986; Dawson, V. et al., 1988; Weiner and Brann, 1989). IA lesions of the CP (resulting in a reduction in the cell bodies of this structure) show an approximately 50% loss of D_2 receptor binding and a complete loss of D_1 receptor binding. Coupled with the AF64A observations above, these experiments indicate that D_1 receptors exist postsynaptically on the noncholinergic interneurons and/or projection neurons of the CP. IA lesion of the CP also eliminates D_1 receptor binding in the ipsilateral SNR. This observation indicates that the D_1 receptors within the pars reticulata are presynaptic heteroreceptors on the terminals of the neurons originating in the CP (Porceddu et al., 1986; Starr, 1987; Filloux et al., 1988a). [Some cells in the CP may have both types of dopamine receptor present (Ohno et al., 1987)].

Lesion of the dopaminergic nigrostriatal projection causes a persistent denervation supersensitivity in both the D_1 and D_2 postsynaptic receptor populations within the CP as shown by a variety of methods (Mishra et al., 1974; Creese et al., 1977; Bounamici et al., 1986; Filloux et al., 1988a; Savasta, et al., 1988; Dawson, T. et al., 1991a). The robust increase in postsynaptic D_2 receptors following denervation appears to mask the loss of D_2 receptors associated with the presynaptic dopaminergic terminals.

Following the same outline for the muscarinic cholinergic receptors separated into two subtypes, the M_1 and M_2 receptor populations, lesion of the cholinergic interneurons results in an apparent loss of choline uptake sites and M_2 receptors (Dawson, V. et al., 1990). The majority of the muscarinic cholinergic receptors probably exist postsynaptically in the CP. Some population of M_2 receptor sites, however, appear to be presynaptic (Potter et al., 1984; Schoffelmeer et al., 1986), since lesion of the cholinergic interneurons with AF64A results in a reduction in the M_2 receptor population (Dawson, V. et al., 1990). The selectivity of AF64A for cholinergic neurons has been questioned (McGurk et al., 1987), but the compound appears to be useful for removal of ACh-containing neurons under appropriate conditions (Sandberg et al., 1984; Dawson, V. et al., 1988; 1990). We, and others, have previously shown that the breakdown of muscarinic cholinergic receptors into two major populations is inadequate to account for the binding of supposedly subtype-specific ligands. The discrepancies are due to the fact that more than just two receptor populations for the muscarinic cholinergic system have been shown to exist. Various molecular biological studies have indicated the presence of five different gene products coding for muscarinic cholinergic receptors (Kubo et al., 1986; Bonner et al., 1987; Buckley et al., 1988). Our description of M_1 and M_2 receptor binding probably relates to ligands recognizing M_1 vs. M_2-M_5 sites. These receptor localizations, as described above, are summarized in Figure 1.

RESULTS OF THE TRANSPLANTS

6-OHDA lesion of the dopaminergic projections to the CP causes an increase in both D_1 and D_2 receptor populations of the caudate (Table II). The extent of this presumed denervation supersensitivity relates to the time elapsed after the lesion. A short postlesion time reveals a slight receptor increase, whereas longer periods of time following the lesion may show a larger increase in receptor number (unpublished observations). Prolonged lesion of the DA neurons with 6-OHDA results in a reduction of both M_1 and M_2 receptor binding in the CP (Dawson, T. et al., 1991b). No change in the number of choline uptake sites, on the basis of [^{3}H]hemicholinium-3 binding, is found following the lesion, indicating

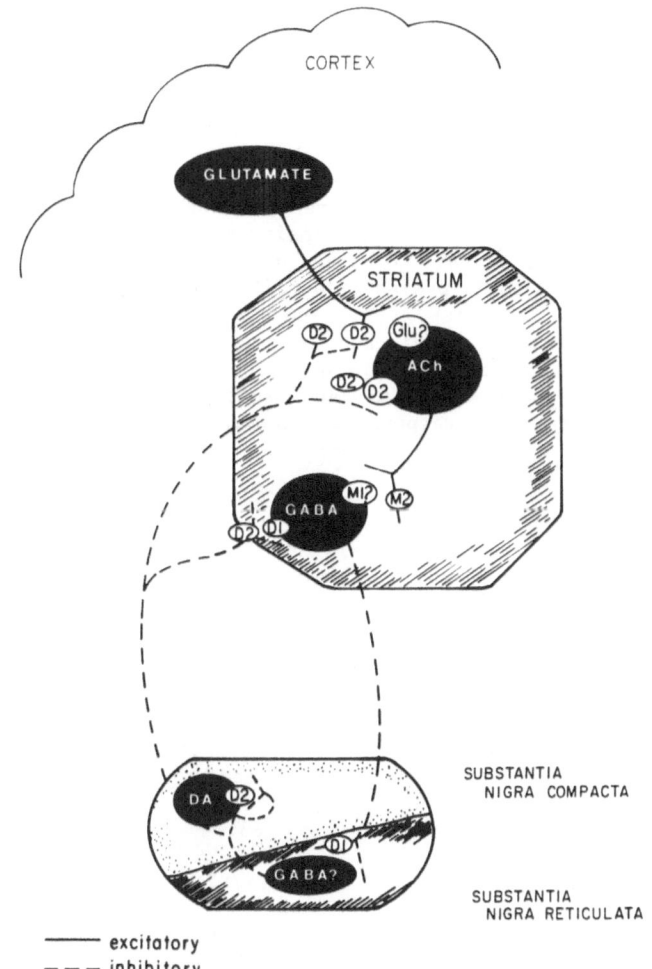

Fig. 1. Schematic representation of neurotransmitter and receptor relationships in striatal pathways.

that the denervation did not alter the cholinergic terminals present in the CP.

The dopaminergic and cholinergic receptor alterations were persistent after the 6-OHDA lesion; they were still present 14 months following the lesion (Dawson, T. et al., 1991a,b). These animals received the 6-OHDA lesion (to the MFB), but did not receive the cell suspension of fetal mesencephalic tissue. However, animals which received the graft approached normal levels of receptor binding with respect to all of the subtypes mentioned above (Table II). These animals could be divided into two groups based on the rotational response to injection of amphetamine. There was a compensated group (which did not rotate, or rotated in a contralateral direction in response to amphetamine) and a noncompensated group, which continued to rotate in an ipsilateral fashion in spite of being grafted. The DA receptors in the compensated group were present in the CP at their prelesion density. The DA uptake sites were present throughout the CP at a level much higher than that in the lesioned animals, but still not back to the "normal" prelesion density. DA terminals are thus apparently spread throughout the CP by a successful graft, but they do not achieve a density as high as normal. This level of DA terminals still appears to be adequate to return the DA receptors (both D_1 and D_2) back to their prelesion density. In the noncompensated animals, no evidence of a viable graft could be found. The DA uptake sites remained low in these animals and the receptor density was similar to that in the lesioned group which did not receive a graft. These transplants were apparently unsuccessful and would account for the continued ipsilateral rotational response and the lack of receptor normalization in this group.

Table II

Effect of Viable Fetal Mesencephalic Cell
Grafts on Ipsilateral Striatal Receptor
Alterations Induced by MFB Lesions

Receptor Type	Change (relative to control)	
	Lesioned	Grafted & Compensated
DA uptake sites	-63.1%	-36.9%
D1 receptors	+26.2%	NS
D2 receptors	+57.7%	NS
choline uptake sites	NS	NS
M1 receptors	-24.6%	NS
M2 receptors	-33.3%	NS

The density of M_1 and M_2 receptors, lowered by the lesion, returned to their normal prelesioned density in the compensated group without a change in the marker for the terminals. The muscarinic receptors in the ipsilateral striatum of animals in the noncompensated group remained low, similar to those in the lesioned animals. The DA receptor alterations caused by the lesion and the restoration induced by the graft would appear to be indicative of functional changes since they indirectly affected the cholinergic receptor system at a cellular level, and may directly relate to the rotational behavior.

DISCUSSION

Persistent alterations in the receptors within the CP after compromising the DA input, on the basis of the connections outlined, would be expected. Dopamine is thought to be inhibitory to cholinergic interneurons and this inhibition is apparently mediated through D_2 receptors (Scatton, 1982). Reducing or eliminating the presence of dopamine-containing terminals within the CP should result in an upregulation of these postsynaptic D_2 receptors. D_1 receptors are also thought to be postsynaptic to these incoming DA terminals, and these receptors are found on the small and medium spiny neurons within the stroma of the CP and also on the projection neurons to the substantia nigra. These receptors would also be expected to upregulate in response to the denervation. Interestingly, D_1 receptors in the SNR also upregulate in response to chronic denervation (shorter periods of denervation may not have this effect; Porceddu et al., 1987) of the DA projections to the striatum. Thus, the same consequence of denervation supersensitivity is taking place both in the region of the deficit (CP) and projections of those affected neurons. The functional consequences of these receptor alterations are unknown.

Dopamine is thought to be inhibitory to cholinergic interneurons, and by removing the source of this inhibition, the cholinergic neurons would be expected to fire more frequently. This would result in an increased release of acetylcholine at the synapse, and postsynaptic (and possibly presynaptic) receptors should downregulate in response to the increased neuronal activity (Kato et al., 1978). In this case, both M_1 and M_2 receptors downregulated, indicating that cholinergic hyperactivity is indeed taking place. The presynaptic M_2 receptors would also be expected to downregulate since they are involved in presynaptic inhibition (James and Cubeddu, 1987), and a chronic presence of acetylcholine in the synaptic cleft would continue to stimulate this receptor population as well. The changes in receptors persist up to 14 months after the lesion. Grafting fetal mesencephalic cells containing DA neurons into the denervated striatum results in the return of these altered receptor populations to normal (Freed et al., 1983; Dawson, T. et al., 1991a,b). The supersensitive D_1 and D_2 receptor populations decrease to their predenervated number both locally within the CP and in the downstream CP projection terminals within the substantia nigra (Dawson, T. et al., 1991a). The M_1 and M_2 receptor populations increase to their "normal" state. The fact that the cholinergic receptors downregulate and subsequently return to normal is intriguing and supports the hypothesis that DA is inhibitory to these neurons.

The prolonged loss of DA containing terminals and their subsequent reappearance as a result of the transplanted tissue was apparent through the localization of [^{3}H]BTCP. The binding of this compound to DA uptake sites was reduced by 63% in lesioned animals even after 14 months had elapsed (Dawson, T. et al., 1991a). Binding of [^{3}H]BTCP, however, was elevated in the direction of the normal predenervated state in animals receiving the transplant. Thus, the DA uptake sites appear spread throughout the CP in the unlesioned control and lesioned animals with transplants, but absent in animals with the 6-OHDA lesion alone.

The animals receiving a successful graft no longer turn in an ipsilateral fashion in response to injected amphetamine. In fact, the animals actually, for the most part, circle in the contralateral direction in response to amphetamine. Thus, it appears that the receptor alterations are involved in mediation of at least some of the behavioral effects of the amphetamine injections, and that the newly innervated system may overcompensate for these effects (Herman et al., 1985). The reasons for the contralateral turning are unknown, but one may surmise that the feedback loop from the substantia nigra to the caudate and back has been severed, so that the return of the receptors to normal compensates in one respect for the behavioral alteration (Arnt and Hyttel, 1984; Robertson and Robertson, 1989), but overcompensates for another.

The DA containing neurons projecting from the SNC to the striatum are thought to be impaired in Parkinson's disease. Postmortem brain tissue from patients with Parkinson's disease who have not received medication (i.e., 1-dopa, which crosses the blood-brain barrier and serves as a precursor to DA synthesis) show both supersensitive D_1 and D_2 receptors in the caudate nucleus (Seeman et al., 1987). Thus, the animal model closely parallels the receptor alterations found in the human condition. The fact that transplants can be used in the rat to bring supersensitive receptors back to their normal predenervated state, and that this level seems to relate to at least one aspect of motor function, is provocative and warrants further research. However, caution must be advised since rushing into an area where not all of the essential problems have been considered or worked out could result in false negatives which would delay future research endeavors in this area.

CONCLUSIONS

(1) 6-OHDA lesions induce changes in postsynaptic receptors within the striatum. The dopaminergic receptors, D_1 and D_2, show increased density or "supersensitivity" while the muscarinic receptors, M_1 and M_2, slightly decrease in density. The denervation-induced supersensitivity of postsynaptic dopamine receptors results from the loss of nigrostriatal dopamine terminals within the CP. There is no evidence that a spontaneous recovery of DA levels occurs within the lesioned CP (i.e. sprouting of DA terminals), as DA uptake sites are significantly depressed and DA receptors remain at a supersensitive level for prolonged periods following a DA denervation. The decrease in muscarinic receptors following the lesion appears to reflect an excess of acetylcholine within the CP resulting from the loss of inhibitory DA control over cholinergic neurons.

(2) Grafts of fetal DA tissue restore postsynaptic receptors to "normal" levels. Both D_1 and D_2 receptors decrease to pre-denervation density levels in animals which display improvements in amphetamine-induced rotational behavior. The "normalization" of DA receptors appears to reflect the increased levels of DA within the CP derived from the DA-containing outgrowths of the graft. Similarly, increased DA within the CP restores inhibitory control over cholinergic neurons, reduces acetylcholine levels and increases muscarinic receptors to control densities.

(3) Changes in DA receptors within the CP following DA-denervation mimic those observed in PD patients prior to drug therapy. Thus, the results obtained from animal models of PD using fetal grafts offer cautious optimism that such techniques may prove useful in a clinical setting.

(4) Different DA receptors are associated with different functions, and dysfunctions, within the brain. Understanding the specific role of these receptors within the brain, possibly through the use of specific receptor agonists or antagonists, may prove useful in developing effective intervention strategies for patients with disorders of the basal ganglia.

ACKNOWLEDGMENTS

The authors acknowledge the secretarial assistance of Karen Meidinger. This work was supported by grants from the Public Health Service (DA05167 and NS22033) and funds from Schering Corporation to J.K.W.; and grants AG6088 and funds from the Margaret and Herbert Hoover Foundation to F.H.G.

LITERATURE CITED

Albin, R.L., Young, A.B., and Penny, J.B., 1989, The functional anatomy of basal ganglia disorders, Trends Neurosci., 12:366-375.

Arnt, J., and Hyttel, J., 1984, Differential inhibition by dopamine D-1 and D-2 antagonists of circling behavior induced by dopamine agonists in rats with unilateral 6-hydroxydopamine lesions, Eur. J. Pharmacol., 102:349-354.

Barnett, A., 1986, Review on dopamine receptors, Drugs Future, 11:49-55.

Barnett, A., Iorio, L.C., McQuade, R.D., and Chipkin, R.E., 1986, Pharmacological and behavioral effects of D1 dopamine antagonists, in: "Central D1 Dopamine Receptors," M. Goldstein, K. Fuxe and I. Tabachnick, eds., Plenum Press, New York, pp. 137-144.

Beckstead, R.M., 1987, Striatal substance P cell clusters coincide with the high density terminal zones of the discontinuous nigrostriatal dopaminergic projection in the cat: A study by combined immunohistochemistry and autoradiographic axon-tracing, Neurosci., 20:557-576.

Bjorklund, A., and Lindvall, O., 1984, Dopamine-containing systems in the CNS, in: "Handbook of Chemical Neuroanatomy", Bjorklund, A. and Hokfelt, T., eds. Elsevier, Amsterdam, 2:55-122.

Bjorklund, A., Lindvall, O., Isacson, O., Brundin, P., Wictorin, K., Strecker, R.E., Clarke, D.J., and Dunnett, S.B., 1987, Mechanisms of action of intracerebral neural implants: Studies on nigral and striatal grafts to the lesioned striatum, Trends Neurosci., 10:509-516.

Bjorklund, A., and Stenevi, U., 1984, Intracerebral neural implants: Neuronal replacement and reconstruction of damaged circuitries, Ann. Rev. Neurosci., 7:279-308.

Bonner, T.I., 1989, The molecular basis of muscarinic receptor diversity, Trends Neurosci., 12:148-151.

Bonner, T.I., Buckley, N.J., Yound, A.C., and Brann, M.R., 1987, Identification of a family of muscarinic acetylcholine receptor genes, Science, 237:527-532.

Bounamici, M., Caccia, D., Carpentieri, M., Pegrassi, L., Rossi, A.C., and Di Chiara, G., 1986, D-1 Receptor supersensitivity in the rat striatum after unilateral 6-hydroxydopamine lesions, Eur. J. Pharmacol., 126:347-348.

Boyson, S.J., McGonigle, P., and Molinoff, P.B., 1986, Quantitative autoradiographic localization of the D_1 and D_2 subtypes of dopamine receptors in rat brain, J. Neurosci., 6(11):3177-3188.

Buckley, N.J., Bonner, T.I., and Brann, M.R., 1988, Localization of a family of muscarinic receptor mRNAs in rat brain, J. Neurosci., 8:4624-4652.

Creese, I., Burt, D.R., Snyder, S.H., 1977, Dopamine receptor binding enhancement accompanies lesion-induced behavioral supersensitivity, Science, 197:596-598.

Creese, I., Sibley, D.R., Hamblin, M.W., and Leff, S.E., 1983, The classification of dopamine receptors: Relationship to radioligand binding, Ann. Rev. Neurosci., 6:43-71.

Dawson, T.M., Barone, P., Sidhu, A., Wamsley, J.K., and Chase, T.N., 1988, The D-1 dopamine receptor in the rat brain: Quantitative autoradiographic localization using an iodinated ligand, Neurosci., 26:83-100.

Dawson, T.M., Dawson, V.L., Gage, F.H., Fisher, L.J., Hunt, M.A., and Wamsley, J.K., 1991a, Functional recovery of supersensitive dopamine receptors after intrastriatal grafts of fetal substantia nigra, Exp. Neurol., [in press].

Dawson, T.M., Dawson, V.L., Gage, F.H., Fisher, L.J., Hunt, M.A., and Wamsley, J.K., 1991b, Downregulation of muscarinic receptors in the rat caudate-putamen after lesioning of the ipsilateral nigrostriatal dopamine pathway with 6-hydroxydopamine (6-OHDA): Normalization by fetal mesencephalic transplants, Brain Res., [in press].

Dawson, T.M., Gehlert, D.R., McCabe, R.T., Barnett, A., and Wamsley, J.K., 1986a, D-1 dopamine receptors in the rat brain: A quantitative autoradiographic analysis, J. Neurosci., 8:2352-2365.

Dawson, T.M., Gehlert, D.R., and Wamsley, J.K., 1986b, Quantitative autoradiographic localization of central dopamine D-1 and D-2 receptors, in: "Neurobiology and Central D1-Dopamine Receptors," G.R. Breese and I. Creese, eds., Plenum Press, New York, pp. 93-118.

Dawson, T.M., Gehlert, D.R., Yamamura, H.I., Barnett, A., and Wamsley, J.K., 1985, D-1 dopamine receptors in the rat brain: Autoradiographic localization using [^{3}H]SCH23390, Eur. J. Pharmacol., 180:323-325.

Dawson, V.L., Dawson, T.M., Filloux, F.M., and Wamsley, J.K., 1988, Evidence for dopamine D-2 receptors on cholinergic interneurons in the rat caudate-putamen, Life Sci., 42:1933-1939.

Dawson, V.L., Dawson, T.M., and Wamsley, J.K., 1990, Muscarinic and dopaminergic receptor subtypes on striatal cholinergic interneurons. Brain Res. Bulletin, 25:903-912.

Fage, D., and Scatton, B., 1986, Opposing effects of D-1 and D-2 receptor antagonists on acetylcholine levels in rat striatum, Eur. J. Pharmacol., 129:359-362.

Filloux, F., Dawson, T.M., and Wamsley, J.K., 1988a, Localization of nigrostriatal dopamine receptor subtypes and adenylate cyclase, Brain Res. Bulletin, 20:447-459.

Filloux, F., Hunt, M.A., and Wamsley, J.K., 1989, Localization of the dopamine uptake complex using [^{3}H]-[1-(2-benzo(B)thiophenyl) cyclohexyl] piperidine ([^{3}H]BTCP) in rat brain, Neurosci. Lett., 100:105-110.

Filloux, F., Liu, T.H., Hsu, C.Y., Hunt, M.A., and Wamsley, J.K., 1988b, Selective cortical infarction reduces [^{3}H]sulpiride binding in rat caudate-putamen: Autoradiographic evidence for presynaptic D2 receptors on corticostriate terminals, Synapse, 2:521-531.

Filloux, F., Wamsley, J.K., and Dawson, T.M., 1987a, Dopamine D-2 auto- and postsynaptic receptors in the nigrostriatal system of the rat brain: localization by quantitative autoradiography with [^{3}H]-sulpiride, Eur. J. Pharmacol., 138:61-68.

Filloux, F., Wamsley, J.K., and Dawson, T.M., 1987b, Presynaptic and postsynaptic D-1 dopamine receptors in the nigrostriatal system of the rat brain: A quantitative autoradiographic study using the selective D-1 antagonist [^{3}H]-SCH 23390, Brain Res., 408:205-209.

Freed, W.J., Ko, G.N., Niehoff, D.L., Kuhar, M.J., Hoffer, B.J., Olson, L., Cannon-Spoor, H.E., Morihisa, J.M., and Wyatt, R.J., 1983, Normalization of spiroperidol binding in the denervated rat striatum by homologous grafts of substantia nigra, Science, 222:937-939.

Freund, T.F., Bolam, J.P., Bjorklund, A., Stenevi, U., Dunnett, S.B., Powell, J.F., and Smith, A.D., 1985, Efferent synaptic connections of grafted dopaminergic neurons reinnervating the host neostriatum: A tyrosine hydroxylase immunocytochemical study, J. Neurosci., 5:603-616 (1985).

Gehlert, D.R., Dawson, T.M., Filloux, F.M., Sanna, E., Hanbauer, I., and Wamsley, J.K., 1987, Evidence that [³H]-forskolin binding in the substantia nigra is intrinsic to a striatal-nigral projection: An autoradiographic study of rat brain, Neurosci. Lett., 73:114-118.

Gehlert, D.R., and Wamsley, J.K., 1984, Autoradiographic localization of [³H]-sulpiride binding sites in the rat brain, Eur. J. Pharmacol., 98:311-312.

Gehlert, D.R. and Wamsley, J.K., 1985, Dopamine receptors in the rat brain: Quantitative autoradiographic localization using [³H]-sulpiride, Neurochem. Int., 7:717-723.

Graybiel, A.M., and Ragsdale, C.W., 1983, Biochemical anatomy of the striatum, in: "Chemical Neuroanatomy", P.C. Emson, ed., Raven Press, New York, pp.427-504.

Hall, M.D., Kelly, J.D., and Marsden, C.D., 1983, Differential anatomical location of [³H]-spiperone binding sites in the striatum and substantia nigra of the rat, Br. J. Pharmac., 79:599-610.

Herman, J.P., Choulli, K., Le Moal, M., 1985, Hyper-reactivity to amphetamine in rats with dopaminergic grafts, Exp. Brain Res., 60:521-526.

Hess, E.J., Bracha, H.S., Kleinman, J.E., and Creese, I., 1987, Dopamine receptor subtype imbalance in schizophrenia, Life Sci., 40:1487-1497.

Izzo, P.N., and Bolam, J.P., 1988, Cholinergic synaptic input to different parts of spiny striatonigral neurons in the rat, J. Comp. Neurol., 269:219-234.

James, M.K., and Cubeddu, L.X., 1987, Pharmacologic characterization and functional role of muscarinic autoreceptors in the rabbit striatum, J. Pharmacol. Exp. Ther., 240:203-215.

Joyce, J.N., and Marshall, J.F., 1985, Striatal topography of D-2 receptors correlates with indexes of cholinergic neuron localization, Neurosci. Lett., 53:127-131.

Kato, G., Carson, S., Kemel, M.L., Glowinski, J., and Giorguieff, M.F., 1978, Changes in striatal specific ³H-atropine binding after unilateral 6-hydroxydopamine lesions of nigrostriatal dopaminergic neurones, Life Sci., 22:1607-1614.

Kebabian, J.W., and Calne, D.B., 1979, Multiple receptors for dopamine, Nature, 277:93-96.

Kubo, T., Maeda, A., Sugimoto, K., Akiba, I., Mikami, A., Takahashi, H., Mishina, H., Haga, T., Haga, K., Ichiyama, A., Kangawa, K., Kojima, M., Matuso, M., Hirose, T., and Numa, S., 1986, Cloning, sequencing and expression of complementary DNA encoding the muscarinic acetylcholine receptor, Nature, 323:411-416.

Kubota, Y., Inagaki, S., Kito, S., and Wu, J.-Y., 1987a, Dopaminergic axons directly make synapses with GABAergic neurons in the rat neostriatum, Brain Res., 406:147-156.

Kubota, Y., Inagaki, S., Shimada, S., Kito, S., Eckenstein, F., and Tohyama, M., 1987b, Neostriatal cholinergic neurons receive direct synaptic inputs from dopaminergic axons, Brain Res., 413:179-184.

Kuhar, M.J., DeSouza, E.B., and Unnerstall, J.R., 1986, Neurotransmitter receptor mapping by autoradiography and other methods, Ann Rev. Neurosci., 9:27-59.

Lehman, J., and Langer, S.Z., 1983, The striatal cholinergic interneuron: Synaptic target of dopaminergic terminals?, Neurosci., 10:1105-1120.

Limberger, N., Spath, L., and Starke, K., 1986, A search for receptors modulating the release of [³H]gamma-aminobutyric acid in rabbit caudate nucleus slices, J. Neurochem., 46:1109-1117.

Lloyd, K.G., 1978, Neurotransmitter interactions related to central dopamine neurons, in: "Essays in Neurochemistry and Neuropharmacology", M.B.H. Youdim, W. Lovenberg, D.F. Sharman and J.R. Lagnado, eds., Wiley, New York, 3:131.

Marshall, J.F., and Ungerstedt, U., 1977, Supersensitivity to apomorphine following destruction of the ascending dopamine neurons: Quantification using the rotational model, Eur. J. Pharmacol., 41:361-367.

McGurk, S.R., Hartgraves, S.L., Kelly, P.H., Gordon, P.H., and Butcher, L.L., 1987, Is ethylcholine mustard aziridinium ion a specific cholinergic neurotoxin, Neurosci., 222:215-224.

Mishra, R.K., Gardner, E.L., Katzman, R., and Makman, M.H., 1974, Enhancement of dopamine-stimulated adenylate cyclase activity in rat caudate after lesions in the substantia nigra: Evidence for denervation supersensitivity, Proc. Natl. Acad. Sci. U.S.A., 71:3883-3887.

Ohno, Y., Sasa, M., and Takaori, S., 1987, Coexistence of inhibitory dopamine D-1 and excitatory D-2 receptors on the same caudate nucleus neurons, Life Sci., 40:1937-1945.

Porceddu, M.L., Giorgi, O., De Montis, G., Mele, S., Cocco, L., Ongini, E., and Biggio, G., 1987, 6-Hydroxydopamine-induced degeneration of nigral dopamine neurons: Differential effect on nigral and striatal D-1 dopamine receptors, Life. Sci., 41:697-706.

Porceddu, M.L., Giorgi, O., Ongini, E., Mele, S., and Biggio, G., 1986, [3H]-SCH 23390 binding sites in the rat substantia nigra: Evidence for a presynaptic localization and innervation by dopamine, Life Sci., 39:321-328.

Potter, L.T., Flynn, D.D., Hanchet, H.E., Kalinoski, D.L., Luber-Narod, J., and Mash, D.C., 1984, Independent M_1 and M_2 receptors, ligands, autoradiography and function, Trends Pharmacol. Sci., Suppl 22-31.

Regenold, W., Araujo, D., and Quirion, R., 1987, Direct visualization of brain M_2 muscarinic receptors using the selective antagonist [3H]AF-DX116, Eur. J. Pharmacol., 144:417-419.

Robertson, G.S., and Robertson, H.A., 1989, Evidence that l-dopa-induced rotational behavior is dependent on both striatal and nigral mechanisms, J. Neurosci., 9:3326-3331.

Sandberg, K., Hanin, I., Fisher, A., and Coyle, J.T., 1984, Selective cholinergic neurotoxin AF64A's effects in rat striatum, Brain Res., 293:49-55.

Savasta, M., Dubois, A., Benavides, J., and Scatton, B., 1986, Different neuronal location of [3H]SCH 23390 binding sites in pars reticulata and pars compacta of the substantia nigra in the rat, Neurosci. Lett., 72:265-271.

Savasta, M., Dubois, A., Benavides, J., and Scatton, B., 1988, Different plasticity changes in D_1 and D_2 receptors in rat striatal subregions following impairment of dopaminergic transmission, Neurosci. Lett., 85:119-124.

Scatton, B., 1982, Further evidence for the involvement of D_2, but not D_1 dopamine receptors in dopaminergic control of striatal cholinergic transmission, Life Sci., 31:2883-2890.

Scheel-Krüger, J., 1986, Dopamine-GABA interactions: Evidence that GABA transmits, modulates, and mediates dopaminergic functions in the basal ganglia and limbic system, Acta. Neurol. Scand., 73:S107.

Schmidt, R.H., Bjorklund, A., Stenevi, U., Dunnett, S.B., Gage, F.H., 1983, Intracerebral grafting of neuronal cell suspensions III. Activity of intrastriatal nigral suspension implants as assessed by measurements of dopamine synthesis and metabolism, Acta Physiol. Scand., Suppl 522:19-28.

Schoffelmeer, A.N.M., VanVliet, B.J., Wardeh, G., and Mulder, A.H., 1986, Muscarinic receptor-mediated modulation of [3H]-dopamine and [14C]-acetylcholine release from rat neostriatal slices: Selective antagonism by gallamine but not pirenzepine, Eur. J. Pharmacol., 128:291-294.

Seeman, P., Bzowej, N.H., Guan, H.C., Bergeron, C., Reynolds, G.P., Bird, E.D., Riederer, P., Jellinger, K., and Tourtellotte, W.W., 1987, Human brain D_1 and D_2 dopamine receptors in schizophrenia, Alzheimer's, Parkinson's, and Huntington's diseases, Neuropsychopharmacol., 1:5-15.

Seiler, M.P., and Markstein, R., 1984, Further characterization of structural requirements for agonists at the striatal dopamine D-2 receptor: Studies with a series of monohydroxyaminotetralins on acetylcholine release from rat striatum, Mol. Pharmacol., 26:452-457.

Sokoloff, P., Giros, B., Martres, M.-P., Bouthenet, M.-L., and Schwartz, J.-C., 1990, Molecular cloning and characterization of a novel dopamine receptor (D_3) as a target for neuroleptics, Nature, 347:146-151.

Starr, M., 1987, Opposing roles of dopamine D_1 and D_2 receptors in nigral gamma-[^{3}H]-aminobutyric acid release, J. Neurochem., 49:1042-1049.

Ungerstedt, U., 1971, Striatal dopamine release after amphetamine or nerve degeneration revealed by rotational behavior, Acta. Physiol. Scand. Suppl., 367:49-68.

Ungerstedt, U., and Arbuthnott, G.W., 1970, Quantitative recording of rotational behavior in rats after 6-hydroxydopamine lesions of the nigrostriatal dopamine system, Brain Res., 24:485-493.

Vickroy, T.W., Roeske, W.R., Gehlert, D.R., Wamsley, J.K., and Yamamura, H.I., 1985, Quantitative light microscopic autoradiography of [^{3}H]-hemicholinium-3 binding sites in the rat central nervous system: A novel biochemical marker for mapping the distribution of cholinergic nerve terminals, Brain Res., 329:368-378.

Waddington, J.L., and O'Boyle, K.M., 1989, Drugs acting on brain dopamine receptors: A conceptual reevaluation five years after the first selective D-1 antagonist, Pharmacol. Ther., 43(1):1-52.

Wamsley, J.K., Gehlert, D.R., Filloux, F.M., and Dawson, T.M., 1989, Comparison of the density and distribution of D-1 and D-2 dopamine receptors in the rat brain, J. Chem. Neuroanat., 2:119-137.

Wamsley, J.K., Gehlert, D.R., Roeske, W.R. amd Yamamura, H.I., 1984, Muscarinic antagonist binding site heterogeneity as evidenced by autoradiography after direct labeling with [^{3}H]-QNB and [^{3}H]-pirenzepine, Life Sci., 34:1395-1402.

Wamsley, J.K., Lewis, M.S., Young III, W.S., and Kuhar, M.J., 1981, Autoradiographic localization of muscarinic cholinergic receptors in rat brainstem, J. Neurosci., 1:176-191.

Wang, J., Roeske, W.R., Hawkins, K.N., Gehlert, D.R., and Yamamura, H.I., 1989, Quantitative autoradiography of M_2 muscarinic receptors in the rat brain identified by using a selective radioligand [^{3}H]AF-DX 116, Brain Research, 477:322-326.

Watson, M., Roeske, W.R., Vickroy, T.W., Smith, T.L., Akiyama, K., Guyla, K., Duckles, S.P., Serra, M., Adem, A., Nordberg, A., Gehlert, D.R., Wamsley, J.K., and Yamamura, H.I., 1986, Biochemical and functional basis of putative muscarinic receptor subtypes and its implications, Trends. Pharmacol. Sci., 7:(Supp. II) 46-55.

Weiner, D.M., and Brann, M.R., 1989, The distribution of a dopamine D2 receptor mRNA in rat brain, FEBS Letters, 253:207-213.

Yamamura, H.I., Wamsley, J.K., Deshmukh, P., and Roeske, W.R., 1983, Differential light microscopic autoradiographic localization of muscarinic cholinergic receptors in the brainstem and spinal cord of the rat using [^{3}H]-pirenzepine, Eur. J. Pharmacol., 91:147-149.

KEY WORDS: D_1, D_2 receptors; M_1, M_2 receptors; cell transplants; grafting; nigrostriatal pathway; dopamine receptor subtypes; fetal cell transplants; dopamine cell transplants; fetal cell grafting; dopamine cell grafting; presynaptic receptors; postsynaptic receptors; 6-OHDA lesion; acetylcholine receptors; muscarinic receptor subtypes

OCTOPAMINE: AN ENDOGENOUS BLOCKER OF

DOPAMINE D-1 RECEPTORS

*Juei-Tang Cheng and *Jui-Ting Tsai*

Department of Pharmacology and *Ophthalmology
College of Medicine, National Cheng Kung University
Tainan City, Taiwan 70101, Republic of China

Introduction

Octopamine, an important neurohormone in invertebrates (Robertson and Juorio, 1976; Axelrod and Saavedra, 1977; David and Coulon, 1985), has been recognized as the false transmitter of adrenergic neurotransmission for a long time (Korol et al., 1968). Octopamine has been identified in mammalian brain (Harmer and Horn, 1976; Duffield et al., 1981) and a deficiency has been found to be associated with depressive states (Sandler et al., 1983). In the animal behaviors, octopamine has been reported to increase motor activity (Hicks, 1977) and avoidance conditioning (Delacour et al., 1983). Recently, we found that octopamine induced the relaxation of mammalian intestine by an activation of dopamine DA-1 receptors (Cheng and Hsieh-Chen, 1988). In the present report we will briefly discuss some recent findings that indicate the possible role of octopamine on dopamine D-1 receptor in mammalian brain.

Effect on the bindings of [^{3}H]SCH23390 and the activity of dopamine D-1 receptors

·In the isolated striatal membrane of rats, we found that octopamine produced a dose-dependent displacement of [^{3}H]SCH23390 bindings (Cheng et al., 1990). Also, octopamine reduced the dopamine-stimulated formation of cyclic AMP in the presence of sulpiride. Thus, the inhibitory effect of octopamine on dopamine D-1 receptors of rat brain can be considered. Then, we analyzed the relationship of octopamine-induced actions between the reduction of cyclic AMP formation and the displacement of [^{3}H]SCH23390 bindings.

As shown in Fig. 1, there is a positive relationship between these two actions where the correlative coefficient is 0.998 ($P < 0.001$). From these results, we suggest that octopamine may bind with dopamine D-1 receptors to reduce the action of dopamine. This view is supported by our previous findings that octopamine did not affect the increase of cyclic AMP induced by forskolin at the concentrations sufficient to block the formation of cyclic AMP stimulated by dopamine (Cheng et al., 1990).

Neuroreceptor Mechanisms in Brain, Edited by S. Kito *et al.*
Plenum Press, New York, 1991

Protection of dopamine D-1 receptor sites

According to the previous reports (Meller et al., 1985; Hess et al., 1987), N-ethoxycarbonyl-2-ethoxy-1,2-dihydroquinoline (EEDQ) produced an inactivation of dopamine receptors. We employed this compound to investigate the possible protective effect of octopamine recently.

We used Balb/C strain mice of both sexes (weighing 20 to 25 g) that were housed in a five to six in a cage, had free access to food and water, and were maintained on a 12-hr light-dark cycle room under constant temperature (26 - 28°C). All animals received intraperitoneal (i.p.) injections of EEDQ(Aldrich) at the dose of 6 mg/kg/day, as previously reported (Meller et al., 1985). Octopamine (Sigma) and SCH23390 (RBI,Natick,MA), at the dose of 10 mg/kg/day, were also i.p. injected at 30 min before the administration of EEDQ. Animals received the i.p. injection of vehicle at same volume were taken as control. After continuous injection for 3 days, mice were sacrificed at 24 hr later of the last injection with EEDQ. The corpus striatum and cortical hemispheres were dissected to prepare the membrane fractions. Receptor binding study was then carried out according to the previous method (Billard et al., 1984).

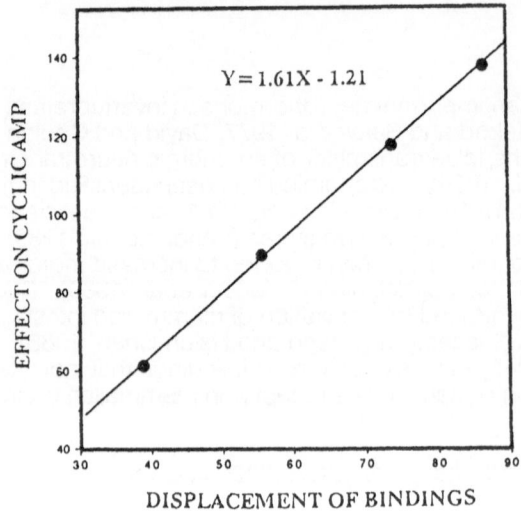

$$Y = 1.61X - 1.21$$

DISPLACEMENT OF BINDINGS

Fig.1 Relationship between displacement of [³H]SCH23390 binding (%) and reduction of dopamine-stimulted formation of cyclic AMP (pmol/min/mg protein) for 4 different concentrations of octopamine. The correlation coefficient is 0.998 (P < 0.001).

Fig. 2 shows the results of this *Ex Vivo* binding of dopamine D-1 receptors using [³H]SCH23390 as radioligand. Because EEDQ functions as the blocker of dopamine receptors, binding of radioligand in the control animals can be considered as the non-specific binding. Then, binding of [³H]SCH23390 in the animals pretreated with octopamine or SCH23390 indicates the combination of specific binding sites, that were *In Vivo* protected from the blockade of EEDQ, and some non-specific bindings. After substraction with the non-specific bindings in control group, the specific binding values can be obtained to represent the active sites that were selectively protected by octopamine or SCH23390 from the inactivation of EEDQ.

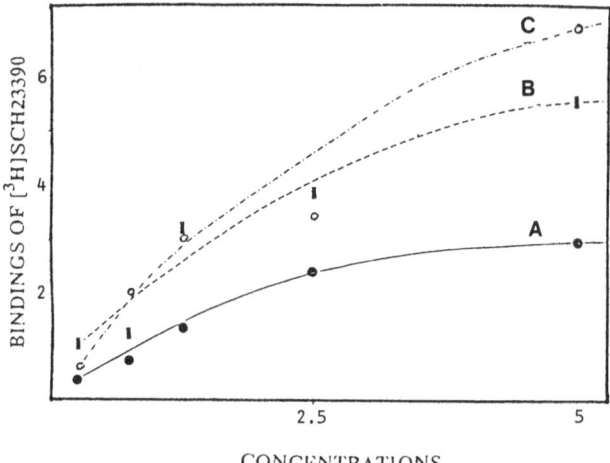

Fig.2 Bindings of [³H]SCH23390 on the striatal membrane isolated from the EEDQ-treated mice (A), ocotpamine + EEDQ-treated mice (B), and SCH23390 + EEDQ-treated mice (C). All values are the mean from 6 mice.

Table 1 shows the values of Kd and Bmax from Scatchard plot of specific [³H]SCH23390 binding in octopamine-protected and SCH23390-pretreated mice. There is no statistical difference ($P > 0.01$) between these two values.

The prevention of EEDQ-induced inactivation of dopamine D-1 receptors by octopamine is likely to be due to site-directed occupancy of the radioligand binding site. Similar results were also obtained in the unlabelled SCH23390-treated mice, indicating that the pharmacokinetic interactions and/or crossed interruptions between octopamine and EEDQ can thus be ruled out.

Blockade of dopamine D-1 receptors by EEDQ was found both *in vivo* and *in vitro* (Hess et al., 1987). The action mechanism of EEDQ is attributed to occur via peptide carboxyl group activation (Meller et al., 1985). Octopamine produced the *in vivo* protection of dopamine receptors from EEDQ-induced inactivation of brain, suggesting that octopamine passed the blood-brain barrier to enter central nervous system. Thus, the active receptor sites occupied by octopamine were reserved for the binding of [³H]SCH23390, the useful profile in the identification of dopamine D-1 receptors (Hyttel, 1983).

There are several factors shall be considered in the *In Vivo* study, however, the obtained results identically suggest that octopamine may bind with dopamine D-1 receptors in striatum to protect the EEDQ-induced inactivations.

Table 1. The activity of dopamine D-1 receptors in striatal membrane protected by octopamine or SCH23390 from the blockade of EEDQ.

	Dopamine D-1 Receptors*	
	Kd (nM)	Bmax(pmol/mg)
Octopamine-induced	9.5	107.4 ± 26.7
SCH23390-induced	9.7	112.9 ± 18.4

Receptor activity was analyzed by Scatchard plot through EBDA computer program.

Conclusion

Octopamine has the affinity to bind with dopamine D-1 receptors. Combined with the findings of the presence of octopamine in mammalian brain (Harman and Horn, 1976; Duffield et al., 1981), our results may have major implications both for characterizating the role of octopamine in neuropyschological disorders and for the future development of compounds available as specific antagonists of dopamine-D-1 receptors.

Acknowledgement

We appreciate Prof. Hal Malmud for reading of this report. Thanks are also due to Miss T.F. Yang and Mr. Peter Lian for excellent technical assistance. The present work was supported in part by a grant from National Science Council of the Republic of China (NSC79-0412-B006-08).

References

Axelrod, J. and Saavedra, J.M. 1977 Octopamine. Nature, 265:501 - 504.

Billard, W., Ruperto, V., Crosby, G., Iorio, L.C. and Barnett, A. 1984 Characterization of the binding of [^{3}H]SCH 23390, a selective D-1 receptor antagonist ligand, in rat striatum. Life Sci., 35:1885 - 1893.

Cheng, J.T. and Hsieh-Chen, S.C. 1988 Octopamine relaxes rabbit jejunal smooth muscle by selective activation of dopamine D1 receptors. Naunyn-Schmiedeberg's Arch. Pharmacol., 338:373 - 378 .

Cheng, J.T., Shen, C.L. and Jou, T.C. 1990 Inhibitory effect of octopamine on the striatal dopamine D-1 receptor of rats. Brain Res. (submitted)

David, J.C. and Coulon, J.F. 1985 Octopamine in invertebrates and vertebrates: a review. Prog. Neurobiol., 24:141 - 185 .

Delacour, J., Coulon, J.F., David, J.C. and Guenaine, C. 1983 Brain octopamine and strain differences in avoidance behavior. Brain Res., 288:169 - 176.

Duffield, P.H., Dougan, D.F.H., Wade, D.N. and Duffield, A.M. 1981 A chemical ionization gas chromatographic mass spectrometric assay for octopamine and tyramine in rat brain. Biomed. Mass Spectrom., 8:170 - 173 .

Harmar, A.J. and Horn, A.S. 1976 Octopamine in mammalian brain: rapid post mortem increase and effects of drugs. J. Neurochem., 26:987 - 993

Hess, E.J., Battaglia, G., Norman, A.B. and Creese, I. 1987 Differential modification of striatal D-1 dopamine receptors and effector moieties by-ethoxycarbonyl-2-ethoxy-1,2-dihydroquinoline in vivo and in vitro. Mol. Pharmacol., 31:50 - 57.

Hicks, T.P.1977 The possible role of octopamine as a synaptic transmitter: a review. Canad. J. Physiol. Pharmacol., 55:137 - 152 .

Hyttel, J. 1983 SCH23390-The first selective dopamine D-1 antagonist. Europ. J. Pharmacol., 19:153 - 154.

Korol, B., Soffer, L., and Brown, M.L. 1968 Some cardiovascular studies on octopamine. Arch. int. Pharmacodyn., 171:415 - 424 .

Meller, E., Bohmaker, K., Goldstein, M. and Friedhoff, A.J. 1985 Inactivation of D-1 and D-2 dopamine receptors by N- ethoxycarbonyl-2-ethoxy-1,2-dihydroquinoline in vivo: selective protection by neuroleptics. J. Pharmacol. Expt. Therap., 233:656 - 662.

Robertson, H.A. and Juorio, A.V. 1976 Octopamine and some related non-catecholic amines in invertebrate nervoussystem. Ann. Rev. Neurobiol., 19: 173 - 224 .

Sandler, M., Ruthven, C.R.J., Goodwin, B.L., Reynolds, G.P., Rao, V.A.R. and Coppen, A. 1979 Deficient production of tyramine and octopamine in cases of depression. Nature, 278:257 - 358 .

DEPRESSION OF NMDA-EVOKED ACETYLCHOLINE RELEASE BY ENDOGENOUS
ADENOSINE IN STRIATUM SLICES

Elena Porsche and Thomas Schwan

Hoechst AG Werk Kalle-Albert, 6200 Wiesbaden, FRG

ABSTRACT

The mechanisms of NMDA-evoked transmitter release are still not clear. We demonstrate for the first time that NMDA-evoked acetylcholine release is depressed by endogenous liberated adenosine in functionally-hypoxic and non-hypoxic brain slices. Adenosine deaminase potentiates the release of ^{3}H-acetylcholine from rat striatum slices in response to N-Methyl-D-Aspartate (NMDA) stimulation, whereas (-)-R-phenylisopropyl-adenosine (R-PIA) decreases the NMDA-evoked transmitter release. The NMDA-evoked transmitter release is potentiated by 8-cyclopentyl-1,3-di-propylxanthine (DPCPX) only in adenosine depleted slices. The increase of extraneuronal adenosine by dipyridamole also depresses the presynaptic transmitter release.

Our results indicate: (i) that stimulation of NMDA-receptors induces a complex neuronal response, e.g., release of acetylcholine and release of adenosine, (ii) that adenosine induces an agonist high affinity con-formation of the A_1-receptors which apparently are insensitive to an A_1-adenosine receptor-antagonist (DPCPX) and (iii) that brain slices 0.5 mm thick can serve as an experimental model for testing antihypoxic substances.

INTRODUCTION

Adenosine is released by K^+ and glutamate from brain slices, as recently shown by Hoehn and White (1990). Previous studies demonstrated that exogenous adenosine depresses neuroexcitation and the evoked trans-mitter release (Phillis and Wu 1981, Reddingten et al. 1982, Stone and Bartrup 1988, Fredholm et al. 1989) via the inhibitory, presynaptic adenosine A_1-receptors. However, there is less information on inhibition by endogenous adenosine (Jackisch et al. 1984, Haas and Greene 1988). Considering the neuroprotective role played by adenosine in hypoxic/ischemic neuronal degeneration where excitatory glutamate receptors of NMDA-type are powerfully involved (Rothmann and Olney 1987), the question emerged, whether endogenous adenosine is able to inhibit neuro-transmitter release evoked by specific NMDA-stimulation.

Neuroreceptor Mechanisms in Brain, Edited by S. Kito *et al.*
Plenum Press, New York, 1991

Elucidation of the interactions between NMDA- and A_1-adenosine receptors as a particular feature of physiological neuroprotection is also important for the understanding of brain vulnerability and regional differences in hypoxia or ischemia, and for therapeutic reasons.

Our purpose was to establish an in vitro model of functional brain hypoxia caused by an NMDA-receptor stimulation. In the present study we determined the time course of acetylcholine release evoked by NMDA-receptor stimulation in 0.3 and 0.5 mm thick striatum slices. Adenosine deaminase and dipyridamole were employed in order to detect the effect of endogenous adenosine on neurotransmitter release.

Further, the effects of (-)-R-phenylisopropyladenosine (PIA) as agonist, and 8-cyclopentyl-1,3-dipropylxanthine (DPCPX) as antagonist of A_1-adenosine receptors on the NMDA-stimulated acetylcholine release were investigated. The results support our assumption that endogenous adenosine is released in response to stimulation with NMDA.

MATERIAL AND METHODS

Striatum slices were obtained from Sprague Dawley female rat brain using a McIlwain slicer. Superfusion experiments were performed based on those previously described by Wichmann et al. (1988). The superfusion chambers were fitted with a special valve for rapid exchange of superfusion fluids in less than 10 s and for decreasing the dead volume ($<$ 0.150 ml). For details see legend of Fig. 1 and Fig. 2. Tritium in solubilized slices and superfusate samples was determined by liquid scintillation counting using a Packard 1900 CA at 50 % efficiency.

(Methyl-^{3}H)-choline chloride, specific activity 80.0 Ci/mmol, was purchased from Amersham (Braunschweig, FRG), (-)-R-N^6-phenylisopropyl-adenosine and adenosine deaminase from Boehringer Mannheim (Mannheim, FRG), 2,6-bis/diaethanolamino)-4,8-dipiperidino-pyrimido(5,4-d)pyrimidine, dipyridamole from Thomae (Biberach, FRG) and 8-cyclopentyl-1,3-dipropyl-xanthine (DPCPX) from Research Biochemicals Incorporated (Natic, MA, USA). (+)--5,5-Methyl-10,11-dihydro-H-dibenzo(a,D)-cyclohept-5,10-imine maleate (MK-801) was supplied by Dr. Faasch from Hoechst AG, Werk Kalle-Albert, (Wiesbaden, FRG). Drugs were dissolved in superfusion medium except DPCPX (dimethylsuphoxide: final concentration in superfusion fluid was 1%).

DATA PRESENTATION AND STATISTICAL EVALUATION

Basal and NMDA-stimulated release of tritium is expressed as fractional rate in percent (Limberger et al. 1988, Wichman et al. 1988). Results are expressed as means $\pm$ SD. Differences between groups were tested for significance with Mann-Whitney test for unpaired samples. Differences with $p < 0.05$ were considered to be statistically significant.

RESULTS

The stimulatory effect of NMDA is dependent on the presence of magnesium ions in superfusion medium (results not shown) and abolished by 10 μM of NMDA-receptor-channel complex antagonist, MK-801. It can be assumed that the evoked transmitter release is mediated by excitatory amino acid receptors of the NMDA-type. The tritium overflow is decreased during the second stimulation with NMDA (S_2, Table 1) suggesting that the NMDA receptor was desensitized, as observed by Fink et al. (1989).

Summarized results:

1. Similar amounts of ACH were released from both 0.3 mm or 0.5 mm striatum slices after NMDA-stimulation, demonstrating that both ticknesses were not hypoxic (Table 1).

2. Adenosine deaminase increases the ACH-release under following conditions.
a) when added together with NMDA during the S_1 or S_2 stimulation. This effect was observed in 0.5 mm thick slices only (Fig. 1,B)
b) when added continuously for 40 min (Fig. 1, C and D)
c) when added for 10 min before slices were stimulated with NMDA + DPCPX (Fig. 2)
d) when slices were previously superfused with dipyridamole (Table 2, experiments 5 and 7).

3. Dipyridamole decreases ($p < 0.05$) the NMDA-mediated ACH-release (Table 2, exp. 4). This effect is partially reversed by adenosine deaminase (see above).

4. R-PIA decreases the NMDA-mediated ACH-release ($p < 0.05$). This effect is not reversed by adenosine deaminase (Table 2, exp. 6 and 7).

5. DPCPX potentiates the NMDA-mediated ACH-release in 0.5 mm thick slices ($p < 0.05$) and 0.3 mm thick slices. The effect was observed in slices previously superfused with adenosine deaminase (Fig. 2, lower left panel and Table 2, exp. B).

Table 1. NMDA induced ^{3}H-acetylcholine release in striatum slices: Comparison between 0.5 and 0.3 mm thick slices

Slices	n (slices)	^{3}H-Overflow (fractional rate: %/min)		S_2/S_1
		S_1	S_2	
0.5 mm	20	1.46 $\pm$ 0.36	0.99 $\pm$ 0.24	0.68 $\pm$ 0.09
0.3 mm	22	1.40 $\pm$ 0.20	0.89 $\pm$ 0.17	0.64 $\pm$ 0.10

The values are means $\pm$ standard deviation; slices were stimulated with 25 µM NMDA for two minutes (S_1 respectively S_2).

DISCUSSION

It has often been demonstrated that the adenosine inhibitory system is activated by hypoxia or by excitation (Phillis and Wu 1981, Reddington et al. 1982, Schubert 1988). The inhibition of neuronal activity by adenosine is presynaptic (Fredholm et al. 1989, Phillis 1989, Stone and Bartrup 1989).

We document that adenosine contributes directly to the depression of NMDA-evoked acetylcholine release. The 0.5 mm thick slices released adenosine in response to the NMDA stimulation as did hypoxic slices. Under NMDA-stimulation thick slices develop a functional hypoxic state (FHS). They can

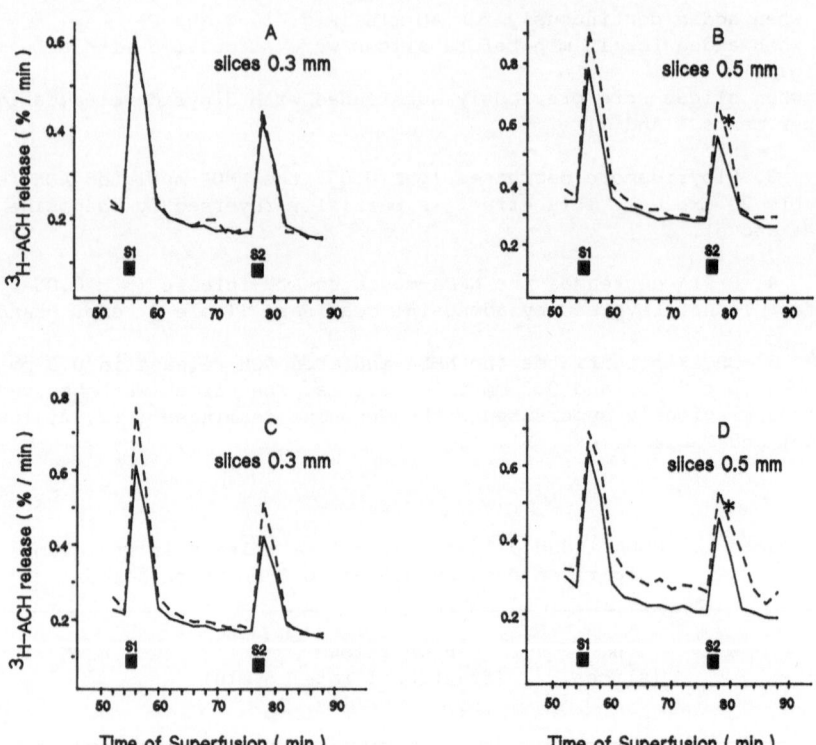

Fig. 1.   Time course of NMDA-evoked ^{3}H-acetylcholine (ACH) release from
striatum slices. Slices 0.3 mm thick and 0.5 mm thick were
preincubated for 30 min. in 2 ml medium containing ^{3}H-choline
0.1 µM, specific activity 80 Ci/mmol. One slice was transferred to
each of six chambers and superfused with medium for 90 min. at a
rate of 1.5 ml/ min. Incubation medium contained (mM) NaCl 118,
KCl 4.8, CaCl$_2$ 1.2, MgSO$_4$ 1.2, NaHCO$_2$ 25, KH$_2$PO$_4$ 1.2, Glucose 10.
Media were saturated with 5 % CO$_2$ + 95 % O$_2$, warmed to 37° C, pH
was adjusted to 7.4. MgSO$_4$ was omitted from superfusion fluid.
Two-minute fractions of the superfusate were collected continuously. 25 µM NMDA was applied to each slice for two minutes (S$_1$, S$_2$,
solid line). 0.2 IU/ml adenosine deaminase were added with NMDA^2
(panel A and B, dashed line) or continuously, for 40 min. (panel C
and D).

Table 2. NMDA induced ^{3}H-acetylcholine release from 0.5 mm thick slices: effect of adenosine deaminase (ADA) and adenosine system effectors

Nr.	Superfusion conditions	n (slices)	^{3}H-overflow (fractional rate: %/min) S_1	S_2	S_2/S_1
1.	A. Control	20	1.46 + 0.36 (100 %)	0.99 + 0.24 (100 %)	0.68 + 0.09 (100 %)
2.	+ ADA (0.2 IU/ml)[a,b]	28	2.08 + 0.55* (+42.5 %)	1.34 + 0.31* (+35.4 %)	0.68 + 0.11
3.	+ ADA (0.2 IU/ml)[c]	12	1.58 + 0.43	0.99 + 0.23	0.65 + 0.17
4.	+ Dipyridamole (6.54 μM)[c]	7	1.42 + 0.27	0.58 + 0.21	0.36 + 0.17* (-44.2 %)
5.	+ Dipyridamole[c] + ADA[b]	9	1.60 + 0.32	0.95 + 0.30	0.55 + 0.11* (-25.7 %)
6.	+R-PIA (10 μM)[c]	10	1.21 + 0.28	0.66 + 0.29	0.52 + 0.13* (-24.6 %)
7.	+R-PIA[c] + ADA[b]	12	1.72 + 0.34	0.83 + 0.17	0.49 + 0.11* (-27.9 %)
8.	B. Control[d]	12	1.58 + 0.43	0.99 + 0.23	0.65 + 0.17 (100 %)
9.	+ DPCPX (0.5 μM)[b,d] + ADA[c]	9	1.42 + 0.38	1.11 + 0.22	0.80 + 0.15 (+23.1 %)

The values are means $\pm$ standard deviation; * values are different at $p < 0.05$

Substance addition: a) for two minutes during S_1 stimulation with 25 μM NMDA; b) for two minutes during S_2 stimulation; c) for ten minutes before S_2 stimulation; d) 1 % dimethylsulphoxide in superfusion medium

245

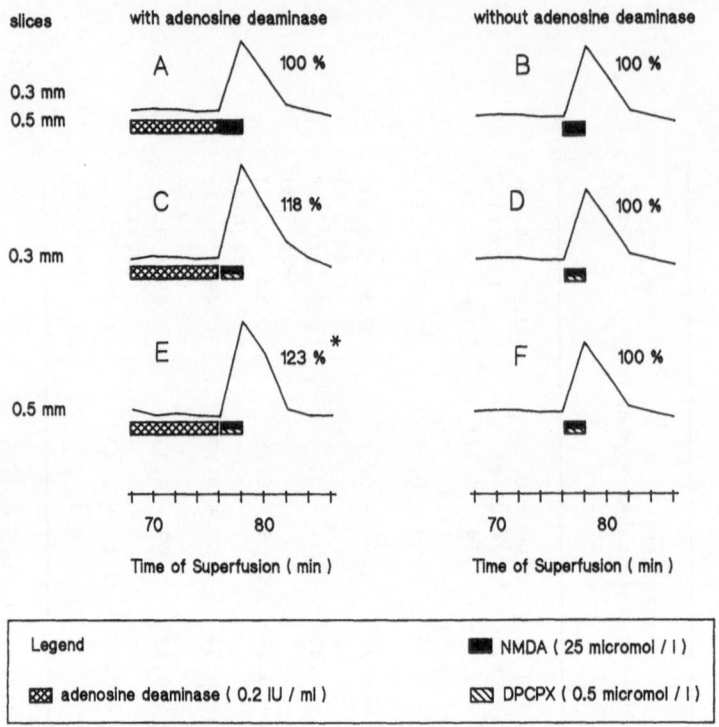

Fig. 2. Time course of NMDA-induced ³H-acetylcholine (ACH) release (S₂-stimulation). ACH fractional release rate (%/min) is represented by solid line. Changes in release are given in percent to control 0.3 mm or 0.5 mm thick slices. Statistical significance (p < 0.05) is marked (*). Additions, as indicated in legend (bottom panel). For experimental details see legend to Fig. 1.

therefore serve as an appropiate model to study interactions between excitatory and inhibitory receptors. However, our results indicate for the first time that 0.3 mm thick slices - often considered as "non hypoxic" slices - obviously release adenosine too. Neurons - and probably glia cells - are the source of adenosine (Schubert 1988). The effects of endogenous adenosine observed in this study are Mg^{2+}-independent, and limited to presynaptic events. NMDA, unlike glutamate has the advantage that it stimulates transmitter release without interfering with the amino acid metabolism in neurons or in glia cells.

With regard to the possible mechanism of A_1-adenosine receptor function, our results with DPCPX (Fig. 2, C, D, E, F) demonstrate that the expected potentiation of NMDA-effects (namely as result of adenosine A_1-receptor antagonism) can be observed when adenosine previously was removed by deamination. These data clearly indicate that in the presence of adenosine, the A_1-adenosine receptor takes the agonist conformation, which has low affinity for the antagonist (Barrington 1989). In other experiments (not shown here) where adenosine was washed out, DPCPX shows a high affinity for the A_1-adenosine receptor. Based on the data with DPCPX and with PIA and dipyrdamole (Table 2) we assume that the adenosine liberated during neuron stimulation first induces an agonist conformation necessary for further activation of the A_1-adenosine receptor and for presynaptic depression.

We suggest that the excitatory impulse is controlled by a multi-factorial process: amino acid receptors, specific receptors, e.g., cholinergic, dopaminergic or noradrenergic, and adenosine A_1-receptors. Each control unit may be separately adjusted by pharmacological intervention. Adenosine antagonists, e.g., theophylline or caffeine, can potentiate the NMDA-effects by diminishing adenosine inhibitory tonus. The use of these pharmacological effects is important for alertness and learning, and conversely, substances which potentiate the adenosine effects in the central nervous system are of interest in post-stroke therapy (Porsche-Wiebking 1989).

REFERENCES

Barrington, W. W., Jacobson, K. A., and Stiles, G. L., 1989, Demonstration of distinct agonist and antagonist conformation of A_1-adenosine receptor, J. Biol. Chem., 264:13157-13164.

Fink, K., Göthert, M., Molderings, G., and Schlicker, E., 1989, N-methyl-D-aspartate (NMDA) receptor-mediated stimulation of noradrenaline release, but not release of other neurotransmitters, in the rat brain cortex: receptor location, characterization and desensitization, Naunyn-Schmiedeberg's Arch. Pharmacol., 339:514-521.

Fredholm, B. B., Fastbom, J., Duner-Engstrom, M., Hu P.-Sch., van der Ploeg, I., and Dunwiddie, T. V., 1989, Mechanism(s) of inhibition of transmitter release by adenosine receptor activation, in: "Adenosine receptors in the nervous system," J. A. Ribeiro, ed., Taylor & Francis, London, New York, Philadelphia, 123-131.

Haas, H. L., and Greene, R. W., 1988, Endogenous adenosine inhibits hippocampal CA_1 neurones: further evidence from extra- and intracellular recording, Naunyn-Schmiedeberg's Arch. Pharmacol., 337:561-565.

Hoehn, K., and White, T., 1990, Role for excitatory amino acid receptors in K^+- and glutamate-evoked release of endogenous adenosine from rat cortical slices, J. Neurochem., 54:256-265.

Jackisch, R., Strittmatter, H., Kasakow, L., and Hertting, G., 1984, Endogenous adenosine as a modulator of hippocampal acetylcholine release, Naunyn-Schmiedeberg's Arch. Pharmacol., 327:319-325.

Limberger, N., Spath, L., and Starke, K., 1988, Presynaptic a_2-adrenoceptor, opioid K-receptor and adenosine A_1-receptor interactions on noradrenaline release in rabbit brain cortex, Naunyn-Schmiedeberg's Arch. Pharmacol., 338:53-61.

Porsche-Wiebking, E., 1989, New N-Methyl-D-Aspartate antagonists for the treatment of stroke, Drug Develop. Res., 17:367-375.

Phillis, J. W., and Wu, P. H., 1981, The role of adenosine and its nucleotides in central synaptic transmission, Prog. Neurobiol., 16:187-239.

Reddington, M., Lee, K. S., and Schubert, P., 1982, An A_1-receptor characterised by (^{3}H)-cyclohexyladenosine binding, mediates the depression of evoked potentials in a rat hippocampus slice preparation, Neurosci. Lett., 28:275-279.

Rothmann, S. M., and Olney, J., 1987, Excitotoxicity and the NMDA receptor, Trends Neurol. Sci., 10:299-302.

Schubert, P., 1988, Modulation of synaptically evoked neuronal caclium fluxes by adenosine, in: "Neurotransmitters and cortical frunction. From molecules to mind," M. Avoli, T. A. Reader, R. W. Dykes, and P, Gloor, eds., Plenum Press, New York, 471-483.

Stone, T. W., and Bartrup, J. T., 1989, Magnesium-dependency of presynaptic inhibition by adenosine, in: "Adenosine receptors in the nervous system," J. A. Ribeiro, ed., Taylor & Francis, London, New York, Philadelphia, 87-95.

Wichman, T., Wictorin, K., Björklund, A., and Starke, K., 1988, Release of acetylcholine and its dopaminergic control in slices from striatal grafts in the ibotenic acid lesioned rat striatum, Naunyn-Schmiedeberg's Arch. Pharmacol., 338:623-631.

Acknowledgement: We thank Miss Kerstin Brühl for her expert secretarial work. We feel in debt to Professor Ulrich Trendelenburg, Institut of Pharmacology, Würzburg and thank him for his friendly criticism.

IMAGING OF DOPAMINE D1 AND D2 RECEPTORS BY A HIGH RESOLUTION POSITRON EMISSION TOMOGRAPHY

Hitoshi Shinotoh,[1,2] Akiyo Aotsuka,[1,2] Osamu Inoue,[1] Kazutoshi Suzuki,[1] Hiroshi Fukuda,[1] Masaomi Iyo,[1] Toshiro Yamazaki,[1] Yukio Tateno,[1] Keizo Hirayama,[2] and Norimasa Nohara[1]

[1]National Institute of Radiological Sciences, Chiba City; and [2]Department of Neurology, School of Medicine, Chiba University, Chiba City, Japan

INTRODUCTION

The recent development of a high resolution positron emission tomography and appropriate radioligands labelled with positron emitter has made it possible to clearly delineate distribution of neuroreceptors in the human brain.

In this study, dopamine D1 and D2 receptors in the human brain were imaged by a newly developed, high resolution positron emission tomography (PET), carbon-11 labelled SCH23390 and N-methylspiperone (C-11 NMSP). And a comparison of the distribution of two types of dopamine receptors in controls and patients with parkinsonism was made.

MATERIALS AND METHODS

A five-ring PET system "SHR-1200" was used to follow the radioactivity in 9 sections of the brain covering an axial distance of 70 mm (Yamashita et al., 1990). The spatial resolution of the reconstructed images was 3.6 mm full width at half maximum (FWHM), and the axial thickness was 6 mm FWHM for direct slices and 8 mm FWHM for cross slices.

SCH23390 is a potential benzazepine antipsychotic and a selective high affinity dopamine D1 receptor antagonist. C-11 SCH23390 was prepared by N-alkylation of SCH24518 with C-11 methyliodide. C-11 NMSP was also prepared by N-alkylation of spiperone with C-11 methyliodide for imaging of dopamine D2 receptors.

The number of PET study with C-11 SCH23390 and the high resolution PET in our institute has been 14 until the present. Two healthy males and three females, ranging in age from 30 to 73 years old (y.o.) were recruited as controls. Five female patients with idiopathic Parkinson's disease (aged 48-70 y.o.; disease duration 0.5-3 years; Hoehn and Yahr stage I-III) and

Neuroreceptor Mechanisms in Brain, Edited by S. Kito *et al.*
Plenum Press, New York, 1991

four patients with clinically diagnosed striato-nigral degeneration (SND) (two males and two females; aged 61-69 y.o.; disease duration 4-9 years; Hoehn and Yahr stage IV-V) were also investigated with C-11 SCH23390. When the patients had been treated, all the neuropsychiatric drugs including levodopa and dopamine agonists were withdrawn 48 hours before the PET study.

The number of PET study with C-11 NMSP and the high resolution PET has been 3 until the present. Two healthy male volunteers, aged 30 and 61 y.o. were recruited as controls, and one female patient with clinically diagnosed SND participated in this study. The neuropsychiatric drugs were also withdrawn 48 hours before the study.

All idiopathic PD patients showed typical parkinsonian tremor and responded well to anti-parkinsonian drugs. The diagnosis of SND was based on the following: 1) the predominance of rigidity and akinesia with absent or minimal tremor; 2) rapid disability compared with PD patients; 3) mild to moderate ponto-cerebellar atrophy on CT scan; 4) parkinsonism was the predominant symptom from the onset of disease until the current study was conducted; and 5) poor response to anti-parkinsonian drugs.

A dose of 20.2 ± 2.2 (mean $\pm$ SD) mCi of the radioligands was injected intravenously and the brain radioacitivy was followd by the PET. The specific activity at the time of injection varied between 250 and 960 Ci/mmol. Each study comprised 15-21 sequential scans during a period of 57.5-87.5 minutes.

RESULTS

C-11 SCH23390

The radioactivity peaked within 6 min, and then decreased gradually to the end of the experiment following injection of C-11 SCH23390 (Fig. 1). The ratio of radioactivity in the striatum to that in the cerebellum (Ast/Acb ratio) reached a peak within 37.5 min following injection and stayed a same level or slightly decreased thereafter (Fig. 1). The radioactivity in the caudate-putamen (striatum) was higher than in the cerebellum and the PET images in controls showed the caudate-putamen clearly 10 min following injection of C-11 SCH23390 (Fig. 2).

No remarkable difference of the kinetics of C-11 SCH23390 in the brain and the radioactivity distribution between controls and PD patients was observed.

PET images in SND patients showed that the accumulation of C-11 radioactivity in the putamen was lower than controls and PD patients and the putamen was delineated smaller than controls and PD patients (Fig. 3).

C-11 NMSP

The radioactivity in the caudate-putamen increased up to 37.5 min following injection and then gradually decreased,

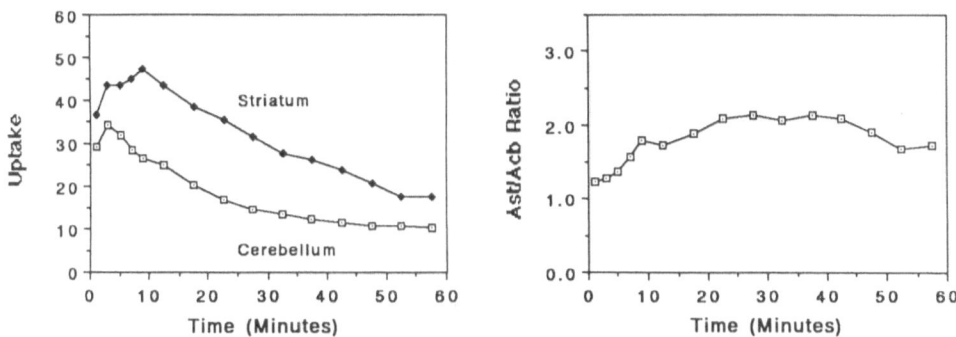

Fig. 1. Kinetics of C-11 SCH23390 in the brain and the
Ast/Acb ratio in a control (70 y.o. female).

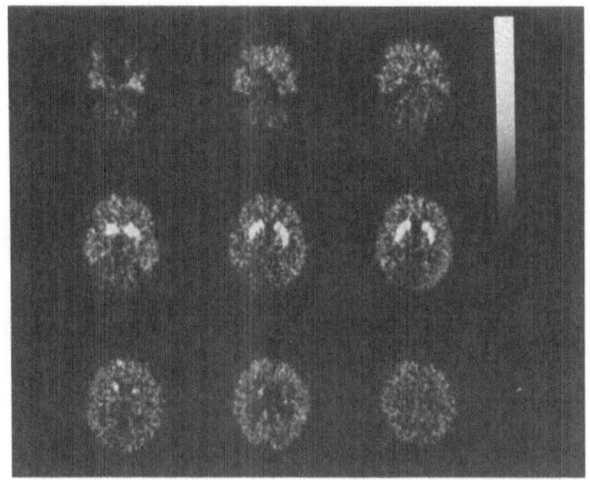

Fig. 2. PET images of the brain of a healthy female
volunteer (66 y.o.) at 27.5-57.5 min following C-11
SCH23390.

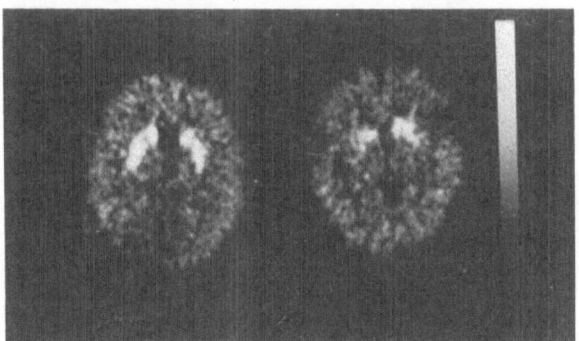

Fig. 3. PET images of a control (Left image; 66 y.o.
female) and of a SND patient (Right image; 68 y.o.
female) following injection of C-11 SCH23390. These
images were normalized to the radioactivity in the
cerebellum.

251

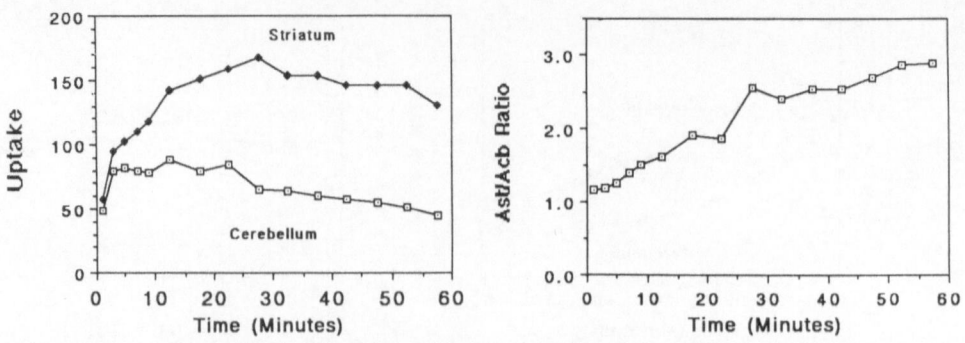

Fig. 4. Kinetics of C-11 NMSP in the brain and the Ast/Acb
ratio in a control (30 y.o. male).

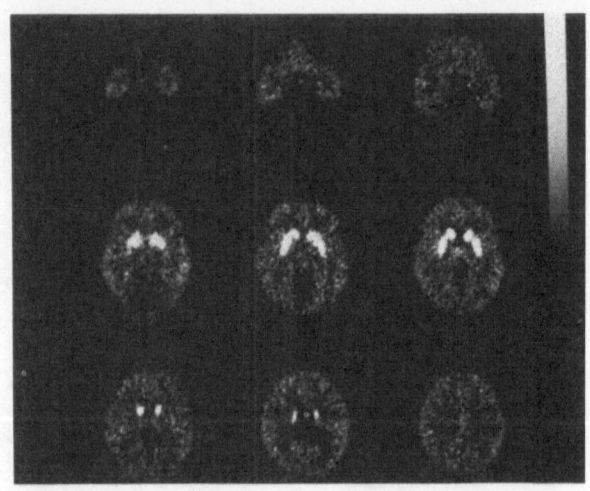

Fig. 5. PET images the brain in a healthy, male volunteer
(30 y.o.) at 57.5-87.5 min following injection of C-11
NMSP.

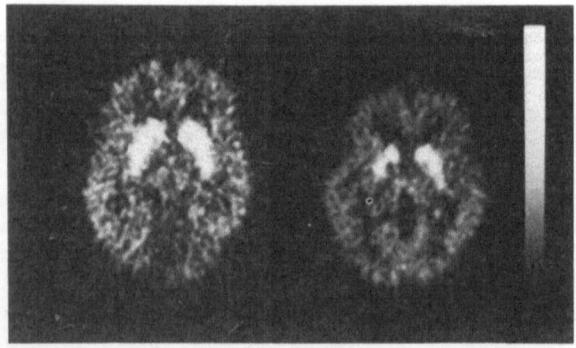

Fig. 6. PET images of a control (Left image; 30 y.o.
male) and a SND patient (Right image; 68 y.o.
female) following injection of C-11 NMSP. These
images were normalized to the radioactivity in the
cerebellum.

whereas the radioactivity in the cerebellum reached a peak within 12.5 min and then gradually decreased thereafter (Fig. 4 a). The Ast/Acb ratio increased linearly to the end of the experiment (Fig. 4 b). PET images in controls showed the caudate-putamen clearly (Fig. 5).

PET images in one SND patient showed that the accumulation of C-11 radioactivity in the putamen was lower and the putamen was delineated smaller than in controls (Fig. 6).

DISCUSSION

PET images following injection of C-11 SCH23390 and C-11 NMSP clearly showed the caudate-putamen, in which both types of dopamine receptors are known to be located in the highest densities in the brain. The high resolution PET could show not only the caudate head but also the caudate tail (Fig. 5). The distribution of both types of dopamine receptors appeared similar in our study.

The Ast/Acb ratio reached a plateau level after 37.5 min following injection of C-11 SCH23390, whereas the Ast/Acb ratio increased linearly to the end of the study following injection of C-11 NMSP. Although the kinetics of C-11 SCH23390 and C-11 NMSP were different, both types of dopamine receptors can be estimated semiquantitatively.

Assuming that radioligand concentration in the cerebellum represents exchangeable radioligand in the striatum, the plateau level after 27.5 min following injection of C-11 SCH23390 indicates a pseudoequibrium has been reached. In a pseudoequibrium, the Ast/Acb ratio reflects the binding potential of dopamine D1 receptors in the striatum. The slope of the Ast/Acb ratio line following C-11 NMSP injection reflects the rate of C-11 NMSP binding to dopamine D2 receptors from exchangeable radioligands.

It is well established that most patients with idiopathic PD respond well, at least temporarily, to levodopa therapy. However, when parkinsonism is associated with SND, the response to levodopa therapy is usually disappointing. PET images in SND patients indicate that the inefficiency of levodopa therapy in this disorder should be attributed at least in part, to the loss of both types of dopamine receptors.

In conclusion, this technique has great potential in investigating the pathophysiology of various neuropsychiatric disorders.

REFERENCE

Yamashita, T., Uchida, H., Okada, H., Kurono, T., Takemori, T., Watanabe, M, Shimizu, K., Yoshikawa, E., Ohmura, T., Satoh, N., Tanaka, E., Nohara, N., Tomitani, T., Yamamoto, M., Murayama, H., Endo, M., 1990, Development of a high resolution PET, IEEE Trans. Nucl. Sci, NS-37 (in press).

THE NICOTINIC ACETYLCHOLINE RECEPTOR GENE FAMILY:

STRUCTURE OF NICOTINIC RECEPTORS FROM MUSCLE AND NEURONS AND NEURONAL

α-BUNGAROTOXIN-BINDING PROTEINS

Jon Lindstrom, Ralf Schoepfer, William Conroy, Paul Whiting,
Manoj Das, Mohammad Saedi, and Rene Anand

The Salk Institute for Biological Studies, P.O. Box 85800
San Diego, CA 92138

INTRODUCTION

Three branches of the ligand-gated ion channel gene superfamily encode proteins that bind cholinergic ligands: 1) nicotinic acetylcholine receptors (AChRs) from skeletal muscle, 2) nicotinic AChRs from neurons, and 3) α-bungarotoxin-binding proteins (αBgtBPs) from neurons. AChRs from vertebrate muscles and nerves differ in subunit composition, and in some cases in functional role, but both appear to be formed from several homologous subunits which form ACh-gated cation channels. αBgtBPs from vertebrate neurons have uncertain subunit compositions, uncertain endogenous ligands, and unknown functions.   The ligand-gated ion channel gene superfamily also includes receptors for GABA and glycine, which are ligand-gated anion channels, and it probably also includes other ligand-gated ion channels (Barnard et al., 1987; Betz and Becker, 1988). The relation of glutamate receptors to this superfamily is less certain (Gregor et al., 1989; Hollmann et al., 1989; Wada et al., 1989).

Here we will briefly review our recent studies of these three types of cholinergic ligand-binding proteins.   First, muscle-type nicotinic AChRs will be considered, because they are the best characterized of these proteins and serve as an archetype for studies of the ligand-gated ion channel family. Second, we will discuss neuronal nicotinic AChRs. In the past few years monoclonal antibody (mAb) and complementary DNA (cDNA) probes initially developed from studies of muscle-type AChRs have permitted molecular studies of neuronal AChRs.   Now, a plethora of previously unrecognized neuronal AChR subtypes have been discovered, and neuronal AChR-specific mAb and cDNA probes are being used to identify which subtypes occur in vivo, where they occur, what their functions are, what subunit structures they have, and which genes encode these subunits. Finally, we will discuss neuronal αBgtBPs.   These were discovered shortly after αBgt was first used as a high-affinity antagonist for muscle-type AChRs. Neuronal αBgtBPs were initially assumed to be neuronal AChRs, but it was soon found that some or all of these did not function as ACh-gated cation channels, and were not located at synapses.   Although these neuronal αBgtBPs outnumber neuronal AChRs, little progress was made on molecular studies of them until we recently succeeded in identifying cDNAs for two of their subunits.

Neuroreceptor Mechanisms in Brain, Edited by S. Kito *et al.*
Plenum Press, New York, 1991

255

AChRs from the electric organs of *Torpedo* are the best characterized neurotransmitter receptors because they are available in large amounts at high concentration (Stroud and Finer-Moore, 1985; Karlin et al., 1986; Changeux et al., 1987; Unwin, 1989). These AChRs are composed of two ACh-binding subunits termed $\alpha 1$, and one copy each of three kinds of structural subunits termed $\beta 1$, γ, and δ. These subunits exhibit sequence homologies which indicate that all of these subunits are part of a gene family and that they have similar transmembrane orientations of their polypeptide chains (Raftery et al., 1980; Noda et al., 1983). This is illustrated by the sequences of $\alpha 1$ and δ subunits shown in Figures 1 and 2. The rod-shaped subunits appear to be organized around a central cation channel in the order $\alpha\beta\alpha\gamma\delta$ (Kubalek et al., 1987). The AChR appears to extend about 6 nm above the extracellular surface of the membrane as a hollow rigid cylinder 9 nm across with about a 2.5 nm diameter channel that narrows at the level of the membrane to a diameter <1 nm before broadening on the cytoplasmic surface within a domain that is more amorphous than the extracellular surface (Toyoshima and Unwin, 1988; Mitra et al., 1989). It has been suggested that binding of ACh triggers opening of the cation channel through a slight tilting of the subunits which produces an iris-like opening of the narrowest part of the channel in the lipid layer region (Unwin et al., 1989).

The sequence of *Torpedo* AChR α subunits exhibits patterns characteristic of all members of the gene superfamily. In the N-terminal 209 amino acids there are several short conserved sequences characteristic of $\alpha 1$ subunits from any species; similarly, in other AChR subunits there are several conserved sequences characteristic of each subunit type in any species. There is a disulfide bond between cysteines at $\alpha 128$ and $\alpha 142$ that forms a loop including a highly conserved sequence with an N-glycosylation site at $\alpha 141$ (Kellaris et al., 1989). This glycosylated disulfide-linked loop is characteristic of all subunits in the gene superfamily. There is a disulfide-linked pair of cysteines at $\alpha 192$, 193 (Kao and Karlin, 1986; Kellaris et al., 1989) which are characteristic of $\alpha 1$ and other ACh-binding subunits in the gene superfamily, but are not found in structural subunits like $\beta 1$, γ, and δ. These amino acids are known to be near the ACh-binding site on $\alpha 1$ subunits because they react with affinity labeling reagents for the ACh binding site (Kao et al., 1984; Dennis et al., 1988). Tyrosine $\alpha 190$ which is characteristic of all ACh-binding subunits is also thought to be near the ACh binding site because it is labeled by lophotoxin (Abramson et al., 1989). The corresponding region of structural subunits is both highly variable between species and hydrophilic, suggesting that it encodes a surface domain of little functional consequence. There are three hydrophobic sequences between $\alpha 210$ and $\alpha 296$ termed M1, M2 and M3 which are thought to comprise α helical transmembrane domains (Claudio et al., 1983; Devillers-Thiery et al., 1983; Noda et al., 1983). These are highly conserved between different AChR subunits and present in all members of the gene superfamily (Barnard et al., 1987; Betz and Becker, 1988; Hucho and Hilgenfeld, 1989; Unwin, 1989). Several lines of evidence suggest that M2 may be a domain contributed by each subunit to the lining of the cation channel (Giraudat et al., 1986; Imoto et al., 1986, 1988; Oiki et al., 1988; Hucho and Hilgenfeld, 1989). Following M3 is an extended sequence between $\alpha 297$ and $\alpha 407$ which is largely unique to subunits from each species. A fourth hydrophobic sequence termed M4 between $\alpha 407$ and $\alpha 427$ is characteristic of all members of the gene family, but has a poorly conserved sequence and seems of limited importance since it can be replaced by other hydrophobic sequences (Tobimatsu et al., 1987). The C-terminus of $\alpha 1$ subunits is at $\alpha 437$, and all subunits in the superfamily terminate shortly after M4.

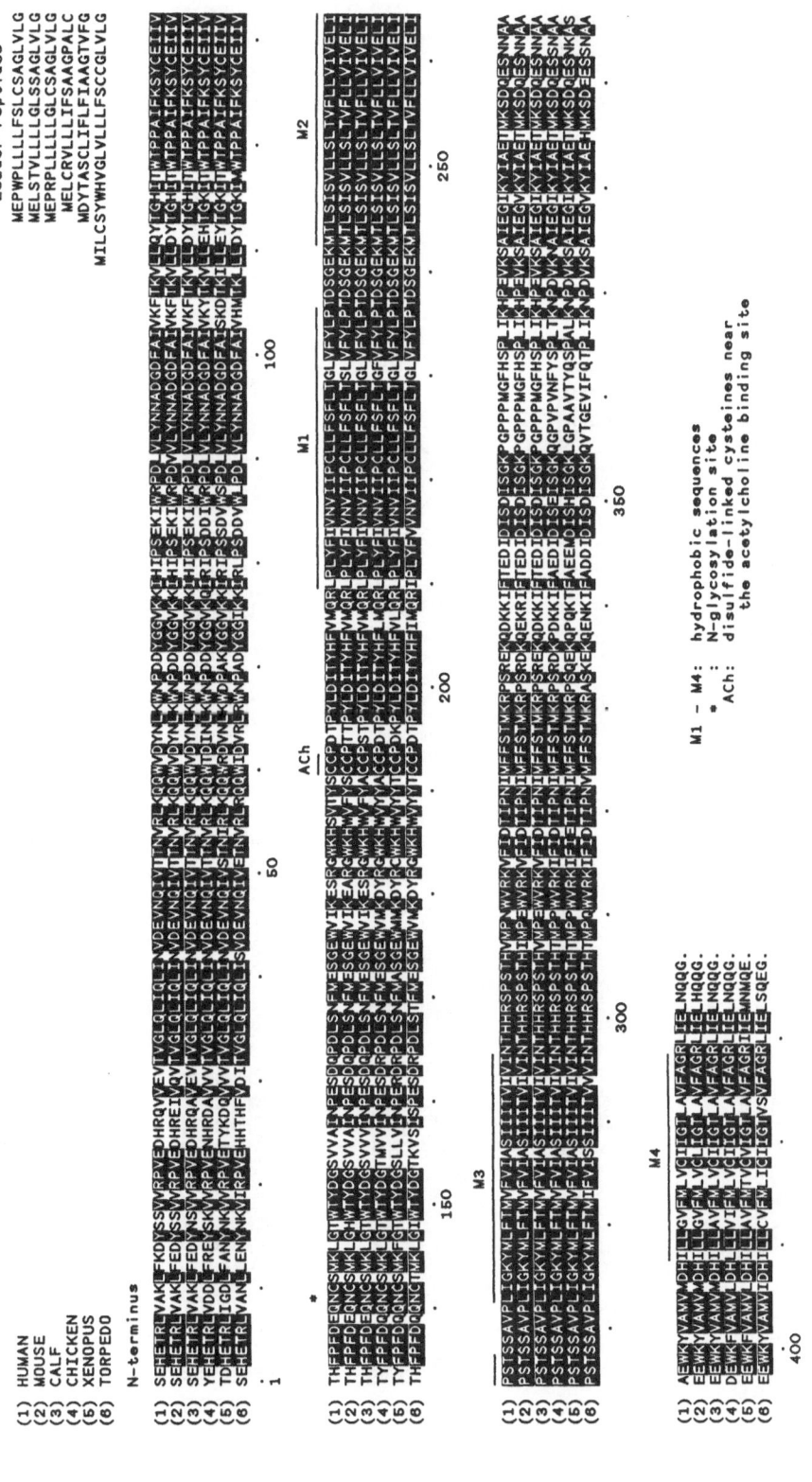

Figure 1. Homologies among AChR α1 subunits. Reproduced from Luther et al. (1989). Sequences are from Schoepfer et al., (1988), Boulter et al. (1987), Noda et al. (1983), Nef et al. (1988), Baldwin et al. (1988), and Noda et al. (1982).

257

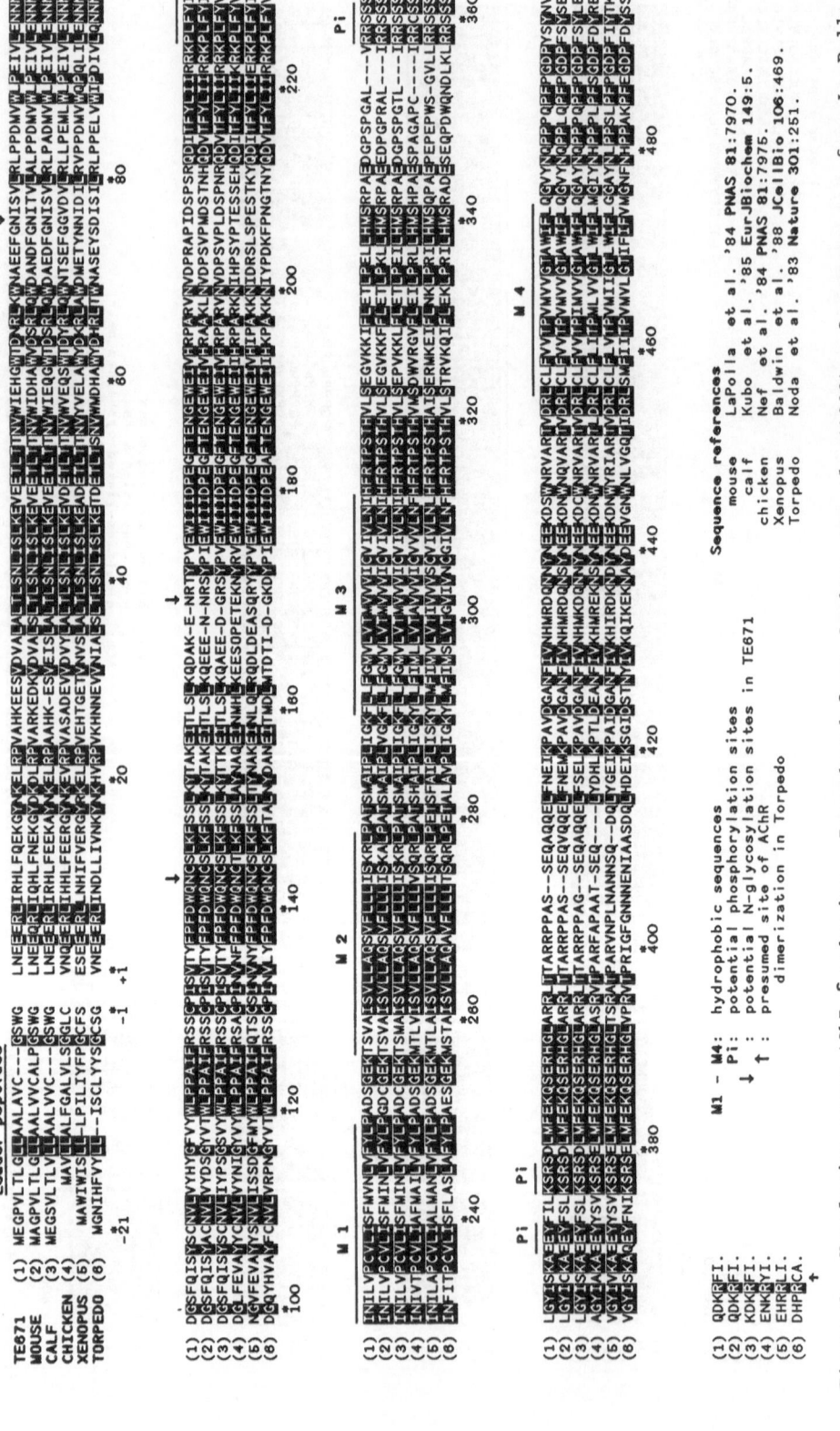

Figure 2. Homologies among AchR δ subunits. Reproduced from Luther et al. (1988). Sequences are from La Polla et al. (1984), Kubo et al. (1985), Nef et al. (1984), Baldwin et al. (1988), and Noda et al. (1983).

We have used mAbs as probes of AChR structure and as model autoanti-bodies (Tzartos and Lindstrom, 1980; Conti-Tronconi et al., 1981, 1985, 1986; Tzartos et al., 1981, 1982, 1983, 1986, 1987; Ratnam et al., 1984, 1986a,b; Criado et al., 1985a,b, 1986; Das and Lindstrom 1989). More than half of the antibodies in an antiserum to AChRs purified from electric organ or muscle or in antisera from patients with myasthenia gravis are directed at the main immunogenic region (MIR) on α subunits (Tzartos et al., 1981, 1982, 1983). mAbs to the MIR are also interesting because these proved to be the first useful mAb probes for neuronal nicotinic AChRs (Swanson et al., 1983; Jacob et al., 1984, 1986; Smith et al., 1985, 1986; Whiting and Lindstrom, 1986). The MIR appears to be located on the outside of the large cylindrical extracellular surface of the AChR (Kubalek et al., 1987). Using synthetic peptides Tzartos et al. (1988) found that amino acids contributing to the MIR were contained within the sequence $\alpha 66$-76. We confirmed this, and showed that $\alpha 68$ and $\alpha 71$ were especially important (Das and Lindstrom, 1989), as others have found as well (Bellone et al., 1989; Tzartos et al., 1990). These two amino acids are replaced by non-conservative amino acids in $\alpha 1$ subunits from *Xenopus* (N$\alpha 68$ and D$\alpha 71$ in *Torpedo* $\alpha 1$ were replaced by D$\alpha 68$ and K$\alpha 71$ in *Xenopus*, see Figure 1). This is interesting because AChRs from *Xenopus* muscle are unique among muscle AChRs tested in not reacting with mAbs to the MIR (Sargent et al., 1983). Synthetic peptides corresponding to $\alpha 66$-76 of *Torpedo* or human AChR were bound by mAbs to the MIR, but the corresponding *Xenopus* peptide was not (Das and Lindstrom, 1989). mAbs to the MIR bind synthetic peptides with much lower affinity than they do native AChRs, and some mAbs to the MIR do not bind at all to denatured α subunits or synthetic peptides. Thus, the question remained whether the weak binding to synthetic peptides reflected real binding to the MIR in native AChRs. To answer this, $\alpha 1$ subunits of *Torpedo* AChR were mutated *in vitro* at $\alpha 68$ and $\alpha 71$ to replace these amino acids with their equivalents in *Xenopus* α subunits, and then the mutated subunits were expressed in *Xenopus* oocytes along with *Torpedo* $\beta 1$, γ, and δ subunits to form intact AChRs with normal αBgt binding properties (Saedi, Anand, and Lindstrom, unpublished). Mutation of either or both amino acids eliminated binding of all MIR mAbs which had bound to the peptides; and in the case of mAb 35, which did not recognize peptides, but only reacted with the MIR in its native conformation, mutation of $\alpha 68$ reduced binding and mutation of $\alpha 71$ eliminated it. Thus, we know that these amino acids are important to the native conformation of the MIR.

We are especially interested in using mAbs as sequence-specific probes to determine the transmembrane orientation of AChR subunit polypeptide chains (Criado et al., 1985a,b; Ratnam et al., 1986a,b). Using the Geysen technique (Geysen et al., 1987) we can precisely map the epitopes to which mAbs bind (Lindstrom et al., 1989; Das and Lindstrom, unpublished). The entire sequence of *Torpedo* $\alpha 1$ subunits was synthesized as overlapping octamers on plastic pegs, $\alpha 1$-8 on the first peg, $\alpha 2$-9 on the second, etc. Antibodies which could bind to these peptides were detected by ELISA. Antisera to denatured α subunits applied to these peptides revealed major epitopes around $\alpha 160$ and several epitopes in the $\alpha 340$-370 region. mAbs to denatured α subunits were mapped to at least four epitopes within $\alpha 330$-$\alpha 370$, and mAbs were made to synthetic peptides corresponding to $\alpha 127$-143, $\alpha 152$-167, $\alpha 172$-189, $\alpha 235$-242, and $\alpha 261$-277 (Criado et al., 1985a,b, 1986; Ratnam et al., 1986a,b; Criado, Das, and Lindstrom, unpublished). mAbs to the denatured α subunits were selected for their ability to bind to native AChR, but mAbs to synthetic peptides were much less effective at binding to native AChR. mAbs to $\alpha 152$-167 were especially interesting because there was some evidence that they could bind to the cytoplasmic surface of the AChR (Criado et al., 1985a). This was unexpected because most models presumed that all of the sequence N-terminal of the first hydrophobic sequence M1 starting at $\alpha 210$ was located in the extracellular part of the AChR, but detection of sequences around $\alpha 160$ on the cytoplasmic surface would require a transmembrane sequence prior to α160 and another afterwards. Epitope mapping by the Geysen method (1987) showed that these mAbs mapped precisely to α158-165 and did not react elsewhere on the sequence of α (Lindstrom et al., 1989, Das and Lindstrom, unpublished). Cohen and coworkers recently confirmed these observations using mAbs to denatured α subunits which bind to epitopes within α156-179 and which bind to the cytoplasmic surface of membrane bound AChRs (Pedersen et al., 1990). A nagging problem with these studies is that these epitopes seem to be obscured in native AChRs, and neither Cohen's nor our mAbs to this region bind well to detergent solubilized AChR. Nonetheless, these studies suggest that simple hydrophobicity plots may not accurately predict the transmembrane orientation of the subunit polypeptide chains. Additional techniques will be required to probe the transmembrane orientation of the subunit polypeptide chains while awaiting the ultimate answer from X-ray crystallographic studies that remains several years in the future.

Much or all of the sequence between M3 and M4 is likely to form a domain on the cytoplasmic surface of the AChR which is loosely organized (Claudio et al., 1983; Devillers-Thiery et al., 1983; Noda et al., 1983; Ratnam et al., 1986a,b). mAbs to epitopes within α330-α370 bind well to native AChRs and denatured α subunits. Binding of these mAbs can be mapped precisely using the Geysen (1987) technique, and, can thereby be shown to occur only at these sequences of α subunits (Ratnam et al., 1986a,b; Das and Lindstrom, unpublished). Electron microscopy shows that they bind to the cytoplasmic surface (Ratnam et al., 1986a,b). The observation that this sequence is highly variable between subunits is consistent with the idea that this region is not highly spatially constrained, and electron microscopy of two dimensional crystals of AChRs suggests that the cytoplasmic domain may be less rigidly structured than is the extracellular domain (Toyoshima and Unwin, 1988; Mitra et al., 1989). Others have also found that this cytoplasmic region is relatively immunogenic (Souroujon et al., 1986). We have taken advantage of the subunit-specific sequences in this region, their immunogenicity, and their antigenicity in both native and denatured AChRs. These properties are ideal for peptides expressed in bacteria from subunit cDNA fragments for use as antigens in making subunit-specific mAbs. We have used this approach effectively in a number of cases, some of which will be described subsequently.

Human muscle AChRs have been characterized from the cell line TE671 (Lindstrom et al., 1987; Luther et al., 1988; Schoepfer et al., 1988a). This cell line was originally reported to be a neuromedulloblastoma (McAllister et al., 1977), but now appears to be a rhabdomyosarcoma (Stratton et al., 1989), consistent with the observed expression of muscle proteins. These AChRs have the single channel electrophysiological properties expected of muscle AChRs (Luther et al., 1988; Sine, 1988). They are composed of α1, β1, γ and δ subunits of molecular weights similar to the corresponding subunits from _Torpedo_ electric organ, and are immunologically crossreactive with these subunits at some epitopes (Luther et al., 1988). cDNAs have been cloned for all four subunits and the sequences reported for α1 and δ (as shown in Figures 1 and 2, Schoepfer et al., 1988a; Luther et al., 1988; Schoepfer, Anand, and Lindstrom, unpublished). α1 subunits from all species are more highly conserved than are β1, γ, or δ subunits, perhaps reflecting their critical role in ACh binding and channel gating. α1 subunits of human AChR have been used in studies of the transmembrane orientation of subunit polypeptides chains (Saedi and Lindstrom, unpublished). α1 subunits from human muscle assemble very efficiently with _Torpedo_ β1, γ, and δ subunits when expressed from synthetic mRNAs in _Xenopus_ oocytes. mAb 142 raised to denatured _Torpedo_ α1 subunits was mapped using immunoprecipitation of synthetic peptides to within α349-365 (Ratnam et al., 1986b) and precisely mapped using the Geysen technique to α359-366 while being shown to bind only to this α sequence (Das and Lindstrom, unpublished). This sequence is not found in human α1 subunits, and mAb 142 does not bind to these AChRs. mAb 142 was shown by electron microscopy to bind to the cytoplasmic surface of *Torpedo* α1 subunits (Ratnam et al., 1986b). Insertion of the epitope for mAb 142 by *in vitro* mutagenesis into the sequence of human α1 subunits at α347 followed by expression in *Xenopus* oocytes with *Torpedo* β1, γ, and δ subunits produced AChRs which could not bind ^{125}I-mAb 142 applied to the outside of the oocytes, but AChRs which, if solubilized with Triton X-100, could then be precipitated using mAb 142 (Saedi and Lindstrom, unpublished). This indicates that the epitope for mAb 142 could be inserted into the sequence of human α1 subunits without disrupting assembly of intact AChRs, that the inserted epitope was still recognizable by the mAb, and that the epitope inserted into the large putative cytoplasmic domain C-terminal of M3 was found, as expected, on the cytoplasmic surface of the AChR. This "reporter epitope" technique, thus, appears capable of probing the transmembrane orientation of the subunit polypeptide chains. It should be useful in studies of putative transmembrane domains not predicted by hydrophobicity plots.

AChR α1 subunits not assembled with other subunits are produced by TE671 cells in amounts nearly equal to their content of assembled AChRs (Conroy and Lindstrom, unpublished). These have the conformation of a synthetic intermediate previously recognized in the mouse muscle cell line BC3H-1 (Merlie and Lindstrom, 1983). In BC3H-1 cells during the first 15 minutes after synthesis α1 subunits partially mature in conformation and acquire moderate affinity for mAbs to the MIR and for αBgt, but they do not acquire substantial affinity for small cholinergic ligands (Blount and Merlie, 1987). The unassembled α1 subunits of TE671 cells, like unassembled α subunits of BC3H-1 cells, are not expressed on the cell surface membrane, consistent with the idea (Klausner, 1989) that only correctly assembled membrane proteins are expected to reach the cell surface.

We can confirm (Saedi, Conroy, and Lindstrom, unpublished) that α1 subunits acquire high affinity for small cholinergic ligands when they assemble with either γ or δ subunits (Kurosaki et al., 1987; Blount and Merlie, 1989; Sumikawa and Miledi, 1989). In addition, we found that the conformation of their MIR matures fully at this time. We studied combinations of *Torpedo* AChR subunits expressed in *Xenopus* oocytes using subunit-specific mAbs and sucrose gradient sedimentation (Saedi, Conroy, and Lindstrom, unpublished). α1 subunits expressed in combination with γ or δ subunits efficiently assembled into complexes of the size expected for two subunits, and coexpression of α1, γ, and δ subunits resulted in efficient assembly of complexes of the size expected for four subunits, perhaps corresponding to $^{\alpha\gamma}_{\alpha\delta}$. These complexes have high affinity for mAbs to the MIR, αBgt, and carbamylcholine. Coexpression of α1 and β1 subunits results in assembly of aggregates of many sizes, perhaps corresponding to αβ, βαβ, αβαβ, etc., and does not result in maturation of the conformation of α1 subunits. Coexpression of α1, β1, γ and δ subunits efficiently results in the assembly of mature AChR monomers. It may be that these processes reflect the normal pathway of AChR subunit assembly. In that case, assembly of subunits would not be concerted, but take place sequentially. The structural homology of all AChR subunits may permit low affinity interactions between many combinations of the subunits, but the conformation changes associated with binding of the proper interfaces of α, γ, and δ subunits may be the driving force for assembly of the subunits into the unique array characteristic of mature AChRs. Perhaps assembly of β subunits with an $^{\alpha}_{\alpha}\!^{\gamma}_{\delta}$ array is the final step in assembly of native $^{\beta\alpha\gamma}_{\alpha\delta}$ AChRs.

Hess and coworkers have investigated expression of *Torpedo* AChR subunits from their cDNAs in transformed yeast as a model system for expressing, hopefully in functional form and large amounts, neurotransmitter receptors from brain which are normally not easily accessible for functional studies and are present in amounts too small for easy biochemical studies (Fujita et al., 1986; Jansen et al., 1989). We found that under some conditions transformed yeast could express small amounts of all four *Torpedo* AChR subunits (Jansen et al., 1989). The *α* subunits exhibited a conformation typical of the synthetic intermediate with moderate affinity for mAbs to the MIR and negligible affinity for small cholinergic ligands (Conroy, Luther, and Lindstrom, unpublished). However, the *α*1 subunits did not efficiently assemble with the other subunits and did not mature into their native conformation. Thus conditions have not been found to use yeast as an effective expression system for neurotransmitter receptors.

NEURONAL NICOTINIC ACHRS

The strategy we initially used to study neuronal nicotinic AChRs was to immunoaffinity purify chicken brain AChRs that crossreacted with mAbs to the MIR of muscle type AChRs (Whiting and Lindstrom, 1986a), then raise a library of mAbs to these AChRs (Whiting et al., 1987a) to obtain probes for other neuronal AChRs (Whiting and Lindstrom, 1987a) which, in turn, were used to produce more mAbs (Whiting and Lindstrom, 1988), and sequences of purified subunits were used to identify subunit cDNAs (Whiting et al., 1987b, Schoepfer et al., 1988b). Simultaneously other groups used muscle AChR subunit cDNA probes in low stringency hybridization screens to identify putative neuronal nicotinic AChR subunit cDNAs (Boulter et al., 1986; Nef et al., 1988) which were then used to identify others (Goldman et al., 1986; Deneris et al., 1988, 1989; Wada et al., 1988; Duvoisin, 1989). Now these approaches are merging and subunit terminologies used by the various groups are becoming more uniform as the cDNAs encoding subunits of purified AChRs are being identified. Also, mAbs are being prepared against bacterially expressed cDNA fragments to aid in identification of the genes encoding subunits of AChR subtypes found *in vivo* (Schoepfer et al., 1989; Whiting et al., submitted; Schoepfer, Conroy, and Lindstrom, unpublished). Figure 3 compares the properties of neuronal AChR subtypes immunoaffinity purified from brains of chickens (Whiting et al., 1987a), rats (Whiting and Lindstrom, 1987a), cattle, and humans (Whiting and Lindstrom, 1988) with the known subtypes of muscle AChRs. Neuronal nicotinic AChR subunit cDNAs reveal a gene family of subunits clearly related to those of muscle AChR, suggesting a common genetic origin and basically similar transmembrane orientation of their polypeptide chains and similar shapes of their subunits. Figures 4 and 5 compare the sequences of some neuronal nicotinic AChR subunits from chickens and rats.

Neuronal nicotinic AChRs are thought to be composed of only two kinds of subunits: 1) ACh-binding subunits termed *α*2, *α*3, *α*4... in various subtypes, and 2) structural subunits termed *β*2, *β*3, *β*4... in various AChR subtypes. Immunoaffinity purified AChR preparations appear to consist of these two kinds of subunits (Whiting and Lindstrom 1986a, 1987a, 1988; Whiting et al., 1987a). Coexpression of *α*2, *α*3, or *α*4 subunits with *β*2 or *β*4 subunits in *Xenopus* oocytes yields efficient expression of ACh-gated cation channels (Boulter et al., 1987; Ballivet et al., 1988; Deneris et al., 1988, 1989; Wada et al., 1988; Duvoisin et al., 1989; Papke et al., 1989). Detailed comparison of the properties of AChRs expressed from defined gene products with native AChRs will be required to firmly establish the identify of the subunits comprising the AChRs which occur *in vivo*.

262

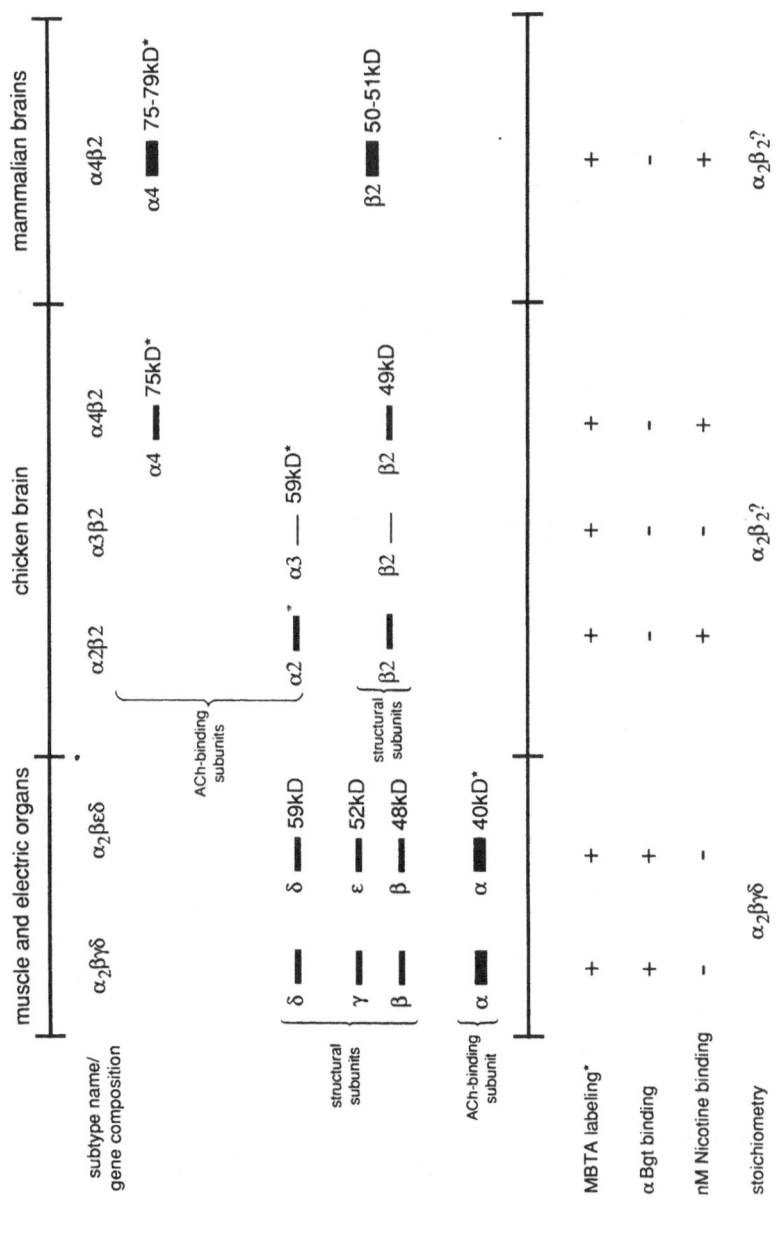

Figure 3. Comparison of some AChR subtypes from muscles and neurons. Reproduced from Lindstrom et al. (1990).

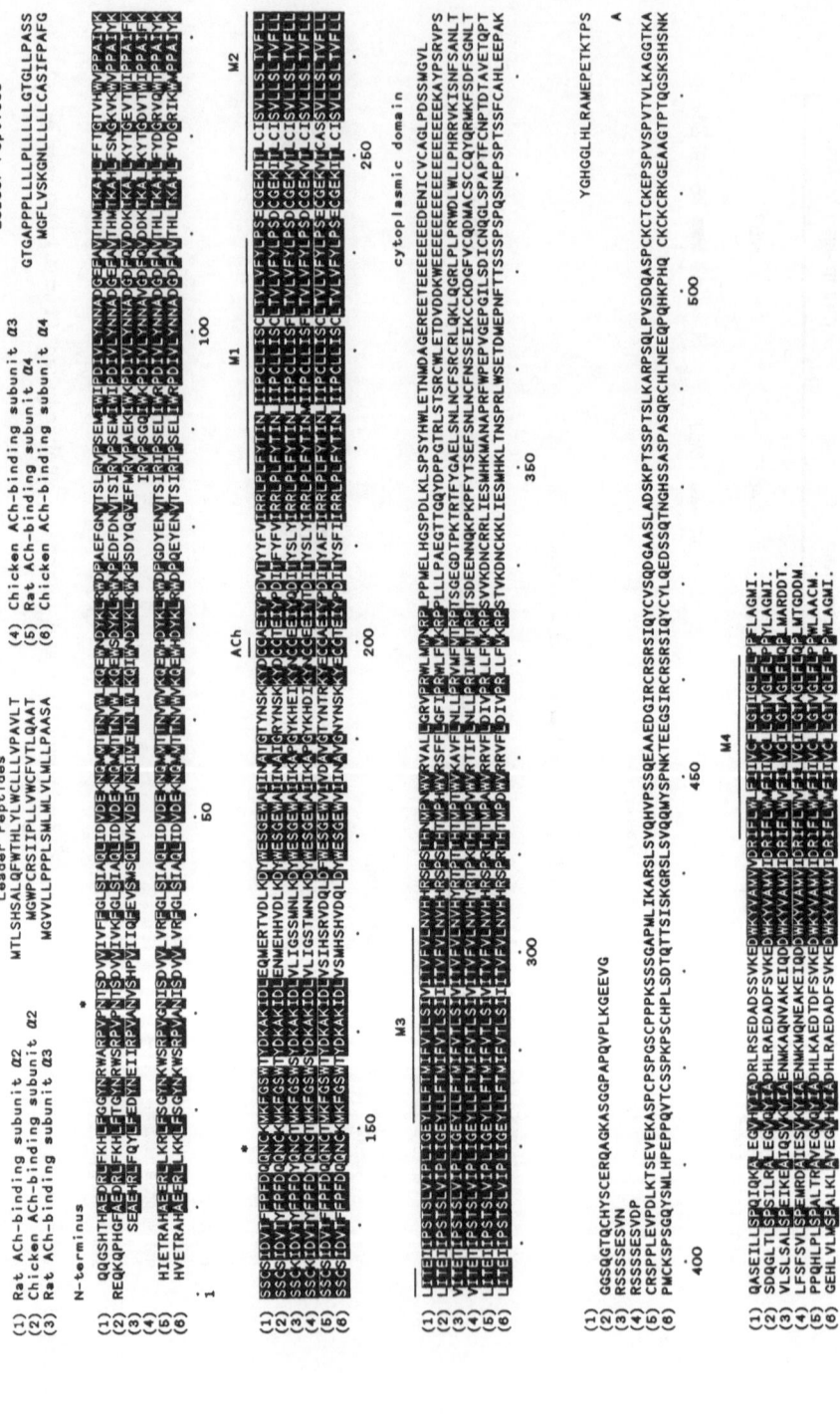

Figure 4. Comparison of ACh-binding subunit sequences from neuronal nicotinic AChRs of chickens and rats. Reproduced from Lindstrom et al. (1990). Sequences are from Boulter et al. (1986), Goldman et al. (1986), Nef et al. (1988), and Wada et al. (1988).

Leader peptides

MALLRVLCLLAALRRSLC
MLACMAGHSNSMALFSFSLLWLCSGVLG

(1) Chicken brain AChR structural subunit   (pCh23.1)
(2) Rat brain AChR structural subunit   (beta2)

N-Terminus

```
                                                                                              M1                      M2
(1) TDTEERLVEVLLDPIRYNKLIRPATNGSQLVTVQLMVSLAQLISVHEREQIMTNVWLTQEWEDYRLTWKPEDFDNMKKVRLPSKHIWLPDVVLYNNADGMYEVSFYSNAVISVDGSIFWLPPAIYKSACKI
(2) TDTEERLVE.LLDPSRYNKLIRPATNGSELVTVQLMVSLAQLISVHEREQIMTNVWLTQEWEDYRLTWKPEDFDNMKKVRLPSKHIWLPDVVLYNNADGMYEVSFYSNAVVSYDGSIFWLPPAIYKSACKI
 1                              .            50                          .            100                           .            250

                                                                                        cytoplasmic domain
(1) EVKHFPFDQQNCTMKFRSWTYDRTETDLVLKSEVASLDDFTPSGEWDIVALPGRRNENPDDSTYVDITYDFIIRRKPLFYTINLIIPCILITSLAILVFYLPSDCGEKMTLCISVLLALTVFLLLISKIVPP
(2) EVKHFPFDQQNCTMKFRSWTYDRTETDLVLKSDVASLDDFTPSGEWDIIALPGRRNENPDDSTYVDITYDFIIRRKPLFYTINLIIPCVLITSLAILVFYLPSDCGEKMTLCISVLLALTVFLLLISKIVPP
                       .            150                           .            200

                    M3                                                                           .            350
(1) TSLDVPLVGKYLMFTMVLVTFSIVTSVCVLNVHHRSPTTHTMPPWVRTLFLRKLPALLFMKQPQQNCARQRLRQRGTQERAAATLFLRAGRACYANPGAAKAEGLNGYRERGGGP  DPPAPCCCG
(2) TSLDVPLVGKYLMFTMVLVTFSIVTSVCVLNVHHRSPTTHTMAPWVKVVFLEKLPTLLFLQQPRHHCARQRLRRRQREREGEAVFFREGPADPCTCFVNPASVQGLAGAFRAEPTAAGPGRSVGPCSCG
                     .            300

                                          M4
(1) LEAVEGVRFIADHMRSEDDDQSVSEDWKYVAMVIDRLFLWIFVFVCVFGTVGWFLQPLFQNYATNSLKQLGQGTPTSK
(2) LREAVDGVRFIADHMRSEDDDQSVREDWKYVAMVIDRLFLWIFVFVCVFGTVGWFLQPLFQNYTATTFLHPDHSAPSSK
 400                           .            450
```

M1-M4: hydrophobic putative transmembrane sequences
cytoplasmic domain: putative cytoplasmic domain
*: putative N-glycosylation sites

Figure 5. Comparison of β2 structural subunit sequences from neuronal nicotinic AChRs of chickens and rats. Reproduced from Lindstrom et al. (1990). Sequences are from Deneris et al. (1988) and Schoepfer et al. (1988b).

The major subtype of nicotinic AChR in brains is thought to be composed of α4 and β2 subunits; this subtype accounts for about 90% of the high affinity nicotine binding sites in mammalian brains (Whiting et al., 1987b; Whiting and Lindstrom, 1987a, 1988; Schoepfer et al., 1988b) and about 60% of these sites in chicken brains (Whiting et al., 1987a; Schoepfer et al., 1988b). In chicken brains a subtype which appears to be composed of α2 and β2 subunits also comprises a substantial fraction of the nicotine binding sites (Whiting et al., 1987a; Schoepfer et al., 1988b; Schoepfer, Whiting, Conroy, and Lindstrom, unpublished). Other subtypes appear to be minor components located in restricted areas. These conclusions are consistent with results both from immunoaffinity purification (Whiting et al., 1987a; Whiting and Lindstrom, 1987a, 1988), immunohistological localization (Swanson et al., 1987; Watson et al., 1988), and *in situ* hybridization (Deneris et al., 1988, 1989; Duvoisin et al., 1989; Wada et al., 1989; Morris et al., 1990). The immunohistological localization of β2 structural subunits in rat brains using mAb 270 (Swanson et al., 1987) closely parallels the localization of high affinity binding sites for ACh and nicotine and is distinct from the pattern of αBgt binding sites (Clarke et al., 1985). This result is consistent with the observations that none of the immunoaffinity purified nicotinic AChRs bind αBgt and that the predominant subtypes have high (nM) affinity for nicotine (Whiting and Lindstrom 1986b, 1987a, 1988; Whiting et al., 1987a). The widespread localization of small amounts of neuronal nicotinic AChRs suggests that these AChRs may be able to exert widespread influences on brain function. The precise roles played by the various neuronal AChR subtypes and their functional properties remain to be identified.

ACh-binding subunits in immunoaffinity purified AChRs have been identified by specific affinity labeling with MBTA (Whiting and Lindstrom , 1987b, 1988; Whiting et al., 1987a). MBTA reacts with the cysteine pair α192, 193 characteristic of α1 subunits (Kao et al., 1984). AChR subunit cDNAs identified by low stringency hybridization which have been found to contain homologues of these cysteines have been considered putative ACh-binding subunits (Boulter et al., 1986; Goldman et al., 1986; Wada et al., 1988). The ACh-binding subunits of the predominant form of AChR immunoaffinity purified from rat brains and the predominant form of the two major subtypes immunoaffinity purified from chicken brains have been identified by N-terminal amino acid sequencing as corresponding to α4 subunit cDNAs as shown in Figure 6 (Whiting et al 1987b, Whiting et al., submitted). mAbs specific for this subunit type react with the ACh-binding subunits of AChRs from cattle and humans, suggesting that these are also encoded by α4 (Whiting and Lindstrom, 1988). Antisera and mAbs raised to a bacterially expressed cDNA fragment of chicken α3 subunits encoding the putative large cytoplasmic domain react with brain AChRs that appear to have β2 structural subunits because of their reaction with mAb 270 (Schoepfer et al., 1989; Schoepfer, Conroy, and Lindstrom, unpublished). These are present in small amounts and have lower affinity for nicotine (μM rather than nM). These mAbs to α3 also react with the AChRs in chicken ciliary ganglia which also have relatively low affinity for ACh (Halvorsen and Berg, 1990). mAbs to a bacterially expressed peptide corresponding to the large putative cytoplasmic domain of chicken α2 subunits bind on western blots to the ACh-binding subunits of AChRs immunoaffinity purified from chicken brains using mAb 35, but these mAbs do not react with the native AChR, presumably because the epitope is buried (Conroy, Schoepfer, and Lindstrom, unpublished).

β2 structural subunits have been identified by N-terminal amino acid sequencing of immunoaffinity purified AChRs from rat and chicken brains, as shown in Figure 6 (Schoepfer et al., 1988b). Both the chicken brain AChR subtype using α4 and the subtype using α2 ACh-binding subunits appear to use β2 structural subunits.

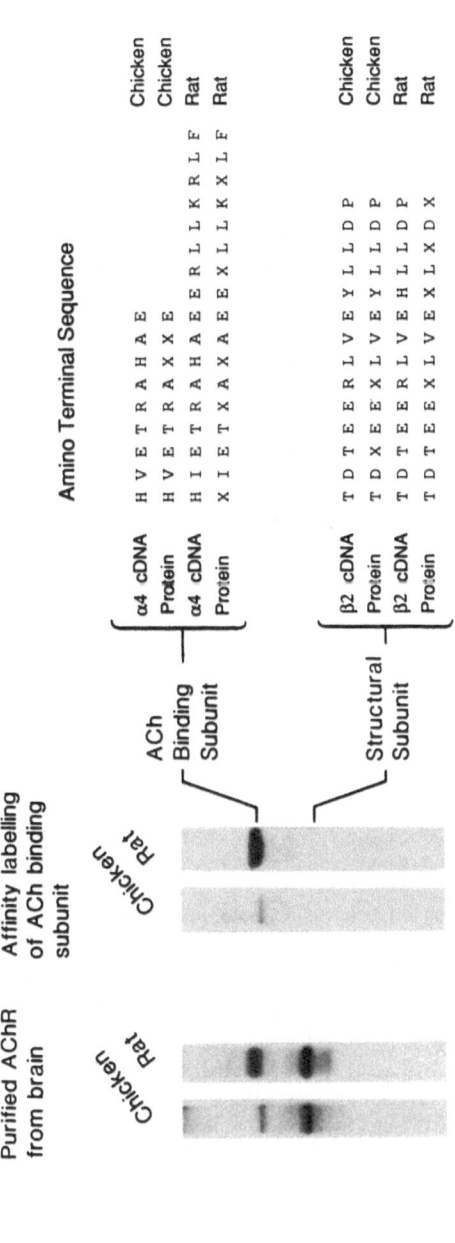

Figure 6. Identification of subunits in the predominant neuronal nicotinic AChR subtype of rats and chickens. Reproduced from Lindstrom et al. (1990). This includes data from Whiting et al. (1987a,b), Schoepfer et al. (1988b), and Whiting et al. (submitted).

Detailed comparison of functional AChRs expressed from cloned subunit cDNAs in *Xenopus* oocytes and transformed cell lines with AChRs from neurons will be required to determine if any other subunits or AChR-associated proteins are required to account for the native properties of neuronal AChR subtypes found *in vivo*. This is complicated by the fact that there are few detailed electrophysiological studies of neuronal AChRs which can be used as a standard to compare with AChRs expressed from cDNAs (Lipton et al., 1987; Margiotta et al., 1987). Pairwise combinations of α2, α3, and α4 subunits with β2 and β4 subunits expressed in oocytes are capable of producing functional AChRs (Ballivet et al., 1988; Duvoisin et al., 1989; Papke et al., 1989). The combinations which exist *in vivo* are a more difficult problem. β2 and α4 mRNAs are the most abundant in brain, and many of these AChR subunit mRNAs are found in overlapping areas (Nef et al., 1988; Wada et al., 1989; Morris et al., 1990). N-terminal amino acid sequence analysis of purified AChRs proves that α4 and β2 subunits are components of the same AChR molecules *in vivo* (Whiting et al., 1987b, submitted; Schoepfer et al., 1988b). However, the possibility of other subunits or associated proteins remains. Developmental considerations add a further variable. During maturation of the neuromuscular junction γ subunits of immature extrajunctional AChRs are exchanged for ε subunits of mature junctional AChRs (Witzemann et al., 1987). It is unknown if similar exchanges are made in neuronal nicotinic AChRs during development or new synapse formation in the adult. It is possible that different subunit combinations may exist in different areas, e.g. α3 subunits might be associated with β2 subunits in one area and β3 or β4 in another. AChR subtypes differ in various areas as suggested by immunohistology (Watson et al., 1988) and *in situ* hybridization (Wada et al., 1989; Morris et al., 1990). For example, α3 is rare at the mRNA and protein level in brain, but relatively abundant in retina (Whiting et al., submitted).

The different neuronal nicotinic AChR subtypes presumably differ in functional roles, but this is not well understood. In ganglia AChRs with α3 subunits play a classic postsynaptic role (Schoepfer et al., 1989; Halvorsen and Berg, 1990). Several lines of evidence suggest that many nicotinic AChRs in brain may be located presynaptically (e.g. Clarke et al., 1986; Rapier et al., 1988). For example, AChR synthesized in retinal ganglion neurons are transported along their axons to the superior colliculus or equivalent structures (Henley et al., 1986; Swanson et al., 1987; Keyser et al., 1988; Sargent et al., 1989). There is evidence that presynaptic AChRs can modulate the release of ACh and other transmitters such as dopamine, and this may be a major functional role (Rapier et al., 1988). AChRs may have functional roles in neurons in addition to synaptic transmission. For example, stimulation of neuronal AChRs can induce rapid gene transcription (Greenberg et al., 1986). Lipton and co-workers (Cheng and Lipton, 1989; Lipton et al., 1989) have shown that AChR activation causes retraction of neurites on retinal ganglion neurons, suggesting a role for AChRs in neuronal plasticity and synapse formation. AChRs with α3 subunits seem to have rather low affinity (μM) for ACh (Smith et al., 1986; Conroy and Lindstrom, unpublished), whereas those with α4 subunits have high affinity (nM), at least in the desensitized state, (Whiting and Lindstrom 1986b, 1988; Whiting et al., 1987a), but the functional significance of this is not clear. As a further complication, the same ACh binding subunit in combination with different structural subunits has been reported to change pharmacological properties (Duvoisin et al., 1989). The ACh binding sites are on the extracellular surface. The most variable part of the subunit sequences is the large putative cytoplasmic domain formed by the sequence between M3 and M4. This could reflect the insignificance and lack of evolutionary constraint on this region, or it could reflect the existence of important properties of this domain unique to each subtype. For example, the cytoplasmic surface would be expected to contain sequences important for targeting AChR transport to axons or dendrites, for targeting

AChR localization in conjunction with associated proteins and the cytoskeleton at synaptic or extrasynaptic locations, and sequences at which phosphorylation might regulate the rate of desensitization or turnover.

The subunit stoichiometry of neuronal nicotinic AChRs appears to differ from muscle AChRs. In neuronal AChRs there appear to be only two kinds of subunits rather than the four kinds observed in muscle nicotinic AChRs (Whiting and Lindstrom 1986a, 1987a, 1988; Whiting et al., 1987a). There appears to be at least two copies of each subunit in neuronal AChRs, because AChRs tethered to agarose beads through a mAb specific for either ACh-binding or structural subunits can bind a second ^{125}I-labeled mAb of the same specificity (Whiting et al., 1987a). There appear to be equal numbers of $\alpha 4$ and $\beta 2$ subunits in neuronal AChRs because equal moles of each subunit type can be ^{125}I-labeled if the AChRs are first denatured to make all of the tyrosines in both subunits equally accessible (Whiting, Cooper, Conroy, and Lindstrom, unpublished). The size of AChR monomers containing $\alpha 4$ and $\beta 2$ subunits from brain and AChRs monomers from *Torpedo* electric organ are approximately the same size as measured by sucrose gradients. This means that the size of neuronal nicotinic AChRs is compatible with two, or perhaps, three copies of each subunit, but is unlikely to accommodate more copies of each subunit. It seems likely that ACh-binding subunits and structural subunits of neuronal AChRs alternate as barrel staves around a central cation channel as do the subunits of muscle AChRs. Neuronal AChRs composed of equal numbers of two kinds of subunits, e.g. $\alpha^{\alpha}_{\beta}\alpha$, would lack the pentagonal symmetry of $\beta^{\alpha\gamma}_{\alpha\delta}$ muscle AChRs. Also, neuronal AChRs with only four subunits forming the central channel might be expected to exhibit lower conductance than would muscle AChRs with a larger channel formed by five subunits. In fact, both chicken and rat $\alpha 4\beta 2$ AChRs expressed in *Xenopus* oocytes exhibit about half the conductance of muscle AChRs expressed in oocytes (Ballivet et al., 1988; Papke et al., 1989).

There is very little neuronal AChR protein available for immunoaffinity purification. Extensive biochemical studies of neuronal nicotinic AChR structure will require the expression of relatively large amounts of AChR from defined subunit cDNAs. AChRs expressed from chicken $\alpha 4$ and $\beta 2$ subunits have been obtained in relatively high concentrations, but small amounts from *Xenopus* oocytes injected with synthetic mRNAs (Anand, Conroy, Schoepfer, and Lindstrom, unpublished). These AChRs have also been expressed in functional form but at somewhat lower concentrations in transfected mouse fibroblast lines (Whiting et al., unpublished) using the same basic approach applied by Claudio et al., (1987) to express *Torpedo* AChRs in fibroblasts or by Blount and Merlie (1989) to express combinations of muscle AChR subunits. Expression of AChRs from defined subunits in transfected cell lines seems a feasible approach for producing relatively large amounts of AChR for further structural studies.

NEURONAL α-BUNGAROTOXIN-BINDING PROTEINS

αBgtBPs are found in brain, retina, and ganglia (Wang et al., 1978; Chiappinelli, 1985; Kemp et al., 1985; Schneider et al., 1985). Like nicotinic AChRs of muscle and neurons, αBgtBPs exhibit nM affinity for αBgt and μM affinity for ACh and nicotine (Marks and Collins, 1982; Lukas, 1986). Like nicotinic AChRs of muscle and neurons, αBgtBP can be affinity labeled by MBTA and bromoacetylcholine (Norman et al., 1982; Leprince, 1983; Kemp et al., 1985). This suggests that αBgtBPs have cysteines homologous to $\alpha 192$, 193 of AChRs from muscle and electric organ which are labeled by these reagents (Kao et al., 1984). Like AChRs of muscle and neurons, αBgtBPs sediment on sucrose gradients at about 10S (Betz et al., 1981; Norman et al., 1982; Conti-Tronconi et al., 1985). In ganglion cells and PC12 cells, αBgtBPs and AChRs are found in the same cells (Patrick and

Stallcup, 1977; Smith et al., 1985, 1986). In brain, the localizations of αBgtBPs and AChRs overlap, but are distinct (Clarke et al., 1986). However, unlike AChRs, these ganglionic αBgtBPs do not appear to function as ACh-gated cation channels (Patrick and Stallcup, 1977; Carbonetto and Fambrough, 1978). Unlike AChRs, αBgtBPs on ganglia are not located at synapses (Jacob and Berg, 1983). Affinity purified αBgtBPs have been found to consist of several polypeptides of apparent molecular weights in the 40,000-60,000 range, but their subunit composition is not well defined (Conti-Tronconi et al., 1985; Kemp et al., 1985; Whiting and Lindstrom, 1987a). The N-terminal amino aid sequence of one subunit of αBgtBP affinity purified from chicken brain was determined and found to exhibit some homology with nicotinic AChR sequences (Conti-Tronconi et al., 1985).

We used the reported N-terminal amino acid sequence of an αBgtBP subunit to construct oligonucleotide probes for screening a chicken brain cDNA library (Schoepfer et al., submitted). In this way a cDNA termed αBgtBP α1 was identified which encoded a protein with a putative N-terminal sequence virtually identical to that reported for the subunit of purified αBgtBP by Conti-Tronconi et al., (1985). Using this cDNA as a probe, a second very similar sequence termed αBgtBPα2 was identified. These cDNAs encoded sequences with evident homology to subunits of AChRs from muscle and nerve, but appear to belong to another branch of the gene family about equally distant from either AChRs of muscle or nerves. These cDNAs exhibit the pattern of four hydrophobic sequences characteristic of all subunits in the ligand gated ion channel gene superfamily. The cDNAs encode subunits which look as though they could function as parts of ligand gated cation channels. The net charge in the sequences N-terminus - M1 and M3 - the C-terminus, and the sequence homology in M1-M3 is much closer to the AChRs, which have cation channels, than to subunits of glycine or GABA receptors, which have anion channels. Both αBgtBPα1 and αBgtBPα2 may be ligand binding subunits. Both exhibit a cysteine pair homologous to α192-193 which is near the ACh binding site of AChRs (Kao et al., 1984; Dennis et al., 1988) and also exhibit other amino acids characteristic of ACh-binding subunits from AChRs of muscles and nerves such as tyrosine α190 (Abramson et al., 1989).

mAbs to bacterially expressed peptides from αBgtBP α1 and α2 were used to show that these subunits are components of authentic αBgtBP solubilized from chicken brains (Schoepfer et al., submitted). Unique sequences corresponding to the large putative cytoplasmic domain between M3 and M4 were expressed in bacteria to produce antigen. mAbs to this sequence of αBgtBPα1 immune precipitated about 85% of ^{125}I-αBgt labeled αBgtBP from chicken brain, whereas mAbs to αBgtBPα2 immune precipitated about 15% of the total. mAbs to both subunits identify subunits of similar apparent molecular weights in western blots of affinity purified αBgtBP. We found mAbs to affinity purified αBgtBP directed at conformation-dependent epitopes which fall into two classes: 1) mAbs that immunoprecipitate all αBgtBP, and 2) mAbs that immunoprecipitate about 15% of αBgtBP. These results are consistent with the results with mAbs to expressed αBgtBP subunit peptides in suggesting that there are subtypes of neuronal αBgtBPs. Reminiscent of the role of α4 and α3 ACh binding subunits in brain AChR subtypes, αBgtBPα1 subunits may be ligand binding subunits of a major subtype of brain αBgtBP, whereas αBgtBPα2 subunits may be ligand binding subunits of a minor subtype of brain αBgtBP. mAbs to purified αBgtBP that immunoprecipitate all αBgtBP may bind to a conformation-dependent epitope on a structural subunit common to all subtypes. mAbs to purified αBgtBP that immunoprecipitate 15% of αBgtBP may bind to a conformation-dependent epitope on αBgtBP α2 subunits. Detailed subunit compositions of αBgtBPs remain to be determined.

In situ hybridization studies suggest that αBgtBPα1 corresponds to a subunit of the major αBgtBP in rat brain (Swanson, Schoepfer, and Lindstrom, unpublished). Using the polymerase chain reaction and primer sequences from chicken αBgtBPα1 cDNA corresponding to the large putative cytoplasmic domain, the corresponding sequence of the rat brain αBgtBPα1 cDNA was identified. *In situ* hybridization using this probe on rat brain sections revealed intense labeling of regions characteristically labeled by ^{125}I-αBgt. Labeling was observed, for example, in the superior colliculus. In particular, there was strong labeling of the hippocampus, which is characteristic of ^{125}I-αBgt labeling, but not of labeling by ^{3}H-ACh or ^{3}H-nicotine (Clarke et al., 1986).

What might be the functions of neuronal αBgtBPs? The sequences of the two cloned subunits available suggest that they could be ligand gated cation channels. The ligand could be ACh, but where they have been studied, αBgtBPs did not appear to function as ACh-gated cation channels (Patrick and Stallcup, 1977; Carbonetto et al., 1978). This could be because the endogenous ligand is not ACh. Thymopoietin (Quick et al., 1989) and substance P (Utkin et al., 1989) have been reported to compete for αBgt binding to αBgtBP, and might be candidates for endogenous ligands. Perhaps to function as gated cation channels αBgtBPs require a cofactor or posttranslational modification not available in the tested systems. Or perhaps they are highly selective for potassium and function as inhibitory channels, causing their activity to have been missed because excitatory AChRs were present in the same cells. In invertebrates there are αBgt-inhibitable inhibitory anion and cation channels (Kehoe et al., 1979; Kristan and French, 1988) and inhibitory AChR gated channels have been reported in vertebrates (Wong and Galleger, 1989). Finally, it remains possible that vertebrate neuronal αBgtBPs are a descendent of ligand gated ion channels, but now function in some other role.

Now we have the mAb and cDNA tools available which should permit immuno-affinity purification of αBgtBP subtypes, determination of their subunit composition, and cloning of all of their subunit cDNAs. Then expression of αBgtBPs from their subunit cDNAs may permit determination of the functional properties of these enigmatic members of the AChR gene family.

REFERENCES

Abramson, S., Y. Culver, P. Taylor, 1989, An analog of lophotoxin reacts covalently with Try^{190} in the α subunit of the nicotinic acetylcholine receptor. *J. Biol. Chem. 264*:1266-1267.

Baldwin, T., C. Yoshihara, K. Blackmer, C. Kinter, and S. Burden, 1988, Regulation of acetylcholine receptor transcript expression during development in *Xenopus laevis. J. Cell Biol. 106*:469-478.

Ballivet M., P. Nef, S. Couturier, D. Rungger, C. Bader, D. Bertrand, and E. Cooper, 1988, Electrophysiology of a chick neuronal nicotinic acetylcholine receptor expressed in *Xenopus* oocytes after cDNA injection. *Neuron 1*:847-852.

Barnard E., M. Darlison, and P. Seeburg, 1987, Molecular biology of the GABA$_A$ receptor: The receptor/channel superfamily. *Trends in Neurosci. 10*:502-509.

Bellone, M., F. Tang, R. Milius, B. Conti-Tronconi, 1989, The main immunogenic region of the nicotinic acetylcholine receptor: Identification of amino acid residues interacting with different antibodies. *J. Immunology 143*:3568-3579.

Betz, H., and C-M. Becker, 1988, The mammalian glycine receptor: Biology and structure of a neuronal chloride channel protein. *Neurochem. Int. 13*:137-146.

Betz, H., D. Graham and H. Rehm, 1981, Identification of polypeptides associated with a putative neuronal nicotinic acetylcholine receptor. *J. Biol. Chem. 257*:11390-11394.

Blount, P., and J. Merlie, 1988, Native folding of an acetylcholine receptor α subunit expressed in the absence of other receptor subunits. *J. Biol. Chem. 262*:4367-4376.

Blount, P,. and J.P. Merlie, 1989, Molecular basis of the two nonequivalent ligand binding sites of the muscle nicotinic acetylcholine receptor. *Neuron 3*:349-357.

Boulter, J., J. Connolly, E. Deneris, D. Goldman, S. Heinemann, and J. Patrick, 1987, Functional expression of two neuronal nicotinic acetylcholine receptors from cDNA clones identifies a gene family. *Proc. Natl. Acad. Sci. USA 84*:7763-7767.

Boulter, J., K. Evans, D. Goldman, G. Martin, D. Treco, S. Heinemann, and J. Patrick, 1986, Isolation of a cDNA clone coding for a possible neural nicotinic acetylcholine receptor α subunit. *Nature 319*:368-374.

Carbonetto, S., D. Fambrough, and K. Muller, 1978, Nonequivalence of α-bungarotoxin receptors and acetylcholine receptors in chick sympathetic neurons. *Proc. Natl. Acad. Sci USA 75*:1016-1020.

Changeux, J-P., J. Giraudat, and M. Dennis, 1987, The nicotinic acetylcholine receptor: Molecular architecture of a ligand-regulated ion channel. *Trends in Pharmacology 8*:459-465.

Cheng, T., and T. Lipton, 1989, Nicotine induces retraction of neuritis in cultured rat retinal ganglion cells. *Neuroscience Society Meeting Abstract* 263.6.

Chiappinelli, V., 1985, Actions of snake venom toxins on neuronal nicotinic receptors and other neuronal receptors. *Pharmac. Ther. 31*:1-32.

Clark, P., G. Hamill, N. Nadi, D. Jacobowitz, and A. Pert, 1986, ^{3}H-Nicotine and ^{125}I-α-bungarotoxin-labeled nicotinic receptors in the interpeduncular nucleus of rats. II. Effects of habenular deafferentation. *J. Comp. Neurol. 251*:407-413.

Clarke, P., R. Schwartz, S. Paul, C. Pert, and A. Pert, 1985, Nicotinic binding in rat brain: Autoradiographic comparison of ^{3}H-acetylcholine, ^{3}H-nicotine, and ^{125}I-α-bungarotoxin. *J. Neurosci. 5*:1307-1315.

Claudio, T., M. Ballivet, J. Patrick, and S. Heinemann, 1983, *Torpedo californica* acetylcholine receptor 60,000 dalton subunit: Nucleotide sequence of cloned cDNA deduced amino acid sequence, subunit structural predictions. *Proc. Natl. Acad. Sci. USA 80*:1111-1115.

Claudio, T., W. Green, D. Hartman, D. Hayden, H. Paulson, F. Sigworth, S. Sine, and A. Swedlund, 1987, Genetic reconstitution of functional acetylcholine receptor channels in mouse fibroblasts. *Science 238*:1688-1694.

Conti-Tronconi, B., S. Dunn, E. Barnard, J. Dolly, F. Lai, N. Ray, and M. Raftery, 1985, Brain and muscle nicotinic acetylcholine receptors are different but homologous proteins. *Proc. Natl. Acad. Sci. USA 82*:5208-5212.

Conti-Tronconi, B., S. Tzartos, and J. Lindstrom, 1981, Monoclonal antibodies as probes of acetylcholine receptor structure: II: Binding to native receptor. *Biochemistry 20*:2181-2191.

Criado, M., S. Hochschwender, V. Sarin, J.L. Fox, and J. Lindstrom, 1985a, Evidence for unpredicted transmembrane domains in acetylcholine receptor subunits. *Proc. Natl. Acad. Sci. USA 82*:2004-2008.

Criado, M., V. Sarin, J.L. Fox, and J. Lindstrom, 1985b, Structural localization of the sequence α235-242 of the nicotinic acetylcholine receptor. *Biochem. Biophys. Res. Commun. 128*:864-871.

Criado, M.,V. Sarin, J. Fox, and J. Lindstrom, 1986, Evidence that the acetylcholine binding site is not formed by the sequence α127-143 of the acetylcholine receptor. *Biochemistry 25*:2839-2846.

Das, M., and J. Lindstrom, 1989, The main immunogenic region of the nicotinic acetylcholine receptor: Interaction of monoclonal antibodies with synthetic peptides. *Biochem. Biophys. Res. Commun. 165*:865-871.

272

Deneris, E.S., J. Boulter, L.W. Swanson, J. Patrick, and S. Heinemann, 1989, β3: A new member of nicotinic acetylcholine receptor gene family is expressed in brain. *J. Biol. Chem. 264:*6268-6272.

Deneris, E., J. Connolly, J. Boulter, E. Wada, K. Wada, L. Swanson, J. Patrick, and S. Heinemann, 1988, Primary structure and expression of β2: A novel subunit of neuronal nicotinic receptors. *Neuron 1:*45-54.

Dennis, M., J. Giraudat, F. Kotzyba-Hibert, M. Goeldner, C. Hirth, J-Y. Chang, C. Lazure, M. Chretien, J-P. Changeux, 1988, Amino acids of the *Torpedo marmorata* acetylcholine receptor α subunit labeled by a photoaffinity ligand for the acetylcholine binding site. *Biochemistry 27:*2346-2358.

Devillers-Thiery, A., J. Giraudat, M. Bentaboulet, and J-P. Changeux, 1983, Complete mRNA coding sequence of the acetylcholine binding α subunit of *Torpedo marmorata* acetylcholine receptor: A model for the transmembrane organization of the polypeptide chain. *Proc. Natl. Acad. Sci. USA 80:*2067-2071.

Duvoisin, R., E. Deneris, J. Patrick, S. Heinemann, 1989, The functional diversity of the neuronal nicotinic receptors is increased by a novel subunit: β4. *Neuron 3:*487-496.

Fujita, N., N. Nelson, T. Fox, T. Claudio, J. Lindstrom, H. Reizman, and G. Hess, 1986, Biosynthesis of the *Torpedo californica* acetylcholine receptor α subunit in yeast. *Science 231:*1284-1287.

Geysen, H., S. Rodda, T. Mason, G. Tribbick, and P. Schoofs, 1987, Strategies for epitope analysis using peptide synthesis. *J. Immunol. Meth. 102:*259-274.

Giraudat, J., M. Dennis, T. Heidmann, J. Chang, and J-P. Changeux, 1986, Structure of the high-affinity binding site for noncompetitive blockers of the acetylcholine receptor: Serine 262 of the δ subunit is labeled by ^{3}H chlorpromazine. *Proc. Natl. Acad. Sci. USA 83:*2719-2723.

Goldman, D., D. Simmons, L. Swanson, J. Patrick, and S. Heinemann, 1986, Mapping of brain areas expressing RNA homologous to two different acetylcholine receptor α subunit cDNAs. *Proc. Natl. Acad. Sci. USA 83:*4076-4080.

Greenberg, M., E. Ziff, and L. Greene, 1986, Stimulation of neuronal acetylcholine receptors induces rapid gene transcription. *Science 234:*80-83.

Gregor, P., I. Mano, I. Maoz, M. McKeown, V. Teichberg, 1989, Molecular structure of the chick cerebellar kainate-binding subunit of a putative glutamate receptor. *Nature 342:* 689-692.

Halvorsen, S., and D. Berg, in press, Subunit composition of nicotinic acetylcholine receptors from chick ciliary ganglia: Correlation with known gene products. *J. Neuroscience.*

Henley, J., J. Lindstrom, and R. Oswald, 1986, Acetylcholine receptor synthesis in retina and transport to the optic tectum in goldfish. *Science 232:* 1627-1629.

Hollmann, M., A. O'Shea-Greenfield, S. Rogers, S. Heinemann, 1989, Cloning by functional expression of a member of the glutamate receptor family. *Nature 342:*643-648.

Hucho, F., and R. Hilgenfeld, 1989, The selectivity filter of a ligand-gated ion channel: The helix-M2 model of the ion channel of the nicotinic acetylcholine receptor. *FEBS Lett. 257:*17-23.

Imoto, K., C. Busch, B. Sakmann, M. Mishina, T. Konno, J. Nakai, H. Bujo, Y. Mori, K. Fukuda, and S. Numa, 1988, Rings of negatively charged amino acids determine the acetylcholine receptor channel conductance. *Nature 335:*645-648.

Imoto, F., C. Methfessel, B. Sakmann, M. Mishina, Y. Mori, T. Konno, F. Fukuda, M. Kurasaki, H. Bujo, Y. Fujita, and S. Numa, 1986, Location of a δ subunit region determining ion transport through the acetylcholine receptor channel. *Nature 324:*670-674.

Jacob, M. and D. Berg, 1983, The ultrastructural localization of α-bungarotoxin binding sites in relation to synapses on chick ciliary ganglion neurons. *J. Neurosci. 3:*260-271.

Jacob, M., D. Berg, and J. Lindstrom, 1984, A shared antigenic determinant between the *Electrophorus* acetylcholine receptor and a synaptic component on chick ciliary ganglion neurons. *Proc. Natl. Acad. Sci. USA* 81:3223-3227.

Jacob, M., J. Lindstrom, and D. Berg, 1986, Surface and intracellular distribution of a putative neuronal nicotinic acetylcholine receptor. *J. Cell Biology*, 103:205-214.

Jansen, K.U, W.G. Conroy, T. Claudio, T.D. Fox, N. Fujita, O. Hamill, J.M. Lindstrom, M. Luther, N. Nelson, K.A. Ryan, M.T. Sweet, and G.P. Hess, 1989, Expression of the four subunits of the *Torpedo californica* nicotinic acetylcholine receptor in *Saccharomyces cerevisiae*. *J. Biol. Chem.* 264:15022-15027.

Kao, P., A. Dwork, R. Kaldany, M. Silver, J. Wideman, S. Stein, and A. Karlin, 1984, Identification of the *α* subunit half cysteine specifically labeled by an affinity reagent for the acetylcholine receptor binding site. *J. Biol. Chem.* 259:11662-11665.

Kao, P. and A. Karlin, 1986, Acetylcholine receptor binding site contains a disulfide crosslink between adjacent half-cystinyl residues. *J. Biol. Chem.* 261:8085-8088.

Karlin, A., R. Cox, M. Di Paola, E. Holtzman, P. Kao, P. Label, L. Wang, and N. Yodh, 1986, Functional domains of the nicotinic acetylcholine receptor. *Annals of the NY Acad. of Sci. 463:* 53-69.

Kehoe, J., 1979, Acetylcholine receptors in molluscan neurons. *Adv. Pharmacol. Ther.* 8:285-298.

Kellaris, K., D. Ware, S. Smith, and J. Kyte, 1989, Assessment of the number of free cysteines and isolation and identification of cysteine-containing peptides from acetylcholine receptor. *Biochemistry* 28:3469-3482.

Kemp, G., L. Bentley, M. McNamee, and B. Morley, 1985, Purification and characterization of the *α*-bungarotoxin binding protein from rat brain. *Brain Res. 347:* 274-283.

Kemp G., and B. Morley, 1986, Ganglionic nAChRs and high affinity nicotinic binding sites are not equivalent. *FEBS Lett.* 205:265-268.

Keyser, K.T., T.E. Hughes, P.J. Whiting, J.M. Lindstrom, and H.J. Karten, 1988, Cholinoceptive neurons in the retina of the chick: An immunohistochemical study of the nicotinic acetylcholine receptors. *Visual Neurosci.* 1:349-366.

Klausner, R., 1989, Architectural editing: Determining the fate of newly synthesized membrane proteins. *The New Biologist* 1:3-8.

Kristan, W. and K. French, 1988, Segment-specific differences in ACh receptors in leech retzins neurons. *Neuroscience Society Meeting Abstracts* 69:11.

Kubalek, E., S. Ralston, J. Lindstrom, and N. Unwin, 1987, Location of subunits within the acetylcholine receptor: Analysis of tubular crystals from *Torpedo marmorata*. *J. Cell. Biol.* 105:9-18.

Kubo, T. M. Noda, T. Takai, T. Tanabe, T. Kayano, S. Shimizu, K. Tanaka, H. Takahashi, T. Hirose, S. Inayama, R. Kikuno, T. Miyata, and S. Numa, 1985, Primary structure of *δ* subunit precursor of calf muscle acetylcholine receptor deduced from cDNA sequence. *EUR. J. Biochem.* 149:5-13.

Kurosaki, T., K. Fukuda, T. Konno, Y. Mori, K. Tanaka, M. Mishina, and S. Numa, 1987, Functional properties of nicotinic acetylcholine receptor subunits expressed in various combinations. *FEBS Lett.* 214:253-258.

La Polla, R., K. Mayne, and N. Davidson, 1984, Isolation and characterization of a cDNA clone for the complete protein coding region of the *δ* subunit of the mouse acetylcholine receptor. *Proc. Natl. Acad. Sci. USA* 81:7970-7974.

Leprince, P., 1983, Chemical modification of the nicotinic cholinergic receptor of PC-12 nerve cell line. *Biochemistry* 22:5551-5556.

Lindstrom, J., M. Criado, M. Ratnam, P. Whiting, S. Ralston, J. Rivier, V. Sarin, and P. Sargent, 1987, Using monoclonal antibodies to determine the structures of acetylcholine receptors from electric organs, muscles, and neurons. *Ann. N.Y. Acad. Sci.* 505:208-225.

Lindstrom, J., R. Schoepfer, W.G. Conroy, P. Whiting, 1990, Structural and functional heterogeneity of nicotinic receptors. *Ciba Foundation Symposium #152*, John Wiley & Sons, New York.

Lindstrom, J., P. Whiting, R. Schoepfer, M. Luther, and M. Das, 1989, Structure of nicotinic acetylcholine receptors from muscle and neurons. In: "Computer-Assisted Modeling of Receptor-Ligand Interactions: Theoretical Aspects and Applications", R. Rein and A. Golombek (eds.), Alan R. Liss, New York.

Lipton, S., E. Aizenman, and R. Loring, 1987, Neural nicotinic acetylcholine responses in solitary mammalian retinal ganglion cells. *Plgers Arch.* 410:37-43.

Lipton, S., M. Frosch, M. Phillips, D. Tauck, and E. Aizenman, 1988, Nicotinic antagonists enhance process outgrowth by rat retinal ganglion cells in culture. *Science* 239:1293-1296.

Lukas, R., 1986, Immunochemical and pharmacological distinctions between curaremimetic neurotoxin binding sites of central, autonomic, and peripheral origin. *Proc. Natl. Acad. Sci. USA* 83:5741-5745.

Luther, M., R. Schoepfer, P. Whiting, Y. Blatt, M.S. Montal, M. Montal, and L. Lindstrom, 1988, A muscle acetylcholine receptor is expressed in the human cerebellar medulloblastoma cell line TE671. *J. Neurosci.* 9:1082-1096.

Margiotta, J., D. Berg, and V. Dionne, 1987, The properties and regulation of functional acetylcholine receptors on chick ciliary ganglion neurons. *J. Neurosci.* 7:3612-3622.

Marks, M., and A. Collins, 1982, Characterization of nicotine binding in mouse brain and comparison with binding of α-bungarotoxin and quinuclidinly benzilate. *Mol. Pharmacol* 22:554-564.

McAllister, R., H. Isaacs, R. Rongey, M. Peer, W. Au, S. Soukup, and M. Gardner, 1977, Establishment of a human medulloblastoma cell line. *Intl. J. Cancer* 20:206-212.

Merlie, J.P., and J. Lindstrom, 1983, Assembly *in vivo* of mouse muscle acetylcholine receptor: Identification of an α subunit species which may be an assembly intermediate. *Cell* 34:747-757.

Mitra, A., M. McCarthy, and R. Stroud, 1989, Three-dimensional structure of the nicotinic acetylcholine receptor and location of the major associated 43kD cytoskeletal protein, determined at 22Å by low-dose electron microscopy and x-ray diffraction to 12.5Å. *J. Cell Biol.* 109:755-774.

Morris, B., A. Hicks, W. Wisden, M. Darlison, S. Hunt, and E. Barnard, in press, Distinct regional expression of nicotinic acetylcholine receptor genes in chick brain. *Mol. Brain Research*

Nef, P., A. Mauron, R. Stalder, C. Alliod, and M. Ballivet, 1984, Structure, linkage, and sequence of the two genes encoding the δ and γ subunits of the nicotinic acetylcholine receptor. *Proc. Natl. Acad. Sci. USA* 81:7975-7979.

Nef, P., C. Oneyser, C. Alliod, S. Couturier, and M. Ballivet, 1988, Genes expressed in the brain define three distinct neuronal nicotinic acetylcholine receptors. *EMBO J.:*595-601.

Noda, M., Y. Furutani, H. Takahashi, M. Toyosato, T. Tanabe, S. Shimizu, S. Kikyotani, T. Kayano, T. Hirose, S. Inayama, and S. Numa, 1983, Cloning and sequence analysis of calf cDNA and human genomic DNA encoding α subunit precursor of muscle acetylcholine receptor. *Nature* 305:818-823.

Noda, M., H. Takahashi, T. Tanabe, M. Toyosato, Y. Furutani, T. Hirose, M. Asai, S. Inayama, T. Miyata, and S. Numa, 1982, Primary structure of α subunit precursor of *Torpedo californica* acetylcholine receptor deduced from cDNA sequence. *Nature* 299:793-797.

Noda, M., H. Takahashi, T. Tanabe, M. Toyosato, S. Kikyotani, T. Hirose, M. Asai, H. Takashima, S. Inayama, T. Miyata, and S. Numa, 1983, Primary structures of β- and δ-subunit precursors of *Torpedo californica* acetylcholine receptor deduced from cDNA sequences. *Nature* 301:251-255.

Noda, M., H. Takahashi, T. Tanabe, M. Toyosato, S. Kikyotani, Y. Furutani, T. Hirose, H. Takashima, S. Inayama, T. Miyata, and S. Numa, 1983, Structural homology of *Torpedo californica* acetylcholine receptor subunits. *Nature 302*:528-532.

Norman, R., F. Mehraban, E. Barnard, and O. Dolly, 1982, Nicotinic acetylcholine receptor from chick optic lobe. *Proc. Natl. Acad. Sci. USA 79*:1321-1325.

Oiki, S., W. Danho, K. Madison, and M. Montal, 1988, M2δ, a candidate for the structure lining the ionic channel of the nicotinic cholinergic receptor. *Proc. Natl. Acad. Sci. USA 85*:8703-8707.

Papke, R., J. Boulter, J. Patrick, and S. Heinemann, 1989, Single-channel currents of rat neuronal nicotinic acetylcholine receptors expressed in *Xenopus laevis* oocytes. *Neuron 3*:589-596.

Patrick, J., and M. Stallcup, 1977, α-Bungarotoxin binding and cholinergic receptor function on a rat sympathetic nerve line. *J. Biol. Chem. 252*:8629-8633.

Pedersen, S., P. Bridgman, S. Sharp, J. Cohen, 1990, Identification of a cytoplasmic region of the *Torpedo* nicotinic acetylcholine receptor α subunit by epitope mapping. *J. Biol. Chem. 265*:569-581.

Quik, R., R. Afar, T. Audhya, and G. Goldstein, 1989, Thymopoietin, a thymic polypeptide, specifically interacts at neuronal nicotinic α-bungarotoxin receptors. *J. Neurochem. 53*:1320-1323.

Raftery, M., M. Hunkapillar, C. Strader, and L. Hood, 1980, Acetylcholine receptor: Complex of homologous subunits. *Science 208*:1454-1457.

Rapier, C., G. Lunt, and S. Wonnacott, 1988, Stereoselective nicotine-induced release of dopamine from striatal synaptosomes: Concentration dependence and repetitive stimulation. *J. Neurochem. 50*:1123-1130.

Ratnam, M. and J. Lindstrom, 1984, Structural features of the nicotinic acetylcholine receptor revealed by antibodies to synthetic peptides. *Biochem. Biophys. Res. Commun. 122*:1225-1233.

Ratnam, M., P.B. Sargent, V. Sarin, J.L. Fox, D. Le Nguyen, J. Rivier, M. Criado, and J. Lindstrom, 1986a, Location of antigenic determinants on primary sequences of subunits of nicotinic acetylcholine receptor by peptide mapping. *Biochemistry 25*:2621-2632.

Ratnam, M. D. Le Nguyen, J. Rivier, P.B. Sargent, and J. Lindstrom, 1986b, Transmembrane topography of nicotinic acetylcholine receptor: Immunochemical tests contradict theoretical predictions based on hydrophobicity profiles. *Biochemistry 25*:2633-2643.

Sargent, P., B. Hedges, L. Tsavaler, L. Clemmons, S. Tzartos, and J. Lindstrom, 1983, The structure and transmembrane nature of the acetylcholine receptor in amphibian skeletal muscles revealed by crossreacting monoclonal antibodies. *J. Cell Biol. 98*:609-618.

Sargent, P.B., S.H. Pike, S.B. Nadel, and J.M. Lindstrom, 1989, Nicotinic acetylcholine receptor-like molecules in the retina, retinotectal pathway, and optic tectum of the frog. *J. Neurosci. 9*:565-573.

Schneider, M., C. Adee, H. Betz, and J. Schmidt, 1985, Biochemical characterization of two nicotinic receptors from the optic lobe of the chick. *J. Biol. Chem. 260*:14505-14512.

Schoepfer, R. W.G. Conroy, P. Whiting, M. Gore, and L. Lindstrom, submitted, cDNA clones define brain α-bungarotoxin-binding proteins as members of the ligand-gated ion channel family.

Schoepfer, R., S. Halvorsen, W.G. Conroy, P. Whiting, and J. Lindstrom, 1989, Antisera against an α-3 fusion protein bind to ganglionic but not to brain nicotinic acetylcholine receptors. *FEBS Lett. 257*:393-399.

Schoepfer, R., M. Luther, and J. Lindstrom, 1988a, The human medulloblastoma cell line TE671 expresses a muscle-like acetylcholine receptor: Cloning of the α subunit cDNA. *FEBS Lett. 226*:235-240.

Schoepfer, R. P. Whiting, F. Esch, R. Blacher, S. Shimasaki, and J. Lindstrom, 1988b, cDNA clone coding for the structural subunit of a chicken brain nicotinic acetylcholine receptor. *Neuron 1*:241-248.

Sine, S., 1988, Functional properties of human skeletal muscle acetylcholine receptors expressed by the TE671 cell line. *J. Biol. Chem. 263*:18052-18062.

Smith, M., J. Margiotta, A. Franco, J. Lindstrom, and D. Berg, 1986, Cholinergic modulation of an acetylcholine receptor-like antigen on the surface of chick ciliary ganglion neurons in cell culture. *J. Neurosci.* 6:946-953.

Smith, M., J. Stollberg, J. Lindstrom, and D.K. Berg, 1985, Characterization of a component in chick ciliary ganglia that cross-reacts with monoclonal antibodies to muscle and electric organ acetylcholine receptor. *J. Neurosci.* 5:2726-2731.

Souroujon, M., D. Neumann, S. Pizzighella, A. Safran, and S. Fuchs, 1986, Localization of a highly immunogenic region on the acetylcholine receptor *a* subunit. *Biochem. Biophys. Res. Commun.* 135:82-89.

Steinbach, J., and C. Ifune, 1989, How many kinds of nicotinic acetylcholine receptor are there? *TINS* 12:3-6.

Stratton, M., B. Reeves, and C. Cooper, 1989, "Scientific Correspondence", *Nature* 337:311-312.

Stroud, R.M. and J. Finer-Moore, 1985, Acetylcholine receptor structure, function, and evolution. *Ann. Rev. Cell Biol.* 1:317-351.

Sumikawa, K., and R. Miledi, 1989, Assembly and N-glycosylation of all ACh receptor subunits are required for their efficient insertion into plasma membranes. *Mol. Brain Res.* 5:183-192.

Swanson, L., J. Lindstrom, S. Tzartos, L. Schmued, D. O'Leary, and W. Cowan, 1983, Immunohistochemical localization of monoclonal antibodies to the nicotinic acetylcholine receptor in the midbrain of the chick. *Proc. Natl. Acad. Sci. USA* 80:4532-4536.

Swanson, L., D. Simmons, P. Whiting, and J. Lindstrom, 1987, Immunohistochemical localization of neuronal nicotinic receptors in the rodent central nervous system. *J. Neurosci.* 7:3334-3342.

Tobimatsu, T., Y. Fujita, K. Fukuda, K. Tanaka, Y. Mori, T. Konno, M. Mishina, and S. Numa, 1987, Effects of substitution of putative transmembrane segments on nicotinic acetylcholine receptor function. *FEBS Lett.* 222:56-62.

Toyoshima, C. and N. Uniwn, 1988, Ion channel of acetylcholine receptor reconstructed from images of postsynaptic membranes. *Nature* 336:247-250.

Tzartos, S. and J. Lindstrom, 1980, Monoclonal antibodies used to probe acetylcholine receptor structure: Localization of the main immunogenic region and detection of similarities between subunits. *Proc. Natl. Acad. Sci. USA* 77:755-759.

Tzartos, S. S. Hochschwender, L. Langeberg, and J. Lindstrom, 1983, Demonstration of a main immunogenic region on acetylcholine receptors from human muscle using monoclonal antibodies to human receptor. *FEBS Lett.* 158:116-118.

Tzartos, S., S. Hochschwender, P. Vasquez, and J. Lindstrom, 1987, Passive transfer of experimental autoimmune myasthenia gravis by monoclonal antibodies to the main immunogenic region of the acetylcholine receptor. *J. Neuroimmunol.* 15:185-194.

Tzartos, S., A. Kokla, S. Walgrave, and B. Conti-Tronconi, 1988, The main immunogenic region of human muscle acetylcholine receptor is localized within residues 63-80 of the *a* subunit. *Proc. Natl. Acad. Sci. USA* 85:2899-2903.

Tzartos, S., L. Langeberg, S. Hochschwender, L. Swanson, and J. Lindstrom, 1986, Characteristics of monoclonal antibodies to denatured *Torpedo* and to native calf acetylcholine receptors: species, subunit, and region specificity. *J. Neuroimmunol.* 10:235-253.

Tzartos, R. H. Loutrari, F. Tang, A. Kokla, S. Walgrave, R. Milius, Conti-Tronconi, B., 1990, Main immunogenic region of *Torpedo* electroplax and human muscle acetylcholine receptor: Localization and microheterogeneity revealed by the use of synthetic peptides. *J. Neurochemistry* 54:51-61.

Tzartos, S., D. Rand, B. Einarson, and J. Lindstrom, 1981, Mapping of surface structures on *Electrophorus* acetylcholine receptor using monoclonal antibodies. *J. Biol. Chem.* 256:8635-8645.

Tzartos, S. M. Seybond, and J. Lindstrom (1982) Specificity of antibodies to acetylcholine receptors in sera from myasthenia gravis patients measured by monoclonal antibodies. *Proc. Natl. Acad. Sci. USA 79:*188-192.

Unwin, N., 1989, The structure of ion channels in membranes of excitable cells. *Neuron 3:*665-675.

Utkin, Y., E. Lazakovich, E.I. Kasheverov, and V. Tsetlin, 1989, α-Bungarotoxin interacts with the rat brain tchykinin receptors. *FEBS Lett. 255:*111-115.

Wada, K., M. Ballivet, J. Boulter, J. Connolly, E. Wada, E. Deneris, L. Swanson, S. Heinemann, and J. Patrick, 1988, Functional expression of a new pharmacological subtype of brain nicotinic acetylcholine receptor. *Science 240:*330-334.

Wada, K. C. Dechesne, S. Shimasaki, R. King, K. Kusano, C. Banner, R. Wenthold, Y. Nakatani, 1989, Sequence and expression of a frog brain complementary DNA encoding a kainate-binding protein. *Nature 342:*684-689.

Wada, E., K. Wada, J. Boulter, E. Deneris, S. Heinemann, J. Patrick, and L. Swanson, 1989, The distribution of $\alpha 2$, $\alpha 3$, $\alpha 4$ and $\beta 2$ neuronal nicotinic receptor subunit mRNAs in the central nervous system: A hybridization histochemical study in the rat. *J. Comp. Neurol. 284:*314-335.

Wang, G., S. Molinaro, and J. Schmidt, 1978, Ligand responses of α-bungarotoxin-binding sites from skeletal muscle and optic lobe of the chick. *J. Biol. Chem. 253:*8507-8512.

Watson, J., E. Adkins-Regan, P. Whiting, J.M. Lindstrom, and T. Podleski, 1988, Autoradiographic localization of nicotinic acetylcholine receptors in the brain of the zebra finch *(Poephila guttata)*. *J. Comp. Neurol. 275:*255-264.

Whiting, P.J. and J.M. Lindstrom, 1986a, Purification and characterization of a nicotinic acetylcholine receptor from chick brain. *Biochemistry, 25:*2082-2093.

Whiting, P. and J. Lindstrom, 1986b, Pharmacological properties of immunoisolated neuronal nicotinic receptors. *J. Neurosci. 6:*3061-3069.

Whiting, P.J., and J.M. Lindstrom 1987a, Purification and characterization of a nicotinic from rat brain. *Proc. Natl. Acad. Sci. USA 84:*595-599.

Whiting, P. and J. Lindstrom, 1987b, Affinity labeling of neuronal acetylcholine receptors localizes the neurotransmitter binding site to the β subunit. *FEBS Lett. 213:*55-60.

Whiting, P.J. and J.M. Lindstrom, 1988, Characterization of bovine an human neuronal nicotinic acetylcholine receptors using monoclonal antibodies. *J. Neurosci. 8:*3395-3404.

Whiting, P. R. Liu, B.J. Morley, and J. Lindstrom, 1987a, Structurally different neuronal nicotinic acetylcholine receptor subtypes purified and characterized using monoclonal antibodies. *J. Neurosci. 7:*4005-4016.

Whiting, P. F. Esch, S. Shimasaki, and J. Lindstrom, 1987b, Neuronal nicotinic acetylcholine receptor β subunit is coded for by the cDNA clone $\alpha 4$. *FEBS Lett. 219:*459-463.

Whiting, P. R. Schoepfer., W.G. Conroy, M.V. Gore, K. Keyser, S. Shimasaki, F. Esch, and J. Lindstrom, submitted, Differential expression of nicotinic acetylcholine receptor subtypes in brain and retina.

Witzemann, V., B. Barg, Y. Nishikawa, B. Sakmann, and S. Numa, 1987, Differential regulation of muscle acetylcholine receptor γ and ϵ subunit mRNAs. *FEBS Lett. 223:*104-112.

Wong, L., and J. Gallagher, 1989, A direct nicotinic receptor-mediated inhibition recorded intracellularly *in vitro*. *Nature 341:*439-442.

REGULATION OF THE NICOTINIC ACETYLCHOLINE RECEPTOR

BY SERINE AND TYROSINE PROTEIN KINASES

Richard L. Huganir

Howard Hughes Medical Institute
Department of Neuroscience
The Johns Hopkins University School of Medicine
Baltimore, Maryland   USA

Regulation of Synaptic Transmission by Protein Phosphorylation

Neurotransmitter receptors play a central role in the process of signal transduction across synapses between neurons. Neurotransmitters released from the presynaptic neuron diffuse across the synaptic cleft and bind to neurotransmitter receptors in the membrane of the postsynaptic neuron. The neurotransmitter receptors then transduce this signal across the postsynaptic membrane either by directly activating ion channels or by regulating the level of intracellular second messengers in the postsynaptic neuron. Because of the essential role of neurotransmitter receptors in synaptic transmission, the short and long term modulation of neurotransmitter receptor function could be an extremely effective mechanism for the regulation of synaptic plasticity. What are the molecular mechanisms that may be involved in the modulation of neurotransmitter receptor function? Studies on the regulation of cellular metabolism over the past four decades have shown that protein phosphorylation is the primary mechanisms in the regulation of almost all cellular processes (Edelman et al, 1987, Nairn et al, 1985, Hunter et al, 1985). Protein phosphorylation has been demonstrated to regulate such diverse functions as glycogen metabolism, lipid metabolism, muscle contraction, protein synthesis and DNA transcription (Edelman et al, 1987, Nairn et al, 1985, Hunter et al, 1985). Protein phosphorylation systems consist of three basic components: a substrate protein, a protein kinase and a phosphoprotein phosphatase. Protein kinases are enzymes that catalyze the covalent transfer of phosphate from ATP to serine, threonine or tyrosine residues of their substrate proteins. The addition of this highly charged phosphate molecule to the substrate protein can affect its structure and thereby regulates its function. Substrate proteins include many cellular elements such as enzymes, cytoskeletal proteins, ion channels or membrane receptors. Phosphoprotein phosphatases are enzymes that reverse the phosphorylation process by catalyzing the hydrolysis of the phosphate from the substrate protein, returning it to its nonphosphorylated state (Cohen, 1990).

The classic protein kinases are the second messenger-regulated protein kinases such as cAMP-dependent protein kinase, cGMP-dependent protein kinase, protein kinase C and calcium/calmodulin-dependent protein kinases (Edelman et al, 1987, Nairn et al, 1985). In addition to these protein kinases which all phosphorylate serine or threonine residues of their substrate proteins, a unique class of protein kinase which phosphorylates only tyrosine residues of their substrate proteins has recently been described (Hunter et al, 1985). These protein tyrosine kinases were initially discovered because they were the product of the oncogenes of transforming retroviruses (Hunter et al, 1985). Most of these viral protein tyrosine kinases have been shown to have normal cellular homologues that are very similar in structure to the viral proteins.

Recently it has become increasingly clear that one of the major mechanisms for the regulation of synaptic transmission is protein phosphorylation. Recent studies have shown that protein phosphorylation is intimately involved in the regulation of both the presynaptic mechanisms for the release of neurotransmitters as well as the postsynaptic receptor mechanisms (Huganir and Greengard, 1987; Hemmings et al, 1989). However, the molecular details of the functional effects of protein phosphorylation on synaptic transmission have been difficult to study.

The Nicotinic Acetylcholine Receptor

One neurotransmitter receptor, the nicotinic acetylcholine receptor (AchR), has been characterized at the molecular level in great detail (Changeux et al, 1984) and has therefore been an excellent model system to study the molecular details of the regulation of synaptic function by protein phosphorylation (Miles and Huganir, 1988). The AChR is a neurotransmitter-gated ion channel that mediates the response to acetylcholine at the postsynaptic membrane of nicotinic cholinergic synapses such as the neuromuscular junction (Changeux et al, 1984). The receptor has been purified and extensively characterized and is a pentameric complex of four types of subunits in the stoichiometry $\alpha_2 \beta \gamma \delta$ (see figure 1) (Reynolds et al, 1978). This pentameric complex has two acetylcholine binding sites, one on each of the α subunits (Changeux et al, 1984). Each subunit spans the membrane and the five subunits are arranged in a pentameric rosette surrounding a central ion channel region (see figure 1). The purified receptor is biologically functional when reconstituted into phospholipid vesicles and displays the known biological properties of the nicotinic receptor in the native membrane (Huganir and Racker, 1982, Tank et al, 1984).

Although each of the subunits is encoded by a different gene the subunits are very similar in structure and show extensive amino acid sequence identity (Raftery et al, 1980; Noda et al, 1982; Noda et al, 1983; Noda et al, 1983, Claudio et al, 1983; Devillers-Thiery et al, 1983). From the hydrophobicity plots of the amino acid sequence of each subunit a number of theoretical models for the transmembrane topology of each subunit have been proposed (Claudio et al, 1983, Noda et al, 1983, Devillers-Thiery et al, 1983, Finer-Moore et al, 1984). An example of one such model is shown in figure 2 (Claudio et al, 1983). In this model there is a large N-terminal region which is extracellular. It is this portion of the α subunit which binds acetylcholine (Kao et al, 1984). Each subunit has four hydrophobic regions which have been proposed to be transmembrane α-helices. In this model the C-terminus is predicted to be on the intracellular side of the membrane and the major intracellular loop would be predicted to be between the M3 and M4 transmembrane α-helices. There is good evidence from photoaffinity labelling studies with open channel blockers (Giraudat et al, 1986, Hucho et al, 1986) and from site specific mutagenesis (Imoto et al, 1988, Leonard et al, 1988) that the M2 transmembrane α-helix is the region of each subunit that lines the wall of the ion channel.

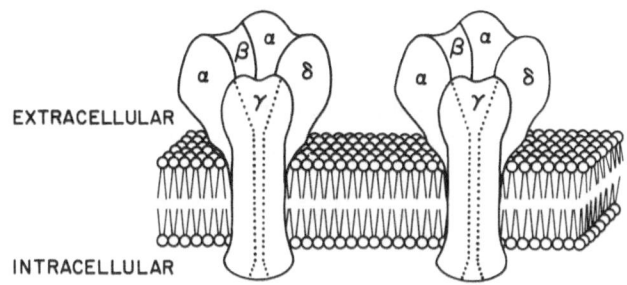

Figure 1 - Schematic model of the structure of the
nicotinic acetylcholine receptor. Arrangement of the
five subunits around the central ion channel as viewed
from a cross section through the plane of the membrane.

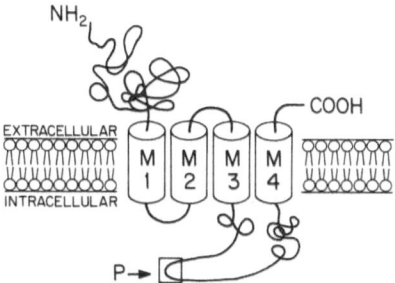

Figure 2 - One proposed model for the transmembrane
topology of each subunit. P indicates the region of
each subunit that contains the proposed phosphorylation
sites for the various protein kinases.

Phosphorylation of the Nicotinic Acetylcholine Receptor

Recent studies have shown that the nicotinic acetylcholine receptor is phosphorylated and regulated in vitro and in vivo by several serine and tyrosine protein kinases (Huganir and Greengard, 1987; Miles and Huganir, 1988). Isolated postsynaptic membranes highly enriched in the nicotinic receptor contain at least four different protein kinases: cAMP-dependent protein kinase (Huganir and Greengard, 1983; Zavoico et al, 1984; Heilbronn et al, 1985), calcium/calmodulin dependent protein kinase (Smilowitz et al, 1982; Huganir and Greengard, 1983), protein kinase C (Huganir et al, 1983; Safran et al, 1987) and a protein tyrosine kinase (Huganir et al, 1984). Three of these protein kinases phosphorylate seven different phosphorylation sites on the various subunits of the nicotinic receptor (see figure 3). The cAMP-dependent protein kinase phosphorylates serine residues on the τ and δ subunits, protein kinase C phosphorylates a serine residue on the δ subunit and to a lesser degree a serine residue on the α subunit. The protein tyrosine kinase phosphorylates a tyrosine

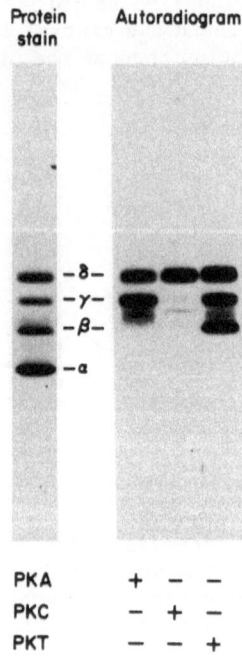

Figure 3 - Subunit specificity of the three different protein kinases that phosphorylate the nicotinic acetylcholine receptor. Polyacrylamide gel electrophoresis of the acetylcholine receptor purified after phosphorylation by cAMP-dependent protein kinase (PKA); protein kinase C (PKC); and the protein tyrosine kinase (PKT).

358   P   P   P 378
ARG-ARG-SER-SER-SER-VAL-GLY-TYR-ILE-SER-LYS-ALA-GLN-GLU-TYR-PHE-ASN-ILE-LYS-SER-ARG

γ-SUBUNIT

350   P   P   370
ARG-ARG-ARG-SER-SER-PHE-GLY-ILE-MET-ILE-LYS-ALA-GLU-GLU-TYR-ILE-LEU-LYS-LYS-PRO-ARG

ß-SUBUNIT

340   P   360
SER-PRO-ASP-SER-LYS-PRO-THR-ILE-ILE-SER-ARG-ALA-ASP-ASP-GLU-TYR-PHE-ILE-ARG-LYS-PRO

α-SUBUNIT

314   P 334
LYS-ILE-PHE-ILE-ASP-THR-ILE-PRO-ASN-VAL-MET-PHE-PHE-SER-THR-MET-LYS-ARG-ALA-SER-LYS

Figure 4 - Proposed location of the phosphorylated amino acid residues on the α, ß, γ and δ subunits of the AChR. The kinases and their proposed phosphorylation sites are: tyrosine-specific protein kinase (ß subunit, Tyr-355; γ subunit, Tyr-364; δ subunit, Tyr-372); cAMP-dependent protein kinase (γ subunit, Ser-353; δ subunit, Ser-361) and protein kinase C (α subunit, Ser-333; δ subunit, Ser-377).

residue on the ß, γ and δ subunits of the AChR. The locations of these phosphorylation sites have been identified (see figure 4) and are all located within twenty amino acids of each other on the major intracellular loop of each subunit (Huganir et al, 1984; Yee and Huganir, 1987) (see figure 2). All of the phosphorylation sites are conserved in the amino acid sequence of the receptor subunits from all species sequenced so far except for the phosphorylation sites on the γ subunit of most mammalian species. It is interesting to note, however, that the cAMP-dependent phosphorylation site is conserved in the adult form of the γ subunit (the ε subunit) in mammalian species (Miles and Huganir, 1988).

Regulation of the Phosphorylation of the Nicotinic Acetylcholine Receptor

Phosphorylation of the nicotinic acetylcholine receptor has also been demonstrated in cultured myotubes and in BC3H1 myocytes (Miles et al, 1987; Ross et al, 1987; Smith et al, 1987). Regulation of acetylcholine receptor phosphorylation in situ has been explored by treating muscle cell cultures with agents that raise second messenger levels or stimulate protein kinases. Forskolin, an activator of adenylyl cyclase, or cAMP analogs have been shown to regulate phosphorylation of the δ and α subunits of the nicotinic receptor in cultured myotubes (Miles et al, 1987, Smith et al, 1987). Acetylcholine receptor phosphorylation has also been shown to be regulated by calcium ionophores, suggesting that the receptor may be a substrate for calcium-dependent protein kinases in situ (Smith et al, 1987), such as phosphorylase kinase (Smith et al, 1989). Recent studies using phorbol esters, activators of protein kinase C, have provided evidence that protein kinase C phosphorylates the δ subunit in myotubes in situ (Ross et al, 1988; Miles et al, 1990). Although an endogenous tyrosine kinase phosphorylates the ß subunit of the AChR in BC3H1 myocytes (Smith et al, 1987) and in rat myotubes (Miles et al, 1989), the β subunit is phosphorylated on tyrosine residues to an extremely low level (< 0.01 mole phosphotyrosine/mole subunit). A variety of agents, including cAMP analogs, forskolin, phorbol esters and calcium ionophores have been used to try to regulate tyrosine phosphorylation of the acetylcholine receptor in muscle cells. However, none of these reagents regulates the tyrosine phosphorylation of the receptor.

The first messengers that are involved in the regulation of the three protein kinases have recently been investigated (Smith et al, 1987; Miles et al, 1989; Ross et al, 1989; Miles et al, 1990). Several neurotransmitters and hormones that are known to stimulate intracellular cAMP levels have been considered. Isoproterenol, the ß-adrenergic receptor agonist, stimulated AChR phosphorylation in BC3H1 myocytes, suggesting that catecholamines may modulate AChR function (Smith et al, 1987). The neuropeptide calcitonin gene-related peptide (CGRP) is a particularly attractive candidate as a first messenger for the regulation of AChR phosphorylation at the neuromuscular junction because of its location in presynaptic large dense core vesicles (Matteoli et al, 1988), its trophic effects on AChR synthesis, and its ability to raise intracellular cAMP levels (New and Mudge, 1986; Fontaine et al, 1986; Takami et al, 1986; Kobayashi et al, 1987; Laufer and Changeux, 1987).

Recent results have demonstrated that CGRP increases phosphorylation of the acetylcholine receptor in rat primary myotubes by a mechanism likely to be mediated by cAMP-dependent protein kinase (Miles et al, 1989). Phosphorylation of the AChR δ subunit in rat myotubes exposed to CGRP reached a maximum within 5 minutes and mimicked the time course of AChR phosphorylation caused by forskolin. The CGRP-induced increase in AChR α subunit phosphorylation was potentiated by phosphodiesterase inhibitors. At maximal doses, in the presence of a phosphodiesterase inhibitor, the effects of CGRP and forskolin on AChR phosphorylation were not additive, suggesting that CGRP and forskolin use the same biochemical mechanism to increase AChR phosphorylation. This argument is supported by phosphopeptide maps of AChR subunits which indicate that the same peptides are phosphorylated in situ by treatment of myotubes with either CGRP or forskolin (Miles et al, 1989). In addition, phosphoamino acid analysis revealed that both CGRP and forskolin increased phosphorylation of the α and δ subunits on serine residues.

One first messenger that may be involved in the regulation of protein kinase C phosphorylation of the acetylcholine receptor is acetylcholine itself (Adamo et al, 1985; Ross et al, 1988; Laufer et al, 1989; Miles et al, 1990). Recent studies have shown that acetylcholine stimulates phosphatidylinositol turnover in muscle cells through both a muscarinic receptor pathway (Miles et al, 1990) and a nicotinic receptor pathway (Adamo et al, 1985). In addition, carbamylcholine stimulates phosphorylation of the AChR in a manner similar to that seen with phorbol ester treatment of myotubes (Ross et al, 1988; Miles et al, 1990). Recent results have shown that this stimulation of protein kinase C phosphorylation of the nicotinic receptor by carbamylcholine occurs through a receptor with nicotinic pharmacology (Miles et al, 1990). These results suggest that acetylcholine, through the nicotinic receptor, stimulates the phosphorylation of the nicotinic receptor by protein kinase C. The molecular mechanism of the activation of protein kinase C by nicotinic receptors is not clear; however, it may be due in part to direct or indirect activation of protein kinase C by calcium ions that permeate the AChR channel (Miles et al, 1990).

The first messengers that are involved in the regulation of the tyrosine phosphorylation of the nicotinic acetylcholine receptor are completely unknown. Several neurotransmitters, hormones, and growth factors, such as acetylcholine, CGRP, isoproterenol, substance P, insulin, FGF, EGF, and PDGF have been tested to see if they regulate the tyrosine phosphorylation of the acetylcholine receptor in muscle cell cultures. None of these agents regulate the tyrosine phosphorylation of the AChR. However, recent results have suggested that tyrosine phosphorylation of the AChR in muscle is regulated by neuronal innervation (Qu et al, 1990). As discussed above, tyrosine phosphorylation of the AChR in rat myotube cultures is barely detectable. To examine whether this low level of tyrosine phosphorylation of the AChR in muscle cell cultures was due to a lack of neuronal innervation, the tyrosine phosphorylation of the AChR in rat diaphragm in vivo was examined (Qu et al, 1990). Immunofluorescent double labelling of cryostat sections of rat diaphragm using antibodies specific for phosphotyrosine or the AChR showed a direct co-localization of phosphotyrosine with the AChR at the neuromuscular junction (See figure 5). In examining well over a thousand endplates that had been double-labelled with anti-phosphotyrosine antibodies and antireceptor antibodies, it was found that all endplates stained for phosphotyrosine in adult rat diaphragm. In addition, using anti-phosphotyrosine antibodies, immunoblots of nicotinic acetylcholine receptor partially purified from rat diaphragm demonstrated that the rat nicotinic acetylcholine receptor contains high levels of phosphotyrosine on its β and δ subunits (Qu et al, 1990).

Denervation of rat diaphragm induced a time-dependent decrease in tyrosine phosphorylation of the AChR as measured by immunocytochemical and immunoblot techniques. The tyrosine phosphorylation of the AChR was detectably lower at 10 days and completely disappeared at most synapses 34 days after denervation (see figure 6). In contrast, endplates still show intense labelling for the AChR (figure 6).

When the developmental time course of tyrosine phosphorylation of the AChR was examined, it was found to occur late in the development of the neuromuscular junction between postnatal days 7 and 14. These studies suggest that muscle innervation regulates tyrosine phosphorylation of the AChR, and that tyrosine phosphorylation may play an important role in the developmental regulation of the AChR.

Functional Effects of Phosphorylation of the Nicotinic Acetylcholine Receptor

The functional effects of phosphorylation of the nicotinic acetylcholine receptor were recently examined directly by analyzing the properties of the purified and reconstituted receptor before and after phosphorylation (Huganir et al, 1986; Hopfield et al, 1988). Stop-flow and quench-flow rapid kinetic techniques were used to measure the properties of ion transport by the AChR before and after cAMP-dependent protein phosphorylation (Huganir et al, 1986). It was found that the rate of desensitization of the AChR was increased several fold after cAMP-dependent phosphorylation. Desensitization is the process by which the receptor is inactivated in the prolonged presence of acetylcholine.

This effect of phosphorylation on the rate of desensitization has more recently been extended to tyrosine phosphorylation of the receptor (Hopfield et al, 1988). The purified receptor phosphorylated to different

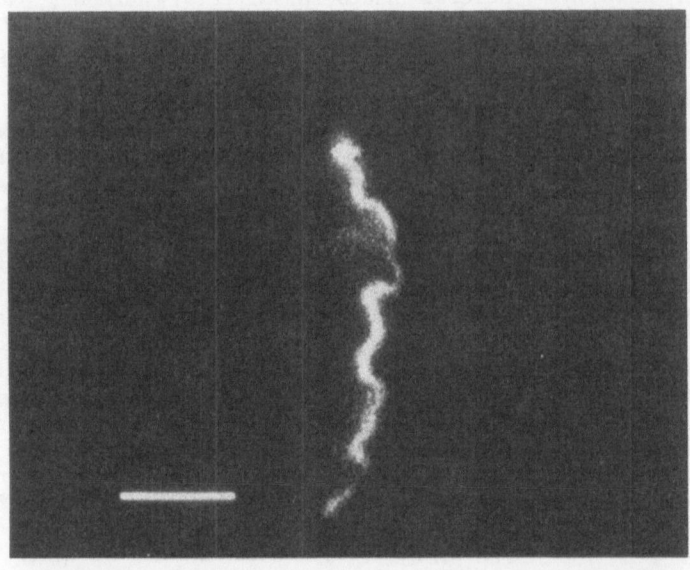

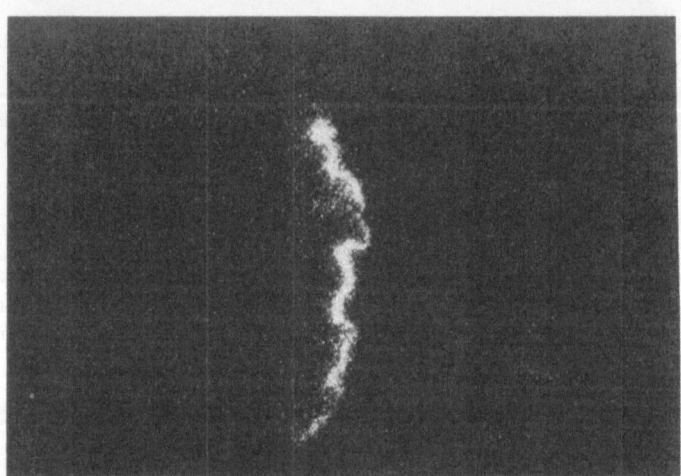

Figure 5 - Co-localization of phosphotyrosine with the
AChR at the rat neuromuscular junction. Cryostat sections
of rat diaphragm were fixed and then double-labelled with
a monoclonal antibody (88b) against AChR (top panel) and
an affinity purified antibody against phosphotyrosine
(bottom panel). Immunofluorescent staining was analyzed by
confocal microscopy. Bar=5μM.

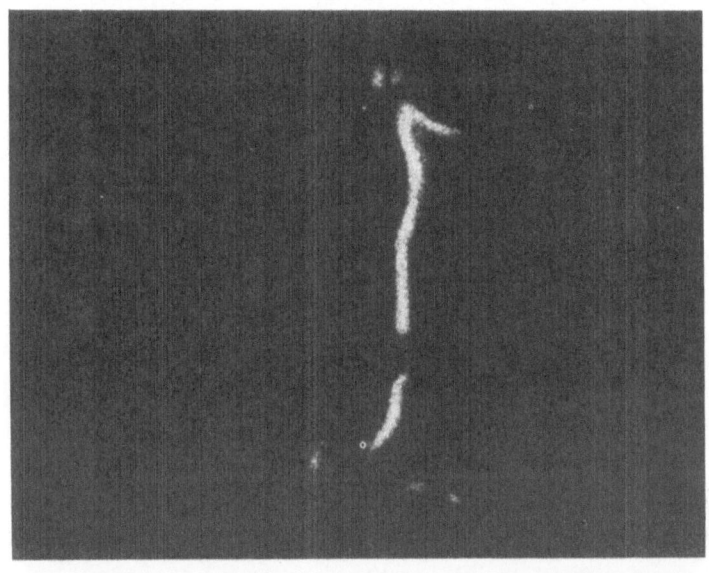

Figure 6 - Regulation of tyrosine phosphorylation of AChR at the rat neuromuscular junction by denervation. Immunofluorescent staining of rat diaphragm with a monoclonal antibody (88b) against the AChR (top panel) or with affinity purified anti-phosphotyrosine antibody (bottom panel) 34 days after muscle denervation analyzed by confocal microscopy.

stoichiometries of tyrosine phosphorylation was reconstituted into lipid vesicles, and then single channel properties of the various AChR preparations were measured by the patch clamp technique. When the patch pipette contained acetylcholine, patches of liposome membrane containing the AChR showed single channel openings and the frequency of these channel openings diminished with time after exposure to acetylcholine (see figure 7). This desensitization behavior could be quantified from histograms of the number of channel openings per unit time and could be fitted to the sum of two exponentials (see figure 7). When the rates of desensitization of various phosphorylated receptor preparations were examined it was found that there is a striking dependence of the rapid component of desensitization on the stoichiometry of tyrosine phosphorylation (see figure 8). These results clearly demonstrate that phosphorylation of the

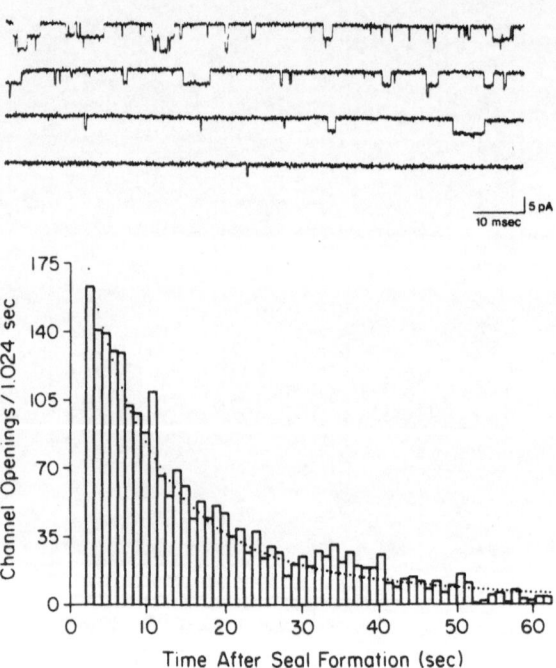

Figure 7 - Desensitization behavior of purified and reconstituted AChR as seen in single patches. Top panel: 0.512 s portions of a single record sampled 0, 15, 30 and 60 s following seal formation and voltage application. Pipette holding potential, +100mV; 1 μM acetylcholine. Bottom panel: Exponential decay with time of the frequency of channel-opening events. The number of channel openings in successive 1.024 s intervals was plotted versus the midpoint of these time intervals. Histograms of records containing at least 100 channel events over the recording period were used to determine the time constants for the rapid and slow phase of desensitization. The dotted curve shows a double exponential distribution with τ_{fast}=7.8 s, τ_{slow}=30.2 s, % fast=50 (Reprinted by permission from NATURE Vol. 336 pp. 677-680. Copyright (c) 1988 Macmillan Magazines Ltd.).

nicotinic acetylcholine receptor by cAMP-dependent protein kinase and by the protein tyrosine kinase _in vitro_ regulates its rate of desensitization.

Recent studies have also suggested that protein phosphorylation of the nicotinic acetylcholine receptor _in vivo_ in muscle cells regulates the rate of desensitization of the receptor. Intracellular recordings from rat soleus muscle have shown an increased rate of receptor desensitization after treatment of the tissue with forskolin (Middleton et al, 1986; Albuquerque et al, 1986). In addition, forskolin or cAMP analogs regulated the rate of AChR desensitization analyzed by intracellular and single channel recordings of rat myotubes in culture (Middleton et al, 1988; Mulle et al, 1988). The time course and concentration of forskolin treatment necessary to elicit an increased rate of desensitization correlated with the forskolin-induced increase in phosphorylation of the AChR δ subunit (Miles et al, 1987). Studies with forskolin have been complicated by the fact that, in addition to its ability to activate adenylyl cyclase, forskolin exerts a direct local anesthetic effort on AChR function (McHugh and McGee, 1986; Wagoner and Pallota, 1988). However, at low concentrations where the direct effects of forskolin were negligible, increases in the AChR desensitization rate were still detected, suggesting that this observed increase in the rate of AChR desensitization was due to activation of adenylyl cyclase (Middleton et al, 1988; Mulle et al, 1988). In addition, phorbol esters, which activate protein kinase C, increased the rate of desensitization and reduced acetylcholine sensitivity in cultured chick myotubes (Eusebi et al, 1985).

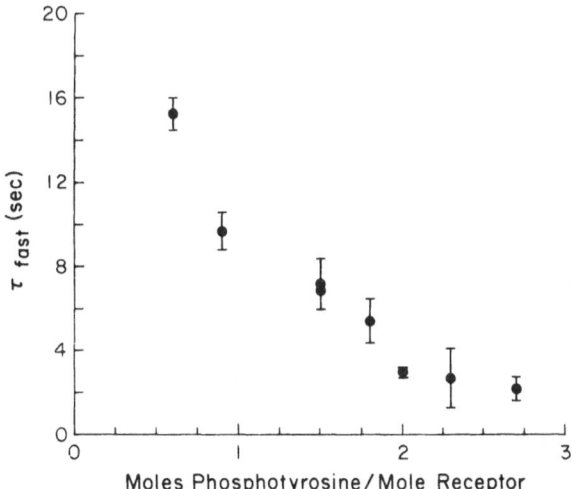

Figure 8 - Increased levels of tyrosine phosphorylation increase the rate of the rapid phase of desensitization of the AChR. Time constants of the rapid phase of desensitization are plotted versus the stoichiometry of tyrosine phosphorylation. Data are plotted as mean values $\pm$ standard deviations (Reprinted by permission from NATURE Vol. 336 pp. 677-680. Copyright (c) 1988 Macmillan Magazines Ltd.).

Recent studies have also demonstrated that the neuropeptide CGRP enhances the rate of the rapid phase of AChR desensitization in mouse myotubes (Mulle et al, 1988). Concentrations of CGRP in the 10^{-7} M range were effective, and the maximal effect was achieved within 5-8 minutes after application of CGRP. Single-channel recordings from myotubes indicated that neither the mean channel open time nor the simple channel conductance was affected by CGRP, but rather that the frequency of channel openings was reduced. This effect of CGRP could be mimicked by forskolin or cAMP. Maximal doses of forskolin and CGRP together were not additive in their ability to increase the AChR desensitization rate, arguing that CGRP and forskolin use the same biochemical pathway to achieve this altered physiological behavior of the ion channel. This data on the physiological effects of CGRP on the desensitization of the AChR, taken together with the biochemical data on the effect of CGRP on AChR phosphorylation by cAMP-dependent protein kinase, strongly suggests that the CGRP-induced phosphorylation of the AChR increases its rate of desensitization.

CONCLUSION

Protein phosphorylation of neurotransmitter receptors is most likely one of the major mechanisms in the regulation of synaptic transmission. The nicotinic acetylcholine receptor has served as excellent model system for the study of the role of protein phosphorylation in the regulation of neurotransmitter receptor function. The nicotinic acetylcholine receptor is multiply phosphorylated by at least three different protein kinases in vitro and in vivo. Cyclic AMP-dependent protein kinase phosphorylates the γ and δ subunits of the receptor, protein kinase C phosphorylates the δ subunit, and a protein tyrosine kinase phosphorylates the β, γ, and δ subunits of the receptor. The phosphorylation of the nicotinic receptor by these three protein kinases is regulated by at least three different first messengers (See figure 9). The cAMP-dependent protein phosphorylation of the receptor appears to be regulated by the neuropeptide CGRP, while the phosphorylation of the nicotinic receptor by protein kinase C appears to be regulated by acetylcholine itself. Although it is not clear what first messengers regulate the tyrosine phosphorylation of the nicotinic receptor, recent results suggest that some factor from neurons regulates the tyrosine phosphorylation.

Phosphorylation of the nicotinic receptor by all three of these protein kinases appears to regulate the rate of desensitization of the receptor in vitro and in vivo. The physiological role of desensitization at synapses is not clearly understood; however, desensitization is a well-conserved property of all neurotransmitter receptors including the $GABA_A$, glycine, and glutamate receptors. Desensitization has been proposed to be a form of short-term regulation in the second-to-minute time range of synaptic efficacy, and protein phosphorylation may be an important way of modulating this process. With the recent cloning of the $GABA_A$ and glycine receptors, it is clear that these neurotransmitter-gated ion channels are extremely similar in structure to the nicotinic acetylcholine receptor. The subunits of these receptors have the same pattern of four hydrophobic transmembrane domains as do those of the nicotinic acetylcholine receptor and are homologous in their amino acid sequence to each other and to the nicotinic acetylcholine receptor. Moreover, consensus sequences for a cAMP-dependent phosphorylation site and a protein tyrosine kinase phosphorylation site are located on the ß and γ subunits, respectively, of the $GABA_A$ receptor. These sites are located on the major intracellular domain between the third and fourth transmembrane α-helices, in a position similar to the phosphorylation sites on AChR. Recent results have also shown that phosphorylation of the $GABA_A$ receptor by cAMP-dependent protein kinase appears to regulate the rate of desensitization of the GABA$_A$ receptor. Protein phosphorylation of postsynaptic neurotransmitter receptors, in general, appears to be an important and well-conserved mechanism of synaptic plasticity.

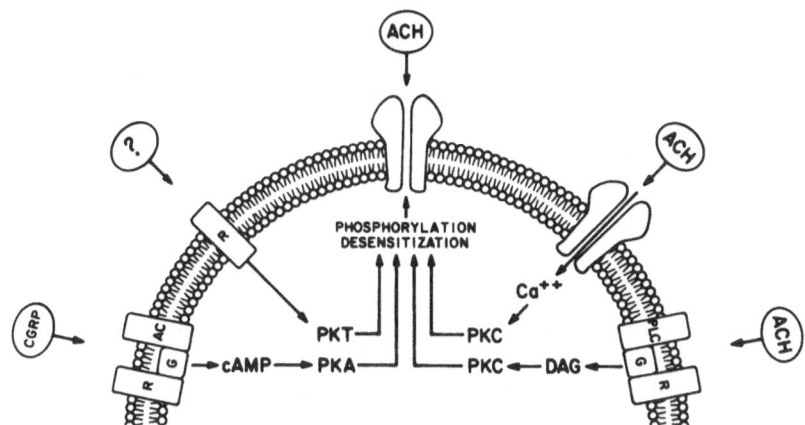

Figure 9 - Schematic diagram illustrating proposed regulation of the acetylcholine receptor by three protein kinase systems. Three neurotransmitters, through the activation of their respective receptors and associated protein kinase systems, bring about the phosphorylation and increased rate of desensitization of the acetylcholine receptor.

REFERENCES

Adamo S, Zani BM, Nerri C, Senni MI, Molinare M, and Eusebi F. (1985). Acetylcholine stimulates phosphatidylinositol turnover at nicotinic receptors of cultured myotubes. FEBS Lett. 190: 161-164.

Albuquerque EX, Deshpande SS, Aracava Y, Alkondon M, Daly JW (1986). A possible involvement of cyclic AMP in the expression of desensitization of the nicotinic acetylcholine receptor: a study with forskolin and its analogs. FEBS Lett 199:113-120.

Changeux J-P, Devillers-Thiery A, Chemouilli P (1984). Acetylcholine receptor: an allosteric protein. Science 225:1333-1345.

Claudio T, Ballivet M, Patrick J, Heinemann S (1983). Nucleotide and deduced amino acid sequences of Torpedo californica acetylcholine receptor τ subunit. Proc Natl Acad Sci USA 80:1111-1115.

Cohen P (1989). The structure and regulation of protein phosphatases, in "Annual Review of Biochemistry," Richardson CC, Abelson JN et al, eds., Annual Reviews Inc., Palo Alto, CA.

Devillers-Thiery A, Giraudat J, Benaboulet M, Changeux J-P (1983). Complete mRNA coding sequence of the acetylcholine-binding α-subunit of Torpedo marmorata acetylcholine receptor: a model for the transmembrane organization of the polypeptide chain. Proc Natl Acad Sci USA 80:2067-2071.

Edelman AM, Blumenthal DK, Krebs EG (1987). Protein serine/threonine kinases. Annu Rev Biochem 56:567-613.

Eusebi F, Molinaro M, Zani BM (1985). Agents that activate protein kinase C reduce acetylcholine sensitivity in cultured myotubes. J Cell Biol 100:1339-1342.

Finer-Moore J, Stroud RM (1984). Amphipathic analysis and possible formation of the ion channel in an acetylcholine receptor. Proc Natl Acad Sci USA 81:155-159.

Fontaine B, Klarsfeld A, Hokfelt T, Changuex J-P (1986). Calcitonin gene-related peptide, a peptide present in spinal cord motoneurons, increases the number of acetylcholine receptors in primary cultures of chick embryo myotubes. Neurosci Lett 71:59-65.

Giraudat J, Dennis M, Heidmann T, Chang J-Y, Changeux J-P (1986). Structure of the high-affinity binding site for noncompetitive blockers of the acetylcholine receptors: serine-262 of the δ is labeled by [3H] chlorpromazine. Proc Natl Acad Sci USA 83:2719-2723.

Heilbronn H, Eriksson R, Salmansson R (1985). Regulation of the nicotinic acetylcholine receptor by phosphorylation. In Changeux, Hucho, Maelicke, and Neumann, (eds): "Molecular Basis of Nerve Activity," Berlin: Walter de Gruyter, pp 237-250.

Hemmings HC, Nairn AC, McGuinness TL, Huganir RL, Greengard P (1989). Role of protein phosphorylation in neuronal signal transduction. FASEB 3:1583-1592.

Hopfield JF, Tank DW, Greengard P, Huganir RL. (1988). Functional modulation of the nicotinic acetylcholine receptor by tyrosine phosphorylation. Nature 336:677-680.

Hucho F, Oberthur W, Lottspeich, F (1986). The ion channel of the nicotinic acetylcholine receptor is formed by the homologous helices M II of the receptor subunits. FEBS Lett 205:137-142.

Huganir RL, Albert KA, Greengard P (1983). Phosphorylation of the nicotinic acetylcholine receptor by Ca^{2+}/phospholipid-dependent protein kinase, and comparison with its phosphorylation by cAMP-dependent protein kinase. Soc Neurosci Abstr 9:578.

Huganir RL, Delcour AH, Greengard P, Hess GP (1986). Phosphorylation of the nicotinic acetylcholine receptor regulates its rate of desensitization. Nature 321:774-776.

Huganir RL, Greengard P (1983). cAMP-dependent protein kinase phosphorylates the nicotinic acetylcholine receptor. Proc Natl Acad Sci USA 80:1130-1134.

Huganir RL, Greengard P (1987). Regulation of receptor function by protein phosphorylation. TIPS 8:472-477.

Huganir RL, Miles K, Greengard P (1984). Phosphorylation of the nicotinic acetylcholine receptor by an endogenous tyrosine-specific protein kinase. Proc Natl Acad Sci USA 81:6963-6972.

Huganir RL, Racker E (1982). Properties of proteoliposimes reconstituted with acetylcholine receptor from Torpedo californica. J Biol Chem 257:9372-9378.

Hunter T, Cooper JA (1985). Protein-tyrosine kinases. Annu Rev Biochem 54:897-930.

Imoto K, Busch C, Sakmann B, Mishina M, Konno T, Nakai J, Bujo H, Mori Y, Fukuda K, Numa S. (1988). Rings of negatively charged amino acids determine the acetylcholine receptor channel conductance. Nature 335:645-648.

Kao PN, Dwork AJ, Kaldany RRJ, Silver ML, Wideman J, Stein S, Karlin A (1984). Identification of the α subunit Half-cystine specifically labeled by an affinity reagent for the acetylcholine receptor binding site. J Biol Chem 259:11662-11665.

Kobayashi H. Hashimoto K, Sakuma J, Takami K, Tohyama M, Izumi F, Yoshida H (1987). Calcitonin gene-related peptide stimulates adenylate cyclase activity in rat striatal muscle. Experientia 43:314-316.

Laufer R, Changeux J-P (1989). Calcitonin gene-related peptide and cyclic AMP stimulate phosphoinositide turnover in skeletal muscle cells: interaction between two second messenger systems. J Biol Chem 264: 2683-2689.

Laufer R, Changeux J-P (1987). Calcitonin gene-related peptide elevates cyclic AMP levels in chick skeletal muscle: possible neurotrophic role for a coexisting neuronal messenger. EMBO J 6:901-906.

Leonard RJ, Labarca CG, Charnet P, Davidson N., Lester HA (1988). Evidence that the M2 Membrane-spanning region lines the ion channel pore of the nicotinic receptor. Science 242:1578-1581.

Matteoli M, Haimann C, Torri-Tarelli F, Polak JM Ceccarelli B, DeCamilli P (1988). Differential effect of α-latrotoxin on exocytosis from small synaptic vesicles and from large dense-core vesicles containing calcitonin gene-related peptide at the frog neuromuscular junction. Proc Natl Acad Sci USA 85:7366-7370.

McHugh EM, McGee, Jr R (1986). Direct anesthetic-like effects of forskolin on the nicotinic acetylcholine receptors of PC12 cells. J Biol Chem 261:3103-3106.

Middleton P, Jaramillo F, Scheutze SM (1986). Forskolin increases the rate of acetylcholine receptor desensitization at rat soleus endplates. Proc Natl Acad Sci USA 83:4967-4971.

Middleton P, Rubin LL, Schuetze SM (1988). Modulation of acetylcholine receptor desensitization in rat myotubes. J Neurosci 8:3405-3412.

Miles K, Anthony DT, Rubin LL, Greengard P, Huganir RL (1987). Regulation of nicotinic acetylcholine receptor phosphorylation in rat myotubes by forskolin and cAMP. Proc Natl Acad Sci USA 84:6591-6595.

Miles K, Greengard P, Huganir RL (1989). Calcitonin gene-related peptide regulates phosphorylation of the nicotinic acetylcholine receptor in rat myotubes. Neuron 2: 1517-1524.

Miles K, Huganir RL (1988). Regulation of Nicotinic Acetylcholine Receptors by Protein Phosphorylation. Molecular Neurobiology 2:91-124.

Miles K, Greengard P, Huganir RL (1990). Manuscript in preparation.

Mulle C, Benoit P, Pinset C, Roa M, Changeux J-P (1988). Calcitonin gene-related peptide enhances the rate of desensitization of the nicotinic acetylcholine receptor in cultured mouse muscle cell. Proc Natl Acad Sci USA 85:5728-5732.

Nairn AC, Hemmings HC, Greengard P (1985). Protein kinases in the brain. Annu Rev Biochem 54:931-976.

New HV, and Mudge AW (1986). Calcitonin gene-related peptide regulates muscle acetylcholine receptor synthesis. Nature 323:809-811.

Noda M, Takahashi H, Tanabe T, Toyosato M, Furutani Y, Hirose T, Asai M, Inayama S, Miyata T, Numa S (1982). Primary structure of α-subunit precursor of Torpedo californica acetylcholine receptor deduced from cDNA sequence. Nature 299:793-797.

Noda M, Takahashi H, Tanabe T, Toyosato M, Kikyotani S, Furutani Y, Hirose T, Takashima H, Inayama S, Miyata T, Numa S (1983). Structural homology of Torpedo californica acetylcholine receptor subunits. Nature 302:528-532.

Noda M, Takahashi H, Tanabe T, Toyosato M, Kikyotani S, Hirose T, Asai M, Takashima H, Inayama S, Miyata T, Numa S (1983). Primary structures of ß and δ subunit precursors of Torpedo californica acetylcholine receptor deduced from cDNA sequences. Nature 301:251-255.

Qu Z, Moritz E, Huganir RL (1990). Regulation of tyrosine phosphorylation of the nicotinic acetylcholine receptor at the rat neuromuscular junction. Neuron, in press.

Raftery MA, Hunkapiller MW, Strader CD, Hood LE (1980). Acetylcholine receptor: complex of homologous subunits. Science 208:1454-1457.

Reynolds JA, Karlin A (1978). Molecular weight in detergent solution of acetylcholine receptor from Torpedo californica. Biochemistry 17:2035-2038.

Ross A, Rapuano M, and Prives J. (1988). Induction of phosphorylation and cell surface redistribution of acetylcholine receptors by phorbol ester and carbamylcholine in cultured chick muscle cells. J Cell Biol 107:1139-1145

Ross A, Rapuano M, Schmidt J, Prives J (1987). Phosphorylation and assembly of nicotinic acetylcholine receptor subunits in cultured chick muscle cells. J Biol Chem 262:14640-14647.

Safran A, Eisenberg RS, Neumann D, Fuchs S (1987). Phosphorylation of the acetylcholine receptor by protein kinase C and identification of the phosphorylation site within the receptor δ subunit. J Biol Chem 262:10506-10510.

Smilowitz H, Hadjian RA, Dwyer J, Feinstein MB (1981). Regulation of acetylcholine receptor phosphorylation by calcium and calmodulin. Proc Natl Acad Sci USA 78:4708-4712.

Smith MM, Merlie JP, Lawrence, Jr JC, (1987). Regulation of phosphorylation of nicotinic acetylcholine receptors in mouse BC3H1 myocytes. Proc Natl Acad Sci USA 84:6601-6605.

Smith MM, Merlie JP, Lawrence Jr. JC (1989). Ca^{+2}-dependent and cAMP-dependent control of nicotinic acetylcholine receptor phophorylation in muscle cells. J Biol Chem 264: 12813-12819.

Takami K, Hashimito K, Uchida S, Tohyama M, Yashida H. (1986). Effect of calcitonin gene-related peptide on the cyclic AMP level of isolated mouse diaphragm. Jap J Pharmacol 42:345-350.

Tank DE, Huganir RL. Greenard P, Webb WW (1983). Patch-recorded single-channel currents of the purified and reconstituted Torpedo acetylcholine receptor. Proc Natl Acad Sci USA 80:5129-5133.

Wagoner PK, Pallotta BS (1988). Modulation of acetylcholine receptor desensitization by forskolin is independent of cAMP. Science 240:1655-1657.

Yee GH, Huganir RL (1987). Determination of the sites of cAMP-dependent phosphorylation on the nicotinic acetylcholine receptor. J Biol Chem 262: 16748-16753.

Zavoico GB, Comerci C, Subers E, Egon JJ, Huang CK, Feinstein MB, Smilowitz H (1984). cAMP, not Ca^{2+}/calmodulin, regulates the phosphorylation of acetylcholine receptor in Torpedo californica electroplax. Biochim Biophys Acta 770:225-229.

AFFINITY PURIFICATION OF NICOTINIC ACETYLCHOLINE RECEPTOR FROM RAT BRAIN

Hitoshi Nakayama*, Jon Lindstrom**, Toshikatsu Nakashima*
and Yutaka Kurogochi*

* Department of Pharmacology, Nara Medical University
   Kashihara 634, Japan
** The Salk Institute for Biological Studies, P.O.Box
   85800, San Diego, California 92138, U.S.A.

INTRODUCTION

Recently, several kinds of neuronal nicotinic acetylcholine receptor
(nAChR) genes have been identified. Complementary DNA expression experiments performed in Xenopus oocytes have shown that both ACh-binding (α
subunit) and another nonACh binding subunits (β subunit) are necessary to
form a functional nAChR. Monoclonal antibody (mAb) column chromatography
has been developed to purify nAChR from chicken, rat, bovine and human
brains (Lindstrom et al., 1987), which have shown that chicken brain contains two ACh-binding subunits and one structural subunit which form two
subtypes of nAChRs, whereas rat and bovine brains contain only one ACh-
binding subunit and one structural subunit. In this case, the immunoaffinity-purified nAChR accounts for more than 90% of the high affinity
^{3}H-nicotine binding sites in a detergent extract of rat brain. To date,
minor components of nAChR in brains have not been purified. Furthermore,
it still remains obscure as to pharmacological roles of each subtype of
neuronal nAChR. To investigate the heterogeneity of neuronal nAChR,
cholinergic ligand affinity purification of the nAChR would be useful.
In this report, we deal with cholinergic ligand affinity purification of
nAChR from rat brain.

PURIFICATION OF nAChR

Purification of nAChR from rat brain was performed as described
elsewhere (Nakayama et al., 1990). The method consisted of Lubrol solubilization followed by a combination of DE-52 and ACh-Affi-Gel column
chromatography (Fig. 1). ACh-Affi-Gel was prepared from bromoacetylcholine and Affi-Gel 401 (Bio-Rad). When Lubrol extract was directly applied to the affinity column, most of nAChR in the extract passed through
the affinity column. We therefore employed DE-52 column chromatography
prior to the affinity chromatography. nAChR was eluted immediately after
the NaCl concentration increased and approximately 50% of the nAChR ac-
itivity was recovered from the column. With the use of the DE-52 column,
efficient adsorption of the nAChR to ACh-Affi-Gel column was obtained,
suggesting that competitive inhibitors were removed. In order to remove
carbachol and to concentrate nAChR eluted from the affinity column, a
hydroxylapatite column was connected to the outflow tube of the affinity

Neuroreceptor Mechanisms in Brain, Edited by S. Kito *et al.*
Plenum Press, New York, 1991

```
Membrane fraction
    ├─solubilized with 1.5% Lubrol PX and then centrifuged at
    │ 60,000 x g for 3 hr.
DE-52 column
    ├─eluted with a linear gradient of NaCl
Elute
    │
ACh-Affi-Gel column
    │  (coupled to a small volume of hydroxylapatite column)
    ├─eluted with carbachol
Hydroxylapatite column
    ├─eluted with 0.2 M phosphate buffer(pH 7.4) containing
    │ 0.1% Lubrol PX, 0.4 mM PMSF and 0.02% NaN3
Elute (round 1)
    │
2nd ACh-Affi-Gel column
    ├─eluted with carbachol or nicotine
Elute (round 2)
```

Fig. 1. Purification of nAChR

column and then nAChR was eluted from the affinity column with carbachol. nAChR adsorbed on the hydroxylapatite was eluted with a high concentration of phosphate buffer. Table 1 shows the results of a purification procedure. In four different preparations, the affinity chromatography gave a 4000–7000 fold purification of the nAChR at this step, and the nAChR was purified 7000–13,000 fold with the recovery of 10–15% from the Lubrol extract. nAChR eluted from the first affinity column was stable at –100°C for at least 4 months. The nAChR activity was liable to be unstable during 2nd affinity column chromatography, probably due to low concentration of the protein. More than 50 µg of the protein are necessary to perform the 2nd affinity chromatography in our experimental conditions. The purified preparation contained no cholinesterase activity. α-Bungarotoxin (7 µM) did not inhibit ^{3}H-ACh binding to the purified preparation.

Table 1. Purification of nAChR from rat brain

Step	Protein (mg)	^{3}H ACh binding activity (pmol)	Specific activity (pmol/mg protein)	Yield (%)	Fold
Lubrol extract	1844	52.3	0.0284	100	1.0
DE-52	385	22.3	0.0579	43	2.0
ACh-Affi-Gel (round 1)	0.022	7.9	359	15	12,641

^{3}H-ACh binding activities were determined at 8 nM ^{3}H-ACh in the assay conditions as described in the text.

CHARACTERIZATION OF AFFINITY-PURIFIED nAChR

Assay of ^{3}H-ACh binding by the purified preparation was performed by filtration under vacuum using Whatman GF/B glass fiber filters presoaked in 0.3% polyethylenimine. The standard assay mixture contained 50 mM Tris-HCl buffer(pH 7.4), 0.1 mM eserine, 2 µM atropine, 0.1–0.4% Lubrol

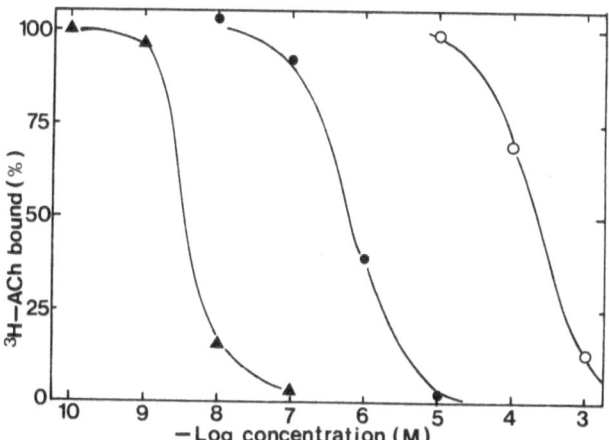

Fig. 2. Displacement of ^{3}H–ACh binding to purified nAChR by cholinergic ligands. d–tubocurarine (▲), carbachol (●), cytisine (○).

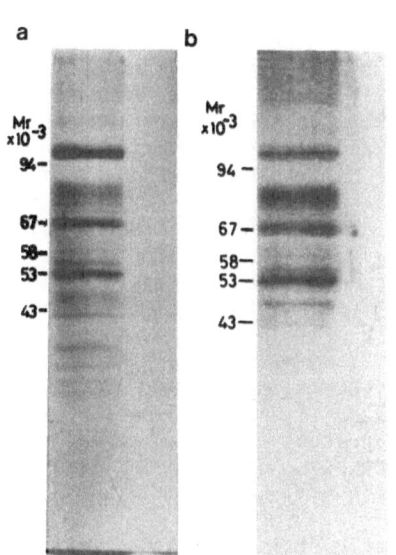

Fig. 3. SDS–PAGE of purified nAChR.
a) the first affinity column,
b) the second affinity column.

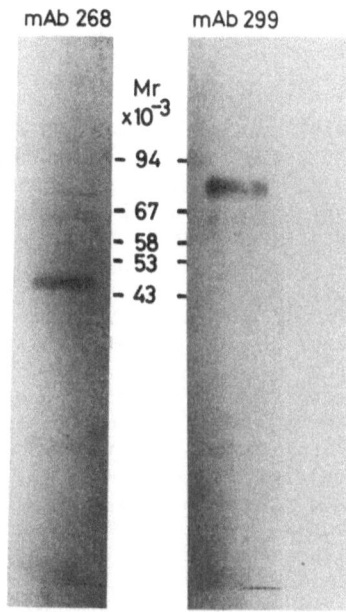

Fig. 4. Immunoblots of purified nAChR.

PX, 8 or 18 nM ^{3}H-ACh (81.7 Ci/mmol), and the purified preparation in the presence or absence of 1 mM carbachol. Specific ^{3}H-ACh binding increased linearly with protein. By this assay method, we found high affinity ^{3}H-ACh binding sites at all purification steps. ^{3}H-ACh binding to the purified nAChR was inhibited by cytisine, carbachol and d-tubocurarine (Fig. 2). Cytisine was most effective and d-tubocurarine was less effective than other cholinergic ligands. These pharmacological properties of the purified nAChR is very similar to that of immunoaffinity-purified nAChR (Whiting and Lindstrom, 1986).

SDS-PAGE of the purified preparation revealed four major protein bands and a few minor bands (Fig. 3). Four major proteins were estimated to have molecular weights of 53,000, 67,000, 80,000 and 108,000. Some of the minor bands could be removed by a second affinity chromatography. When nAChR was eluted with nicotine from the affinity column, similar results were obtained on SDS-PAGE. Since carbachol is more easily removed from nAChR than nicotine, we usually use carbachol.

nAChR immunoaffinity-purified from rat brain has been found to contain an ACh-binding and a strucural subunits with apparent molecular weights of 79,000 and 51,000, respectively (Whiting and Lindstrom, 1987). mAb 299 binds to the α_4 subunit (Mr 79,000) on western blots and native nAChR. mAb 268 was first reported to bind a structural subunit on western blots, but this component has recently found not to be β_2 (Lindstrom et al., unpublished). Proteins eluted from the first affinity column were subjected to immunoblot analysis using mAbs 268 and 299 (Fig. 4). For western blots, purified preparations were electrophoresed on 10% acrylamide gels in SDS and then transfered onto polyvinylidine difluoride papers. The detection system included biotinylated-anti-rat IgG, alkaline phosphatase-conjugated Streptoavidine, 5-bromo-4-chloro-3-indoylphosphate-p-toluidine and nitro blue tetorazolium. mAb 299 bound to a Mr 80,000 protein, and mAb 268 bound to a Mr 48,000 protein which was found to be a minor band on SDS-PAGE gel. The similar results were also obtained from the 2nd affinity column preparation. These results strongly suggest that Mr 80,000 protein is α_4 of ACh-binding subunit. The identity of the component identified by mAb268 remains unclear.

To investigate depletion of ^{3}H-ACh binding sites of the purified nAChR by mAb 299, mAb 299-Affi-Gel-Protein A complex was incubated with the purified preparation and proteins bound to mAb 299 were precipitated by centrifugation, and then the supernatant was assayed for ^{3}H-ACh binding. mAb 299 depleted more than 85% of ^{3}H-ACh binding activity of the purified preparation.

The ACh-binding subunit (Mr 79,000) of the nAChR immunoaffinity-purified from rat brain is encoded by α_4 gene and the structural subunit (Mr 51,000) corresponds to β_2. Based on immunoprecipitation and immunoblot analysis, most of nAChR eluted from ACh-Affi-Gel column is probably composed of $\alpha_4\beta_2$. The Mr 48,000 protein bound to mAb 268 was found to be a minor component on SDS-PAGE and was smaller than that of the structural subunit of immunoaffinity-purified nAChR. Mr 48,000 protein may be partially degraded during purification. Abood et al. have purified a nicotine-binding protein of apparent molecular weight 56,000 from rat brain by affinity chromatography using nicotine analogue (Abood et al., 1983), which is different from ACh-Affi-Gel chromatography. This discrepancy remains unclear.

The present study advances cholinergic ligand affinity purification of nAChR in rat brain. This approach has the virtue that its competitive elution does not denature nAChR, whereas the acid elution of immunoaffinity purification is likely to denature nAChR. ACh-Affi-Gel column would be expected to bind other types of neuronal nAChR. Whether the purified preparation eluted from the affinity column contains the minor subtypes of nAChR is to be examine. Whether Mr 53,000 and 67,000 proteins we have found are connected with nAChR is under investigation.

The results in this study would be of use in the further purification and characterization of neuronal nAChR.

REFERENCES

Abood, L. G., Latham, W. and Grassi, S., 1983, Isolation of a nicotine binding site from rat brain by affinity chromatography, Proc. Natl. Acad. Sci. U.S.A., 80:3536-3539.

Lindstrom, J., Schoepfer, R. and Whiting, P., 1987, Molecular studies of the neuronal nicotinic acetylcholine receptor family, Mol. Neubiol., 1:281-337.

Nakayama, H., Shirase, M., Nakashima, T., Kurogochi, Y. and Lindstrom, J 1990, Affinity purification of nicotinic acetylcholine receptor from rat brain, Mol. Brain Res., in press.

Whiting, P. and Lindstrom, J., 1986, Pharmacological properties of immuno-isolated neuronal nicotinic receptors, J. Neurosci., 6:3061-3069.

Whiting, P. and Lindstrom, J., 1987, Purification and characterization of a nicotinic acetylcholine receptor from rat brain, Proc. Natl. Acad. Sci. U.S.A., 84:595-599.

DIFFERENTIAL COUPLING OF SUBTYPES OF THE MUSCARINIC RECEPTOR TO

SIGNALING MECHANISMS IN BRAIN AND PERIPHERAL TISSUES

F. J. Ehlert, F. M. Delen, S. H. Yun and P. Tran

Department of Pharmacology
College of Medicine
University of California, Irvine
Irvine, California 92717

INTRODUCTION

It has been possible to classify muscarinic receptors on the basis of their specificity for selective muscarinic agents for quite some time. However more recently, considerable progress has been made in characterizing muscarinic receptors, and there is now unequivocal evidence for subtypes of the muscarinic receptor. In this chapter, the relationship between the pharmacological and molecular biological classifications of subtypes of the muscarinic receptor will be discussed briefly, and some recent data on the coupling of subtypes of the muscarinic receptor to adenylate cyclase and phosphoinositide hydrolysis in central and peripheral tissues will also be described.

PHARMACOLOGICAL AND MOLECULAR DISTINCTIONS AMONG SUBTYPES OF THE MUSCARINIC RECEPTOR

The first evidence for subtypes of the muscarinic receptors came in the early 1950's just after the introduction of gallamine as an adjunct in general anesthesia. In addition to its neuromuscular blocking activity, gallamine caused sinus tachycardia as a side effect during anesthesia (Walton, 1950). This action was attributed to a blockade of the muscarinic effects of ACh (Riker and Wescoe, 1951; Rathbun and Hamilton, 1970) and to an enhancement of the release of catecholamines from cardiac sympathetic nerves (Brown and Crout, 1970). The latter could be due to a blockade of presynaptic muscarinic receptors modulating the release of norepinephrine (Loffelholz et al., 1967). Interestingly, gallamine was shown to be a much less potent antagonist at muscarinic sites in the ileum, bladder and salivary glands (Riker and Wescoe, 1951). The cardioselective action of gallamine is shared by some other neuromuscular blocking agents including pancuronium (Saxena and Bonta, 1970) and stercuronium (Li and Mitchelson, 1980), but not by analogs of gallamine in which the N-ethyl groups have been replaced by N-methyl groups (Riker and Wescoe, 1951).

The basis of the specificity of gallamine rests on the nature of its interaction with a secondary allosteric site on the muscarinic receptor and not on its binding to the primary recognition site (i.e. ACh binding site). In experiments on isolated atria, Clark and Mitchelson (1976) showed that gallamine antagonized the effects of ACh and carbachol in a manner which was clearly different from competitive antagonism, and suggested that gallamine altered the affinity of agonists for the muscarinic receptor allosterically. Similarly, Stockton et al. (1983) have demonstrated that gallamine modifies the binding of the muscarinic antagonist [^{3}H]N-methylscopolamine ([^{3}H]NMS) by an allosteric mechanism, and that the potency and negatively cooperative effects of gallamine are greatest in the heart, but much less in brain and salivary glands. These studies with gallamine provide clear implications for the development of novel agonists and antagonists which tune up or tune down muscarinic neurotransmission by acting at the gallamine site. Thus, the issue of muscarinic receptor specificity is one which entails not only the primary recognition site but the allosteric site as well.

Further evidence for subtypes of the muscarinic receptor came in the early 1960's when Roszkowski (1961) reported on a unique compound (McN A343; 4-(m-chlorophenylcarbamoyloxy)-2-butynyltrimethylammonium chloride) which had characteristics of a muscarinic ganglionic stimulant. When injected into cats, McN A343 caused a large pressor response that was preceded by a small depressor response. Both responses were blocked by atropine or bilateral sympathectomy and adrenalectomy, but not by hexamethonium. McN A343 was remarkably inactive on the heart and gut, so that its predominant effect *in vivo* is a pressor response due to ganglionic stimulation. Nonselective muscarinic agonists also stimulate sympathetic ganglia, but their profound peripheral effects on the heart and blood vessels cause a net reduction in blood pressure when administered *in vivo*. Other compounds have been described which appear to be selective muscarinic ganglionic stimulants (Franko et al., 1963; Jones, 1963; Roszkowski and Yelnowsky, 1967).

In retrospect, it can be seen that as early as the 1960's there was pharmacological evidence for three subtypes of the muscarinic receptor - 1) ganglionic muscarinic receptors, 2) cardiac muscarinic receptors, and 3) muscarinic receptors in exocrine glands and smooth muscle. Presently, it is still only possible to divide muscarinic receptors into three categories using pharmacological criteria. Nevertheless, much progress has been made in identifying new selective antagonists useful in classifying muscarinic receptors. One of the most important of these is pirenzepine, a competitive muscarinic antagonist used to treat gastric and duodenal ulcer (Hammer et al., 1980). It blocks gastric acid and pepsinogen secretion at doses which do not produce typical atropine-like effects including tachycardia, dry mouth, urinary retention and constipation. Pirenzepine also blocks the pressor response to McN A343 at doses 30-fold lower than those required to block the bradycardia caused by vagal stimulation (Hammer and Giachetti, 1982). In contrast, atropine has nearly equal potency at blocking these two effects. The selective antimuscarinic effects of pirenzepine are clearly evident in its competitive binding properties. A high affinity pirenzepine binding site ($K_D \cong 10^{-8}$ M) has been identified in brain and sympathetic ganglia and has been designated as an M_1 muscarinic receptor (Hammer et al., 1980; Hammer and Giachetti, 1982; Watson et al., 1983). In contrast, pirenzepine binds weakly to muscarinic receptors in the heart, glands and smooth muscle (Hammer et al., 1980).

Additional compounds which exhibit a pharmacological specificity for muscarinic receptors similar to that of gallamine have also been described and include AF-DX 116 and methoctramine (Giachetti et al., 1986; Hammer et al., 1986; Giraldo et al., 1988a). AF-DX 116 is a competitive antagonist structurally related to pirenzepine. It potently antagonizes muscarinic responses in the heart, but only weakly antagonizes muscarinic responses in ganglia, smooth muscles and exocrine glands (Giachetti et al., 1986). Methoctramine is a relatively long, linear compound containing four secondary amines, and it is structurally unrelated to other muscarinic antagonists. Its unusual tetraamino structure suggests that it might bind at the gallamine site on the muscarinic receptor. Although allosteric effects of methoctramine on the rate of dissociation of [^{3}H]NMS from the muscarinic receptor have been described (Giraldo et al., 1988a), it is not clear whether the most potent action of methoctramine on cardiac muscarinic receptors is a direct competition or an allosteric inhibition. The cardioselective antimuscarinic actions of AF-DX 116 and methoctramine have been verified in radioligand binding studies (Hammer et al., 1986; Giraldo et al., 1988a), and muscarinic receptors which have high affinity for the cardioselective antagonists are defined as M_2 muscarinic receptors.

Compounds which exhibit a selectivity for muscarinic receptors in exocrine glands and smooth muscle have also been described, and prominent among these are derivatives of difenidol including hexahydro-sila-difenidol (HHSiD) and its *para*-fluoro derivative

Table 1. Nomenclature for Subtypes of the Muscarinic Receptor

Pharmacological characterization

	M_1	M_2	non-M1, non-M2		
			(M_3)	-	-
Selective antagonists	pirenzepine	AF-DX 116 methoctramine	p-F-HHSiD HHSiD	-	-

Molecular characterization

Sequences	m1	m2	m3	m4	m5

(p-F-HHSiD) (Lambrecht et al., 1988). These compounds have been shown to block muscarinic responses in smooth muscle more potently than responses elicited by muscarinic receptors in the heart and sympathetic ganglia.

The use of molecular biological techniques to investigate the muscarinic receptor has confirmed the existence of receptor subtypes and has provided evidence for additional heterogeneity not apparent in previous pharmacological studies. At least five subtypes of the muscarinic receptor have been cloned from pig, rat and human brain and expressed in Xenopus oocytes and various mammalian cells (Kubo et al., 1986; Peralta et al., 1987a; Bonner et al., 1987; Peralta et al., 1987b; Bonner et al., 1988). These five subtypes have been designated m1 - m5 by Bonner and coworkers (1987, 1988) The expression studies of several investigators (Akiba et al., 1988; Lai et el., 1988; Mei et al., 1989; Buckley et al., 1989) indicate that the cloned m1 receptor exhibits high affinity for pirenzepine ($K_D \cong 10^{-8}$ M) and low affinities for AF-DX 116 and HHSiD. Also, mRNA for the m1 receptor is abundant in forebrain where high affinity pirenzepine binding sites predominate (Brann et al., 1988). Consequently, the high affinity pirenzepine site identified by pharmacological procedures (M_1 site) most likely represents the cloned m1 receptor. The pharmacologically defined M_2 muscarinic receptor, which exhibits high affinity for the cardioselective antagonist AF-DX 116, most likely represents the cloned m2 receptor because when expressed in Xenopus oocytes (Akiba et al., 1988) and CHO-K1 cells (Buckley et al., 1989), the m2 receptor exhibited high affinity for AF-DX 116 and low affinity for pirenzepine and HHSiD. Moreover, its mRNA is low in brain, but abundant in heart relative to the total amount of mRNA for muscarinic receptors in these two tissues (Kubo et al., 1986, Maeda et al., 1988). The remainder of the subtypes of the muscarinic receptor (m3, m4 and m5) exhibit low affinity for pirenzepine and AF-DX 116 when expressed in CHO-K1 cells (Buckley et al., 1989); consequently, these subtypes can be combined under the same implicitly heterogeneous pharmacological class of non-M_1, non-M_2. It has been shown, however, that compound HHSiD exhibits selectivity for the m3 receptor when this subtype is expressed in CHO-K1 cells (Buckley et al., 1989), and HHSiD and p-F-HHSiD antagonize the contraction of smooth muscle in the guinea pig ileum, selectively (Lambrecht et al., 1988). Thus, the cloned m3 muscarinic receptor could be designated pharmacologically as M_3 on the basis of its selectivity for HHSiD and p-F-HHSiD. In this report, however, only the pharmacological classes of M_1, M_2 and non-M_1, non-M_2 will be used because the selectivity of HHSiD and its para-fluoro derivative is not sufficiently characterized to allow an unambiguous classification of the m3 receptor. Table 1 summarizes the nomenclature for subtypes of the muscarinic receptor.

CARDIAC MUSCARINIC RECEPTORS AND ADENYLATE CYCLASE

Our studies on the coupling of subtypes of the muscarinic receptor to second messenger systems were initiated in the heart. The binding properties of muscarinic receptors in the heart are well characterized, and this tissue contains a relatively homogeneous population of M_2 muscarinic receptors (Waelbroek et al., 1987). Our strategy was to estimate the dissociation constants of a series of selective muscarinic antagonists using two different methods: 1) competitive binding assays with [^{3}H]NMS and 2) pharmacological antagonism of muscarinic receptor-mediated inhibition of adenylate cyclase. Provided that the experiments are carried out under similar conditions, the two methods should yield the same estimate for the dissociation constant of the antagonist. Thus, agreement between the two estimates should provide strong evidence that a particular binding site was mediating a given response. By examining a series of antagonists, it should be possible to determine the subtype of the muscarinic receptor mediating the response.

To determine the dissociation constant of a muscarinic antagonist from pharmacological data, we measured the dose-response curve of the highly efficacious agonist oxotremorine-M in the absence and presence of the antagonist. It has been shown that the dissociation constant of a competitive antagonist (K_I) can be estimated from the following null equation:

$$K_I = \frac{[I]}{DR - 1}$$

in which *[I]* denotes the concentration of the antagonist and *DR* (dose ratio) denotes the ratio of the EC_{50} value of the agonist in the presence of the antagonist divided by that measured in the absence of the antagonist (note: the EC_{50} value indicates the concentration of oxotremorine-M causing a half-maximal response). In most instances, we estimated the K_I value of an antagonist from a single dose ratio because most of the antagonists that we investigated have been shown previously to be competitive antagonists. Thus, it seemed unnecessary to measure the dose ratios at several concentrations of antagonist and prove whether the nature of the antagonism is competitive. Although our single dose ratio

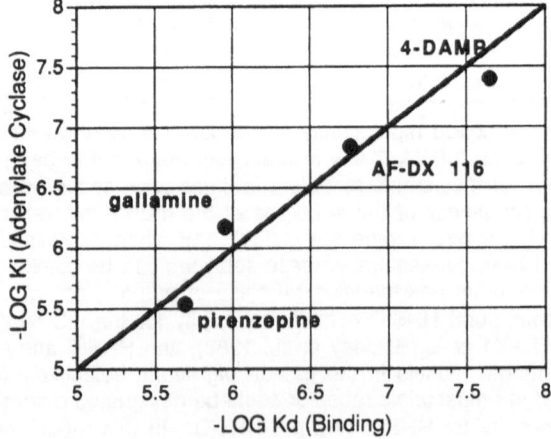

Figure 1. Comparison of the dissociation constants of muscarinic antagonists determined by antagonism of the adenylate cyclase response (Ki) and by inhibition of the binding of [^{3}H]NMS (Kd). These experiments were carried out on homogenates of the rat myocardium. The data are from Ehlert et al. (1989).

measurements could be flawed if more than one receptor subtype contributes to a response, verification that the estimated K_I value agrees with that measured by radioligand binding would appear to justify the pharmacological estimate of K_I. If more than one receptor subtype contributed to a response, then the estimated K_I value would approximate the geometric average of those measured by radioligand binding. For our experiments with the allosteric muscarinic antagonist gallamine, we examined the effects produced by several concentrations of this antagonist and estimated its dissociation constant using the allosteric method of Ehlert (1988).

The results of our studies on the heart are summarized in figure 1, which shows a comparison of the dissociation constants of various selective muscarinic antagonists measured by antagonism of the adenylate cyclase response (K_I) and by competitive inhibition of the binding of [^{3}H]NMS (K_D). In these studies, the binding properties of the various antagonists were consistent with a simple one-site model having dissociation constants which were in agreement with those measured by antagonism of the adenylate cyclase response. The agreement between the two estimates provides strong evidence that the site labeled by [^{3}H]NMS is the same receptor which mediates an inhibition of adenylate cyclase. This site exhibits relatively high affinity for AF-DX 116 ($K_D \cong 10^{-7}$ M) and gallamine ($K_D \cong 10^{-6}$ M) and relatively low affinity for pirenzepine ($K_D \cong 10^{-6}$ M); consequently, it can be concluded that M_2 muscarinic receptors mediate an inhibition of adenylate cyclase in the rat heart. Our results are consistent with those of other investigators who previously the characterized the binding properties of cardiac muscarinic receptors and their inhibition of adenylate cyclase activity (Gil and Wolfe, 1985; Nomura et al., 1987).

CENTRAL MUSCARINIC RECEPTORS AND ADENYLATE CYCLASE

The good agreement between binding affinity and pharmacological antagonism noted in the heart (see figure 1) suggested that the same approach might be feasible for investigating receptor coupling mechanisms in a tissue containing a mixture of subtypes of the muscarinic receptor. Consequently, we investigated the binding properties of selective muscarinic antagonists in various regions of the brain and compared binding affinity with that measured by antagonism of oxotremorine-M-mediated inhibition of adenylate cyclase activity in the corpus striatum.

Our competitive binding experiments with the nonselective muscarinic antagonist [^{3}H]NMS indicated that muscarinic receptors in the brain could be divided into three categories on the basis of their specificity for selective muscarinic antagonists (Ehlert et al., 1989). For example, in the cerebral cortex, the results of pirenzepine/[^{3}H]NMS competition experiments indicated that 50% of the receptors exhibited high affinity for pirenzepine ($K_H \cong 10^{-8}$ M), whereas the remainder of the sites exhibited low affinity ($K_L \cong 10^{-7}$ M). These results are consistent with the postulate that 50% of the receptors labeled by [^{3}H]NMS in the cerebral cortex are M_1 muscarinic receptors. The results of AF-DX 116/[^{3}H]NMS competition experiments in the cerebral cortex indicated that approximately 18% of the sites exhibited high affinity for AF-DX 116 ($K_H \cong 3 \times 10^{-8}$ M), whereas the remainder of the sites exhibited low affinity ($K_L \cong 10^{-6}$ M). These results are consistent with the postulate that approximately 18% of the sites labeled by [^{3}H]NMS in the cerebral cortex are M_2 muscarinic receptors. It can be seen that the sum of the densities of the M_1 and M_2 receptors only amount to 68% of the total sites, indicating that there must be a substantial amount of receptors lacking high affinity for both pirenzepine and AF-DX 116. These latter sites have been designated as non-M_1, non-M_2 muscarinic receptors. This three-site model for the interaction of selective antagonists with muscarinic receptors in the cerebral cortex is shown in figure 2. According to this model, the low affinity component of the pirenzepine/[^{3}H]NMS competition curve actually represents two sites (M_2 and non-M_1, non-M_2) which have similar, but perhaps not identical, affinities for pirenzepine. Similarly, this model indicates that the low affinity component of the AF-DX 116/[^{3}H]NMS competition curves represents two sites (M_1 and non-M_1, non-M_2) having similar low affinities for AF-DX 116. We compared the distribution of these site in various brain regions and found that the proportions of M_1 and non-M1, non-M2 subtypes were greatest in

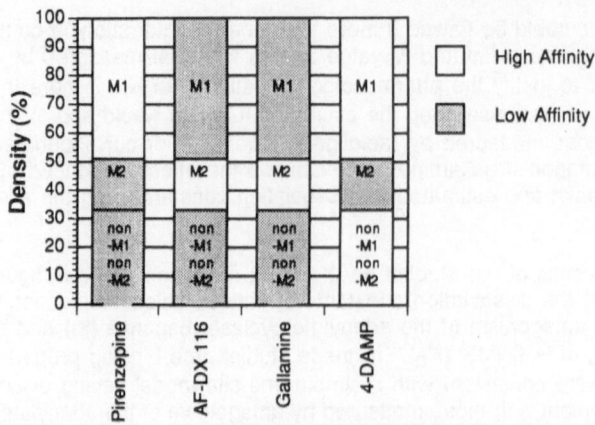

Figure 2. Three-site model for the analysis of the competitive inhibition of [³H]NMS binding by selective muscarinic antagonists in the cerebral cortex. The competitive inhibition of the specific binding of [³H]NMS by pirenzepine and AF-DX 116 yielded competition curves which were consistent with a two-site model. The high affinity component of the pirenzepine competition curve is designated as an M_1 receptor, whereas the low affinity component consists of two receptor populations designated as M_2 and non-M_1, non-M_2. The high affinity component of the AF-DX 116/[³H]NMS competition curve is designated as an M_2 receptor whereas the low affinity component consists of two receptor populations designated as M_1 and non-M_1, non-M_2. Gallamine exhibits a spectrum of affinities similar to that of AF-DX 116, whereas 4-DAMP has high affinity for both M_1 and non-M_1, non-M_2 receptors and low affinity for M_2 receptors. The data are from Ehlert et al. (1989).

cerebral cortex, hippocampus and corpus striatum; intermediate in the thalamus and hypothalamus; and lowest in the midbrain, medulla-pons and cerebellum (Ehlert and Tran, 1989). The relative distribution of the M_2 subtype was exactly the converse. However, when the absolute densities of the subtypes were calculated, it was seen that the M_1 and non-M_1, non-M_2 sites exhibited the greatest regional variation in a manner consistent with their relative abundances. In contrast, the density of the M_2 subtype remained relatively constant throughout the brain, except in the hippocampus, where it was very low. A comparison of the relative densities of the pharmacological subtypes of the muscarinic receptor in the cerebral cortex and corpus striatum is shown in figure 3. The results of our binding experiments with pirenzepine and AF-DX 116 are generally consistent with those of Giraldo et al. (1987) and Waelbroek et al. (1987) who previously divided muscarinic receptors in brain into three categories.

In contrast to the results shown in figure 3, numerous other investigators have calculated a much greater proportion (65 - 100%) of M_1 receptors in the cerebral cortex and corpus striatum by running pirenzepine/[³H]QNB competition experiments or by comparing the binding capacity of [³H]pirenzepine with those of [³H]QNB or [³H]NMS (Birdsall et al., 1983; Fisher and Bartus, 1985; Watson et al., 1986). One potential reason for this discrepancy is that the binding capacity of [³H]QNB is greater than that of [³H]NMS. Lee and El-Fakahany (1985) have shown that the tertiary amine [³H]QNB labels more binding sites than the quaternary ammonium derivative [³H]NMS, and have suggested that some muscarinic receptors are sequestered in a membrane compartment which is inaccessible to fully charged quaternary ammonium compounds but not to partially charged tertiary amines. Strong evidence for this postulate has been obtained in studies of the binding of tertiary amine and quaternary ammonium radioligands to intact cells (Brown and Goldstein, 1986; Fisher, 1988). To obtain additional evidence for the compartmentalization

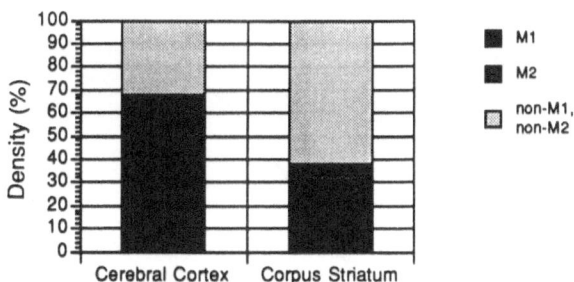

Figure 3. Relative distribution of muscarinic receptor binding sites in the cerebral cortex and corpus striatum. The data are from Ehlert et al. (1989).

model in brain homogenates, we synthesized [³H]N-methylQNB from [³H]QNB and nonradioactive methyliodide, so that the specific activity of our quaternary ligand([³H]N-methylQNB) would be identical to that of the tertiary ligand ([³H]QNB). Using this approach, the estimate of the relative difference in the binding capacities of the two radioligands is unaffected by an incorrect estimate of the specific activity. Figure 4 shows the results of saturation experiments with [³H]QNB and [³H]N-methylQNB in homogenates prepared from frozen rat cerebral cortex. The theoretical curves represent the least squares fit to the data assuming a simple one-site model. It can be seen that the binding curve for [³H]QNB reaches a higher plateau than that of [³H]N-methylQNB. Regression analysis showed that the binding capacity (B_{max}) of [³H]N-methylQNB was approximately 30% lower than that of [³H]QNB. In fresh tissue we have observed a 20% difference in the binding capacities of the two radioligands. The results shown in figure 4 are consistent with the postulate of Lee and El-Fakahany (1985) that the tertiary amine radioligand [³H]QNB binds with more receptors than the quaternary ammonium radioligands [³H]N-methylQNB and [³H]NMS.

To determine whether differences in the B_{max} values of [³H]QNB and [³H]NMS could account for the apparent differences in high affinity component of the pirenzepine/[³H]NMS and pirenzepine/[³H]QNB competition curves, we investigated the effects of benzilylcholine mustard (BCM) on the competitive inhibition of [³H]QNB binding by pirenzepine. BCM is a 2-chloroethylamine derivative of benzilylcholine which cyclizes spontaneously at neutral pH into a quaternary aziridinium ion which binds covalently to the muscarinic receptor (Gill and Rang, 1966). Because the aziridinium ion is a quaternary ammonium compound, it should only interact with the [³H]NMS sites and not the additional sequestered sites which can bind [³H]QNB. By allowing sufficient time for the complete conversion of BCM into its aziridinium ion before incubation with brain homogenate, we were able to alkylate practically all of the [³H]NMS sites but only 75% of the [³H]QNB sites in homogenates of the cerebral cortex. When pirenzepine/[³H]QNB competition experiments were carried out on the residual [³H]QNB binding sites, the competition curves represented a nearly homogeneous population of high affinity pirenzepine sites compared to control curves which exhibited 75% high affinity pirenzepine sites (unpublished observations). These results are consistent with the postulate that the extra sites labeled by [³H]QNB are predominantly M_1 receptors, which can account for the greater apparent proportion of M_1 sites when measured by pirenzepine/[³H]QNB competition experiments as compared to that measured by pirenzepine/[³H]NMS competition.

The results of our studies characterizing the distribution of binding sites for selective muscarinic antagonists in brain are consistent with the distribution of mRNA's for subtypes of the muscarinic receptor. The results of northern blot analysis and *in situ* hybridization studies show that m1 mRNA is abundant in various forebrain regions but low in caudal regions of the brain which agrees with the distribution of the M_1 site described above (Maeda et al., 1988; Brann et al., 1988). Also, it is now clear that the cerebral cortex and corpus striatum express substantial amounts of m3 and m4 mRNA, respectively (Brann et al.,

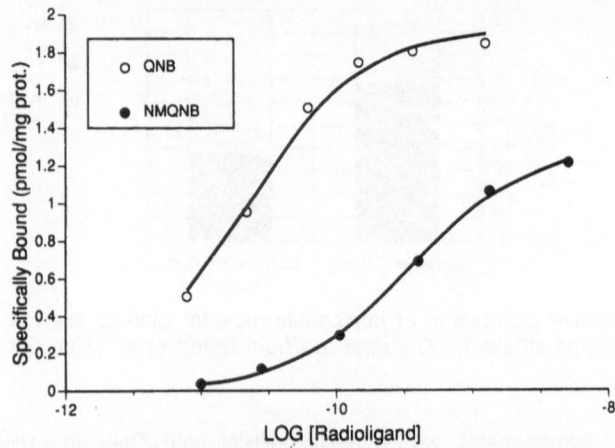

Figure 4. Comparison of the specific binding of [³H]QNB and [³H]N-methylQNB to homogenates of the rat cerebral cortex. The specific binding of [³H]QNB and [³H]N-methylQNB was measured by the a filtration method similar to that described by Ehlert et al. (1989). The incubations were carried out for 60 min at 37°C. The data points represent the mean binding values from six experiments, each done in triplicate. The theoretical curve represents the least squares fit to the data assuming a simple one-site model. The dissociation constant and binding capacity of [³H]QNB were 20.5 pM and 1.93 fmol/mg tissue, whereas those for [³H]N-methylQNB were 351 pM and 1.31 fmol/mg tissue.

1988), which could account for the large abundance of non-M$_1$, non-M$_2$ sites in these brain regions (see figure 3).

The results of our studies characterizing the subtype of the muscarinic receptor which inhibits adenylate cyclase activity in the corpus striatum are summarized in figure 5. Oxotremorine-M-mediated inhibition of adenylate cyclase activity was relatively insensitive to both pirenzepine and AF-DX 116 indicating that a non-M$_1$, non-M$_2$ muscarinic receptor mediates inhibition of adenylate cyclase activity in the corpus striatum. The excellent agreement between the K$_D$ values of antagonists for the non-M$_1$, non-M$_2$ site and the K$_I$ value measured by antagonism of the adenylate cyclase response is clearly seen in figure 5. There was also good agreement between the K$_D$ values of AF-DX 116, 4-DAMP and gallamine for the M$_1$ receptor and their respective K$_I$ values for antagonizing the adenylate cyclase response; however, the relatively low potency of pirenzepine for blocking the adenylate cyclase response (K$_I \cong 10^{-7}$ M) indicates a non-M$_1$ mechanism. The potencies of AF-DX 116 and gallamine for antagonizing the adenylate cyclase response in the corpus striatum were approximately 100 to 360-fold lower than that expected for an M$_2$ response. However, because the relative density of the M$_2$ site in the corpus striatum (7%) was much less than that of the non-M$_1$, non-M$_2$ site (62%; see figure 3), it would be difficult to measure a small M$_2$ component of the adenylate cyclase response in the face of such a large non-M$_1$, non-M$_2$ response. Consequently, we cannot rule out the possibility that the small number of M$_2$ receptors in the corpus striatum are coupled to adenylate cyclase.

Our studies on the pharmacological antagonism of the adenylate cyclase response by pirenzepine in the corpus striatum are consistent with the previous observations of Gil and Wolfe (1985). In contrast, our results differ from those of McKinney et al. (1989) who found that the potency of AF-DX 116 for antagonizing the adenylate cyclase response in the corpus striatum was approximately 10-fold greater than that reported here. Nevertheless,

McKinney et al. (1989) also concluded that a subtype of the muscarinic receptor different from the M_1 and cardiac M_2 subtypes mediates inhibition of adenylate cyclase activity in the corpus striatum.

The observation that a muscarinic receptor belonging to the non-M_1, non-M_2 class mediates inhibition of adenylate cyclase activity in the corpus striatum raises the question of which one of the three subtypes of the muscarinic receptor (i.e. m3, m4 and m5) that contribute to the non-M_1, non-M_2 binding site in brain actually mediates the inhibition of adenylate cyclase activity. The results of studies in which the individual subtypes of the muscarinic receptor were expressed in cells previously lacking muscarinic receptors now indicate that the m1, m3 and m5 receptors couple preferentially to phosphoinositide hydrolysis whereas the m2 and the m4 receptors preferentially inhibit adenylate cyclase activity (Peralta et al., 1988; Liao et al., 1989). Thus, the most likely hypothesis is that m4 muscarinic receptors mediate inhibition of adenylate cyclase activity in the corpus striatum. This conclusion is consistence with the great abundance of m4 mRNA in the corpus striatum (Brann et al., 1988).

MUSCARINIC RECEPTOR COUPLING MECHANISMS IN SMOOTH MUSCLE

In preliminary studies on the longitudinal muscle of the rat ileum, we found evidence for two types of muscarinic receptors (Candell et al., 1990). The majority of the sites (approximately 80%) exhibit high affinity for the cardioselective antagonists AF-DX 116 and methoctramine and have been designated as M_2 muscarinic receptors. The remainder of the sites lack appreciable affinity for pirenzepine as well as the cardioselective antagonists; consequently, we have designated these sites as non-M_1, non-M_2 muscarinic receptors. These results are consistent with the prior observations of Giraldo et al. (1988a) who found that the guinea pig ileum contains predominantly M_2 receptors and a small amount of sites lacking selectivity for M_1 and M_2 antagonists. The studies of Giraldo et al. (1988b) on the guinea pig ileum indicate that the small population of non-M1, non-M2 site exhibit high affinity for the M_3 selective antagonist HHSiD. However, in our studies on the rat, we have been unable to demonstrate a high affinity binding site for HHSiD even though this compound antagonizes oxotremorine-M-induced contractions of the gut with a K_I value consistent with that of an M_3

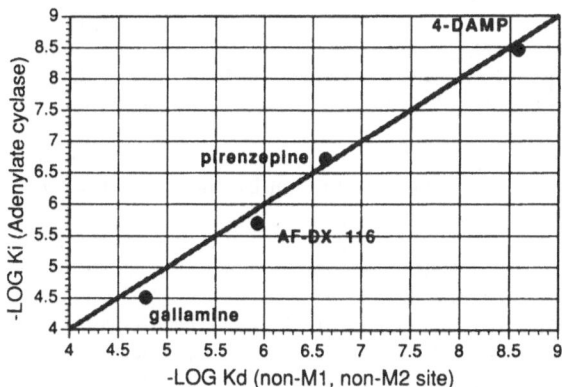

Figure 5. Comparison of the dissociation constants of muscarinic antagonists determined in the corpus striatum by antagonism of the adenylate cyclase response (Ki) and by inhibition of the binding of [³H]NMS to non-M1, non-M2 receptors (Kd). The data are from Ehlert et al. (1989).

(m3) subtype (i.e. $K_I \cong 2 \times 10^{-8}$ M; unpublished observations). In preliminary studies, we have obtained results consistent with the postulate that the major site in the ileum (M_2 receptor) mediates inhibition of adenylate cyclase activity whereas the minor site (non-M_1, non-M_2 site) stimulates phosphoinositide hydrolysis (Candell et al., 1990).

CONCLUSIONS

Our results indicate that the subtypes of the muscarinic receptor couple to phosphoinositide hydrolysis and adenylate cyclase activity differentially. The pattern of coupling seen in the brain, heart and ileum is consistent with that observed in studies where the individual recombinant subtypes of the muscarinic receptor were transfected into cells previously lacking muscarinic receptors. Our results also indicate that it is possible to characterize the subtype of the muscarinic receptor mediating a given response in a tissue containing a mixed population of receptors by measuring the potency with which a series of antagonists interfere with the response.

ACKNOWLEDGEMENTS

This work was supported by National Institute of Health Grant NS 26511, a Research Starter Grant from the Pharmaceutical Manufacturers Association Foundation, a Grant-in-Aid from the American Heart Association, California Affiliate and with funds contributed by the American Heart Association, Chapters. F. J. E. is a recipient of a U. S. Public Health Service Research Career Development Award (NS 01396) from the National Institute of Neurological Disorders and Stroke.

REFERENCES

Akiba, I., Kubo, T., Maeda, A., Bujo, H., Nakai, J., Mishina, M. and Numa, S., 1988, Primary structure of porcine muscarinic acetylcholine receptor III and antagonist binding studies. FEBS Letts. 235:257-261.

Birdsall, N.J.M., Hulme, E. C. and Stockton, J. M., 1983, Muscarinic receptor heterogeneity. Trends Pharmacol. Sci. Suppl., pp 4-8.

Bonner, T. I., Buckley, N. J., Young, A. C. and Brann, M. R., 1987, Identification of a family of muscarinic acetylcholine receptor genes. Science 237:527-532.

Bonner, T. I., Young, A. C., Brann, M. R. and Buckley, N J., 1988, Cloning and expression of the human and rat m5 muscarinic acetylcholine receptor genes. Neuron 1:403-410.

Brann, M.R., Buckley, N. J. and Bonner, T I., 1988, The striatum and cerebral cortex express different muscarinic receptor mRNA's. FEBS Letts. 230:90-94.

Brown, B. R. and Crout, J. R., 1970, The sympathomimetic effect of gallamine on the heart, J. Pharmacol. Ex. Ther. 172:266-273.

Brown, J. H. and Goldstein, D., 1986, Analysis of cardiac muscarinic receptors recognized selectively by nonquaternary but not by quaternary ligands, J. Pharmacol. Ex. Ther. 238:580-586.

Buckley, N.J., Bonner, T. I., Buckley, N. J. and Brann, M. R., 1989, Antagonist binding properties of five cloned muscarinic receptors expressed in CHO-K1 cells. Mol. Pharmacol. 35:469-476.

Candell, L. M., Yun, S. H. and Ehlert, F. J., 1990, Differential coupling of subtypes of the muscarinic receptor to adenylate cyclase and phosphoinositide hydrolysis in the longitudinal muscle of the rat ileum, FASEB J. (in press).

Clark, A. L. and Mitchelson, F., 1976, The inhibitory effect of gallamine on muscarinic receptors, Brit. J. Pharmacol. 58:323-331.

Ehlert, F. J., 1988, Estimation of the affinities of allosteric ligands using radioligand binding and pharmacological null methods, Mol. Pharmacol. 33:187-194.

Ehlert, F.J., 1988, Gallamine allosterically antagonizes muscarinic receptor-mediated inhibition of adenylate cyclase activity in the rat myocardium, J. Pharmacol. Ex. Ther. 247:596-602.

Ehlert, F. J., Delen, F. M., Yun, S. H., Friedman, D. J. and Self, D. W., 1989, Coupling of subtypes of the muscarinic receptor to adenylate cyclase in the corpus striatum and heart, J. Pharmacol. Ex. Ther. 251:660-670.

Ehlert, F. J. and Tran, P., 1989, Regional distribution of binding sites for selective muscarinic antagonists in rat brain, Soc. Neurosci. Abs. 15:229.

Fisher, S. K., 1988, Recognition of muscarinic cholinergic receptors in human SK-N-SH neuroblastoma cells by quaternary and tertiary ligands is dependent upon temperature, cell integrity, and the presence of agonists, Mol. Pharmacol. 33:414-422.

Fisher, S.K. and Bartus, R. T., 1985, Regional differences in the coupling of muscarinic receptors to inositol phospholipid hydrolysis in guinea pig brain, J. Neurochem. 45:1085-1095.

Franko, B. V., Ward, J. W., Alphin, R. S., 1963, Pharmacological studies of N-benzyl-3-pyrrolidyl acetate methobromide (AHR-602), a ganglion stimulating agent, J. Pharmacol. Exp. Ther. 139:25-30.

Giachetti, A., Micheletti, R. and Montagna, E.,1986, Cardioselective profile of AF-DX 116, a muscarinic M2-receptor antagonist. Life Sci. 38:1663-1672.

Gill, E. W. and Rang, H. P., 1966, An alkylating derivative of benzilylcholine with specific and long-lasting parasympatholytic activity, Mol. Pharmacol. 2:284-297.

Gil, D.W. and Wolfe, B. B., 1985, Pirenzepine distinguishes between muscarinic receptor-mediated phosphoinositide breakdown and inhibition of adenylate cyclase, J. Pharmacol. Ex. Ther. 232:608-616.

Giraldo, E., Hammer, R., and Ladinsky, H., 1987, Distribution of muscarinic receptors subtypes in rat brain as determined in binding studies with AF-DX 116 and pirenzepine, Life Sci. 40:833-840.

Giraldo, E., Vigano, M. A., Hammer, R. and Ladinsky, H., 1988a, Characterization of muscarinic receptors in guinea pig ileum longitudinal smooth muscle, J. Pharmacol. Ex. Ther. 33:617-625.

Giraldo, E., Micheletti, R., Montagna, E., Giachetti, A., Vigano, M. A., Ladinsky, H. and Melchiorre, C., 1988a, Binding and functional characterization of the cardioselective muscarinic antagonist methoctramine, J. Pharmacol. Ex. Ther. 244:1016-1020.

Hammer, R., Berrie, C. P., Birdsall, N. J. M., Burgen, A. S. V. and Hulme, E. C., 1980, Pirenzepine distinguishes between different subclasses of muscarinic receptors, Nature 183:90-91.

Hammer, R. and Giachetti, A., 1982, Muscarinic receptor subtypes: M1 and M2 biochemical and functional characterization. Life Sci. 31:2991-2998.

Hammer, R., Giraldo, E., Schiavi, G. B., Monferini, E. and Ladinski, H., 1986, Binding profile of a novel cardioselective muscarine receptor antagonist, AF-DX 116, to membranes of peripheral tissues and brain in the rat. Life Sci. 38:1653-1662.

Jones, A., 1963, Ganglionic actions of muscarinic substances, J. Pharmacol. Ex. Ther. 141:195-205.

Kubo, T., Fukuda, K., Mikami, A., Maeda, A., Takahashi, H., Mishina, M., Haga, T., Haga, K., Ichyama, A., Kangawa, K., Kojima, M., Matsuo, H., Hirose, T. and Numa, S., 1986, Cloning, sequencing and expression of complementary DNA encoding the muscarinic acetylcholine receptor. Nature 323:411-416.

Lai, J., Mei, L., Roeske, W. R., Chung, F.-Z., Yamamura, H. I. and Venter, J, C., 1988, The cloned murine M1 muscarinic receptor is associated with the hydrolysis of phosphatidylinositols in transfected murine B82 cells. Life Sci. 42:2489-2502.

Lambrecht, G., Feifel, R., Forth, B., Strohmann, C., Tacke, R. and Mutschler, E., 1988, p-Fluro-hexahydro-sila-difenidol: The first $M_{2\beta}$-selective muscarinic antagonist, Eur. J. Pharmacol. 152:193-194.

Lee J.-H. El-Fakahany, E. E., 1985, Heterogeneity of binding of muscarinic receptor antagonists in rat brain homogenates, J. Pharmacol. Ex. Ther. 233:707-714.

Li, C. K. and Mitchelson, F., 1980, The selective antimuscarinic action of stercuronium, Brit. J. Pharmacol. 70:313-321.

Liao, C.-F., Themmen, A. P. N., Joho, R., Barberis, C., Birnbaumer, M. and Birnbaumer, L., 1989, Molecular cloning and expression of a fifth muscarinic acetylcholine receptor, J. Biol. Chem. 265:7328-7337.

Loffelholz, K., Lindmark, R. and Muscholl, E., 1967, Der Einfluss von atropin auf die noradrenalinfreisetzung durch acetylcholin, Arch. Pharmak. Exp. Path. 257:308.

Maeda, A., Kubo, T., Mishina, M. and Numa, S., 1988, Tissue distribution of mRNAs encoding muscarinic acetylcholine receptor subtypes. FEBS Lett. 239:339-342.

McKinney, M., Anderson, D., Forray, C. and El-Fakahany, E. E., 1989, Characterization of the striatal M_2 muscarinic receptor mediating inhibition of cyclic AMP using selective antagonists: A comparison with the brainstem M_2 receptor, J. Pharmacol. Ex. Ther. 250:565-572.

Mei, L., Lai, J., Roeske, W. R., Fraser, C. M., Venter, J. C. and Yamamura, H. I., 1989, Pharmacological characterization of the M1 muscarinic receptors expressed in murine fibroblast B82 cells. J. Pharmacol. Ex. Ther. 248:661-670.

Nomura, S., Zorn, S. H. and Enna, S. J., 1987, Selective interaction of tricyclic antidepressants with a subclass of rat brain cholinergic muscarinic receptors. Life Sci. 40:1751-1760.

Peralta, E. G., Winslow, J. W., Peterson, G. L., Smith, D. H., Ashkenazi, A., Ramachandran, J., Schimerlik, M. H. and Capon, D. J., 1987a, Primary structure and biochemical properties of an M_2 muscarinic receptor. Science 236:600-605.

Peralta, E. G., Ashkenazi, A., Winslow, J. W., Smith, D. H., Ramachandran, J. and Capon, D. J., 1987b, Distinct primary structures, ligand-binding properties and tissue-specific expression of four human muscarinic acetylcholine receptors. EMBO J. 6:3923-3929.

Peralta, E.G., Ashkenazi, A., Winslow, J. W., Ramachandran, J. and Capon, D. J., 1988, Differential regulation of PI hydrolysis and adenylate cyclase by muscarinic receptor subtypes. Nature 334:434-437.

Rathbun, F.J. and J.T. Hamilton. Effect of gallamine on cholinergic receptors. Can. Anaes. Soc. J. 17:574-590 (1970).

Riker, W. F. and Wescoe, W. C., 1951, The pharmacology of flaxedil with observations on certain analogs, Ann. N.Y. Acad. Sci. 54:373-394.

Roszkowski, A. P., 1961, An unusual type of sympathetic ganglionic stimulant, J. Pharmacol. Ex. Ther. 132:156-170.

Roszkowski, A. P. and Yelnoski, J., 1967, Structure-activity relationships among a series of acetylenic carbamates related to McN-A-343, J. Pharmacol. Ex. Ther. 156:238-245.

Saxena, P. R. and Bonta, I. L., 1970, Mechanism of selective cardiac vagolytic action of pancuronium bromide. Specific blockade of cardiac muscarinic receptors, Eur. J. Pharmacol. 11:332-341.

Stockton, J.M., Birdsall, N. J. M., Burgen, A. S. V. and Hulme, E. C., 1983, Modification of the binding properties of muscarinic receptors by gallamine. Mol. Pharmacol. 23:551-557.

Waelbroek, M., Gillard, M., Robberecht, P. and Christophe, J., 1987, Muscarinic receptor heterogeneity in rat central nervous system. I. Binding of four selective antagonists to three muscarinic receptor subclasses: A comparison with M2 cardiac muscarinic receptors of the C type. Mol. Pharmacol. 32:91-99.

Walton, F. A., 1950, Flaxedil: A new curarizing agent, Can. Med. Assoc. J. 63:123-129.

Watson, M., Roeske, W. R. and Yamamura, H. I., 1986, [^{3}H]Pirenzepine and (-)-[^{3}H]quinuclidinyl benzilate binding to rat cerebral cortical and cardiac muscarinic cholinergic sites. II. Characterization and regulation of antagonist binding to putative muscarinic subtypes. J. Pharmacol. Ex. Ther. 237:419-427.

Watson, M., Yamamura, H. I. and Roeske, W. R. 1983, A unique regulatory profile and regional distribution of [^{3}H]pirenzepine binding in the rat provide evidence for distinct M1 and M2 muscarinic receptor subtypes. Life Sci. 32:3001-3011.

THE MOLECULAR PROPERTIES OF THE M_1 MUSCARINIC RECEPTOR AND ITS

REGULATION OF CYTOSOLIC CALCIUM IN A EUKARYOTIC GENE EXPRESSION SYSTEM

Josephine Lai, Thomas L. Smith, Lin Mei, Masaaki Ikeda, Yutaka
Fujiwara, Jorge Gomez, Marilyn Halonen, William R. Roeske and Henry
I. Yamamura

Departments of Pharmacology (J.L., T.L.S., L.M., M.I., Y.F., M.H.
W.R.R., H.I.Y.), Psychiatry (H.I.Y.), Internal Medicine (W.R.R.)
Biochemistry (H.I.Y.) and Immunology (J.G., M.H.), University of
Arizona, Tucson, AZ 85724; Research Service (T.L.S.), Veterans
Administration Medical Center, Tucson, AZ 85723

INTRODUCTION

Muscarinic receptor heterogeneity has been implicated in the
selective binding characteristics of non-classical muscarinic
antagonists such as pirenzepine (PZ) and AF-DX 116 (11-[[2-[(diethyl-
amino)methyl]-1-piperi-dinyl]acetyl]-5,11-dihydro-6H-pyrido [2,3-b]-
[1,4]benzodiazepine-6-one) (Watson et al., 1986). Pirenzepine labels a
class of muscarinic receptor binding sites with high affinity, termed
the M_1 type, which is predominant in the brain. AF-DX 116, on the
other hand, distinguishes a second class of binding sites by its high
affinity, known as the M_2 type, which is predominant in cardiac
tissues (Giachetti et al., 1986). A third class of muscarinic
receptor sites, noted for its low affinity for both PZ and AF-DX 116,
is represented in a number of glandular tissues (Hammer et al., 1986;
Korc et al., 1987). The heterogeneity of the muscarinic receptors was
proven unequivocally by the identification of several genes (m_1-m_5,
genotypic definition established by Bonner et al., 1987, 1988) which
encode distinct polypeptides that show muscarinic cholinergic receptor
properties (Bonner et al., 1987, 1988; Peralta et al., 1987; Liao et
al., 1989). These structurally distinct types of the muscarinic
receptors could be individually analyzed, by means of in vitro
expression in eukaryotic cell lines, in order to correlate the
structural diversity of the muscarinic receptors with their multiple
physiological activities. Recent studies have thus been devoted to
compare and contrast the ligand binding characteristics (Bonner et al.,
1987, 1988; Peralta et al., 1987; Fukuda et al., 1987; Akiba et al.,
1988; Liao et al., 1989), and second messenger coupling mechanisms
(Fukuda et al., 1987, 1988; Akiba et al., 1988; Bujo et al., 1988;
Jones et al., 1988; Neher et al., 1988, Peralta et al., 1988; Liao et
al., 1989) of these receptor types.

Ligand binding studies show that receptors encoded by the m_1 gene
assimilate the M_1 type binding characteristics while those encoded by
the m_2 gene product resemble the M_2 type (Bonner et al., 1988;
Akiba et al., 1988). A recent study analyzed the binding affinities of

Neuroreceptor Mechanisms in Brain, Edited by S. Kito *et al.*
Plenum Press, New York, 1991

an array of antagonists to the five muscarinic receptor types encoded by the m_1-m_5 genes (Buckley et al., 1989). The data substantiate earlier evidence that compounds such as PZ and AF-DX 116 also distinguish the receptors encoded by m_3 and m_4 genes from the M_1 and M_2 types. Characterization of these receptors by their functional coupling to signal transduction pathways shows that receptors from m_1, m_3 and m_5 activate phospholipase C in the hydrolysis of polyphosphoinositides (PPI) (Lai et al., 1988; Peralta et al., 1988; Liao et al., 1989) and mediate an intracellular calcium response which modulates calcium dependent potassium conductances (Fukuda et al., 1988; Jones et al., 1988). The receptors encoded by m_1 and m_3 also activate phospholipase A_2 (Conklin et al., 1988). On the other hand, both m_2 and m_4 encoded receptors are inversely coupled to adenylyl cyclase activity (Ashkenazi et al., 1987; Peralta et al., 1988).

We reported an in vitro expression system using a murine fibroblast L cell line, B82, in the expression of the muscarinic receptor encoded by the rat m_1 gene (Lai et al., 1988). B82 is an enzyme deficient clonal subline of the L line of mouse cells (Earle, 1943). These cells are heteroploid fibroblast cells lacking in thymidine kinase activity (available from the Institute for Medical Research, Camden, N.J.). The receptor expressed from m_1 exhibits M_1 type characteristics in radioligand binding studies and the activation of this receptor regulates PPI hydrolysis via a pertussis toxin sensitive guanine nucleotide regulatory protein (G protein) in the B82 cells (Lai et al., 1988; Mei et al., 1989a). In vitro expression systems as such provide a vehicle for biochemical and pharmacological analyses of structurally defined polypeptides, as illustrated by our study of the M_1 receptor which exhibits molecular characteristics similar to the endogeneous M_1 receptor and furthermore regulates intracellular calcium concentration possibly via the G protein coupled PPI hydrolysis pathway.

RESULTS

Transfection of the m_1 muscarinic receptor gene into murine fibroblast B82 cells. The rat m_1 muscarinic receptor gene was transfected into the murine fibroblast B82 cells via the expression plasmid $pH_{\beta}APr$-1-neo (fig. 1). The m_1 gene was originally identified in a rat genomic DNA library (Clontech, Palo Alto, CA) by its hybridization with a consensus sequence for the putative second transmembrane domain of the muscarinic and adrenergic receptors (Lai et al., 1988). The genomic DNA clone, c71, which carried this gene, contained a 2.2 kb TaqI-BamHI fragment which included the entire open reading frame (1.38 kb) for m_1 as determined by DNA sequencing analysis using the standard chain termination method. The deduced amino acid sequence from this genomic m_1 clone (fig. 2) was identical to that of a cDNA clone for the m_1 gene from rat brain (Bonner et al., 1987). There was no sequence divergence between the 5'-untranslated region of the cDNA and the region upstream of the genomic m_1 coding region. This fragment had previously been subcloned into the expression vector, pMSV-neo, and transfection of the recombinant, pMSV-neo-m_1, yielded 2 clones, cTB9 and cTB10 (Lai et al., 1988). The latter of the two was also employed in some of the present analysis. However, higher level of expression of the receptor was achieved by inserting the coding region of the m_1 gene (a 1.6 kb HpaII-BamHI fragment, which contained the 1.38 kb open reading frame for the rat m_1 muscarinic receptor gene and 0.25 kb of its

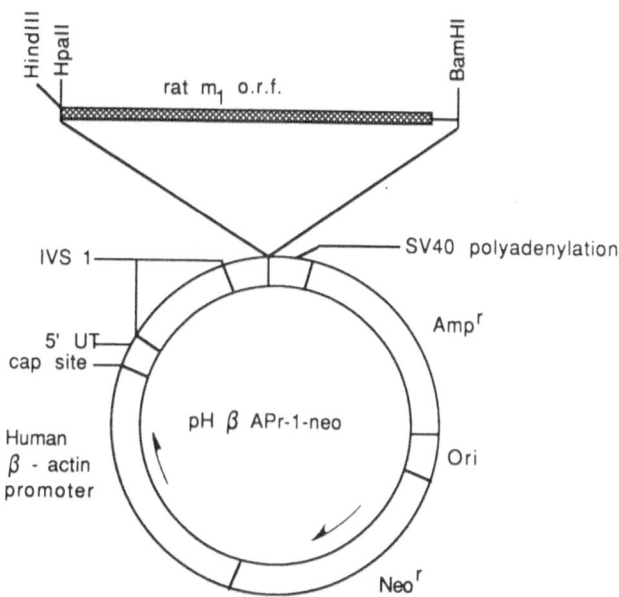

Fig. 1 Construction of the m_1 muscarinic receptor
gene into the eukaryotic expression vector pH β
APr-1-neo. A 1.6 kb HpaII-BamHI restriction
fragment from a rat genomic DNA clone, c71, which
consisted of the 1.38 kb open reading frame (o.r.f.)
and 0.25 kb of the 3'-untranslated region of the
m_1 gene was subcloned into pIBI24 and excised with
HindIII (5') and BamHI (3') (IBI, New Haven, CT).
This fragment was inserted into the unique HindIII
and BamHI sites downstream from the human β-actin
promoter. 5'-UT, 5'-untranslated region of the
human β-actin gene; IVS1, intervening sequence 1 of
the human β-actin gene; Amp^r, ampicillin
resistance gene; Neo^r, neomycin resistance gene;
Ori, SV40 origin of replication.

3'-untranslated region) downstream from the human β-actin promoter
construct, which has been identified as a strong eukaryotic promoter
(Gunning et al., 1987). B82 cells transfected with this recombinant
plasmid, pHβAPr-1-neo-m_1, by the calcium phosphate precipitation
method (Graham and Van Der Eb, 1973), were selected by both their
expression of neomycin resistance (in 500 µg/ml G418, Gibco, Grand
Island, NY) as well as a stable expression of the m_1 gene. Resistant
clones were maintained in a non-selecting growth medium (5 percent
fetal calf serum, 5 percent new born calf serum, 45 percent Ham's F-12,
45 percent Dulbecco's modified Eagle's medium, 100 U/ml penicillin and
100 µg/ml streptomycin) in a humidified atmosphere with 5 percent
CO_2, and periodically screened with 300 µg/ml of G418.

In vitro expression of the M_1 muscarinic receptor in the
transfected murine fibroblast B82 cells. The stable expression of the
muscarinic receptor encoded by the m_1 gene in transfected B82 cells
was detected by specific binding of [^{3}H](-)MQNB ([^{3}H](-)methyl-3-
quinuclidinyl benzilate) to intact cells (Mei et al., 1989b). This

315

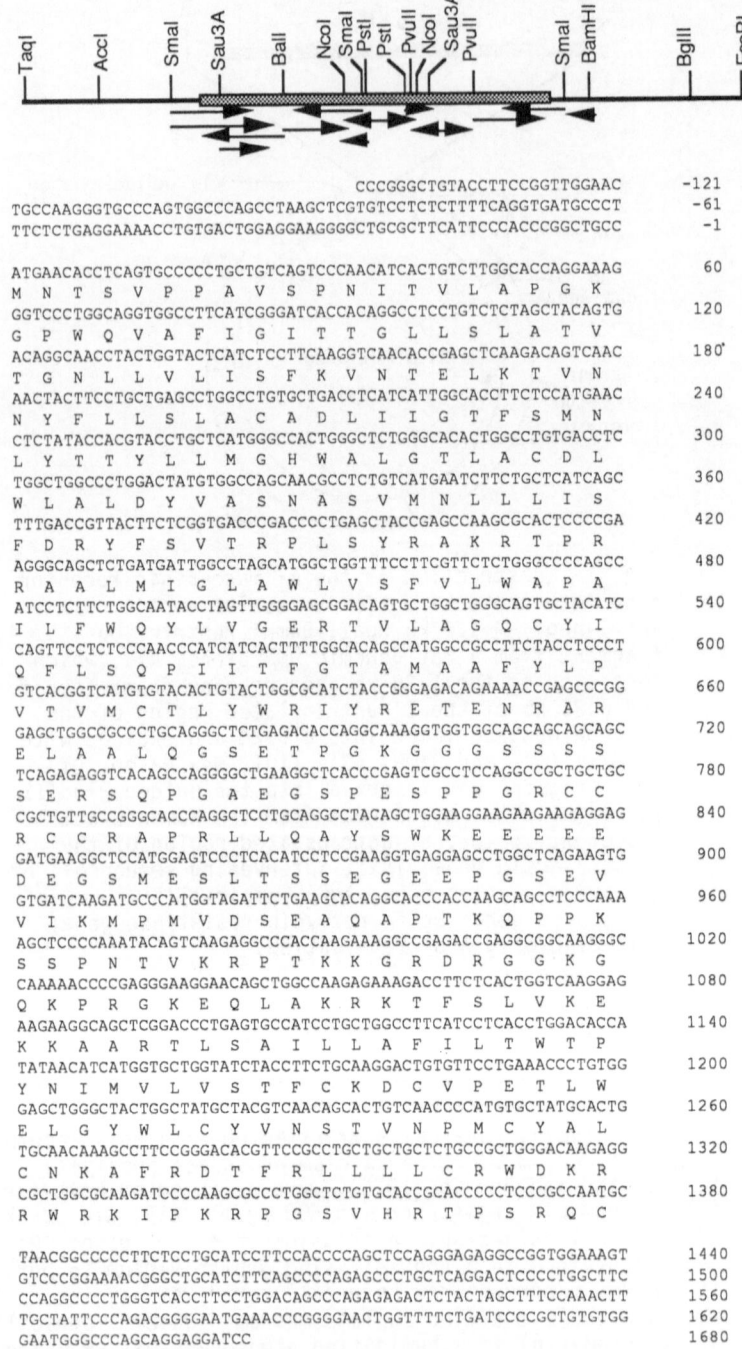

```
                                        CCCGGGCTGTACCTTCCGGTTGGAAC        -121
TGCCAAGGGTGCCCAGTGGCCCAGCCTAAGCTCGTGTCCTCTCTTTTCAGGTGATGCCCT              -61
TTCTCTGAGGAAAACCTGTGACTGGAGGAAGGGGCTGCGCTTCATTCCCACCCGGCTGCC              -1

ATGAACACCTCAGTGCCCCCTGCTGTCAGTCCCAACATCACTGTCTTGGCACCAGGAAAG             60
M   N   T   S   V   P   P   A   V   S   P   N   I   T   V   L   A   P   G   K
GGTCCCTGGCAGGTGGCCTTCATCGGGATCACCACAGGCCTCCTGTCTCTAGCTACAGTG             120
G   P   W   Q   V   A   F   I   G   I   T   T   G   L   L   S   L   A   T   V
ACAGGCAACCTACTGGTACTCATCTCCTTCAAGGTCAACACCGAGCTCAAGACAGTCAAC             180
T   G   N   L   L   V   L   I   S   F   K   V   N   T   E   L   K   T   V   N
AACTACTTCCTGCTGAGCCTGGCCTGTGCTGACCTCATCATTGGCACCTTCTCCATGAAC             240
N   Y   F   L   L   S   L   A   C   A   D   L   I   I   G   T   F   S   M   N
CTCTATACCACGTACCTGCTCATGGGCCACTGGGCTCTGGGCACACTGGCCTGTGACCTC             300
L   Y   T   T   Y   L   L   M   G   H   W   A   L   G   T   L   A   C   D   L
TGGCTGGCCCTGGACTATGTGGCCAGCAACGCCTCTGTCATGAATCTTCTGCTCATCAGC             360
W   L   A   L   D   Y   V   A   S   N   A   S   V   M   N   L   L   L   I   S
TTTGACCGTTACTTCTCGGTGACCCGACCCCTGAGCTACCGAGCCAAGCGCACTCCCCGA             420
F   D   R   Y   F   S   V   T   R   P   L   S   Y   R   A   K   R   T   P   R
AGGGCAGCTCTGATGATTGGCCTAGCATGGCTGGTTTCCTTCGTTCTCTGGGCCCCAGCC             480
R   A   A   L   M   I   G   L   A   W   L   V   S   F   V   L   W   A   P   A
ATCCTCTTCTGGCAATACCTAGTTGGGGAGCGGACAGTGCTGGCTGGGCAGTGCTACATC             540
I   L   F   W   Q   Y   L   V   G   E   R   T   V   L   A   G   Q   C   Y   I
CAGTTCCTCTCCCAACCCATCATCACTTTTGGCACAGCCATGGCCGCCTTCTACCTCCCT             600
Q   F   L   S   Q   P   I   I   T   F   G   T   A   M   A   A   F   Y   L   P
GTCACGGTCATGTGTACACTGTACTGGCGCATCTACCGGGAGACAGAAAACCGAGCCCGG             660
V   T   V   M   C   T   L   Y   W   R   I   Y   R   E   T   E   N   R   A   R
GAGCTGGCCGCCCTGCAGGGCTCTGAGACACCAGGCAAAGGTGGTGGCAGCAGCAGCAGC             720
E   L   A   A   L   Q   G   S   E   T   P   G   K   G   G   G   S   S   S   S
TCAGAGAGGTCACAGCCAGGGGCTGAAGGCTCACCCGAGTCGCCTCCAGGCCGCTGCTGC             780
S   E   R   S   Q   P   G   A   E   G   S   P   E   S   P   P   G   R   C   C
CGCTGTTGCCGGGCACCCAGGCTCCTGCAGGCCTACAGCTGGAAGGAAGAAGAAGAGGAG             840
R   C   C   R   A   P   R   L   L   Q   A   Y   S   W   K   E   E   E   E
GATGAAGGCTCCATGGAGTCCCTCACATCCTCCGAAGGTGAGGAGCCTGGCTCAGAAGTG             900
D   E   G   S   M   E   S   L   T   S   S   E   G   E   E   P   G   S   E   V
GTGATCAAGATGCCCATGGTAGATTCTGAAGCACAGGCACCCACCAAGCAGCCTCCCAAA             960
V   I   K   M   P   M   V   D   S   E   A   Q   A   P   T   K   Q   P   P   K
AGCTCCCCAAATACAGTCAAGAGGCCCACCAAGAAAGGCCGAGACCGAGGCGGCAAGGGC             1020
S   S   P   N   T   V   K   R   P   T   K   K   G   R   D   R   G   G   K   G
CAAAAACCCGAGGGAAGGAAGCAGCTGGCCAAGAGAAAGACCTTCTCACTGGTCAAGGAG             1080
Q   K   P   R   G   K   E   Q   L   A   K   R   K   T   F   S   L   V   K   E
AAGAAGGCAGCTCGGACCCTGAGTGCCATCCTGCTGGCCTTCATCCTCACCTGGACACCA             1140
K   K   A   A   R   T   L   S   A   I   L   L   A   F   I   L   T   W   T   P
TATAACATCATGGTGCTGGTATCTACCTTCTGCAAGGACTGTGTTCCTGAAACCCTGTGG             1200
Y   N   I   M   V   L   V   S   T   F   C   K   D   C   V   P   E   T   L   W
GAGCTGGGCTACTGGCTATGCTACGTCAACAGCACTGTCAACCCCATGTGCTATGCACTG             1260
E   L   G   Y   W   L   C   Y   V   N   S   T   V   N   P   M   C   Y   A   L
TGCAACAAAGCCTTCCGGGACACGTTCCGCCTGCTGCTGCTCTGCCGCTGGGACAAGAGG             1320
C   N   K   A   F   R   D   T   F   R   L   L   L   L   C   R   W   D   K   R
CGGCTGGCGCAAGATCCCCAAGCGCCCTGGCTCTGTGCACCGCACCCCCTCCCGCCAATGC             1380
R   W   R   K   I   P   K   R   P   G   S   V   H   R   T   P   S   R   Q   C

TAACGGCCCCCTTCTCCTGCATCCTTCCACCCCAGCTCCAGGGAGAGGCCGGTGGAAAGT             1440
GTCCCGGAAAACGGGCTGCATCTTCAGCCCCAGAGCCCTGCTCAGGACTCCCCTGGCTTC             1500
CCAGGCCCCTGGGTCACCTTCCTGGACAGCCCAGAGAGACTCTACTAGCTTTCCAAACTT             1560
TGCTATTCCCAGACGGGGAATGAAACCCGGGGAACTGGTTTTCTGATCCCCGCTGTGTGG             1620
GAATGGGCCCAGCAGGAGGATCC                                                  1680
```

Fig 2. Restriction map and partial sequence of rat genomic
clone c71. Arrows indicate direction and positions of
sequencing reactions from which the nucleotide sequence of the
coding region and part of the 5' and 3'-untranslated region of
the m_1 gene were determined. The deduced amino acid sequence
of the coding region is shown below the nucleotide sequence.

receptor has been characterized previously to be of the M_1 type (Mei et al., 1989a). Of fourteen randomly selected clones, three did not show any significant [³H](-)MQNB binding, four exhibited binding of <20 fmol/10⁶ cells, and seven showed significant binding in a range of 20-264 fmol/10⁶ cells. One of these clones, LK3-3, had a K_d value of 160 (140-180) pM and a B_{max} value of 240±15 fmol/10⁶ cells for [³H](-)MQNB and was one of the two clones which expressed the highest muscarinic receptor densities from the transfection experiment. In agreement with the previous studies, the receptor expressed in LK3-3 exhibited high affinity for atropine and PZ, with K_i values of 0.51(0.28-0.74) nM and 13(5.1-28) nM respectively, and low affinity for AF-DX 116, with a K_i value of 540(410-720) nM. The receptor densities expressed in these transfected cells remain stable to date, representing over a year of continual cell passage. Non-transfected B82 cells, or those transfected with pHβAPr-1-neo alone had no specific [³H](-)MQNB binding.

The molecular weight of the M_1 muscarinic receptor in transfected B82 cells. Covalent labeling of cell membranes from both cTB10 and LK3-3 cells with [³H]PrBCM ([³H](-)propylbenzilycholine mustard) showed that the ligand labeled predominantly a protein which migrated with an apparent molecular weight of 79,000±1,000 daltons on sodium dodecyl sulfate (SDS) polyacrylamide gels (figs. 3A, 4). The binding of [³H]PrBCM to this protein could be completely blocked by 10 μM atropine sulfate which showed that [³H]PrBCM labeled specifically the muscarinic receptors expressed in these cells. Non-transfected B82 cells, on the other hand, did not contain any protein which could be specifically labeled by [³H]PrBCM (fig. 3B). Furthermore, the [³H]PrBCM labeled, 80,000 daltons protein in transfected cells was converted to significantly lower molecular weight species, predominantly 53,000 daltons after endoglycosidase F treatment (fig. 4).

Determination of the isoelectric point of the M_1 muscarinic receptor in the transfected B82 cells. Preliminary analysis by isoelectric focusing of the [³H]PrBCM labeled, digitonin solubilized M_1 receptor from transfected B82 cells (cTB10) showed that the [³H]PrBCM labeled proteins resided within pH 4-6 along a pH 3.5-9.5 gradient (data not shown). The labeled membrane proteins were therefore subjected to electrophoresis along a pH 4-6 gradient (fig. 5); analysis of proteins from both the cTB10 and the LK3-3 cells showed that the [³H]PrBCM labeled receptor had an isoelectric point of 5.0±0.05. The focused profile also includes several minor peaks at pH 4.2, 5.7 and 5.9, which probably correspond to the minor aggregate or degraded products of the [³H]PrBCM labeled receptor observed in SDS-PAGE analysis.

Carbachol stimulated an increase in cytosolic free calcium in transfected B82 cells. [Ca^{2+}]$_i$ (cytosolic free calcium concentration) in the LK3-3 cells was measured by its interaction with fluo-3, a fluorescein-like fluorophore which contains the 8-coordinate tetracarboxylate chelating site of BAPTA (Minta et al., 1989). The dye was made permeable to the cell membrane by generating its acetoxymethyl ester derivative, fluo-3/AM, which was used to load the LK3-3 cells. The active component, fluo-3, was regenerated by the endogenous esterase activity of the cells during loading. Optimal loading time for fluo-3/AM was determined experimentally to be 40 min, after which the basal level of fluorescence in the loaded cells remained stable for at least 60 min, when measurements were conducted. Fluo-3 was efficiently loaded and hydrolyzed in the LK3-3 cells as evidenced by a greater than 20 fold increase in fluorescence of these cells after

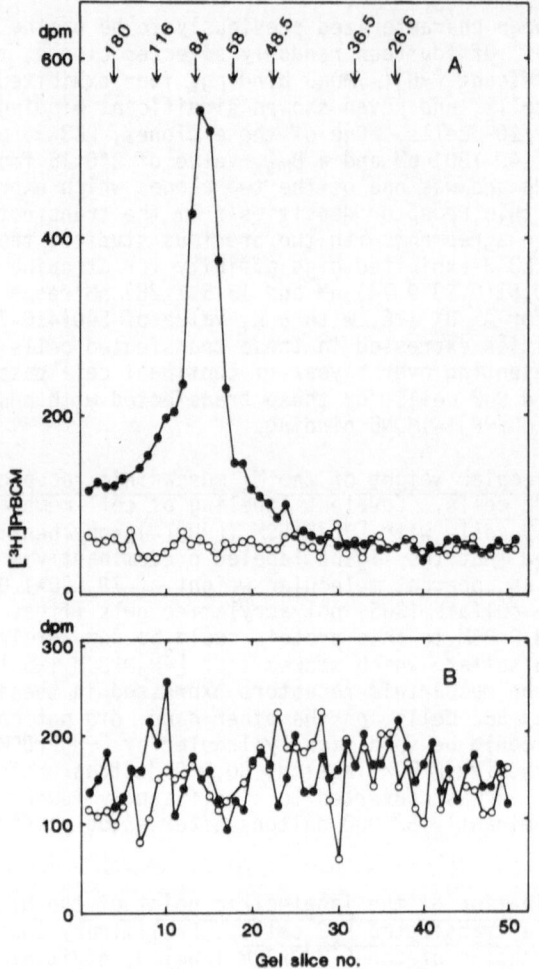

Fig. 3 SDS-PAGE (SDS-polyacrylamide gel electrophoresis)
analysis of [^{3}H]PrBCM labeled cell membranes from B82 cells
transfected with the m_1 gene (A) or from non-transfected B82
cells (B). Membrane preparations were resuspended at 0.2 mg
protein/ml in 10 mM sodium phosphate, 1 mM EDTA, 0.1 mM
phenylmethylsulfonyl fluoride (PMSF) , pH 7.4 and covalently
labeled PrBCM (Burgen et al., 1974) at 1 nM by incubation at
30°C for 30 min. The membranes were recovered by centrifugation
and rinsed with an equal volume of the buffer above. The
membrane proteins were then solubilized by heating at 90°C for 3
min in 2 percent SDS, 1 percent β-mercaptoethanol, 8 M urea, 65
mM Tris, pH 6.8 and analyzed by SDS-PAGE according to Laemmli
(1977). Membrane proteins (0.2 mg) from LK3-3 (A) and 2 mg of
membrane proteins from non-transfected B82 cells (B) were
labeled in the absence (●) or in the presence (○) of 10 μM
atropine sulfate prior to electrophoresis. Positions of the
molecular weight markers are as indicated (in kDa). Profile in
(A) is representative of 4 experiments, using either 2 mg of
membranes from cTB10 or 0.2 mg of membranes from LK3-3.

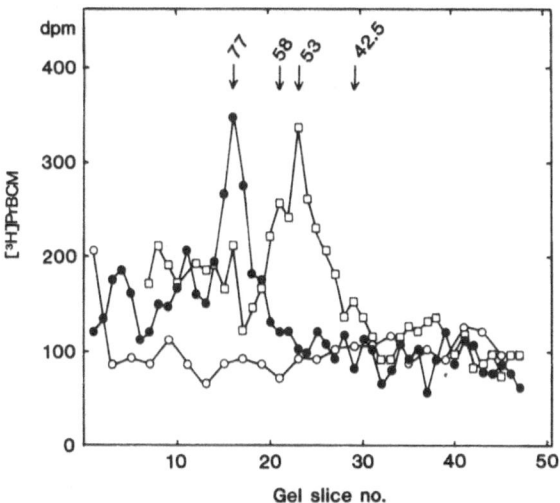

Fig. 4 Endoglycosidase F treatment of [³H]PrBCM
labeled membrane proteins in transfected B82 cells.
Cell membranes from cTB10 cells (2 mg/sample) were
labeled with 1 nM [³H]PrBCM in the absence (●,□)
or in the presence (○) of 10 μM atropine sulfate
and solubilized with 0.5 percent digitonin. These
proteins were either (□) treated with
endoglycosidase F (0.3 u/ml of endoglycosidase F in
0.5 percent digitonin, 0.2 percent SDS, 1 percent
β-mercaptoethanol, 100 mM EDTA, 0.1 mM PMSF, 100 mM
sodium acetate, pH 5.2 at 37°C for 14-16 h with
gentle agitation) or (●) untreated control. The
proteins were analyzed by SDS-PAGE after heat
denaturation (90°C, 3 min) and carboxymethylation
(Dadi and Morris, 1984). The calibrated molecular
weights of the major peaks are shown in kDa.

loading and the 6-10 fold enhancement of fluorescence in response to
calcium. A single measurement of agonist stimulation was made from
each aliquot of 3×10^6 cells, since rapid desensitization occurred
after an initial transient response to the agonist. Typically, the
peak fluorescence observed, usually within 10 to 15 s after the
addition of carbachol to the cell suspension, was taken as the
stimulated value. The fluorescence declined gradually thereafter. The
magnitude of the fluorescence signal upon agonist stimulation appeared
to vary among three different batches of fluo-3. However, the potency
of the agonist and the estimated resting $[Ca^{2+}]_i$ were consistent
among the different dye batches.

The addition of carbachol to a suspension of LK3-3 cells caused a
transient, dose dependent, saturable increase in the cytosolic calcium
concentrations of these cells (fig. 6). The maximum elevation of this
calcium concentration was about 3 fold above basal values, which
averaged 77±2.2 nM (n=96). This transient rise in $[Ca^{2+}]_i$

stimulated by various concentrations of carbachol was expressed as a net increase above their respective resting $[Ca^{2+}]_i$ (stimulated minus basal values). The EC_{50} value of carbachol was 2.8 (2.0-3.1) μM and its E_{max} value was 176 nM $[Ca^{2+}]_i$. The Hill coefficient (n_H) of this saturation isotherm was 1.1. Carbachol had no effect on the $[Ca^{2+}]_i$ in non-transfected B82 cells: $[Ca^{2+}]_i$ was 92.5±9.5 nM in the absence or presence of 100 μM carbachol. A $[Ca^{2+}]_i$ flux was only barely detectable in the cTB10 cells (a net increase in $[Ca^{2+}]_i$ of only 6 percent of that observed in the LK3-3 cells when stimulated with 1 mM carbachol), probably due to the much lower density (19 fmol/10^6 cells compared to 240 fmol/10^6 cells for LK3-3) of muscarinic receptors expressed in this cell line.

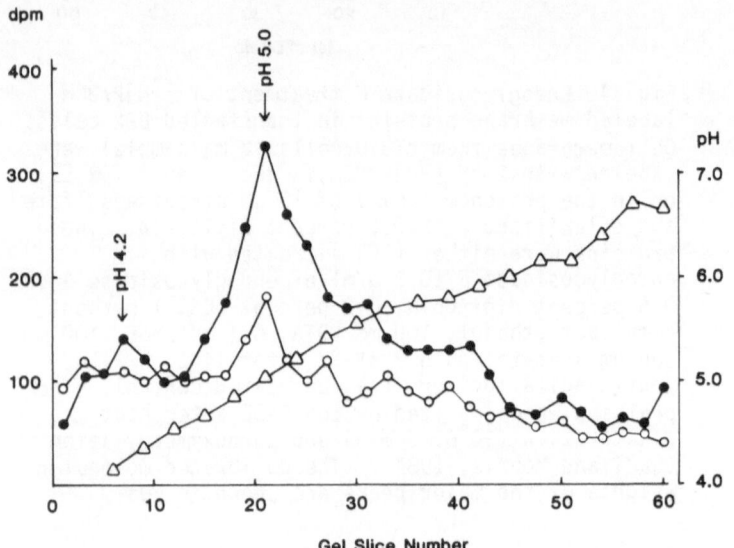

Fig. 5 Isoelectric focusing analysis of solubilized [^{3}H]PrBCM labeled membrane proteins from transfected B82 cells. LK3-3 cell membranes (1 mg) were labeled with 1 nM [^{3}H]PrBCM in the absence (●) or in the presence (O) of 10 μM atropine sulfate prior to digitonin solubilization and electrophoresis. ($\triangle$), pH gradient established along the agarose gel after electrophoresis at 7 watts for 2 h at 5°C. The profile is representative of 5 experiments. 2 mg of cTB10 membranes were used to achieve similar magnitude of labeling. Total [^{3}H]PrBCM detected varied with the efficiency of digitonin solubilization.

Sensitivity of carbachol mediated rise in $[Ca^{2+}]_i$ to external calcium and muscarinic antagonists. Elimination of calcium from the external medium substantially reduced the $[Ca^{2+}]_i$ response to carbachol. If calcium was omitted from the external medium immediately prior to measurement, subsequent stimulation by maximal effective doses

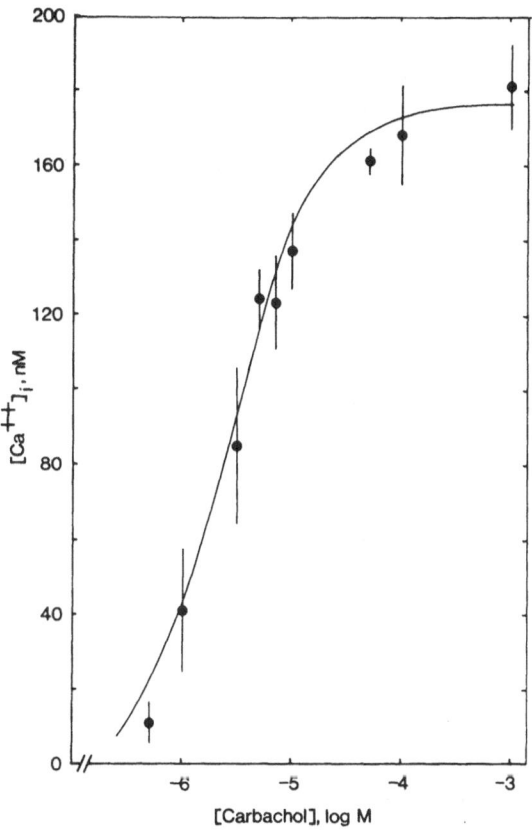

Fig. 6 Carbachol mediated increase in $[Ca^{2+}]_i$
above resting values in LK3-3 cells. Transfected B82
cells cultured to confluency were suspended at 4 x
10^6 cells/ml in buffer A (5 mM HEPES, pH 7.4,
containing NaCl, 144 mM; KCl, 5 mM; MgSO$_4$, 1.2 mM and
glucose, 5 mM) supplemented with 0.2 mM CaCl$_2$ and 10
µM of fluo-3/AM, cell permeant form (Molecular Probes,
Eugene, OR) and incubated with constant shaking at 37°C
for 40 min. Aliquots of the cell suspension ($2x10^6$
cells) loaded with fluo-3/AM were washed two times with
buffer A supplemented with 1 mM CaCl$_2$ and resuspended
in 3 ml of the same buffer immediately prior to
fluorescence determination. Fluorescence of fluo-3 in
the cells was monitored in a T-format series 300
Spectrofluorometer (H and L Instruments, Burlingame,
CA) equipped with a stirring device and an
electronically thermostated cuvette chamber. Data
processing was achieved with an on-line IBM
microcomputer. Samples were excited at 506 nm and the
spectral emission of the dye was monitored at 532 nm
with a narrow band pass filter (Microcoatings, Inc.,
Westford, MA). Resting $[Ca^{2+}]_i$ levels were
measured at 37°C. Stimulation of $[Ca^{2+}]_i$ by the
muscarinic agonist, carbachol, was monitored
continuously for 10 s after its addition. Basal and
stimulated $[Ca^{2+}]_i$ were calculated after correction
for autofluorescence as described previously for

(continued)

(100 μM and 1mM) of carbachol yielded a rise in $[Ca^{2+}]_i$ of only
51±9 nM (n=3) compared to 148±13 nM (n=5) in a medium containing 1 mM
$CaCl_2$; the resting $[Ca^{2+}]_i$ observed in the absence of $[Ca^{2+}]_o$
(external calcium concentration) was within the normal range (85±9
nM). If this calcium free medium was supplemented with 20 μM EGTA, the
basal $[Ca^{2+}]_i$ levels in these cells were reduced to 30 nM, and the
addition of 100 μM carbachol increased $[Ca^{2+}]_i$ by 30 nM above
basal. Hence, removal of $[Ca^{2+}]_o$ diminished the carbachol induced
rise in $[Ca^{2+}]_i$, however, the ratios of carbachol stimulated
$[Ca^{2+}]_i$ to basal $[Ca^{2+}]_i$ in a calcium-free medium show that the
stimulated $[Ca^{2+}]_i$ in LK3-3 cells is still 1.6 fold above basal
levels, compared to 3 fold in a medium containing 1 mM $CaCl_2$. These
data indicate that the magnitude of the carbachol mediated rise in
$[Ca^{2+}]_i$ in LK3-3 cells is very sensitive to the presence of calcium
in the extracellular medium. A transient removal of extracellular
calcium appeared to inhibit a major source of calcium influx into the
cytosol without abolishing the ability of carbachol to elicit a rise in
$[Ca^{2+}]_i$. Moreover, the addition of 50 mM KC1 to the extracellular
medium to depolarize the LK3-3 cells did not have any effect on the
$[Ca^{2+}]_i$ in these cells. This suggests that the LK3-3 cells do not
contain voltage sensitive calcium channels.

Table 1 illustrates inhibitory activities of atropine, PZ and AF-DX
116 on the carbachol induced rise in $[Ca^{2+}]_i$ in LK3-3 cells. The
functional inhibition constants for atropine (0.17 nM), PZ (7.3 nM) and
AF-DX 116 (397 nM) are in the same order of magnitude and rank order of
potency as their K_i values in inhibiting $[^3H](-)MQNB$ binding to the
LK3-3 cells.

Sensitivity of carbachol mediated rise in $[Ca^{2+}]_i$ to pertussis
toxin. This M_1 receptor mediated rise in $[Ca^{2+}]_i$ was also found
to be susceptible to inhibition by pertussis toxin (fig. 7).
Preincubation of the LK3-3 cells with various concentrations of
pertussis toxin 22-24 h prior to stimulating the cells with 100 μM
carbachol resulted in a dose dependent diminution of the transient
increase in $[Ca^{2+}]_i$. This inhibitory effect of pertussis toxin was
apparent at 10 ng/ml and 100 ng/ml, which resulted in a 45±7.5 (P<0.05)
percent and 69±8 (P<0.05) percent inhibition of the rise in
$[Ca^{2+}]_i$, respectively. The toxin did not affect the growth rate,
the morphology or the cells' ability to accumulate fluo-3. The effect
of pertussis toxin on this M_1 receptor mediated response suggests a
role of a pertussis toxin sensitive G protein in this signal
transduction pathway.

Fig. 6 quin-2, using a K_d for fluo-3 of 450 nM (Tsien et
al., 1982). A separate calibration was performed on
each sample essentially as described previously
(Komulainen and Bondy, 1987). Thus, maximal
fluorescence of fluo-3(F_{max}) was determined in the
presence of 0.1 percent SDS and a saturating
concentration of calcium. Minimal fluorescence of
fluo-3 (F_{min}) was determined in the presence of
excess EGTA. Total fluorescence was corrected for both
autofluorescence and dye leakage prior to calculation
of $[Ca^{2+}]_i$ values. The EC_{50} value for carbachol
is 2.8 (2.0-3.1) μM and the Hill coefficient was 1.1.
Data points shown are mean values from three
experiments ± S.E.M. A non-linear least squares
regression analysis gives the line of best fit where
the maximum is 176 nM.

Table 1. The effects of atropine, PZ and AF-DX 116 on carbachol mediated increase in $[Ca^{2+}]_i$ in LK3-3 cells.

Drug	Percent of maximal response[a]
1 mM carbachol	100
+ 10 nM atropine	71.3±7.0
+ 0.1 μM PZ	81.0±5.4
+ 10 μM AF-DX 116	83.6±13.3
0.1 mM carbachol	96±10.8
+ 10 nM atropine	24.0±3.8
+ 0.1 μM PZ	54.2±1.0
+ 10 μM AF-DX 116	39.7±4.0
0.01 mM carbachol	62±1.0
+ 10 nM atropine	10.7±2.9
+ 0.1 μM PZ	16.0±5.9
+ 10 μM AF-DX 116	8.3±0.6

[a]The maximum increase in $[Ca^{2+}]_i$ stimulated by 1 mM carbachol is taken as 100 percent. The rise in $[Ca^{2+}]_i$ at other concentrations of carbachol in the absence or presence of antagonists (added 2-3 min prior to stimulation with carbachol) is expressed as a percentage of the value obtained with 1 mM carbachol. Data points are collected from 4 independent experiments ± S.E.M. These data were analyzed by the method of Tallarida and Murray (1987) using a non-linear least squares regression program and assuming a Hill coefficient of 1.0. The EC_{50} value of carbachol was 6 μM in the absence of antagonist, 350 μM in the presence of 10 nM atropine sulfate, 88 μM in the presence of 0.1 μM PZ and 157 μM in the presence of 10 μM AF-DX 116. The functional inhibition constants for the antagonists were converted from the EC_{50} values by the equation: $[A]/(K_b/K_a-1)$ where [A] is the concentration of the antagonist, K_b is the EC_{50} value of the agonist in the presence of the antagonist and K_a is the EC_{50} value of the agonist in the absence of the antagonist. The functional inhibition constants for atropine, PZ and AF-DX 116 obtained from this analysis were 0.2 nM, 7.3 nM and 397 nM, respectively.

DISCUSSION

The present study demonstrates several salient features of the M_1 muscarinic receptor whose biosynthesis is propagated in a eukaryotic in vitro expression system. Firstly, the stable expression of the receptor is established through a random integration of the recombinant m_1 expression vector into the host cell genome. Secondly, the receptor polypeptide translated from the integrated m_1 undergoes similar post-translational modifications as does the native receptor in vivo. Thirdly, the receptor mediated a rise in the cytosolic calcium concentration through a pertussis toxin sensitive pathway upon agonist activation.

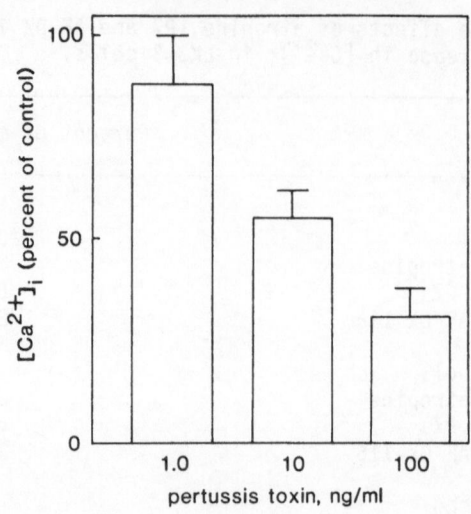

Fig. 7 The effect of pertussis toxin on the
carbachol mediated increase in $[Ca^{2+}]_i$ above the
resting values in the LK3-3 cells. The cells were
preincubated for 22-24 h in IMDM containing various
concentrations of pertussis toxin (List Biological
Laboratories, CA) prior to dye loading and
stimulation with a maximal effective dose of 100 µM
carbachol. The duration of exposure to pertussis
toxin was based on previous studies (Mei et al.,
1989a). Basal and stimulated $[Ca^{2+}]_i$ were
calculated after correction for the maximal increase
in $[Ca^{2+}]_i$ in control LK3-3 cells in the
presence of 100 µM carbachol is taken as 100
percent. Data represent the arithmetic means of
three measurements ± S.E.M.

The in vitro expression of the M_1 muscarinic receptor was
accomplished using a genomic fragment of the m_1 gene. The
constitutive expression of the m_1 gene in the B82 cells in this study
was achieved by a recombination of the coding region of the m_1 gene
and the human β-actin promoter which was then transfected into B82
cells. Neomycin resistant, transfected clonal B82 cells exhibited a
range of maximal binding values for $[^3H](-)$MQNB due to a random
integration of the vector into the host genome, as observed previously
for the β-actin promoter (Gunning et al., 1987). Stable expression of
the receptor encoded by the m_1 in these clonal cell lines exhibited
consistent pharmacological and biochemical properties in the
transfected B82 cells. Specific binding of $[^3H](-)$MQNB was only
observed in cells transfected with the m_1 gene. The specific binding
was accompanied by a comparable, novel expression of an 80,000 daltons
polypeptide s ifically labeled by the muscarinic ligand, $[^3H]$PrBCM
(fig. 3A). Deglycosylation of this labeled polypeptide yielded a
smaller polypeptide (53,000 daltons) (fig. 4) whose molecular mass was
similar to the calculated molecular mass (51,352 daltons) from the

deduced amino acid sequence encoded by the m_1 gene. A similar
molecular mass for the native muscarinic receptor has been observed in
brain (Haga, 1980; Venter et al., 1984; Dadi and Morris, 1984; Haga and
Haga, 1985). The apparent molecular weight of this glycosylated
protein labeled by [^{3}H]PrBCM in transfected B82 cells indicates that
the biosynthesis and post-translational modifications of the muscarinic
receptor expressed in these cells are similar to that of native
receptors in vivo. Taking into account a 10-20 percent over-estimation
of the size of the receptor polypeptide by SDS-PAGE due to its unusual
charge and shape as previously suggested (Peterson et al., 1986), the
glycosylation of this muscarinic receptor polypeptide accounts for
26-30 percent of its molecular mass, similar to that estimated for the
predominant species of muscarinic receptors in the brain (Curtis et
al., 1989).

The surface charge characteristic of this muscarinic receptor
polypeptide after its solubilization with the non-ionic detergent,
digitonin, was observed to have an isoelectric point (pI) of 5.0±0.05
(fig. 5). The pI for this muscarinic receptor deviates significantly
from that reported for the muscarinic receptor in rat brain (pI=4.2 in
digitonin) (Venter et al., 1984). This difference may be due to the
heterogeneous nature of the muscarinic receptors in rat brain, well
established by ligand binding analysis (Watson et al., 1986) and the
detection of m_1, m_3 and m_4 transcripts (Brann et al., 1988).
Subtle differences in the charge properties of the various receptor
polypeptides within 1 pH unit may not be adequately resolved, thereby
resulting in a broad peak which gives a different estimate of the
predominant pI.

We have shown that the M_1 muscarinic receptor expressed in these
transfected clonal cell lines is coupled to the hydrolysis of
polyphosphoinositides (PPI) (Lai et al., 1988; Mei et al., 1989a, b).
Activation of the receptor by the muscarinic agonist, carbachol, also
led to a transient elevation of intracellular free calcium
concentration as demonstrated in this study (fig. 6). That this
cellular response could be inhibited by PZ and AF-DX 116 with the same
order of potency as their binding affinities for the receptor confirmed
that this response was mediated by the expressed M_1 receptor.
Carbachol exhibited similar EC_{50} values for the stimulation of PPI
hydrolysis in the LK3-3 cells (4.0 μM, Mei et al., 1989b) and the rise
in cytosolic calcium (2.8 μM) (fig. 6). Furthermore, both second
messenger pathways could be significantly inhibited by pertussis toxin
at a concentration of 10 ng/ml (fig. 7, Mei et al., 1989a). These
results imply that the two signal transduction pathways activated by
carbachol were closely linked events, possibly mediated by the same
pertussis toxin sensitive G protein upon muscarinic receptor
activation. The relationship between inositol lipid hydrolysis and
calcium mobilization in many tissues is well documented (Berridge and
Irvine, 1984). The intermediate product of PPI hydrolysis, inositol
trisphosphate (1,4,5 IP_3), produced by phospholipase C activity,
stimulates the release of Ca^{2+} from intracellular stores by its
interaction with a specific receptor in the endoplasmic reticulum (Spat
et al., 1986; Ehrlich and Watras, 1988). This signal transduction
mechanism has been demonstrated to mediate muscarinic cholinergic
function in a variety of systems, such as calcium dependent secretory
activity in PC12 cells (Inoue and Kenimer, 1988), adrenal medulla
(Malhotra et al., 1988), parotid acinar cells (Merritt and Rink, 1987),
pancreatic B-cells (Henquin et al., 1988; Sanchez-Andres et al., 1988;

Wolf et al., 1988) or neurotransmitter release in synaptosomes (Audigier et al., 1988) and cortical slices (Diamant et al., 1988). A direct demonstration of inositol phosphate stimulated calcium flux by muscarinic cholinergic activation was recently reported on the human neuroblastoma SK-N-SH cells (Fisher et al., 1989).

Our data show that this receptor-regulated rise in $[Ca^{2+}]_i$ in transfected B82 cells is sensitive to the presence of external calcium, the elimination of which attenuates the magnitude of this response. This study rules out the possibility of calcium influx via voltage dependent calcium channels since they are absent in the B82 cells. This carbachol stimulated increase in cytosolic calcium must be in part due to a mobilization of internal calcium stores, indicated by the significant rise in $[Ca^{2+}]_i$ even in the absence of $[Ca^{2+}]_o$. According to the model proposed by Putney (1986) release of internal calcium could be activated by 1,4,5 IP$_3$ production upon carbachol stimulation accompanied by a simultaneous influx of external calcium via the endoplasmic reticulum to replenish the internal calcium stores. This model could account for the diminished cytosolic calcium accumulation in the absence of external calcium when the intracellular calcium stores become limiting. On the other hand, it may also be speculated that the sensitivity of this carbachol induced rise in $[Ca^{2+}]_i$ to $[Ca^{2+}]_o$ could be contributed by a cell surface muscarinic receptor-mediated calcium conductance. The existence of such a 'receptor-operated' calcium channel regulated by the transfected muscarinic receptors remains to be established. It should be noted that this carbachol mediated $[Ca^{2+}]_i$ response in these transfected B82 cells differ from the NG108-15 cells which were used to express the porcine m$_1$ gene (mAChR I): the increase of $[Ca^{2+}]_i$ upon the activation of the M$_1$ receptors in these transfected NG108-15 cells was independent of extracellular calcium (Neher et al., 1988). A similar absence of extracellular calcium requirement was also observed for the mAChR I activated calcium dependent chloride conductance using Xenopus oocytes as the host (Fukuda et al., 1987). Unlike our system, the G proteins which mediated the mAChR I activation of PPI hydrolysis and a calcium dependent potassium conductance in transfected NG108-15 cells were insensitive to pertussis toxin (100 ng/ml) (Fukuda et al., 1988). These differences are probably attributed to the endogenous proteins characteristic of the host systems. Despite these differences, however, the M$_1$ muscarinic receptor expressed from the m$_1$ gene exhibits comparable ligand binding properties as well as signal transduction processes in oocytes, fibroblast and neuronal cell lines. The fact that the M$_1$ receptor can couple to PPI hydrolysis via different G proteins, as suggested by their different sensitivities to pertussis toxin, implies that this M$_1$ receptor-second messenger coupling can be mediated by more than one type of G proteins.

The eukaryotic in vitro expression systems such as that described for the M$_1$ receptor in this communication provide reliable sources of functional muscarinic receptor types defined by their primary structures. The M$_1$ receptor mediated regulation of $[Ca^{2+}]_i$ in this expression system provides important physiological implications for the M$_1$ receptor in the context of the variety of muscarinic cholinergic activities in various tissues. Any unique function of a particular receptor type is probably not entirely attributable to its intrinsic molecular identity but may also be determined by its tissue localization and tissue specific factors (Wang et al., 1987; Berstein et al., 1989). On the other hand, similarities in the biological activities between two or more receptor polypeptides, in this case ligand binding affinities and localization of the m$_1$ and m$_3$, and effector coupling between the m$_1$, m$_3$ and m$_5$ gene products, may also imply functional convergence between receptor types. It remains a challenge, therefore, to establish a correlation between the structural and physiological heterogeneity of the class of muscarinic receptors. In vitro expression systems developed for this purpose should prove to be highly adaptable vehicles for the structural and functional analysis of signal transduction mechanisms.

Acknowledgement

The authors would like to thank Mrs. Elizabeth Hamlin for technical assistance, Mrs. Carol A. Haussler and Mr. Alfred C. Gallegos for their assistance in cell culture and Ms. Pam Davis for the preparation of this manuscript.

This work was supported in part by U.S. Public Health Service Grants DK-36289 and HL-20984, a research grant from the Department of Veteran Affairs, a grant from the American Heart Association, Arizona Affiliate and a grant from the Arizona Disease Research Commission.

References

Akiba, I., Kubo, T., Maeda, A., Bujo, H., Nakai, J., Mishina, M. and Numa, S., 1988, Primary structure of porcine muscarinic acetylcholine receptor III and antagonist binding studies. FEBS lett. 235:257-261.

Ashkenazi, A., Winslow, J.W., Peralta, E.G., Peterson, G.L., Schimerlik, M.I., Capon, D.J. and Ramachandran, J., 1987, An M_2 muscarinic receptor subtype coupled to both adenylyl cyclase and phosphoinositide turnover. Science 238:672-675.

Audigier, S.M.P., Wang, J.K.T. and Greengard, P., 1988, Membrane depolarization and carbamoylcholine stimulate phosphatidylinositol turnover in intact nerve terminals. Proc. Natl. Acad. Sci. 85:2859-2863.

Berridge, M.J. and Irvine, R.F., 1984, Inositol phosphate, a novel second messenger in cellular signal transduction. Nature 312:315-321.

Berstein, G., Haga, T. and Ichiyama, A., 1989, Effect of the lipid environment on the differenial affinity of purified cerebral and atrial muscarinic acetylcholine receptors for pirenzepine. Mol. Pharmacol. 36:601-607.

Bonner, T.I., Buckley, N.J., Young, A.C. and Brann, M.R., 1987, Identification of a family of muscarinic acetylcholine receptor genes. Science 237:527-532.

Bonner, T.I., Young, A.C., Brann, M.R. and Buckley, N.J., 1988, Cloning and expression of the human and rat m_5 muscarinic acetylcholine receptor genes. Neuron 1:403-410.

Brann, M.R., Buckley, N.J. and Bonner, T.I., 1988, The striatum and cerebral cortex express different muscarinic receptor mRNAs. FEBS lett. 230:90-94.

Buckley, N.J., Bonner, T.I., Buckley, C.M. and Brann, M.R., 1989, Antagonist binding properties of five cloned muscarinic receptors expressed in CHO-K1 cells. Mol. Pharmacol. 35:469-476.

Bujo, H., Nakai, J., Kubo, T., Fukuda, K., Akiba, I., Maeda, A., Mishina, M. and Numa, S., 1988, Different sensitivities to agonist of muscarinic acetylcholine receptor subtypes. FEBS lett. 240:95-100.

Burgen, A.S.V., Hiley, C.R. and Young, J.M., 1974, The properties of muscarinic receptors in mammalian cerebral cortex. Br. J. Pharmacol. 51:279-285.

Conklin, B.R., Brann, M.R., Buckley, N.J., Ma, A.L., Bonner, T.I. and
    Axelrod, J., 1988, Stimulation of arachidonic acid release and
    inhibition of mitogenesis by cloned genes for muscarinic receptor
    subtypes stably expressed in A9L cells. Proc. Natl. Acad. Sci.
    (U.S.A.) 85:8698-8702.
Curtis, C.A.M., Wheatley, M., Bansal, S., Birdsall, N.J.M., Eveleigh,
    P., Pedder, E.K., Poyner, D. and Hulme, E.C., 1989,
    Propylbenzilylcholine mustard labels an acidic residue in
    transmembrane helix 3 of the muscarinic receptor. J. Biol. Chem.
    264:489-495.
Dadi, H.K. and Morris, R.J., 1984, Muscarinic cholinergic receptor of
    rat brain: factors influencing migration in electrophoresis and
    gel filtration in sodium dodecyl sulphate. Eur. J. Biochem.
    144:617-628.
Diamant, S., Lev-Ari, I., Uzielli, I. and Atlas, D., 1988, Muscarinic
    agonists evoke neurotransmitter release: possible roles for
    phosphatidyl inositol bisphosphate breakdown products in
    neuromodulation. J. Neurochem. 51:795-802.
Earle, W.R., 1943, Production of malignancy in vitro; mouse fibroblast
    cultures and changes seen in living cells. J. Natl. Cancer Inst.
    4:165-212.
Ehrlich, B.E. and Watras, J., 1988, Inositol 1,4,5-trisphosphate
    activates a channel from smooth muscle sarcoplasmic reticulum.
    Nature 336:583-586.
Fisher, S.K., Domask, L.M. and Roland, R.M., 1989, Muscarinic receptor
    regulation of cytoplasmic Ca^{2+} concentrations in human SK-N-SH
    neuroblastoma cells: Ca^{2+} requirements for phospholipase C
    activation. Mol. Pharmacol. 35:195-204.
Fukuda, K., Kubo, T., Akiba, I., Maeda, A., Mishina, M. and Numa, S.,
    1987, Molecular distinction between muscarinic acetylcholine
    receptor subtypes. Nature 327:623-625.
Fukuda, K., Higashida, H., Kubo, T., Maeda, A., Akiba, I., Bujo, H.,
    Mishina, M. and Numa, S., 1988, Selective coupling with K^{+}
    currents of muscarinic acetylcholine receptor subtypes in NG 108-15
    cells. Nature 335:355-358.
Giachetti, A., Micheletti, R. and Montagna, E., 1986, Cardioselective
    profile of AF-DX 116, a muscarinic M_2 receptor antagonist. Life
    Sci. 38:1663-1672.
Graham, F. and Van Der Eb, A., 1973, A new technique for the assay of
    infectivity of human adenovirus 5 DNA. Virology 52:456-467.
Gunning, P., Leavitt, J., Muscat, G., Ng, S.Y. and Kedes, L., 1987, A
    human β-actin expression vector system directs high-level
    accumulation of antisense transcripts. Proc. Natl. Acad. Sci.
    (U.S.A.), 84:4831-4835.
Haga, T., 1980, Molecular size of muscarinic acetylcholine receptors
    of rat brain. FEBS lett. 113:68-72.
Haga, K. and Haga, T., 1985, Purification of the muscarinic
    acetylcholine receptor from porcine brain. J. Biol. Chem.
    260:7927-7935.
Hammer, R., Giraldo, E., Schiavi, G.B., Monferini, E. and Ladinsky,
    H., 1986, Binding profile of a novel cardioselective muscarinic
    receptor antagonist, AF-DX 116, to membranes of peripheral tissues
    and brain in the rat. Life Sci. 38:1657-1662.
Henquin, J.C., Garcia, M.C., Bozem, M., Hermans, M.P. and Nenquin, M.,
    1988, Muscarinic control of pancreatic B cell function involves
    sodium dependent depolarization and calcium influx. Endocrinology
    122:2134-2142.
Inoue, K. and Kenimer, J.G., 1988, Muscarinic stimulation of calcium
    influx and norepinephrine release in PC12 cells. J. Biol. Chem.
    263:8157-8161.

Jones, S.V.P., Barker, J.L., Buckley, N.J., Bonner, T.I., Collins, R.M. and Brann, M.R., 1988, Electrophysiological characterization of cloned m_1 muscarinic receptors expressed in A9L cells. Mol. Pharmacol. 34:421-426.

Komulainen, H. and Bondy, S.C., 1987, The estimation of free calcium within synaptosomes and mitochondria with fura-2; comparison to quin-2. Neurochem. Int. 10:55-64.

Korc, M., Ackerman, M.S. and Roeske, W.R., 1987, A cholinergic antagonist identifies a subclass of muscarinic receptors in isolated rat pancreatic acini. J. Pharmacol. Exp. Ther. 240:118-122.

Laemmli, U.K., 1977, Cleavage of structural proteins during the assembly of the head of bacteriophage T_4. Nature 227:680-685.

Lai, J., Mei, L., Roeske, W.R., Chung, F.-Z., Yamamura, H.I. and Venter, J.C., 1988, The cloned murine M_1 muscarinic receptor is associated with the hydrolysis of phosphatidylinositols in transfected murine B82 cells. Life Sci. 42:2489-2502.

Liao, C.F., Themmen, A.P.N., Joho, R., Barberis, C., Birnbaumer, M. and Birnbaumer, L., 1989, Molecular cloning and expression of a fifth muscarinic acetylcholine receptor. J. Biol. Chem. 264:7328-7337.

Malhotra, R.K., Wakade, T.D. and Wakade, A.R., 1988, Vasoactive intestinal polypeptide and muscarine mobilize intracellular Ca^{2+} through breakdown of phosphoinositides to induce catecholamine secretion. J. Biol. Chem. 263:2123-2126.

Mei, L., Lai, J., Roeske, W.R., Fraser, C.M., Venter, J.C. and Yamamura, H.I., 1989a, Pharmacological characterization of the M_1 muscarinic receptors expressed in murine fibroblast B82 cells. J. Pharmacol. Exp. Ther. 248:661-670.

Mei, L., Lai, J., Yamamura, H.I. and Roeske, W.R., 1989b, The relationship between agonist states of the M_1 muscarinic receptor and the hydrolysis of inositol lipids in transfected murine fibroblast cells (B82) expressing different receptor densities. J. Pharmacol. Exp. Ther. 251:90-97.

Merritt, J.E. and Rink, T.J., 1987, Regulation of cytosolic free calcium in fura-2-loaded rat parotid acinar cells. J. Biol. Chem. 262:17362-17369.

Minta, A., Kao, J.P. and Tsien, R.Y., 1989, Fluorescent indicators for cytosolic calcium based on rhodamine and fluorescein chromophores. J. Biol. Chem. 264:8171-8178.

Neher, E., Marty, A., Fukuda, K., Kubo, T. and Numa, S., 1988, Intracellular calcium release mediated by two muscarinic receptor subtypes. FEBS lett. 240:88-94.

Peralta, E.G., Ashkenazi, A., Winslow, J.W., Smith, D.H., Ramachandran, J. and Capon, D.J., 1987, Distinct primary structures, ligand-binding properties and tissue specific expression of four human muscarinic acetylcholine receptors. EMBO J. 6:3923-3929.

Peralta, E.G., Ashkenazi, A., Winslow, J.W., Ramachandran, J. and Capon, D.J., 1988, Differential regulation of PI hydrolysis and adenylyl cyclase by muscarinic receptor subtypes. Nature 334:434-437.

Peterson, G.L., Rosenbaum, L.C., Broderick, D.J. and Schimerlik, M.I., 1986, Physical properties of the purified cardiac muscarinic acetylcholine receptor. Biochemistry 25:3189-3202.

Putney, J.W., Jr., 1986, A model for receptor-regulated calcium entry. Cell Calcium 7:1-12.

Sanchez-Andres, J.V., Ripoll, C. and Soria, B., 1988, Evidence that muscarinic potentiation of insulin release is initiated by an early transient calcium entry. FEBS lett. 231:143-147.

Spat, A., Bradford, P.G., McKinney, J.S., Rubin, R.P. and Putney, J.W., Jr., 1986, A saturable receptor for [^{32}P]-inositol-1,4,5-trisphosphate in hepatocytes and neutrophils. Nature 319:514-516.

Tallarida, R.J. and Murray, R.B., 1987, Manual of Pharmacologic Calculations with Computer Programs, 2nd ed., pp. 53-56, Spring-Verlag, N.Y.

Tsien, R.Y., Poggan, T. and Rink, T.J., 1982, Calcium homeostasis in intact lymphocytes: cytoplasmic free calcium monitored with a new, intracellularly trapped fluorescent indicator. J. Cell Biol. 94:325-334.

Venter, J.C., Eddy, B., Hall, L.M. and Fraser, C.M., 1984, Monoclonal antibodies detect the conservation of muscarinic cholinergic receptor structure from Drosophila to human brain and detect possible structural homology with α_1-adrenergic receptors. Proc. Natl. Acad. Sci. 81:272-276.

Wang, J.X., Mei, L., Yamamura, H.I. and Roeske, W.R., 1987, Solubilization with digitonin alters the kinetics of pirenzepine binding to muscarinic receptors from rat forebrain and heart. J. Pharm. Exp. Ther. 242:981-990.

Watson, M., Roeske, W.R. and Yamamura, H.I., 1986, [^{3}H]Pirenzepine and (-)-[^{3}H]quinuclidinyl benzilate binding to rat cerebral cortical and cardiac muscarinic cholinergic sites. II. Characterization and regulation of antagonist binding to putative muscarinic subtypes. J. Pharmacol. Exp. Ther. 237:419-427.

Wolf, B.A., Florholmen, J., Turk, J. and McDaniel, M.L., 1988, Studies of the Ca^{2+} requirements for glucose-and carbachol-induced augmentation of inositol trisphosphate and inositol tetrakisphosphate accumulation in digitonin-permeabilized islets. J. Biol. Chem. 263:3565-3575.

MOLECULAR MECHANISM UNDERLYING THE OCCURRENCE OF SUPER-
SENSITIVITY AT MUSCARINIC RECEPTORS : ANALYSIS USING
CEREBRAL CORTICAL NEURONS IN PRIMARY CULTURE

Kinya Kuriyama and Seitaro Ohkuma

Department of Pharmacology, Kyoto Prefectural
University of Medicine, Kamikyo-ku, Kyoto 602
Japan

INTRODUCTION

Chronic treatment with muscarinic antagonists is known
to induce an up-regulation at muscarinic receptors in
various organs (Takeyasu et al., 1979; Barak et al., 1981;
Hedlund, 1986; Goodbar and Bartifai, 1988). There is,
however, no actual information on the alteration in
intracellular biosignaling systems associated with
muscarinic receptors under these conditions. Recently,
Goodbar and Bartfai (1988) have reported that the increase
of muscarinic receptor induced by a long-term administration
of a muscarinic antagonist, atropine, leads to a decrease in
the responsiveness of phosphoinositide (PI) hydrolysis to
carbachol stimulation, although the mechanism of the
reduction in carbachol stimulated PI turnover in the brain
with up-regulated muscarinic receptor is not demonstrated.
In this study, we have, therefore, attempted to clarify the
mechanism underlying such a reduction of PI turnover in
response to muscarinic stimulation using mouse cerebral
cortical neurons in primary culture, which possess metabolic
and functional activity as cholinergic neurons (Ohkuma et
al., 1987).

MATERIALS AND METHODS

Primary culture of neurons

Dissociation and primary culture of neurons were
carried out by the method (Ohkuma et al., 1986) previously
described using neopallium of 15 days old fetus of ddy
strain mouse. Dissociated cells by a trypsin treatment were
inoculated on poly-L-lysine coated dishes with modified
Eagle's minimum essential medium (MEM) containing 15 % fetal
calf serum, and these cells were cultured for 3 days at 37° C
in humidified 95 % O_2 - 5 % CO_2. After the exposure of
cells to 20 M cytosine arabinoside for 24 hr to suppress
the proliferation of glial cells, the culture of neurons was
continued up to two weeks with MEM supplemented with 15 %

Neuroreceptor Mechanisms in Brain, Edited by S. Kito *et al.*
Plenum Press, New York, 1991

horse serum. Culture medium was exchanged to fresh medium every four days.

Treatment with atropine

Neurons were exposed to 10^{-8} M atropine for 5 days prior to each experiment. Atropine was dissolved in Hank's solution.

Muscarinic receptor binding

For the determination of muscarinic receptor binding, [^{3}H]quinuclidinyl benzilate ([^{3}H]QNB; 444 GBq/mmol, New England Nuclear, Boston, U.S.A.) was used as a radiolabeled ligand and the binding assay was carried out according to the method (Yamamura and Snyder, 1974) previously reported using the particulate fraction prepared from cultured cerebral cortical neurons. An aliquot (100 - 150 µg protein /assay) of particulate fraction suspended in sodium potassium phosphate buffer (Na-K PB; pH 7.4) (Yamamura and Snyder, 1974) was incubated at 25° C for 1 hr with [^{3}H]QNB at concentrations ranging from 0.05 nM to 4 nM in the presence or absence of 10 µM atropine. The reaction was terminated by filtrating the reaction mixture under vacuum through Whatman GF/B filter and the filter was subsequently washed three times with ice-cold Na-K PB. The filter was then transferred to a scintillation vial containing Triton-toluene scintillator, and it was subjected to the measurement of radioactivity by a liquid scintillation spectrometer.

Measurement of [^{3}H]inositol accumulation

Neurons were incubated with 1 µCi of [^{3}H]inositol (736 GBq/mmol, Amersham International, Amersham, U.K.), added into culture medium, at 37° C for 24 hr prior to the determination of [^{3}H]inositol accumulation. Neurons were then rinsed three times with ice-cold Krebs-Ringer bicarbonate buffer (KRB: pH 7.4) and scraped off with 0.1 M NaOH. An aliquot of the alkaline digested cells, neutralized with 0.1 M acetic acid, was mixed with Triton-toluene scintillator for the measurment of the radioactivity accumulated in these neurons. Net amount of the accumulated [^{3}H]inositol was calculated as described previously (Kishi et al., 1988).

Measurement of [^{3}H]phosphatidylinositides formed in neurons

For examining the capacity to synthesize phosphatidyl-inositides from inositol, neurons were incubated with 1 µCi of [^{3}H]inositol for 24 hr. After the incubation, neurons were scraped off from dishes with 0.2 M KCl/methanol (v/v; 50:50) following five times washing with ice-cold KRB, and then mixed with chloroform/methanol/concentrated HCl (v/v; 2:1:0.02). The organic phase obtained following centrifugation was washed with 0.2 M KCl/methanol and lyophilized. The dried residue was dissolved with chloroform/methanol/H_2O (v/v; 75:25:2) and an aliquot of the dissolved sample with authentic carriers was applied to TLC plates (Merck, silica gel G) to separate [^{3}H]phosphatidyl-inositides formed according to the method (Martin, 1983) previously reported.

Measurement of phosphoinositide (PI) hydrolysis

After prelabeling neurons with 1 μCi [^{3}H]inositol for 24 hr, these neurons were washed five times with ice-cold KRB and incubated with 3 ml of KRB containing 10 mM LiCl at 37°C for 1 hr. Following this preincubation, carbachol (3 x 10^{-4} M), an agonist for muscarinic receptors, was added into the medium and neurons were further incubated at 37°C for 45 min. The reaction was terminated by transferring the medium to a test tube and by the subsequent addition of 3.6 ml of methanol. Neurons detached were then combined with the pooled reaction mixture described above and sonicated following the addition of 3.6 ml of chloroform and 1.0 ml of distilled water. Finally, an aliquot of the organic phase obtained by centrifugation was applied to an ion-exchange column of Dowex 1 X 8 in formate form to separate [^{3}H]inositol phosphate formed (Berridge et al., 1982).

Measurement of GTPase activity

GTPase activity was determined by the methods previously reported (Cassel and Selinger, 1876; Brandt et al., 1983) with a minor modification. Neurons were placed on ice with distilled water for 15 min and then homogenized by a Teflon homogenizer followed by centrifugation. Pellet was resuspended with Hepes buffer (50 mM Hepes, 2 mM MgCl$_2$, 100 mM NaCl; pH 8.0) and recentrifuged. Pellet thus obtained was resuspended in Hepes buffer followed by homogenization with Polytron homogenizer, and this suspension was used as "membrane fraction" for the measurement of GTPase activity. An aliquot of membrane fraction was preincubated with the reaction mixture composed of 1 mM EDTA, 1 mM dithiothreitol, 0.1 mM ascorbic acid and 0.1 mM adenyl-5'-yl-imidodiphosphate (AppNHp) at 30°C for 10 min. After the preincubation, 2 μM [^{3}H]GTP (259 GBq/mmol, Amersham International) and various concentrations of carbachol were added followed by further incubation at 30°C for 15 min. To terminate the reaction, ice-cold methanol was added and centrifuged. Supernatant obtained was then mixed with authentic GTP and GDP, and lyophilized. This lyophilized sample was dissolved in 50 % ethanol and an aliquot of the dissolved sample was used to separate [^{3}H]GDP formed by thin layer chromatography according to the method (Salomon and Rodbell, 1975) previously described.

Measurement of protein

Protein content in neurons were determined by the method of Lowry et al (1951) with bovine serum albumin as a standard.

RESULTS AND DISCUSSION

The exposure of neurons to 10^{-8} M atropine for 5 days induced increases in both Kd and Bmax values for [^{3}H]QNB binding to particulate fraction of neurons as shown in Table 1. [^{3}H]QNB binding to neurons exposed to atropine was dose-dependently inhibited by atropine added in the binding assay system (data not shown), indicating that the increased

Table 1. Changes in kinetic parameters for [³H]quinuclidinyl benzilate (QNB) binding to particulate fraction of mouse cerebral cortical neurons in primary culture following long-term (5 days) exposure to atropine.

	No. of Experiments	[³H]QNB Binding	
		Kd(nM)	Bmax(fmol/mg protein)
Control	6	0.117 ± 0.014	395 ± 7
Atropine (10⁻⁸ M)	7	0.182 ± 0.018*	577 ± 23**

*p<0.05, **p<0.001, compared with control.

binding sites for [³H]QNB following atropine treatment are muscarinic receptors having pharmacological characteristics as the receptor. Similar changes in muscarinic receptors following a long-term treatment with muscarinic antagonists have been reported (Pedigo and Polk, 1985; Hedlund, 1986; Goodbar and Bartifai, 1988). In addition, the incubation of neurons with both atropine and cycloheximide, an inhibitor of protein synthesis, resulted in a significant decrease of [³H]QNB binding, as compared with the level of [³H]QNB binding observed in neurons exposed to atropine alone (Fig.1). These results strongly suggest that the

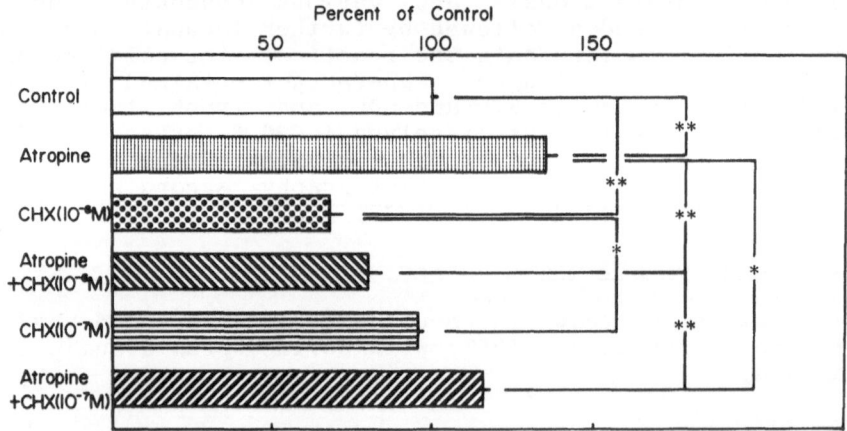

Fig.1. Effect of cycloheximide (CHX) on [³H]quinuclidinyl benzilate ([³H]QNB) binding to particulate fraction from mouse cerebral cortical neurons in primary culture following long-term (5 days) exposure to atropine. Neurons were exposed to CHX for 3 days prior to the preparation of particulate fractions of neurons. Control value:230.2 ± 5.0 fmol/mg protein. *p<0.01, **p<0.001.

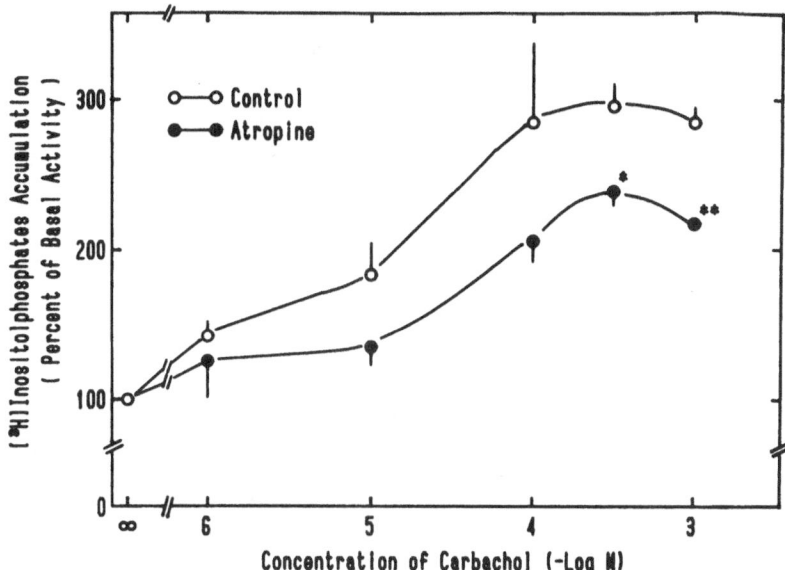

Fig.2. Effect of long-term (5 days) exposure to atropine
(10^{-8} M) on carbachol-stimulated [^{3}H]inositol-
phosphates accumulation in mouse cerebral cortical
neurons in primary culture. *p<0.05, **p<0.01,
compared with each control value.

increase in [^{3}H]QNB binding sites induced by atropine may be
attributed to an increase in the synthesis of new receptor
molecules.

In non-treated neurons, carbachol stimulated PI
turnover in a dose dependent manner. The maximal response
was observed in the presence of 3.0 X 10^{-4} M of
carbachol(Fig.2), which was similar to the data
(Gonzales et al., 1985b; Weiss et al., 1988) previously
reported. Although the response curve of PI turnover to
muscarinic stimulation in atropine-treated neurons was dose-
dependent and exhibited the maximal response at 3.0 X 10^{-4} M
of carbachol, the extent of responsiveness was found to be
significantly lower than that found in non-treated neurons
(Fig.2). Such a reduction in muscarinic agonist-stimulated
PI turnover has also been demonstrated in the cerebral
cortex of rats following a long-term treatment with atropine
(Goodbar and Bartifai, 1988). In the present study, we
have, therefore, attempted to clarify the mechanism
underlying this functional dissociation between the up-
regulated muscarinic receptor and the decreased
responsiveness of PI turnover to muscarinic stimulation.

We have first examined the transport of [^{3}H]inositol
into neurons and the activity to synthesize [^{3}H]phospha-
tidylinositides from [^{3}H]inositol, since we have used
[^{3}H]inositol as a metabolic substrate for phosphatidyl-
inositide formation. As shown in Fig.3, no difference in
the capacity to accumulate [^{3}H]inositol was noted between

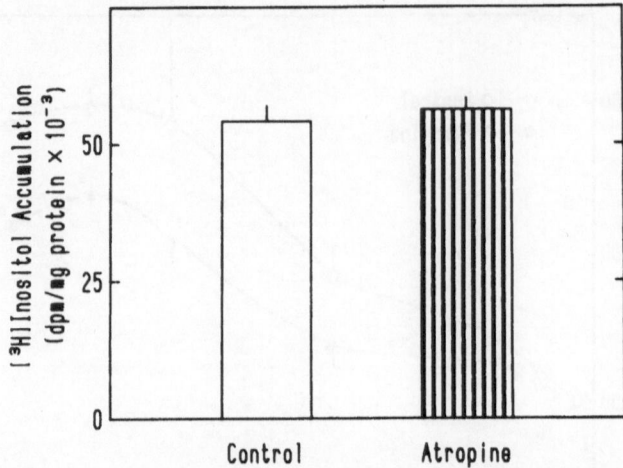

Fig.3. Effect of long-term (5 days) exposure to atropine
(10^{-8} M) on [³H]inositol accumulation in mouse
cerebral cortical neurons in primary culture.

atropine-treated and non-treated neurons. In addition, it
was found that the extent of [³H]inositol incorporation into
phosphatidylinositides in atropine-treated neurons was
similar to that in non-treated neurons (Table 2). These
results clearly indicate that the treatment with atropine
does not affect the metabolic capacities for accumulating
[³H]inositol and for incorporating [³H]inositol into
[³H]phosphatidylinositides in primary cultured neurons used
in this study.

Table 2. [³H]Phosphatidylinositide formation in mouse
cerebral cortical neurons in primary culture
following long-term (5 days) exposure to atropine
(10^{-8} M).

	PI[a]	PIP[a]	PIP$_2$[a]
Control	18,994 ± 944	376 ± 91	115 ± 32
Atropine	17,895 ± 639	432 ± 166	113 ± 69

a) dpm/mg protein.
PI:phosphatidylinositol, PIP:phosphatidylinositol 4-phos-
phate,
PIP$_2$:phosphatidylinositol 4,5-biphosphate

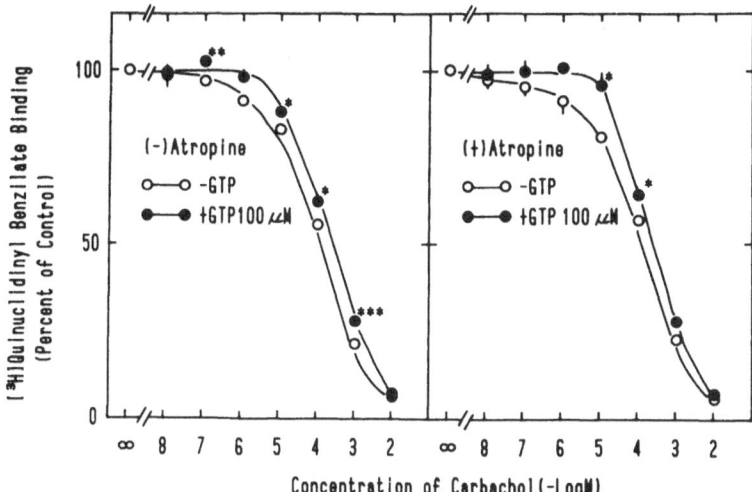

Fig.4. Effect of GTP on agonist binding to muscarinic
receptor in mouse cerebral cortical neurons in
primary culture following long-term (5 days) exposure
to atropine (10^{-8} M). *p<0.05, **p<0.02, ***p<0.01,
compared with the value obtained in the absence of
GTP.

It has been also well documented that the muscarinic
receptor has an interaction with a GTP-binding protein
(G-protein), and this coupling is essential for PI
hydrolysis mediated by muscarinic stimulation (Evans et al.,
1985; Neer and Clapham, 1988).

Therefore, we have also investigated whether or not
this interaction is deteriorated in neurons exposed to
atropine. The [³H]QNB bindings in both atropine-treated and
non-treated neurons were similarly inhibited by carbachol
added into the receptor binding assay system in a dose-
dependent manner (Fig.4). The addition of GTP shifted the
carbachol competition curve of [³H]QNB binding to the right
in both types of neurons and the extent of shift was found
to be similar in both neurons (Fig.4). In addition,
guanylylimido diphosphate (GppNHp) and guanosine 5'-(γ -
thio)triphosphate (GTP γ S), non-hydrolyzable analogues of
GTP, also showed similar effects on carbachol competition
curve in both types of neurons (data not shown). These
evidences strongly suggest that the long-term exposure of
neurons to atropine may have no influence on the functional
interaction between muscarinic receptor and G protein.

It has been also known that G protein is directly
coupled with phospholipase C, which catalyzes PI hydrolysis (Cockcrost and Gomparts, 1985; Gonzales et al., 1985a)
In non-treated neurons, the addition of GTP γ S dose-
dependently stimulated the formation of [³H]inositol
phosphates. On the other hand, GTP γ S-stimulated PI
hydrolysis in atropine-treated neurons was found to be

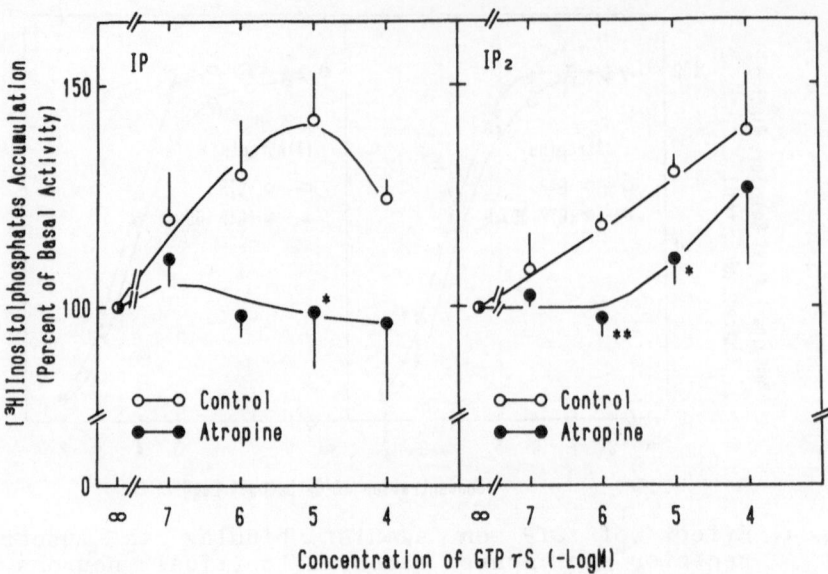

Fig.5. Effect of long-term (5 days) exposure to atropine (10^{-8} M) on GTP γS-stimulated [^{3}H]inositolphosphates accumulation in mouse cerebral cortical neurons in primary culture. *p<0.05, **p<0.02, compared with each control value.

significantly suppressed as compared with that in non-treated neurons, although the stimulation by GTP γS was also dose-dependent in atropine-treated neurons (Fig.5). In contrast, membrane-bound phospholipase C activities in both atropine-treated and non-treated neurons were identical (Table 3). These results, therefore, suggest a possibility that the decreased responsiveness of PI turnover to muscarinic stimulation in neurons exposed to atropine may be due to the functional deterioration of the G protein itself and/or the malfunction of coupling mechanism between the G protein and phospholipase C.

Table 3. Effect of long-term (5 days) exposure to atropine on membrane-bound phospholipase C activity in mouse cerebral cortical neurons in primary culture.

	IP (dpm/mg protein)	IP₂ (dpm/mg protein)
Control	4471 ± 338	975 ± 82
Atropine(10^{-8} M)	5495 ± 663	1157 ± 173

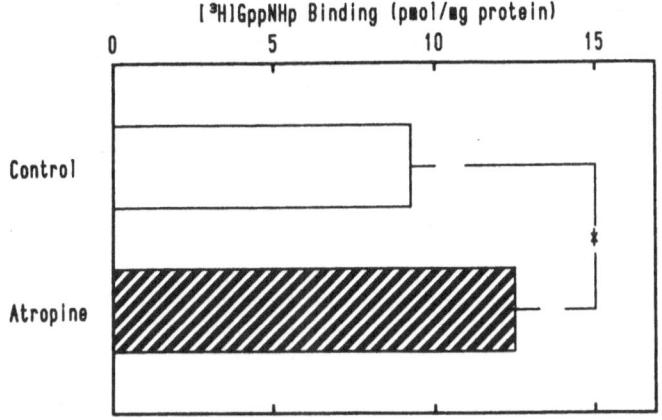

Fig.6. Effect of long-term (5 days) exposure to atropine (10^{-8} M) on [^{3}H]guanylylimido diphosphate (GppNHp) binding to particulate fraction of mouse cerebral cortical neurons in primary culture. *p<0.02

To examine possible changes in functions of G proteins in atropine-treated neurons, we have investigated [^{3}H]GppNHp binding to membrane fraction of neurons and GTPase activity, since the G protein, especially its α-subunit, possesses GTP binding sites and GTPase activity (Gilman, 1987). The binding of [^{3}H]GppNHp, a radiolabeled ligand specific

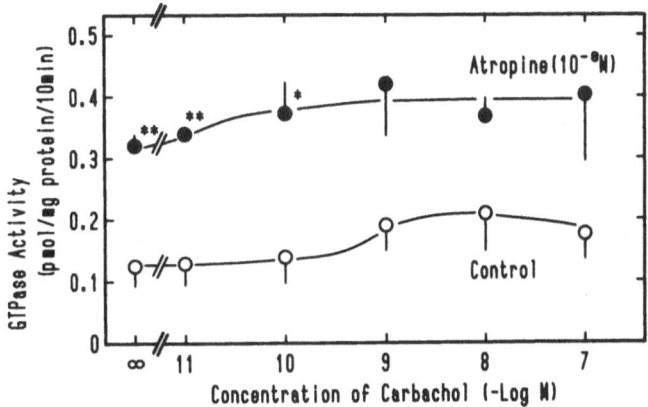

Fig.7. Effect of long-term (5 days) exposure to atropine (10^{-8} M) on GTPase activity in mouse cerebral cortical neurons in primary culture. *p<0.05, **p<0.02.

to G proteins (Salomon and Rodbell, 1975), in atropine-treated neurons was significantly increased as compared with that in non-treated neurons (Fig.6). This evidence indicates that the GTP binding site in G proteins increases in neurons with up-regulation of muscarinic receptor induced by atropine exposure. Moreover, as shown in Fig.7, GTPase activity in atropine-treated neurons showed a significant increase in the presence and absence of carbachol. These results, namely the increase of [^{3}H]GppNHp binding and the elevation of GTPase activity found in atropine-treated neurons, strongly suggest that the number of G protein molecules may be increased during the up-regulation of muscarinic receptors. In spite of the increases in both muscarinic receptors and G proteins and no significant alterations in inositol accumulation and phospholipase C activity, the present results have clearly indicated the occurrence of the attenuation of PI turnover in the presence of carbachol. Concerning these results, it seems reasonable to assume that the decrease in the response of PI turnover to muscarinic stimulation in atropine-treated neurons may be attributed to the functional deterioration of the coupling mechanism between the G protein and phospholipase C.

CONCLUSION

Long-term exposure of neurons to atropine induced not only the increase of muscarinic receptors but also that of G proteins coupled with the receptor. In spite of such increases, PI turnover in the presence of muscarinic agonist was rather suppressed, possibly due to the deterioration of functional coupling between muscarinic receptors and G-proteins. Although molecular mechanisms underlying such a deterioration as well as pharmacological significance of this alteration should be clarified in future studies, possible misleading conclusions on the occurrence of so-called "up-regulation" at muscarinic receptor should be emphasized, if it was judged only by the increase of receptor binding.

REFERENCES

Barak, J. B., Gazit, H., Silman, I. and Dudai, Y., In vivo modulation of the number of muscarinic receptors in rat brain by cholinergic ligands, Eur. J. Pharmacol. 74:73-81 (1981).

Berridge, M. J., Downes, C. P. and Hanley, M. R., Lithium amplifies agonists-dependent phosphatidylinositol responses in brain and salivary glands, Biochem. J. 206:587-595 (1982).

Brandt, D. R., Asano, T., Pedersen, S. E. and Ross, E. M., Reconstitution of catecholamine-stimulated guanosinetri-phosphatase, Biochemistry 22:4357-4362 (1983).

Cassel, D. and Selinger, Z., Catecholamine-stimulated GTPase activity in turkey erythrocyte membrane, Biochim. Biophys. Acta 452:538-551 (1976).

Cockcroft, S. and Gomperts, B. D., Role of guanine nucleotide binding protein in the activation of polyphosphoinositide phosphodiesterase, <u>Nature</u> 314:534-536 (1985).

Evans, T., Hepler, J. R., Masters, S. B., Brown, J. H. and Harden, T. K., Guanine nucleotide regulation of agonist binding to muscarinic cholinergic receptors, <u>Biochem. J.</u> 232:751-757 (1985).

Gilman, A. G., G proteins : Transducers of receptor-generated signals, <u>Ann. Rev. Biochem.</u> 56:615-649 (1987).

Goodbar, L. and Bartifai, T., Long-term atropine treatment lower the efficacy of carbachol to stimulate phosphatidylinositol breakdown in the cerebral cortex and hippocampus of rats, <u>Biochem. J.</u> 250:727-734 (1988).

Gonzales, R. A. and Crews, F. T., Guanine nucleotides stimulate production of inositol triphosphate in rat cortical membrane, <u>Biochem. J.</u> 232:799-804 (1985a).

Gonzales, R. A., Feldstein, J. B., Crews, F. T. and Raizada, M., Receptor-mediated inositide hydrolysis is a neuronal response : comparison of primary neuronal and glial cultures, <u>Brain Res.</u> 345:350-355 (1985b).

Hedlund, B., Long time treatment of NIE-115 neuroblastoma cells with atropine-induces changes in markers of muscarinic cholinergic functions, <u>Neurosci. Lett.</u> 64:196-200 (1986).

Kishi, M., Ohkuma, S., Kimori, M. and Kuriyama, K., Characteristics of taurine transport system and its developmental pattern in mouse cerebral cortical neurons in primary culture, <u>Biochim, Biophys. Acta</u> 939:615-623 (1988).

Lowry, O. H., Rosenbrough, N. J., Farr, A. L. and Randall, R. J.; Protein measurement with the Folin phenol reagent, <u>J. Biol. Chem.</u> 193:265-275 (1951).

Martin, T. F. J., Thyrotropin-releasing hormone rapidly activates the phosphodiester hydrolysis of polyphospho-inositides in GH_3 pituitary cells. Evidence for the role of a polyphosphoinositide-specific phospholipase C in hormone action, <u>J. Biol. Chem.</u> 258:14816-14822 (1983).

Neer, E. J. and Clapham, D. E., Roles of G protein subunits in transmembrane signalling, <u>Nature</u> 333:129-134 (1988).

Ohkuma, S., Ma, F.-H., Tomono, S., Kishi, M. and Kuriyama, K., Development of cerebral cholinergic neurons in primary culture, <u>Japan. J. Pharmacol.</u> 43(Suppl.):148p (1987).

Ohkuma, S., Tomono, S., Tanaka, Y., Kuriyama, K. and Mukainaka, T., Development of taurine biosynthesizing system in cerebral cortical neurons in primary culture, <u>Int. J. Dev. Neurosci.</u> 4:383-395 (1986).

Pedigo, N. N. Jr. and Polk, D. M., Reduced muscarinic receptor plasticity in frontal cortex of aged rats after chronic administration of cholinergic drugs, <u>Life Sci.</u> 37:1443-1449 (1985).

Salomon, Y. and Rodbell, M., Evidence for specific binding sites for guanine nucleotides in adipocyte and hepatocyte plasma membranes. A difference in fate of GTP and guanosine 5'-(β ,γ-imino) triphosphate, <u>J. Biol. Chem.</u> 250:7245-7250 (1975).

Takeyasu, K., Uchida, S., Noguchi, Y., Fujita, N., Saito, K., Hata, F. and Yoshida, H., Changes in brain muscarinic acetylcholine receptors and behavioral responses to atropine and apomorphine in chronic atropine-treated rats, <u>Life Sci.</u> 25:585-592 (1979).

Weiss, S., Schmidt, B. H., Sebben, M., Kemp, D. E., Bockaert, J. and Sladeczek, F., Neurotransmitter-induced inositol phosphate formation in neurons in primary culture, <u>J. Neurochem.</u> 50:1425-1433 (1988).

Yamamura H. I. and Snyder, S. H., Muscarinic cholinergic binding in rat brain, <u>Proc. Natl. Acad. Sci. U.S.A.</u> 71:1725-1729 (1974).

SUPER-DELAYED CHANGES OF MUSCARINIC ACETYLCHOLINE RECEPTOR IN THE

GERBIL HIPPOCAMPUS FOLLOWING TRANSIENT ISCHEMIA

N. Ogawa, K. Haba, M. Asanuma, K. Mizukawa* and A. Mori

Departemnt of Neurochemistry, Institute for Neurobiology
and *Department of Anatomy, Okayama University Medical
School, 2-5-1, Shikata-cho, Okayama 700, Japan

INTRODUCTION

Cases of sequelae of cerebrovascular disease such as vascular dementia due to death of many neurons have been gradually increasing. Such neuronal death following brain ischemia has been considered to be due to energy deficiency resulting from an impaired respiratory chain. However, analysis of the delayed neuronal death showed that neuronal death is not caused by mere energy deficiency. Most studies on delayed neuronal death focused on the morphological changes and energy metabolism in the acute to subacute stage. There are few reports concerning biochemical changes in the chronic stage, especially in neurotransmitter receptors.

Brief bilateral carotid occlusion in the gerbil produces forebrain ischemia resulting in almost complete neuronal destruction in the hippocampal CA1 pyramidal cell layer after a few days, even though metabolic and functional activity recovers during the early recirculation period. Several mechanisms have been considered to be responsible, including synaptically released excitatory neurotransmitters such as glutamate, cytosolic calcium overload, release of excitotoxins, or free radical damage of cell membranes. Since delayed neuronal death was thought to be complete within 7 days, almost all biochemical analysis were carried out up to 7 days after transient ischemia. In the present study, we examined long-term morphological and biochemical changes in the hippocampal region after ischemia.

MATERIALS AND METHODS

Animal experiment

Gerbils weighing 60–85g were used. Under ketamine anesthesia (100 mg/kg body weight, i.p.), silk threads were placed around both common carotid arteries without interrupting carotid blood flow. On the following day, bilateral common carotid arteries were exposed and then occluded with surgical clips after light ether anesthesia (Ogawa et al., 1988). Carotid artery blood flow was restored by releasing the clips following 5 min of occlusion. Gerbils used as controls were treated in the same way, but without occlusion. Then the gerbils were killed by decapitation for receptor binding assay and thiobarbituric acid reacting substances (TBARS)

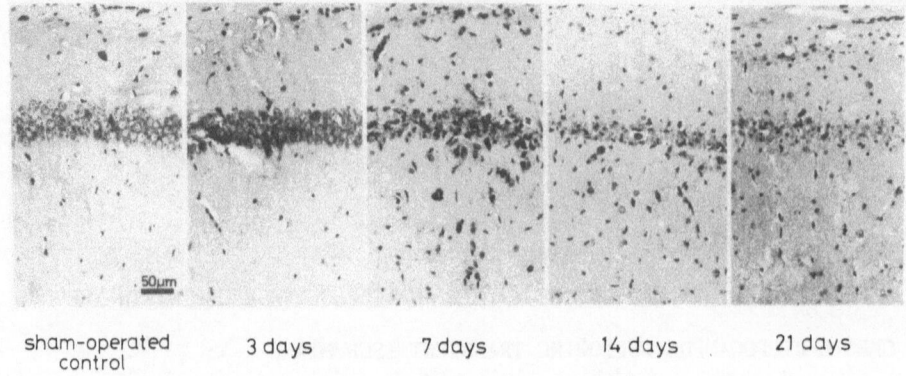

| sham-operated control | 3 days | 7 days | 14 days | 21 days |

Fig. 1.　Histological changes in pyramidal cells in the CA1 area of the gerbil hippocampus after 5-min transient ischemia

determination, and by microwave irradiation (4.8 kw, 1.1 sec) for ACh analysis at the end of, 1, 2, 3, 4, 7, 14, and 21 days after recirculation following cerebral ischemia. The brains were immediately removed and hippocampus was dissected.

Histological examination

Three, 7, 14 and 21 days after ischemia, animals were sacrificed and their brains were processed for histological examination. Briefly, the gerbils were anesthetized with pentobarbital (100 mg/kg, i.p.) then fixed with 10% formalin by whole body perfusion. The brain was carefully removed and fixed in 10% formalin. The brain was sectioned coronally and dehydrated with series of ethanol (30-100%) and epone baths prior to paraffin embedding. Five micron slices containing the hippocampus were cut and slide-mounted prior to staining with cresyl-violet.

Biochemical determinations

The concentration of ACh in the hippocampus was determined as previously reported (Ogawa et al., 1988). Saturation experiments of muscarinic acetylcholine receptor (mACh-R) binding were done by a method previously are described by using [^{3}H]QNB(Ogawa et al., 1983), and N-methyl-D-aspartate receptor (NMDA-R) binding, that is [^{3}H]CPP binding was measured by a slight modification of the method by Murphy et al(1987). TBARS as an indicator of lipid peroxidation was determined by a thiobarbiturate method (Ohkawa et al., 1976).

RESULTS

Histological changes after ishcmeia

Histological findings of pyramidal cell layer in CA1 of the dorsal hippocampal formation of sham-operated gerbils and ischemic gerbils on days 3, 7, 14, and 21 of the survival period (Fig. 1). In the sham-operated gerbils, pyramidal cells were clear, round and moderate to large in size. Their nucleus was clear with a dot-like nucleolus. The pyramidal cell layer consisted of 3-4 rows of these cells. In the stratum radiatum, vertically arranged neuropile bundles noted in the cells. On the 3rd day

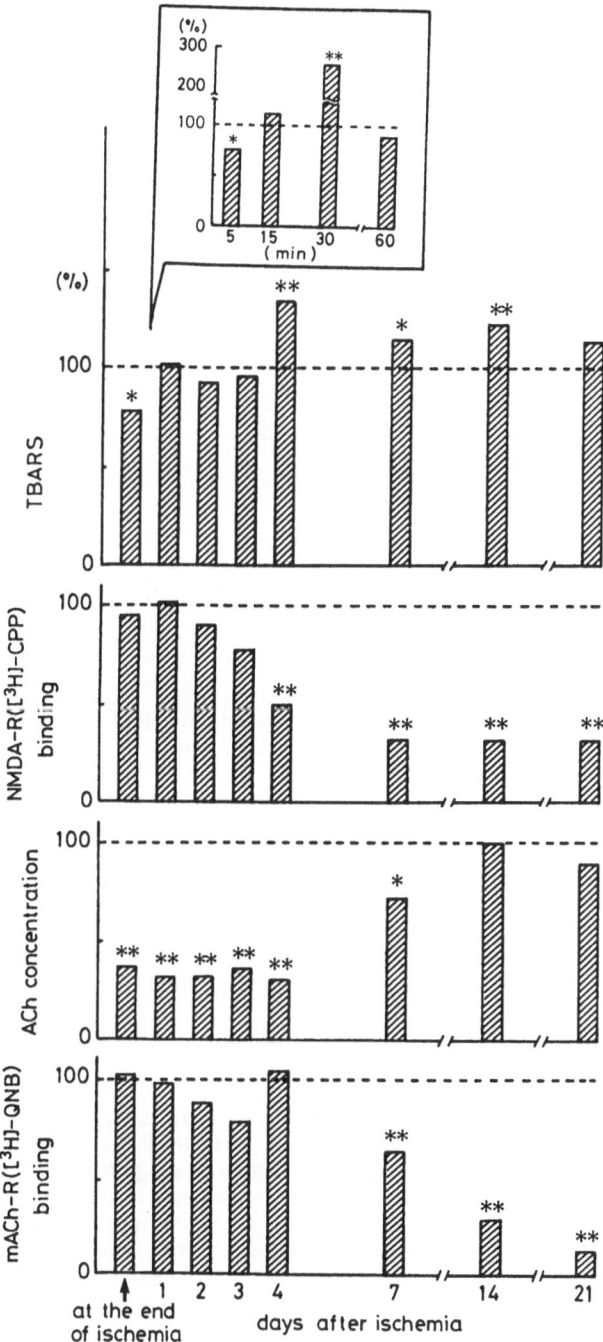

Fig. 2. TBARS levels, NMDA-R binding, ACh concentrations and mACh-R
binding in the gerbil hippocampus after 5-min transient ischemia

after ischemia, pyramidal cells appeared dark but in generally showed findings similar to those in sham-operated gerbils. On day 7, pyramidal cell layers were disordered most of the pyramidal cells in CA1 were destroyed. The pyramidal cells were dark and pyknotic. In these cell layers, the reactive glia cells were very prominent with astroglia and microgrial cells. A large number of phagocytotic cells and neuronal debrils were found and scattered. In the stratum radiata, small to moderate size cells were observed and the vertically arranged neuropil bundles were not prominent. Similar findings were observed on days 14 and 21. In addition, there were phagocytotic reactive cells.

TBARS level

Although the TBARS level was significantly decreased at the end of transient ischemia, it rapidly increased 30 min after recirculation of blood flow (Fig.2, inserted), then it returned to the normal levels. However, the TBARS levels increased by 40-50% between the 4th and 14th day after transient iscehmia. On the 21st day, the TBARS level in ischemic animal returned to the control levels.

NMDA-R binding

The specific NMDA-R binding was slightly decreased 3 days after transient ischemia and markedly decreased at 4 days, remained low thereafter. Scatchard analysis revealed a significant decrease in Bmax on day 21, without any change of Kd.

ACh concentration

The ACh concentration in the hippocampus decreased 30-40% compared with the sham-operated controls from the end of ischemia to 4 days after transient ischemia. On day 7, the ACh concentration in the ischemic group had recovered to 70% of sham-operated group and had returned to the control level on day 14.

mACh-R binding

$[^3H]$-QNB binding was normal within 4 days after transient ischemia, but gradually decreased to 10 % of that in the sham-operated group on day 21. Kd values did not change during the observation.

DISCUSSION

Short time occlusion of the bilateral carotid artery in the gerbil results in a delayed progression of neuronal damage, and hippocampal CA1 neurons were completely destroyed by one week post occlusion at the latest (Kirino, 1982; Suzuki et al., 1983). However, the molecular basis for the permanent neuronal destruction with ischemia remains unclear. Just after ischemia, a large amount of glutamate, an exitatory neurotoxic neurotransmitter, released (Benveniste et al., 1984), and this transient increase is thought to lead to calcium influx with resultant cell death (Rothman et al., 1986).

In the present study, the hippocampal neurons were completely lost 7 days after ischemia as reported elswhere (Fig. 1)(Kirino, 1982; Suzuki et al., 1983). The time course of appearance of morphological damage in hippocampal region paralleled that for depletion of NMDA-R binding . In addition, these two phenomena probably correspond to early surge of TBARS (Fig. 2, inserted).

On the other hand, mACh-R binding was decreased on day 7 and destruction of mACh-R containing neurons gradually increased. This corresponds to the second increase in TBARS between the 4 thard 14th days after ischemia. Although the reason for this super-delayed loss of mACh-R is unclear, some degenerative events may have occured duringthese periods.

The hippocampus is thought to be crucial for learning and memory (Brierley et al., 1962; Adams et al., 1966). Our findings support the hypothesis that the decreases in mACh-R and NMDA-R in postischemic brain as a consequence of irreversible damage to neurons containing the receptors. Although a reduction of NMDA-R occured in parallel with cell destruction, mACh-R decrease more slowly than cell death. Super-delayed destruction of mACh-R in the hippocampus may cause deficits of learning and memory seen in patients with cerebrovascular disease.

ACKNOWLEDGMENTS

This work was supported in part by grants-in-aid for Scientific Research on Priority Areas, Co-operative Research and Scientific Research from the Japanese Ministry of Education, Science and Culture.

REFERENCES

Adams, J. H., Brierley, J. B., Connor R. C. T., and Triep, C., 1966, The effects of systemic hypotension on the human brain-Clinical and neuropahological observation in 11 cases, Brain, 89:235-254.
Benveniste, H., Drejer, J., Schousboe, A., and Diemer, N. H., 1984, Elevation of the extracellular concentrations of glutamate and aspartate in rat hippocampus during transient cerebral ischemia monitored by intracerebral microdialysis, J. Neurochem., 43;1369-1374.
Brierley, J. B., and Cooper, J.E., 1962, Cerebral complecation of hypotensive anesthesia in a healthy adult, J. Neurol. Neurosing Psychiat., 25;24-30.
Kirino, T., 1982, Delayed neuronal death in the gerbil hippocampus following ischemia, Brain Res., 239; 57-69.
Murphy, D. E., Schneider, J., Boehm, C., Iehmann, J., and Williams, M., 1987, Binding of [^{3}H]3-(2-carboxypiperazin-4-y1)propyl-1-phosphonic acid to rat brain membranes: A selective, high affinity ligand for N-methyl-D-aspartate receptors, J. Pharmacol. Exp. Ther., 240;778-784.
Ohkawa, H., Ohishi, N., and Yagi, K., 1976, A new sensitive assay for the measurement of hydroperoxide, Anal. Biochem., 76;184-191.
Ogawa, N., Haba, K., Yoshikawa, H., Ono, T., and Mizukawa, K., 1988, Comparison of the effects of bifemelane hydrochloride, idebenone and indeloxazine hydrochloride on ischemia-induced depletion of brain acetylcholine levels in gerbils, Res. Commun. Chem. Pathol. Pharm., 61;285-288.
Ogawa, N., Mizuno, S., Nukina, I., Tsukamoto, S., and Mori, A., 1983, Chronic thyrotropin releasing hormone (TRH) administration on TRH receptors and muscarinic cholinergic receptors in CNS, Brain Res., 263;348-350.
Rothman, S. M., and Olney, J. W., 1986, Glutamate and the pathophysiology of hypoxic ischemic brain damage, Ann. Neurol., 19;105-111.
Suzuki, T., Yamaguchi, T., Kirino, T., Orzi, F., and Klatzo, I., 1983, The effects of 5-minutes ischemia in Mongolian gerbils I: blood-brain barrier, cerebral blood flow, and local cerebral glucose utilization changes, Acta Neuropahol., 60;207-216.

INTRAMEMBRANE PARTICLES IN THE POSTSYNAPTIC MEMBRANES OF THE S-, F-,
AND C-TYPE SYNAPSES BY FREEZE-FRACTURING, AND DEEP-ETCHING STUDIES
ON THE XENOPUS SPINAL CORD

Hiroshi Watanabe, Hiroshi Washioka and Akira Tonosaki

Dept. of Anatomy, Yamagata Univ. School of Medicine
Yamagata 990-23, Japan

INTRODUCTION

     In the CNS, the S-, F- and C-type of synapses used to be classified
with TEM based on the appearance of synaptic vesicles and the specializa-
tion of synaptic membranes (Charlton and Gray, 1966; Uchizono, 1975; Watanabe, 1981): The S-type, or Gray's type I, synapse has spherical synaptic
vesicles and more or less electron-dense substances in the subsynaptic
cytoplasm.   The F-type, or Gray's type II, has flattened vesicles and
symmetrically thickened pre- and postsynaptic membranes. The C-type is
remarkable because of the subsurface cistern in the subsynaptic cytoplasm.

     Many authors have regarded the S-type as excitatory and F-type as
inhibitory, respectively (Raviola and Gilula, 1975; Uchizono, 1975; Gulley
et al., 1978). On the other hand, the C-type synapse has been assumed as
cholinergic, although its neurological part is yet undetermined (Davidoff
and Irintchev, 1986).   To obtain more detailed criteria to specialize
active zones of those different type synapses, the distribution of intramembrane particles (IMPs) has been investigated mainly by the freeze-fracturing (Heuser and Reese, 1977; Watanabe and Yamamoto, 1981; Sandri, 1982).
However, the quantitative analysis of IMPs remains to be made about an
appropriate object.

     We here obtained the distribution and size of IMPs of the postsynaptic
membranes by the complementary freeze-fracturing from the Xenopus spinal
cord tissue in which S-, F-, and C-type of synapses occurred together and
distinguished from each other unequivocally under TEM.   Deep-etching and
freeze-substitution techniques were also applied to examine the surface
appearance of IMPs on the true surfaces of the membrane.

MATERIALS AND METHODS

     Adult Xenopus toads were perfused with the fixative containing 2.5%
glutaraldehyde under anesthesia with urethane. Spinal cord was excised and
minced for continued fixation with the same fixative. For freeze-fracturing, the specimen was incubated with the fixative mixed with 25% glycerol
for 1 h and frozen in the liquid nitrogen slush at the melting temperature,
-210°C. For deep-etching, the fixed specimen was incubated with 1% osmium

tertroxide for 1 h, infiltrated with 40% methanol for 1 h, and frozen similarly.

Complementary freeze-fracturing, deep-etching and replication were carried out with a JFD-7000 vacuum evaporator (Tonosaki and Yamamoto, 1974; Watanabe et al., 1986). For direct examination of the freeze-fractured postsynaptic membranes, some fractured specimens were further processed by freeze-substitution through acetone containing 2% osmium tetroxide, and embedded with Epon 812 for ultrathin sectioning.

RESULTS

Size and aggregation of IMPs in the postsynaptic membranes

S-type Synapse. The E-face of the postsynaptic membrane showed aggregations of larger IMPs ($2600/\mu m^2$, 12-14 nm in diameter), whereas its complementary P-face only scattering IMPs (Fig. 1).

F-type synapse. The E-face of the postsynaptic membrane showed less dense aggregation of smaller-sized IMPs ($1500/\mu m^2$, 8-9 nm), whereas its P-face denser aggregation of medium-sized IMPs ($3100/\mu m^2$, 10-12 nm) (Fig. 2).

C-type synapse. The E-face of the postsynaptic membrane showed mixed aggregation of smaller- and medium-sized IMPs ($1500/\mu m^2$), whereas its P-face denser aggregation of larger- and medium-sized IMPs ($2700/\mu m^2$).

Deep-etching

By deep-etching replication, the both extracellular and protoplasmic surfaces of the postsynaptic membranes were visualized to be attached with particulate structures (8-15 nm), which were regionally concentrated to the active zones. These surface elements were, however, failed to be distingushed type-specifically from each other on TEM pictures of the replicas at this stage of study (Fig. 3).

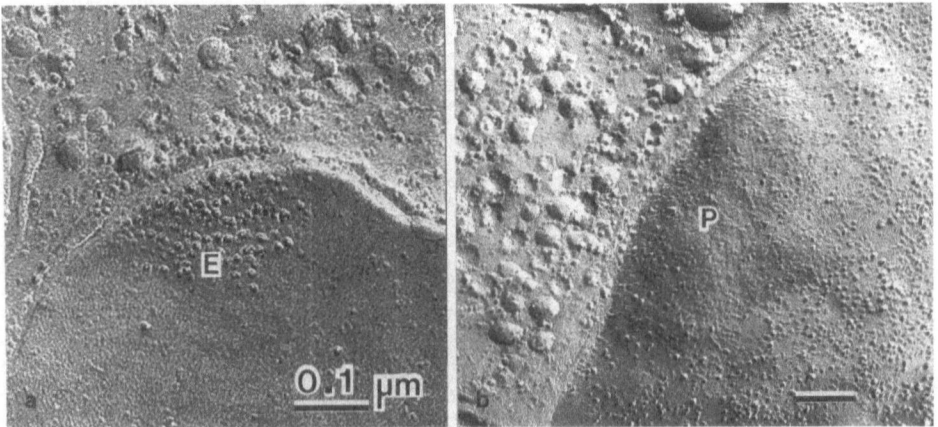

Fig. 1. a, EM of replicas showing E-face (E) of S-type postsynaptic membrane. The active zone is characterized by an aggregation of larger IMPs. b, P-face (P) is conspicuously lacking IMP aggregation.

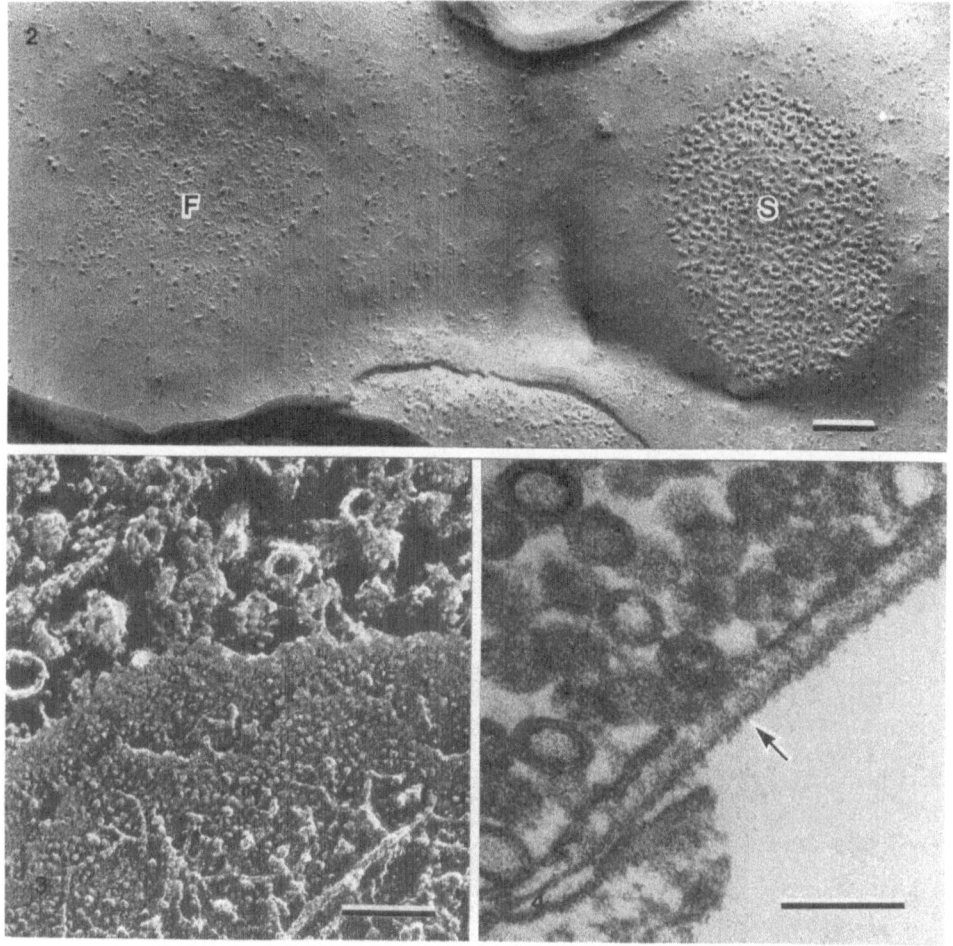

Fig. 2. EM of replica of E-face of postsynaptic membrane. The active zones are differentiated from each other in respect to aggregations of larger (S-type) and smaller (F-type) IMPs.

Fig. 3. EM of deep-etching replica of S-type synapse. Cytoplasmic surface of the postsynaptic membrane is characterized by granular elements mixed with cytoskeletal fibers.

Fig. 4. EM of ultrathin section of freeze-fracture-substituted S-type synapse. The outer (arrow) and inner leaflets, and the topology of E- and P-faces, of fractured postsynaptic membrane are evident.

Ultrathin section of freeze-fracture-substituted materials

This method enabled us to investigate the lateral view of the separated phase of outer and inner leaflets of the postsynaptic membranes. In the S-type postsynaptic active zone, IMPs seemed to protrude into the synaptic cleft and also to be bound with filamentous elements which was conceivably traversing the cleft (Fig. 4). A similar feature was obtained from the F-type postsynaptic active zones.

Fig. 5 shows diagrams of the distribution pattern of IMPs in the postsynaptic active zones from different synapses.

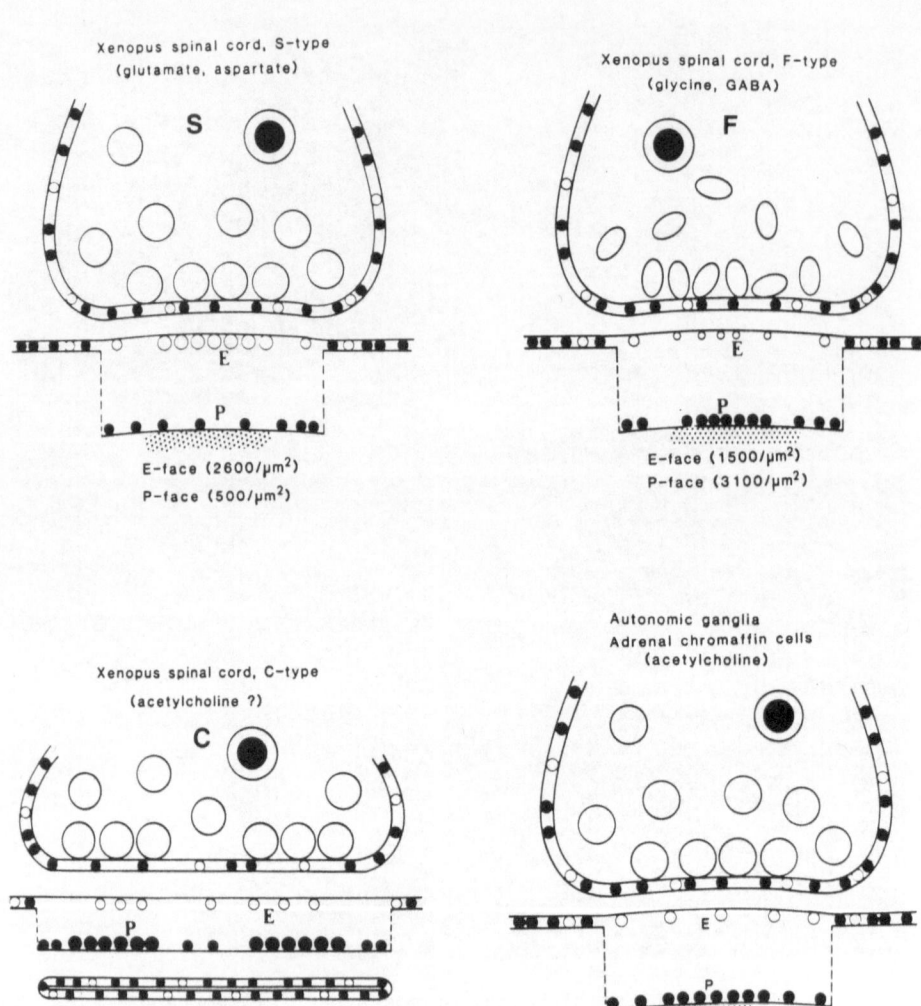

Fig. 5. Diagrams showing the distribution pattern of IMPs in the postsynaptic active zones from the S-, F-, C-type synapses and the "cholinergic" autonomic nervous system. Open and solid circles represent IMPs studded on the E- and P-faces, respectively.

DISCUSSION

S- versus F-type synapses

The postsynaptic membranes of the active zone is obviously specialized for the S- versus F-type synapse in terms of the IMPs. Based on the deep-etching and freeze-fracture-substitution studies, some of the IMPs, if not all, were likely to traverse the postsynaptic membrane of the S- as well as F-type synapse. Consequently, their outer and inner extremities may project externally into the synaptic cleft and internally into the subsynaptic cytoplasm.

Previous physiological studies have suggested that putative transmitters of the excitatory and inhibitory synapses in the amphibian spinal cords will be glutamate or aspartate, and GABA or glycine, respectively (Simpson, 1976; Nistri and Berti, 1983; Bixby and Spitzer, 1984; Antonov et al., 1990). Moreover, recent immunohistochemical studies have confirmed that the specific receptors for the excitatory and inhibitory transmitters are aggregated on the postsynaptic active zones (Möhler et al., 1987; Wenthold et al., 1988).

Taken together, their and our observation will allow us to speculate that the local IMPs will represent unitary or polymerized forms of receptor molecules, each of which is specialized for glutamate or aspartate in the S-type, for GABA or glycine in the F-type synapses.

C-type synapse

Although the functional significance is still uncertain, the localization of ACh-esterase activity has been proved at the C-type synaptic cleft in the spinal cord (Davidoff and Irintchev, 1986). Interestingly, the aggregation and sidedness we obtained of the IMPs in the C-type postsynaptic active zone has close resembrance to those in the bullfrog sympathetic ganglion cells and adrenal chromaffin cells, where the synaptic transmission is established to be mediated via nicotinic ACh receptors (Watanabe et al., 1986; Fig. 5). IMPs aggregating on the C-type postsynaptic active zone will presumably represent unitary or mass of the ACh receptors or related structures.

REFERENCES

Antonov, S, M., Kalinina, N. I., Kurchavyj, G. G., Magazanik, L. G., Shupliakov, O. V., and Vesselkin, N. P., 1990, Identification of two types of excitatory monosynaptic inputs in frog spinal motoneurons, Neurosci. Let., 109:82-87.

Bixby, J. L., and Spitzer, N. C., 1984, The appearance and development of neurotransmitter sensitivity in Xenopus embryonic spinal neurons in vitro, J. Physiol., 353:143-155.

Charlton, B. T., and Gray, E. G., 1966, Comparative electron microscopy of synapses in the vertebral spinal cord, J. Cell Sci., 1:67-80.

Davidoff, M. S., and Irintchev, A. P., 1986, Acetylcholinesterase activity and type C synapses in the hypoglossal, facial and spinal-cord motor nuclei of rats, An electron-microscopy study, Histochemistry, 84:515-524.

Gulley, R. L., Landis, D. M. D., and Reese, T. S., 1978, Internal organization of membranes at end bulbs of Held in the anteroventral cochlear nucleus, J. Comp. Neurol., 180:707-742.

Heuser, J. E., and Reese, T. S., 1977, Structure of synapse, in: "The handbook of physiology, Section 1; The nervous system", Americal Physiological Society, Bethesda, vol 1(1), pp.261-294.

Möhler, H., Schoch, P., Richards, J. G., Häring, P., and Takacs, B., 1987, Structure and location of a GABA-A receptor complex in the central nervous system, J. Receptor Res., 7:617-628.

Nistri, A., and Berti, C., 1983, Caffeine-induced potentiation of GABA effects on frog spinal cord: An electrophysiological study, Brain Res., 258:263-270.

Raviola, E., and Gilula, N. B., 1975, Intramembrane organization of specialized contacts in the outer plexiform layer of retina, J. Cell Biol., 65:192-222.

Sandri, C., Van Buren, J. M., and Akert, K., 1982, Membrane morphology of the vertebrate nervous system, A study with freeze-etch technique, Prog. Brain Res., (2nd revised edition), 46:121-155.

Simpson, J. I., 1976, Functional synaptology of the spinal cord, in: "Frog Neurobiology", R. Llinás and W. Precht, eds., Springer-Verlag, Berlin, pp.728-749.

Tonosaki, A., and Yamamoto, T.-Y., 1974, Double-replicating method for the freeze-fractured retina, J. Ultrastruct. Res., 47:86-94.

Uchizono, K., 1975, "Excitation and inhibition synaptic morphology", Igaku Shoin Ltd., Tokyo.

Watanabe, H., 1981, Development of axosomatic synapses of the Xenopus spinal cord with special reference to subsurface cisterns and C-type synapses, J. Comp. Neurol., 200:323-338.

Watanabe, H., and Yamamoto, T.-Y., 1981, Freeze-fracture study on three types of synapses in the Xenopus spinal cord, J. Comp. Neurol., 198:249-263.

Watanabe, H., Washioka, H., and Tonosaki, A., 1986, "Cholinergic" postsynaptic membranes of bullfrog sympathetic ganglia: Electron microscopy of thin sections and freeze-fracture replicas, Anat. Rec., 214:82-88.

Wenthold, R. J., Parakkal, M. H., Oberdorfer, M. D., and Altschuler, R. A., 1988, Glycine receptor immunoreactivity in the ventral cochlear nucleus of the guinea pig, J. Comp. Neurol., 276:423-435.

GABA$_A$-BENZODIAZEPINE RECEPTORS: DEMONSTRATION OF PHARMACOLOGICAL SUBTYPES

IN THE BRAIN

Richard W. Olsen, Michel Bureau, Shuichi Endo, Geoffrey Smith, Lynn Deng, Douglas Sapp, and Allan J. Tobin[*]

Departments of Pharmacology and [*]Biology, University of California Los Angeles, California, 90024 USA

INTRODUCTION

The GABA$_A$ receptor is a ligand-gated chloride ion channel that mediates the majority of rapid-acting inhibitory synapses in the central nervous system (Olsen and Venter, 1986). The GABA$_A$ receptors are also the target of numerous clinically important depressant and excitatory drugs (Olsen, 1981; Tallman and Gallager, 1985; Biggio and Costa, 1988). The convulsant drug bicuculline acts as a competitive antagonist at the GABA recognition site, beta-carbolines block GABA function as 'inverse agonists' at the benzodiazepine recognition site, and picrotoxin and cage convulsants inhibit the chloride channel function at a site on the receptor complex distinct from the GABA and benzodiazepine receptor sites. Clinically important depressant benzodiazepines enhance GABA-mediated inhibition via their own binding sites on the receptor complex. Still additional sites on the receptor-ion channel complex mediate the action of barbiturates, steroid anesthetics, and possibly ethanol to enhance GABA$_A$ receptor function at the membrane level (Olsen and Venter, 1986; Biggio and Costa, 1988).

As described in the previous ISNR held in Hiroshima in 1987, the GABA$_A$-benzodiazepine receptor-chloride channel complex has been purified from the brain of several mammalian species (Olsen et al., 1988; Barnard et al., 1988; and Kuriyama and Taguchi, 1988). The purified protein contains the binding sites for GABA, benzodiazepines, barbiturates, picrotoxin-like convulsants, and steroid anesthetics, all associated with two major poly-peptide subunits, α (51-53 kDa) and β (56-58 kDa) (Barnard et al., 1988; Deng et al., 1986; Stauber et al., 1987; King et al., 1987; Olsen et al., 1988).

The two major subunits were cloned and expressed in <u>Xenopus</u> oocytes to produce GABA-regulated chloride channels (Schofield et al., 1987; Barnard et al., 1988). Subsequently, multiple GABA$_A$ receptor subunits called α, β, γ, and δ, and subtypes of these have been cloned (Levitan et al., 1988; Pritchett et al., 1989; Shivers et al., 1989; Khrestchatisky et al., 1989; reviewed in Olsen and Tobin, 1990). This multiplicity of clones appears to

Neuroreceptor Mechanisms in Brain, Edited by S. Kito *et al.*
Plenum Press, New York, 1991

reflect a heterogeneity of GABA$_A$ receptors *in vivo*, produced by a tissue-dependent expression of these multiple genes. These isoreceptors show differential physiological functions, mechanisms of regulation, and susceptibility to disease processes, but also differential sensitivity to pharmacological manipulation. In this chapter we describe biochemical evidence for multiple pharmacological subtypes of GABA$_A$ receptors <u>in the brain</u>; these can be related to the multiple cDNA species that have been isolated (Tobin et al., chapter in this volume).

HETEROGENEITY OF GABA$_A$-BENZODIAZEPINE RECEPTORS OBSERVED BY RADIOLIGAND BINDING IN BRAIN SECTIONS WITH AUTORADIOGRAPHY

GABA synapses and GABA$_A$ receptors are wide-spread and abundant throughout the nervous system. Some evidence for regional heterogeneity in pharmacological specificity (e.g., Johnston, 1986) and in ligand binding has been reported (reviewed in Olsen and Tobin, 1990; Biggio and Costa, 1990). Binding studies with brain sections using quantitative autoradiography showed in a particularly clear manner the existence of apparent subtypes of benzodiazepine receptors that varied in brain regional distribution, as well as discrepancies in the distribution of high affinity GABA agonist binding and benzodiazepine binding (e.g., Unnerstall et al., 1981; Young et al., 1981). This heterogeneity can be interpreted as either: [a] multiple conformational states of a single receptor (Yang and Olsen, 1987); [b] regional variation in the amount of interfering endogenous ligands for one or more sites on the receptor complex; [c] oligomeric complexes with multiple components, all components of which are not necessarily present in all complexes (Leeb-Lundberg and Olsen, 1983); or [d] distinct isoreceptor proteins/gene products. This last possibility appears to be supported by several lines of evidence.

While the GABA$_A$ receptor is believed to be an allosteric protein whose function involves ligand-regulated conformational changes, different affinity states do not appear to account for all of the observed binding heterogeneity. For example, the distribution of GABA antagonist binding (bicuculline: Olsen et al., 1984; or SR-95531: McCabe et al., 1988) does not agree with that of agonist binding (muscimol), although it does agree with that of benzodiazepine binding (flunitrazepam or Ro15-1788). The autoradiography technique detects only rather high affinity binding sites due to the necessity for tissue rinsing in nonradioactive buffer for 1-2 seconds. Nevertheless, attempts to measure intermediate-to-low affinity muscimol binding by employing short rinse times and high ligand concentrations increased detectable agonist binding but did not change significantly the distribution. Because of the differential localization, GABA binding sites that can be detected by agonist binding with the autoradiography technique do not appear to be the same as those labeled by GABA antagonists or benzodiazepines (Olsen et al., 1990a).

Indeed, closer examination of seven ligands for the three receptor sites detectable on the GABA$_A$ receptor complex by radioligand binding revealed that comparison of any pair of ligands shows considerable discrepancies in densities of binding across brain regions. This situation occurs despite the dogma that all these binding sites are supposed to be present on the same macromoleoular complex. Ligands compared included [3H]flunitrazepam, the 'BZ1-selective' ligand [3H]2-oxo-quazepam (Yezuita et al., 1988), [3H]muscimol, [3H]bicuculline methochloride, [3H]SR-95531, and two cage convulsants [^{35}S]TBPS and [^{3}H]TBOB. In addition to differences mentioned for GABA agonists versus antagonists and benzodiazepines, and for BZ1 versus BZ2 sites, there were significant differences in binding densities in some brain regions between any pair of the seven ligands including the two GABA antagonists and two cage convulsants! These data strongly suggest that subpopulations of receptors exist, and a minimum of four subtypes are required to explain the data in which some brain regions favor some ligands and disfavor others (Olsen et al., 1990a). Thus, at least four subtypes of GABA$_A$-benzodiazepine receptors with different pharmacological properties are demonstrated by binding data alone. Protein chemistry and molecular cloning studies support this interpretation.

In addition to the brain regional variations in ligand binding densities, regional variations in allosteric interactions between the various sites on the receptor complex(es) were observed. For example, we compared the allosteric modulation of [^{3}H]muscimol, [^{3}H]flunitrazepam, and [^{35}S]TBPS binding by barbiturates and anesthetic steroids across brain regions by quantitative autoradiography. The steroid anesthetic alphaxalone and naturally-occurring analogs enhance muscimol and flunitrazepam binding and inhibit TBPS binding in well-washed brain membranes (Turner et al., 1989). This compound showed a quite remarkable regional variation in such allosteric modulations. Alphaxalone at 1-10 μM completely inhibited TBPS in cerebellar granule layer, while showing no inhibition or enhancement in the molecular layer of cerebellum, as well as in the hippocampus and cerebral cortex (D. Sapp, U. Witte, D.M. Turner, N. Kokka and R.W. Olsen, manuscript in preparation). Under similar conditions, muscimol binding was enhanced by alphaxalone more than two-fold in several brain regions, including the cerebellar molecular layer and hippocampal field CA1, but virtually unaffected in other areas, including the cerebellar granule cell layer and thalamus (M. Bureau and R.W. Olsen, manuscript in preparation).

HETEROGENEITY OF GABA$_A$-BENZODIAZEPINE RECEPTOR SUBUNITS IDENTIFIED BY PROTEIN ISOLATION AND PHOTOAFFINITY LABELING

Multiple polypeptide subunits can be isolated in the purified GABA$_A$ receptor and these polypeptides show differential affinity for various ligands (Bureau and Olsen, 1990; Olsen et al., 1990b). Early studies on purified receptor preparations suggested a homogeneous single protein (Barnard et al., 1988; Möhler et al., 1980; Deng et al., 1986; King et al., 1987; Stauber et al., 1987). However, possible protein heterogeneity was suggested by polyphasic heat inactivation (Squires and Saederup, 1982), differential ontogeny (Lippa et al., 1981), differential loss following brain lesions (Guidotti et al., 1979), and differential solubilization (Lo et al., 1982) of benzodiazepine binding subpopulations.

Even more suggestive was the observation of multiple polypeptide bands on SDS gels following photoaffinity labeling of crude membrane homogenates with [^{3}H]flunitrazepam. In addition to the major polypeptide band at Mr 51 kDa labeled in all brain regions (Möhler et al., 1980), minor labeled species were detected with Mr of 53, 55, and 59 kDa (Sieghart and Karobath, 1980). These minor benzodiazepine binding polypeptides varied in brain regional distribution, ontogeny, tryptic digestion products, and in pharmacological specificity, indicating that they represented distinct gene products (Sieghart et al., 1983; Sieghart and Drexler, 1983). The major Mr 51 kDa band had a higher affinity for the BZ1-specific ligand CL-218872 and for β-carbolines than the higher Mr peptides, and was the only band observed in the cerebellum; the other bands were present in brain regions such as the hippocampus where lower affinities for these ligands were observed, i.e., BZ2 receptors were present. This idea that distinct binding subpopulations could be due to distinct proteins now appears to be borne out by cloning studies, but also requires further demonstration that distinct polypeptides with distinct binding properties are present in the brain.

As first described by Barnard et al. (1988), we also purified the GABA$_A$ receptor complex by benzodiazepine affinity chromatography, and obtained four bands on SDS gels for the rat·protein (Stauber et al., 1987). These included the two bands detected by Barnard et al. (1988), the α subunit at Mr 52 kDa and the β subunit at Mr 57 kDa, as well as a variable band at 47 kDa that we believe is a break-down product of the other two. The fourth band at about 32 kDa has not been identified and may be a contaminant or a receptor-associated protein (Stauber et al., 1987). The two major bands were further identified as receptor subunits by immunoblotting and photoaffinity labeling. Western blotting (Stauber et al., 1987) demonstrated that the 52 kDa band was recognized by an α-specific antibody, and the 57 kDa band by a β-specific antibody (Schoch et al., 1985). Rabbit antiserum raised against the purified rat receptor recognized at high dilution the α subunit in purified preparations and in crude membranes. The 52 kDa band was heavily labeled, and the 57 kDa band lightly photolabeled with [^{3}H]flunitrazepam, while the opposite was true (heavy labeling of the 57 kDa band, light labeling of the 52 kDa band) for [^{3}H]muscimol photolabeling (Stauber et al., 1987).

[^{3}H]Flunitrazepam photoaffinity labeling of purified bovine or rat cortical receptor revealed a microheterogeneity similar to that seen in crude membrane homogenates by Sieghart and Karobath (1980). A labeled doublet at 51 and 53 kDa was evident, and two small peaks at 56 and 58 kDa were also detected (Bureau and Olsen, 1988). These bands could represent cross-labeling of β subunits and the presence of a benzodiazepine binding site on β subunits. Alternatively, they could represent minor species of distinct α polypeptides that have been demonstrated by cloning -- multiple α polypeptides have been detected by immunoblotting (Fuchs et al., 1988; Sato and Neale, 1989; Vitorica et al., 1990; Duggan and Stephenson, 1990; Endo and Olsen, chapter in this volume). The possibility of cross-labeling was also suggested by [^{3}H]muscimol photoaffinity labeling, which indicated incorporation apparently into the same four polypeptides: a major doublet at 56 and 58 kDa and a minor doublet at 51 and 53 kDa (Bureau and Olsen, 1988; 1990). The latter radioactive bands could represent minor species of β subunits, or breakdown products of the 56 and 58 kDa bands, or labeling of the α subunits. The results are not conclusive at this time. Nevertheless, all of the subunit types including α do carry GABA binding sites as indicated by the expression of homo-oligomeric GABA-activated chloride channels from the cDNA or mRNA of any given individual subunit subtype (Schofield, 1989; Pritchett et al., 1989; Shivers et al., 1989; Olsen and Tobin, 1990). Thus, α subunits might be photolabeled with muscimol, possibly with an efficiency differing from β subunits. In addition, there is some possibility that subunits that do not actually carry a binding site can be photoaffinity labeled due to their proximity to the ligand binding site on a nearly subunit; the detailed nature of the photosensitive muscimol breakdown is not understood and thus allows for such a possiblity.

Protein staining of gels revealed at least four polypeptide bands in cow, rat, and human cortical receptor as shown in Figure 1. These appear to correspond to the photolabeled bands, but there may actually be more than

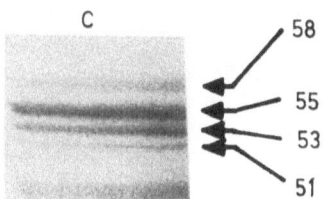

Figure 1. Microheterogeneity of GABA$_A$ Receptor on SDS Gel Electrophoresis Visualized by Protein Staining. The GABA$_A$-benzodiazepine receptor activity was solubilized in Triton X-100, monitored by [^{3}H]muscimol binding, and purified by benzodiazepine affinity chromatography as described in Stauber et al. (1987) and Bureau and Olsen (1990). This preparation from cow (C) was subjected to SDS polyacrylamide gel electrophoresis and stained with Coomassie Blue.

four polypeptides present. The four bovine polypeptide bands were isolated by electro-elution and subjected to N-terminal sequencing, but apparently were blocked, either biologically, or as the result of purification treatments. The isolated bands ran as single bands upon re-electrophoresis in SDS gels or isoelectric focussing, they gave unique amino acid compositions, and they produced unique proteolytic peptide fragments on one-dimensional peptide mapping gels (Bureau and Olsen, 1990). This indicates that they are not likely due to heterogeneity in proteolysis or glycosylation, but more likely represent distinct sequences corresponding to distinct gene products. The proteolytic fragments could be sequenced and provided evidence that the isolated proteins corresponded with the isolated cDNAs (Bureau, Smith, Olsen and Tobin, unpublished). Sequencing of photolabeled protein fragments provides information about ligand binding domains within the sequence (Smith and Olsen, manuscript in preparation).

Consistent with the conclusion that multiple gene products exist for benzodiazepine binding peptides in mammalian brain is our observation that the cerebellum contains only a single α band at Mr 51 kDa, instead of the doublet at 51 and 53 seen in cortex, while the hippocampus contains at least three α bands, at 51, 53, and 56 kDa (Bureau and Olsen, 1990). It is likely that additional polypeptides (α, β, γ, and δ) are present in the purified preparations, depending upon the tissue from which the receptor is prepared, the yield of purification on benzodiazepine affinity chromatography, and the stability to proteolysis during purification. We have produced antisera to subunit subtype-specific synthetic peptides based on several of these clones and shown them to react with distinct bands in the purified receptor on Western blots (Endo and Olsen, chapter in this volume).

Looking at different animal species, however, one finds a rather variable pattern of photoaffinity labeled subunits (e.g., Hebebrand et al., 1987). Benzodiazepine binding activity appears to develop phylogenetically at the time of sharks, with the bony fishes looking rather similar to, but not identical to, the higher vertebrates (Hebebrand et al., 1988; Nielsen et al., 1978). There is also some evidence for benzodiazepine receptors linked to GABA$_A$ receptors in the invertebrates such as insects (Robinson et al., 1986). We found that the Atlantic codfish brain showed a novel pattern of photolabeling with [^{3}H]flunitrazepam, yielding a single benzodiazepine binding polypeptide at about 56 kDa (Deng et al., 1988). Purification of the codfish receptor to homogeneity revealed that the protein was composed of primarily one single polypeptide of Mr 56 kDa, and this polypeptide was photolabeled with both flunitrazepam and muscimol. The codfish GABA-benzodiazepine receptor binding properties appeared to reflect a homogeneous population with specificity similar to that of the mammalian subtype BZ2. This suggests that the codfish receptor is a homo-oligomer with GABA and benzodiazepine binding sites residing on the same polypeptide subunit within the oligomer. This polypeptide is coded for by a gene closely related to the ancestral gene from which the multiple mammalian genes for different subunits and subunit sub-types arose by gene duplication and mutation during evolution (Deng, Nielsen, and Olsen, manuscript submitted).

The multiple mammalian polypeptide bands do differ in their affinity for various ligands, as shown by photoaffinity labeling. When varying concentrations of nonradioactive analogs of flunitrazepam or muscimol, or allosteric modulators, were included in the preincubation with the radioligands prior to photolabeling, differential effects on labeling of the various bands on SDS gels were observed for the rat receptor. For example, flunitrazepam primarily labeled a doublet at 51 and 53 kDa. The 51 kDa band had a higher affinity for displacement by the BZ1-selective ligand CL-218872, and for allosteric enhancement by pentobarbital and steroids, while the 53 kDa band was more enhanced by GABA than the 51 kDa band (Table 1). On the other hand, muscimol photolabeled the 51-53 kDa α doublet and more so the 56-58 kDa β doublet. The 58 kDa β band had a higher affinity for the GABA analog THIP and was a better substrate for phosphorylation by the cyclic AMP-dependent protein kinase than the 56 kDa band. The 56 kDa β and the 51 kDa (α?) bands were more sensitive to enhancement by the steroids and pentobarbital, and the 56 kDa β was a better substrate for phosphorylation by protein kinase C (Table 1; Bureau and Olsen, 1990; Browning et al., 1990). The differential binding properties of the different polypeptides strongly suggest that these bands are constituents of oligomeric receptors that differ in pharmacological properties in the purified receptors *in vitro*; this pharmacological heterogeneity also is likely to exist *in vivo*.

CONCLUSIONS

This chapter summarizes experimental evidence that the multiple subunit clones recently isolated for the GABA$_A$ receptor proteins probably result in isoreceptors in the brain that depend on the subunit composition. Not unexpectedly, these isoreceptors differ in their pharmacological as well as physiological properties. The subtype present in a given cell results from a tissue-specific expression of the various subunit and subunit subtype genes, resulting in one of many possible hetero-oligomeric complexes (Olsen and Tobin, 1990). While all of the clones may not be expressed in sufficient abundance to contribute significantly to any tissue's properties, multiple subtypes (8 or so) appear to exist, and even the minor species appear to have unique localization and presumably other unique properties. At this time, the subunit composition and stoichiometry are not known for any cell. In order to analyze this situation, we have produced subtype-specific antibodies against synthetic peptides based on subtype-specific and immunogenic regions of the deduced amino acid sequence of the cloned polypeptide subtypes (Endo and Olsen, chapter in this volume). These are useful in determining which polypeptide bands on gels are related to which clones, where they are localized, and with which other subunits they are associated. This information can be combined with heterologous cell expression studies to determine the

Table 1. Variable Pharmacology for Different Polypeptides in Purified Rat Brain GABA$_A$-Benzodiazepine Receptor Identified by Photoaffinity Labeling

Binding Site: Ligand	Mr Band (kDa)			
	51	53	56	58
[^{3}H]Muscimol:				
*THIP	ND	ND	+	+++
*Barbiturates	+++	+	+++	+
Steroids	+++	+	+++	+
[^{3}H]Flunitrazepam:				
GABA	+	+++	+	ND
CL-218872	+++	+	ND	ND
Steroids	+++	+	++	+
†Phosphorylation:				
PKA	+	?	+	+++
PKC	+	+	+++	+

Purified receptor was subjected to photoaffinity labeling with either [^{3}H]muscimol or [^{3}H]flunitrazepam, following equilibration at 4°C in the presence of varying concentrations of displacer analogs or enhancing modulatory substances. The samples were subjected to SDS-PAGE, gels sliced and counted for radioactivity. The (+) signs indicate degree of inhibition or enhancement (*Bureau and Olsen, 1990), or phosphorylation (†Browning et al., 1990). ND = not determined.

pharmacological properties of the various oligomers one can construct from the supermarket of available subunits and subtypes (Schofield, 1989). One can then compare this information with what is known about pharmacological subtypes observed in vivo (Johnston, 1986) or by radioligand binding both in brain sections and in isolated polypeptides. In this manner, the jigsaw puzzle of GABA$_A$ receptor heterogeneity begins to be solved.

Acknowledgements: Supported by NIH grants NS 22071; NS 21908; NS 22256; and NS 28772. We thank Lesley Jenkin for preparing the manuscript.

REFERENCES

Barnard, E.A., Darlison, M.G., Fujita, N., Glencorse, T.A., Levitan, E.S., Reale, V., Schofield, P.R., Seeburg, P.H., Squire, M.D., and Stephenson, F.A., 1988, Molecular biology of the GABA$_A$ receptor, in: "Neuroreceptors and Signal Transduction", S. Kito, T. Segawa, K. Kuriyama, M. Tohyama, and R.W. Olsen, eds., Plenum Press, New York.
Biggio, G., and Costa, E., eds., 1988, "Chloride Channels and Their Modulation by Neurotransmitters and Drugs", Raven Press, New York.

Biggio, G., and Costa, E., eds., 1990, "GABA and Benzodiazepine Receptor Subtypes: From Molecular Biology to Clinical Practice", Raven Press, New York.

Browning, M.D., Bureau, M., Dudek, E.M., and Olsen, R.W., 1990, Protein kinase C and cAMP-dependent protein kinase phosphorylate the β-subunit of the purified GABA$_A$ receptor. Natl. Acad. Sci. USA, 87:1315-1318.

Bureau, M., and Olsen, R.W., 1988, γ-Aminobutyric acid/benzodiazepine receptor protein carries binding sites for both ligands on both two major peptide subunits, Biochem. Biophys. Res. Commun., 153:1006-1011.

Bureau, M., and Olsen, R.W., 1990, Multiple distinct subunits of the γ-aminobutyric acid-A receptor protein show different ligand-binding affinities, Mol. Pharmacol., 37:497-502.

Deng, L., Olsen, R.W., and Nielsen, M., 1988, [^{3}H]Muscimol and [^{3}H]flunitrazepam photoaffinity label the same molecular weight band in codfish brain GABA/BZ receptor. Abstr. Soc. Neurosci., 14:168.

Deng, L., Ransom, R.W., and Olsen, R.W., 1986, [^{3}H]Muscimol photolabels the γ-aminobutyric acid receptor binding site on a peptide subunit distinct from that labeled with benzodiazepines, Biochem. Biophys. Res. Commun., 138:1308-1314.

Duggan, M.J., and Stephenson, F.A., 1990, Biochemical evidence for the existence of γ-aminobutyrate$_A$ receptor iso-oligomers, J. Biol. Chem., 265:3831-3835.

Fuchs, K., Möhler, H., and Sieghart, W., 1988, Various proteins from rat brain, specifically and irreversibly labeled by [^{3}H]flunitrazepam, are distinct α-subunits of the GABA-benzodiazepine receptor complex, Neurosci. Lett., 90:314-319.

Guidotti, A., Gale, K., Suria, A., and Toffano, G., 1979, Biochemical evidence for two classes of GABA receptors in rat brain, Brain Res., 172:566-571.

Hebebrand, J., Friedl, W., Breidenbach, B., and Propping, P., 1987, Phylogenetic comparison of the photoaffinity-labeled benzodiazepine receptor subunits, J. Neurochem., 48:1103-1108.

Hebebrand, J., Friedl, W., Reichelt, R., Schmitz, E., Moller, P., and Propping, P., 1988, The shark GABA/BZ receptor: Further evidence for a not so late phylogenetic appearance of the benzodiazepine receptor, Brain Res., 446:251-261.

Johnston, G.A.R., 1986, Multiplicity of GABA receptors, in: "Benzodiazepine/GABA Receptors and Chloride Channels: Structural and Functional Properties", R.W. Olsen and J.C. Venter, eds., Alan R. Liss, Inc., New York.

Khrestchatisky, M., MacLennan, A.J., Chiang, M.-Y., Xu, W., Jackson, M.B., Brecha, N., Sternini, C., Olsen, R.W., and Tobin, A.J., 1989, A novel alpha-subunit in rat brain GABA$_A$ receptors, Neuron, 3:745-753.

King, R.G., Nielsen, M., Stauber, G.B., and Olsen, R.W., 1987, Convulsant/barbiturate activities on the soluble GABA/benzodiazepine receptor complex, Eur. J. Biochem., 169:555-562.

Kuriyama, K., and Taguchi, J., 1988, Biochemical and functional properties of purified GABA receptor/benzodiazepine, receptor/chloride channel complex, and application of its antibody for immunohistochemical studies, in: "Neuroreceptors and Signal Transduction", S. Kito, T. Segawa, K. Kuriyama, M. Tohyama, and R. Olsen, eds., Plenum Press, New York.

Leeb-Lundberg, L.M.F., and Olsen, R.W., 1983, Heterogeneity of benzodiazepine receptor interactions with GABA and barbiturate receptors, Mol. Pharmacol., 23:315-325.

Levitan, E.S., Schofield, P.R., Burt, D.R., Rhee, L.M., Wisden, W., Köhler, M., Fujita, N., Rodriguez, H.F., Stephenson, F.A., Darlison, M.G., Barnard, E.A., and Seeburg, P.H., 1988, Structural and functional basis for GABA$_A$ receptor heterogeneity, Nature, 335:76-79.

Lippa, A.S., Beer, B., Sano, M.C., Vogel, R.A., and Meyerson, L.R., 1981, Differential ontogeny of type 1 and type 2 benzodiazepine receptors, Life Sci., 28:2343-2347.

Lo, M.M.S., Strittmatter, S.M., and Snyder, S.H., 1982, Physical separation and characterization of two types of benzodiazepine receptors, Proc. Natl. Acad. Sci. USA, 79:680-684.

McCabe, R.T., Wamsley, J.K., Yezuita, J.P., and Olsen, R.W., 1988, A novel GABA$_A$ antagonist [^{3}H]SR 95531: Microscopic analysis of binding in the rat brain and allosteric modulation by several benzodiazepine and barbiturate receptor ligands, Synapse, 2:163-173.

Möhler, H., Battersby, M.K., and Richards, J.G., 1980, Benzodiazepine receptor protein identified and visualized in brain tissue by a photoaffinity label, Proc. Natl. Acad. Sci. USA, 77:1666-1670.

Nielsen, M., Braestrup, C., and Squires, R.F., 1978, Evidence for a late evolutionary appearance of brain specific benzodiazepine receptors: An investigation of 18 vertebrate and 5 invertebrate species, Brain Res., 141:342-346.

Olsen, R.W., 1981, GABA-benzodiazepine-barbiturate receptor interactions, J. Neurochem., 37:1-13.

Olsen, R.W., Bureau, M., Khrestchatisky, M., MacLennan, A.J., Chiang, M.-Y, Tobin, A.J., Xu, W., Jackson, M., Sternini, C., and Brecha, N., 1990b, Isolation of pharmacologically distinct GABA-benzodiazepine receptors by protein chemistry and cloning, in: "GABA and Benzodiazepine Receptor Subtypes: From Molecular Biology to Clinical Practice", G. Biggio and E. Costa, eds., Raven Press, New York.

Olsen, R.W., Bureau, M., Ransom, R.W., Deng, L., Dilber, A., Smith, G., Khrestchatisky, M., and Tobin, A.J., 1988, The GABA receptor-chloride ion channel protein complex, in: "Neuroreceptors and Signal Transduction", S. Kito, T. Segawa, K. Kuriyama, M. Tohyama, and R.W. Olsen, eds., Plenum Press, New York.

Olsen, R.W., McCabe, R.T., and Wamsley, J.K., 1990a, GABA$_A$ receptor subtypes: Autoradiographic comparison of GABA, benzodiazepines, and convulsant binding sites in the rat central nervous system, J. Chem. Neuroanat., 3:59-76.

Olsen, R.W., Snowhill, E.W., and Wamsley, J.K., 1984, Autoradiographic localization of low affinity GABA receptors with [^{3}H]bicuculline methochloride, Eur. J. Pharmacol., 99:247-248.

Olsen, R.W., and Tobin, A.J., 1990, Molecular biology of GABA$_A$ receptors, FASEB J., 4:1469-1480.

Olsen, R.W., and Venter, J.C., eds., 1986, "Benzodiazepine/GABA Receptors and Chloride Channels: Structural and Functional Properties", Receptor Biochemistry and Methodology, Volume 5, Alan R. Liss, New York.

Pritchett, D.B., Lüddens, H., and Seeburg, P.H., 1989, Type I and type II GABA$_A$-receptors produced in transfected cells, Science, 245:1389-1392.

Robinson, T.N., MacAllan, D., Lunt, G.G., and Battersby, M., 1986, The GABA receptor complex of insect central nervous system: Characterization of a benzodiazepine binding site. J. Neurochem., 47:1955-1962.

Sato, T.N., and Neale, J.H., 1989, Immunological identification of multiple α-like subunits of the γ-aminobutyric acid$_A$ receptor complex purified from neonatal rat cortex, J. Neurochem., 53:1089-1095.

Schoch, P., Richards, J.G., Haring, P., Takas, B., Stahli, C., Staehelin, T., Haefely, W., and Möhler, H., 1985, Co-localization of GABA$_A$ receptors and benzodiazepine receptors in the brain shown by monoclonal antibodies, Nature, 314:168-171.

Schofield, P.R., 1989, The GABA$_A$ receptor: Molecular biology reveals a complex picture, Trends Pharmacol. Sci., 10:476-478.

Schofield, P.R., Darlison, M.G., Fujita, N., Burt, D.R., Stephenson, F.A., Rodriguez, H., Rhee, L.M., Ramachandran, J., Reale, V.,

Glencorse, T.A., Seeburg, P.H., and Barnard, E.A., 1987, Sequence and functional expression of the GABA-A receptor shows a ligand-gated receptor super-family, Nature 328:221-227.

Shivers, B.D., Killisch, I., Sprengl, R., Sontheimer, H., Köhler, M., Schofield, P.R., and Seeburg, P.H., 1989, Two novel GABA$_A$ receptor subunits exist in distinct neuronal subpopulations, Neuron, 3:327-337.

Sieghart, W., and Drexler, G., 1983, Irreversible binding of [^{3}H]flunitrazepam to different proteins in various brain regions, J. Neurochem., 41:47-55.

Sieghart, W., and Karobath, M., 1980, Molecular heterogeneity of benzodiazepine receptors, Nature, 286:285-287.

Sieghart, W., Mayer, A., and Drexler, G., 1983, Properties of [^{3}H]flunitrazepam binding to different benzodiazepine binding proteins, Eur. J. Pharmacol., 88:291-299.

Squires, R.F., and Saederup, E., 1982, γ-Aminobutyric acid receptors modulate cation binding sites coupled to independent benzodiazepine, picrotoxinin, and anion binding sites, Mol. Pharmacol., 22:327-334.

Stauber, G.B., Ransom, R.W., Dilber, A.I., and Olsen, R.W., 1987, The γ-aminobutyric acid-benzodiazepine receptor protein from rat brain: Large-scale purification and preparation of antibodies, Eur. J. Biochem., 167:125-133.

Tallman, J., and Gallager, D., 1985, The GABAergic system: A locus of benzodiazepine action, Annu. Rev. Neurosci., 8:21-44.

Turner, D.M., Ransom, R.W., Yang, J.S., and Olsen, R.W., 1989, Steroid anesthetics and naturally-occurring analogs modulate the γ-aminobutyric acid receptor complex at a site distinct from barbiturates, J. Pharmacol. Exp. Ther., 248:960-966.

Unnerstall, J.R., Kuhar, M.J., Niehoff, D.L., and Palacios, J.M., 1981, Benzodiazepine receptors are coupled to a subpopulation of GABA receptors: Evidence from a quantitative autoradiographic study, J. Pharmacol. Exp. Ther., 218:797-804.

Vitorica, J., Park, D., Chin, G., and deBlas, A.L., 1990, Characterization with antibodies of the γ-aminobutyric acid$_A$ benzodiazepine receptor complex during development of the rat brain, J. Neurochem., 54:187-194.

Yang, J., and Olsen, R.W., 1987, γ-Aminobutyric acid receptor binding in fresh mouse brain membranes at 22°C: Ligand-induced changes in affinity, Mol. Pharmacol., 32:266-277.

Yezuita, J.P., McCabe, R.T., Barnett, A., Iorio, L.C., and Wamsley, J.K., 1988, Use of the selective benzodiazepine-1 (BZ-1) ligand [^{3}H]2-oxo-quazepam (SCH 15-725) to localize BZ-1 receptors in the rat brain, Neurosci. Lett., 88:86-92.

Young, W.S. III, Niehoff, D.L., Kuhar, M.J., Beer, B., and Lippa, A.S., 1981, Multiple benzodiazepine receptor localization by light microscopic radiohistochemistry, J. Pharmacol. Exp. Ther., 216:425-430.

STRUCTURAL, DEVELOPMENTAL AND FUNCTIONAL HETEROGENEITY OF RAT GABA$_A$

RECEPTORS

Allan J. Tobin,[1,2,3] Michel Khrestchatisky,[1,8]
A. John MacLennan,[1,9] Ming-Yi Chiang,[1]
Niranjala J.K. Tillakaratne,[1] Wentao Xu,[1]
Meyer B. Jackson, [1,2] Nicholas Brecha,[2,4,7]
Catia Sternini,[5,7] and Richard W. Olsen[2,6]

[1]Department of Biology, [2]Brain Research Institute
[3]Molecular Biology Institute, [4]Department of Anatomy
[5]Department of Medicine, [6]Department of Pharmacology
University of California, Los Angeles, CA 90024
[7]Center for Ulcer Research, Veterans Administration Medical
Center, Los Angeles, California 90073
[8]Present address: Unite INSERM 29, 123 Boulevard de Port Royal
75014 Paris, France
[9]Present address: Department of Neuroscience, University of
Florida College of Medicine, Box J-244, J.H. Miller Health
Center, Gainesville, Florida 32610

INTRODUCTION

My laboratory has had a long interest in multigene families,
starting with our earlier work on the developmental regulation of
hemoglobin switching. I have especially been impressed both by the
number of multigene families in the vertebrate genome and by their
widespread developmental regulation, as exemplified by switching within
the globin gene families. Even the lamprey, whose adult hemoglobin,
unlike those of higher vertebrates, contains only one type of globin
chain, has different forms in its larval and adult stages.

The existence of multigene families reflects evolutionary history.
Such families almost certainly arise by gene duplication, unequal
crossing over, and sequence divergence. For many gene family members,
sequence divergence occurred long before the evolution of distinct
species (Hood, et al., 1979). In many cases the different polypeptides
appear to be susceptible to intense natural selection, so that cognate
genes in different species are for more alike than different members of
a multiple gene family. The α_1 and α_2 GABA receptor genes are examples,
with 99% and 96% identity between the corresponding genes of rat
(Khrestchatisky et al., 1989) and cow (Levitan et al., 1988), a pattern
similar to that of the α globin gene family. In other cases, such as α_3
(cow) and α_4 (rat), no cognates have yet been reported in other species.
Perhaps this pattern is similar to that of the evolution of β-globin
family, where vertebrate species have different types and even different
members of β-like globin genes expressed in embryonic or fetal life.

Even in this case, however, individual members may serve distinct functional roles.

The persistence of multiple gene families in many cases reflects differing physiological needs in different cells and at different developmental stages. It may also reflect a need for coordinate regulation. In addition to information specifying an amino acid sequence, each gene contains DNA sequences that establish its pattern of expression, specifying both the cell types in which it is to be expressed and its regulation in response to environmental changes. For genes expressed in the brain, such regulation may be extremely complicated. From the standpoint of a molecular biologist, the two extraordinary characteristics of the vertebrate brain are (1) its huge numbers of component cell types, and (2) its ability to alter its state -- including its pattern of gene expression -- in response to natural sensory input or to experimentally induced changes in neural activity.

In this context, I would like to tell you about our work on the α-polypeptide family of GABA$_A$ receptors. In this paper, I present data on the structural, functional, and developmental heterogeneity of three α polypeptides from rat brain, which contain the major benzodiazepine-binding sites in a heterooligimeric ligand-gated ion channel.

SEQUENCE HETEROGENEITY OF GABA$_A$ RECEPTORS

We used the published bovine α_1 cDNA sequence of Schofield et al. (1987) to isolate three distinct rat brain cDNAs (Khrestchatisky et al., 1989). These encode three α polypeptides, which we call α_1, α_2, and α_4. α_4 represents a previously undescribed sequence and contributes to a receptor with distinct physiological properties and cellular distribution. We have not found a rat counterpart to the bovine α_3 cDNA, originally reported by Levitan et al., (1988).

The deduced amino acid sequences of rat α_1 and α_2 are nearly identical to their bovine cognates. The rat α_1 clone is 1536 nucleotides long, and its sequence is 90% identical to the nucleotide sequence in the coding region of bovine α_1. The sequence specifies an open reading frame of 455 codons, of which 27 are in a putative signal peptide and 428 are present in the mature polypeptide. At the amino acid level, the rat α_1 polypeptide is 99% identical to the bovine α_1 polypeptide. Surprisingly, codon 4 of the bovine α_1 sequence is missing in all of the α_1 cDNAs that we have sequenced.

The nucleotide sequence of the rat α_2 cDNA reveals an open reading frame of 481 codons in length, of which 28 may specify a signal peptide (Khrestchatisky et al., manuscript submitted). This cDNA is the rat counterpart of the bovine α_2 subunit (Levitan et al., 1988), since the deduced rant and cow amino acid sequences are 96% identical.

The other GABA$_A$ receptor cDNA whose sequence we have reported, which we call α_4, contains an open reading frame of 464 codons of which are 433 are likely to represent the mature polypeptide. The deduced amino acid sequence has significant similarities to other rat and bovine α polypeptides. Comparing the rat α_4 amino acid sequence with those to rat α_1 and bovine α_2 and α_3, we find that about 70%, the residues are identical. If we include conservative amino acid substitutions in our estimates, the fraction rises to about 80%.

Differences in amino acid sequences are not evenly distributed throughout the polypeptides, however, as illustrated in Figure 1, which shows the common and variant amino acid residues among the published GABA$_A$ receptor polypeptides. Presumed transmembrane domains and immediate flanking regions are highly conserved. Thus, comparing the rat α_4 polypeptide sequence with that of rat α_1, we find 99% conservation between residues 215-319 and 84% between residues 382-419. Presumed extracellular domains diverge more, with 78% identity to rat α_1 and bovine α_2. Presumed intracellular domains of α_4 show only poor similarity, with 22% identity between residues 320-381.

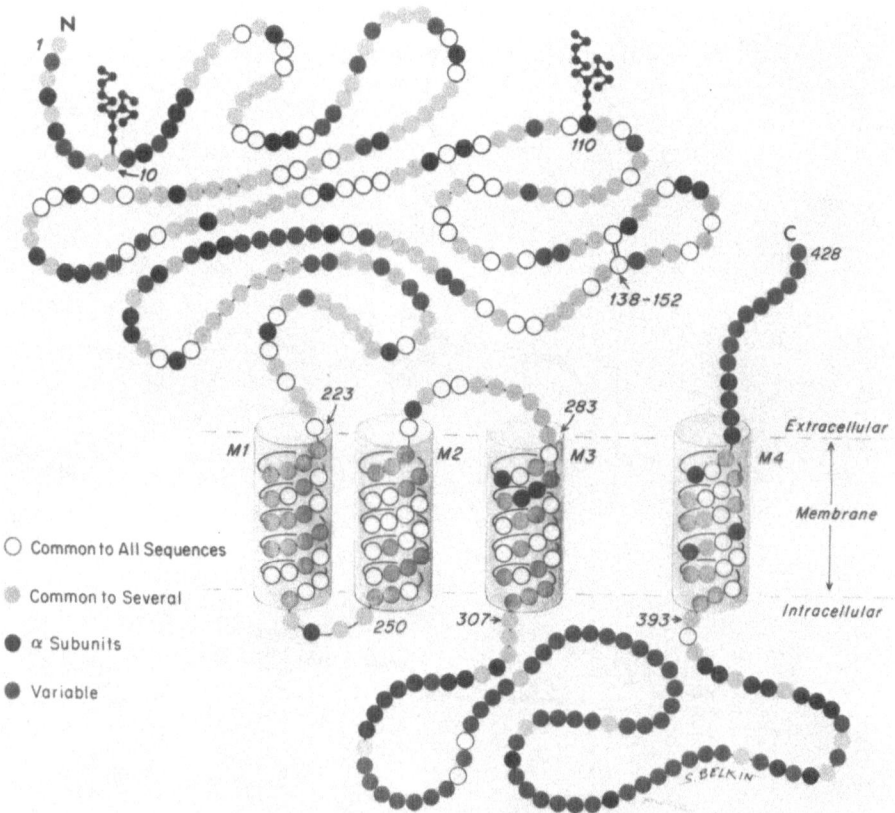

Figure 1. Comparison of published amino acid sequences of GABA$_A$ receptors. M1, M2, M3, and M4 are putative transmembrane domains.

In collaboration with Dr. Meyer Jackson at UCLA, we have injected mRNAs for individual GABA$_A$ receptor polypeptides into Xenopus oocytes to procure GABA-dependent inward currents (presumably chloride channels). These RNAs were each derived from cloned cDNAs. We observed a dose-dependent response to GABA (EC$_{50}$ = 5 micromolar), a reversal potential of 27 mV, and blockade by bicuculline and picrotoxin.

In comparing the properties of channels expressed from the rat α_1 plus β_1 with those produced from α_4 plus β_1, we found that the EC$_{50}$s of GABA were about the same, but that the maximal response to GABA was much bigger with α_1. As shown in Figure 2, both α_1 plus β_1 and α_4 plus β_1

channels were inhibited by picrotoxin. The α_4 plus β_1 channels, however, produced outward currents when picrotoxin alone was added (as also shown in Figure 2), suggesting a block of spontaneously opening channels produced by the GABA receptor mRNAs. Such picrotoxin response have not been observed in neurons, but spontaneous opening of GABA$_A$ chloride channels can occur, and has also been seen in oocytes injected with mRNAs for α_1 plus β_1.

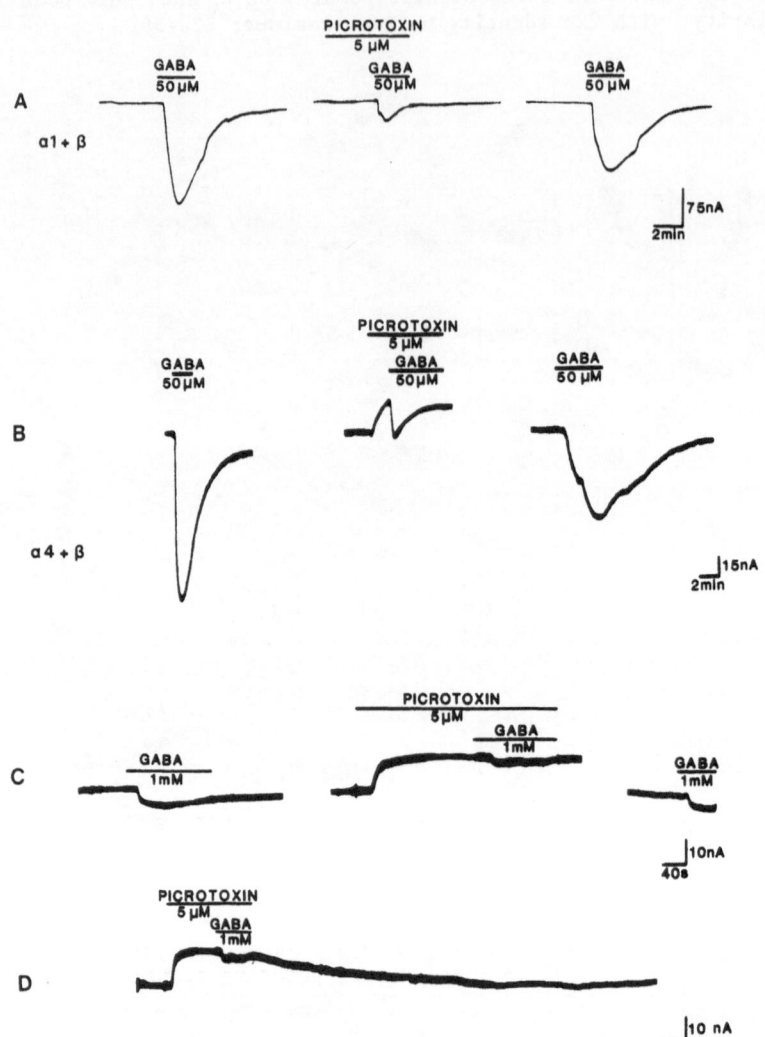

Figure 2. Expression of GABA$_A$ receptor polypeptides in <u>Xenopus</u> oocytes. Oocytes injected with individual or combinations of α_1, α_4 and β_1 RNAs were voltage clamped at -60 mV and challenged with GABA, picrotoxin, or both. The time of application is indicated by the labeled bars. Downward deflections indicate inward currents. (A) an oocyte expressing α_1 plus β_1 polypeptides, challenged first with GABA, then with GABA and picrotoxin, and again with GABA alone; (B) oocyte expressing α_4 plus β_1, challenged as in (A); (C) a second example of the magnitudes of picrotoxin-elicited currents different from that in (B); (D) an oocyte expressing α_4 plus β_1 polypeptides, challenged first with picrotoxin and then with GABA and picrotoxin, showing reversal of the picrotoxin response.

DEVELOPMENTAL REGULATION OF GABA RECEPTOR mRNAs

We have used our cDNA clones to study the distribution of α_1, α_2, and α_4 mRNAs in the brain, both by Northern blot analysis and by <u>in situ</u> hybridization histochemistry. As illustrated in Figure 3, poly(A)$^+$ RNA from the frontal cortex of adult rats shown two distinct α_1 mRNAs, with lengths of 3.8 and 4.3 kb. The mRNAs for α_2 were 3.6 and 6.6 kb long, while that for α_4 mRNA was 2.8 kb long. Total RNA from cortex and hippocampus contained all three mRNAs, but cerebellum contained only α_1 mRNA. In the substantia nigra-ventral tegmental area and in the ventral forebrain, the α_1 messengers were clearly present, whereas the α_2 and α_4 messengers were barely detectable. Neither α_1 nor α_4 mRNAs were detectable in striatum, which, however, did contain α_2 mRNA.

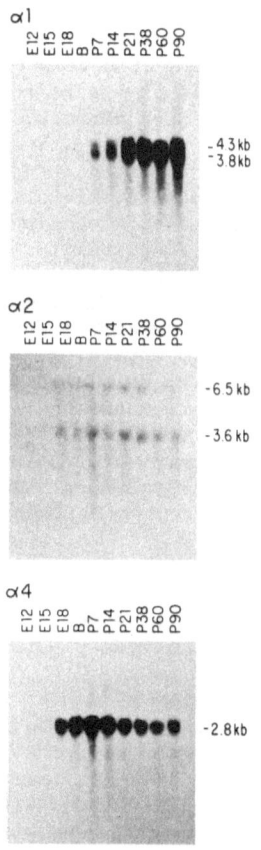

Figure 3. Northern blot analysis with 10 ug poly (A)$^+$ RNA extracted from rat brain at different developmental times.

α_1, α_2, and α_4 mRNAS display different patterns of ontogenetic regulation (MacLennan, et al., manuscript submitted). Northern blot analysis revealed that α_1 mRNA is first expressed in substantial quantities at postnatal day 7. Much lower quantities were detected at embryonic day 18 and at birth, when longer autoradiographic exposure intervals were employed. The concentration of α_1 mRNA dramatically increases between postnatal day 7 and postnatal day 38. From postnatal day 21 to postnatal day 90 the α_1 mRNA levels continue to rise, albeit at a slower rate.

In contrast, α_2 mRNAs were first detected at embryonic day 18.
Maximal levels were found between postnatal days 7 and 38. The α_4 mRNA
is also first seen at embryonic day 18. Its concentration rapidly rises
to its highest level at postnatal day 7 and then just as rapidly falls
to adult levels by postnatal day 21.

In situ hybridization experiments done in collaboration with Dr.
Nicholas Brecha and Catia Sternini confirmed the distribution seen in
Northern blots and showed the cellular distribution of the α mRNAs, as
shown in Figure 4. Overall, a differential expression pattern of α_1, α_2
and α_4 mRNAs is seen in neurons distributed along the neuraxis. α_1 RNAs
are more abundant than either α_2 or α_4 RNAs. Glial cells did not appear
to express these α RNAs.

In many regions, all three mRNA species are present in the same
structure, but often in different cell types and at different levels.
In other structures, one or two of these mRNA species predominate. A_2
and α_4 cRNA probes do not cross-hybridize with α_1 mRNA, which is abundant
in the Purkinje and granule cell layers of the cerebellum, as shown by
our published in situ hybridization studies and by Northern blot
analysis.

Structures with high to moderate levels of α_1 mRNA include the
olfactory bulb, olfactory tubercle, globus pallidus, amygdala,
hippocampal complex, and cerebral cortex (MacLennan et al., manuscript
submitted). Many hypothalamic, thalamic and pretectal nuclei also
contain high to moderate levels of α_1 RNAs. Many midbrain structures
including the substantia nigra, superior and inferior colliculi are
labeled, some prominently. High levels of α_1 RNAs are present in the
cerebellar cortex and in all deep cerebellar nuclei. In the pons and
medulla, many nuclei have high to moderate levels of α_1 RNAs.
Structures with low to undetectable levels of α_1 transcripts include the
nucleus accumbens, caudate-putamen, red nucleus of stria terminalis, and
many other nuclei located within the hypothalamus, thalamus, midbrain,
pons, medulla, and spinal cord.

α_2 RNAs are less abundant and have a more limited distribution
than α_1 RNAs. The highest labeling is observed in the olfactory bulbs,
hippocampal complex and the cranial motor nuclei. Moderate labeling is
found in the olfactory tubercle, caudate-putamen, cerebral cortex, and
amygdala.

α_4 RNAs are also less abundant than α_1 RNAs and are mainly
expressed in neurons present in the olfactory bulbs and the hippocampal
complex. Moderate levels of α_4 RNAs are seen in olfactory tubercle,
cerebral cortex, and amygdala. Structures with low or non-detectable
levels of either α_2 or α_4 RNAs include the hypothalamus, thalamus,
pretectum, midbrain, cerebellum, pons and medulla.

α_1, α_2 and α_4 RNAs are present in all regions of the cerebral
cortex, with α_1 RNAs generally more abundant than α_2 and α_4 RNAs. α_1
mRNAs are distributed in all cortical layers (I-VI). In contrast, α_2
mRNAs are preferentially distributed in superficial cortical layers (II
& III), and α_4 RNAs are distributed most prominently to the deeper
cortical layers (V & VI). Both α_2 and α_4 RNAs are also expressed by
cells in the other cortical layers.

As illustrated in Figure 4, the overall level of α mRNAs in the
hippocampal formation (dentate gyrus and hippocampus) is high to

370

moderate and comparable to other structures including the olfactory
bulbs, cerebral cortex and cerebellum. In the dentate gyrus, all three α
RNAs are expressed by numerous, and perhaps all, granule cells of the
granule cell layer. The relative level of α_2 RNAs appears to be higher
than either α_1 or α_4 RNAs in this layer. A prominent population of
neurons containing high levels of α_1 mRNAs are located in the hilus,
immediately adjacent to the granule cell layer, scattered within the
hilus and located adjacent to the pyramidal cell layer in area CA3 near
the border of area CA3-CA4. Likewise, moderate to low levels of α_2 RNA
are found in hilar neurons. On the basis of their distribution and
density, many of these cells are likely to be hilar interneurons,
perhaps those that express somatostatin-or GABA-RNAs.

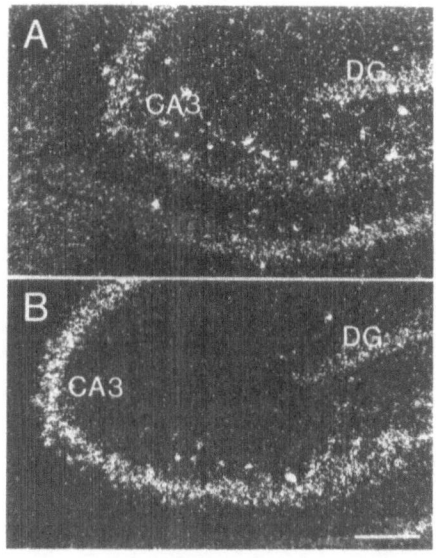

Figure 4. <u>In situ</u> hybridization histochemistry of hippocampal
formation probed with [^{35}S]-RNAs transcribed from α_1 and α_4 cDNAs. (A)
α_1 probe; (B) α_4 probe. Dark-field photographs. DG = dentate gyrus;
CA3 = field CA3 of Ammon's horn. Some cells outside the pyramidal
cell layer are heavily labeled with α_1 probe. α_4 probe appears to
label the vast majority of the cells in the granular and pyramidal
layers. bar = 200 μm.

     In the hippocampus itself, all three α subunits are localized to
neurons of the pyramidal cell layer of Ammon's horn, although there are
marked regional differences in their distribution and level of
expression. There is a gradient of α_1 RNA expression such that the
highest levels are present in areas CA1 and CA2, a lower level in area
CA3 and the lowest level in area CA4. In contrast, α_2 RNAs are highest
in areas CA2, CA3 and CA4, with somewhat lower levels in area CA1.
Similarly, α_4 transcripts are highest in areas CA3 and CA4, lower in
area CA2, and lowest in area CA1. The continuous distribution of label

over the pyramidal cell layer in areas CA1 to CA4 suggests that α RNAs are expressed by most and perhaps all neurons in this layer.

Table 1 summarizes what we have learned about the family of α polypeptides of the GABA$_A$ receptors. We have documented differences in structure, function, regional and cellular distribution within the brain, and have also shown differential regulation during ontogeny.

TABLE 1

COMPARISON OF α POLYPEPTIDES OF THE RAT GABA$_A$ RECEPTOR

Sequence

	α_1	α_2	α_4
Number of amino acids	428	421	433
Leader sequence	27	30	31
Identity in amino acids with bovine:			
α_1	99%	70%	70%
α_2	76%	96%	71%
α_4	70%	70%	66%

Messenger RNA

	α_1	α_2	α_4
sizes (kb)	3.8	3.6	2.8
	4.3	6.5	

Distribution

	α_1	α_2	α_4
Hippocampus	++	+	++
Cerebral cortex	+++	+	+
Cerebellum	+++	-	-
Striatum	±	+	-
Substantia nigra	++	-	±
Olfactory bulb	++	+	+
Thalamus	+	-	-

Expression in oocytes

	α_1	α_2	α_4
GABA current	+	*	*
EC50 (μM)	~5	~5	
Average maximal responses (nA)	284	80	
(to 50 μM GABA)			
Sensitive to picrotoxin	+		+
Picrotoxin currents	+		++

*Not Tested

THE GABA SYNTHETIC ENZYME, GLUTAMATE DECARBOXYLASE, IS ALSO PART OF A MULTIGENE FAMILY

Finally, I want to present some new and surprising data, not on the GABA receptor, but concerning the GABA synthetic enzyme, glutamate decarboxylase (GAD). Several years ago we isolated the first cDNA for GAD, by screening a λgt-11 bacterial expression library, made from feline brain mRNA, with an anti-GAD antibody (Kaufman, et al., 1986). The antibody we used in this screening, that of Oertel et al. (1981) has been widely used to identify GABA neurons by immunohistochemistry. We were able to confirm the identity of our cDNA because -- rather remarkably -- the beta-galactosidase - GAD fusion protein whose synthesis was directed by the recombinant DNA was enzymatically active. That is, it stoichiometrically converted glutamate to GABA$_A$ and CO_2.

Recently we prepared a new antiserum to GAD by directing the synthesis of feline GAD in a bacterial expression system. Programmed bacteria then produced feline GAD cDNA, to the extent of 40% of their protein. We then used the new antiserum to detect GAD an Western blots.

While the conventional antisera to GAD recognized two polypeptides on Western blots, with M_rs of 65,000 and 67,000, our new antiserum recognizes only the larger of these polypeptides. In cotrast, the GAD-6 monoclonal antibody prepared by Chang and Gottlieb (1988) recognizes only the smaller polypeptide. We then examined the distribution of the two polypeptides in the cerebellar cortex, in immunohistochemical experiments performed by Dr. Carolyn Houser, in the Department of Anatomy and Cell Biology at UCLA. The results with the new antiserum were surprising: two forms of GAD have strikingly different intracellular distributions. The smaller polypeptide apparently lies only in axon terminals, while the larger form is also present in cell bodies and even in the primary and secondary dendrites of Purkinje cells. The two forms also differ in their interactions with the GAD cofactor pyridoxal phosphate.

In the last few months we have used the polymerase chain reaction and subsequent subcloning to show that the two forms of GAD derive from separate mRNAs, which are the products of separate genes. We have obtained full length coding regions for these two polypeptides, determined their sequences, and shown that the polypeptides where synthesis they direct in bacterial are enzymatically active. Thus GAD, like the $GABA_A$ receptor, derives from a multiple gene family; Here the significance of the multigene family may lie in the different distributions of the polypeptide products within a single cell.

CONCLUSION

Multigene families encode polypeptides with parallel, but often distinctive functional properties. Even in cases where there are no apparent functional differences among related polypeptides, however, multigene families allow organisms to regulate the total amounts and the distributions of functional molecules. Such regulatory flexibility appears to be heavily used in the brain and may underlie the well-known plasticity of mammalian central nervous system.

Acknowledgements: This work was supported by NIH grants to AJT (NS 22256), to NB (EYO4067), to CS (DK38752 and DK40469), and to AV Delgado-Escueta (NS21908). MK was partially supported by a fellowship from the Fondation de l'Industrie Pharmaceutique pour la Recherche and AJM was funded by a fellowship from the Medical Research Council of Canada.

REFERENCES

Chang, Y.-C., and Gottlieb, D.I. (1988). Characterization of proteins purified with monoclonal antibodies to glutamic acid decarboxylase. J. Neurosci. 8, 2123-2310.
Hood, L., Campbell, J.H., and Elgin, S.C.R (1975). The organization, expression, and evolution of antibody genes and other multigene families. Ann. Rev. Genetics 9, 305-353.
Kaufman, D.L., McGinnis, J.F., Krieger, N.R., and Tobin, A.J. (1986). Brain glutamate decarboxylase gloned in λgt-11: fusion protein produces γ-aminobutyric acid. Science 232, 1138-1140.

Khrestchatisky, M., Maclennan, A.J., Chiang, M.Y., Xu, W., Jackson, M.,
     Brecha, N., Sternini, C., Olsen, R.W., and Tobin, A.J. (1989) A
     novel alpha subunit in rat brain GABA$_A$ receptors. <u>Neuron 3</u>, 745-
     753.
Levitan, E.S., Schofield, P.R., Burt, D.R., Rhee, L.M., Wisden, W.,
     Kohler, M., Fugita, N., Rogriguez, H., Stephenson, F.A., Darlison,
     M.G., Barnard, E.A., and Seeburg, P.H. (1988). Structural and
     functional basis for GABA$_A$ receptor heterogeneity. <u>Nature 335</u>,
     76-79.
Oertel, W.H., Schmechel, D.E., Tappaz, M.L., and Kopin, I.J. (1981).
     Production of a specific antiserum to rat brain glutamic acid
     decarboxylase by injection of an antigen-antibody complex.
     <u>Neuroscience 6</u>, 2689-2700.
Olsen, R.W. and Tobin, A.J. (1990). Molecular biology of GABA$_A$
     receptors. <u>FASEB J. 4</u>, 1469-1480.
Olsen, R.W., and Venter, J.C. (1986) Benzodiazapine/GABA$_A$ receptors and
     chloride channels: Structural and functional properties.
     Receptor Biochemistry and Methodology, Vol. 5 (New York: Alan R.
     Liss).
Olsen, R.W., McCabe, R.T., and Walmsley, J.K. (1990) GABA$_A$ receptor
     subtypes: Autoradiographic comparison of GABA$_A$, benzodiazepine,
     and convulsant binding sites in the rat central nervous system.
     <u>J. Chem. Neuroanatomy 3</u>, 59-76.
Pritchett, D.B., Luddens, H., and Seeburg, P.H. (1989). Type I and Type
     II GABA$_A$-benzodiazepine receptors produced in transfected cells.
     <u>Science 245</u>, 1389-1392.
Pritchett, D.B., Sontheimer, H., Gorman, C.M., Kettenmann, H., Seeburg,
     P.H., and Schofield, P.R. 1988). Transient expression shows
     ligand gating and allosteric potentiation of GABAa receptor
     subunits. <u>Science 242</u>, 1306-1308.
Schofield, P.R., Darlison, M.G., Fujita, N., Burt, D.R., Stephenson,
     R.A., Rodriguez, H., Rhee, L.M., Ramachandran, J., Reale, V.,
     Glencorse, T.A., Seeburg, P.G., and Barnard, E.A. (1987).
     Sequence and functional expression of the GABAa receptor shows a
     ligand-gated receptor superfamily. <u>Nature 328</u>, 221-227.
Shivers, B.D., Killishch, I., Sprengel, R., Sontheimer, H., Kohler, M.,
     Schofield, P.R., and Seeburg, P.H. (1989). Two novel GABAa
     receptor subunits exist in distinct neruonal subpopulations.
     <u>Neuron 3</u>, 327-337.
Siegel, R.E. (1988). The mRNAs encoding GABAa/benzodiazepine receptor
     subunits are localized in different cell populations of the bovine
     cerebellum. <u>Neuron 1</u>, 579-584.

SUBUNIT AND SUBTYPE-SPECIFIC ANTIBODIES TO THE GABA-A/BENZODIAZEPINE RECEPTOR COMPLEX

Shuichi Endo and Richard W. Olsen

Department of Pharmacology and Brain Research Institute, UCLA School of Medicine, Los Angeles, CA 90024 USA

INTRODUCTION

The GABA-A/benzodiazepine (BZ) receptor complex consists of a chloride channel which is gated by the inhibitory neurotransmitter, GABA. A number of therapeutically important drugs, such as benzodiazepines, barbiturates, and steroid anesthetics interact with the receptor complex and allosterically modulate the receptor (Olsen et al., 1988).

The GABA-A/BZ receptor complex was purified from bovine brain with BZ affinity column chromatography and was composed of two major polypeptide bands on SDS-polyacrylamide gel. The smaller molecular weight ($Mr = 52$ kDa) band, termed α, is photoaffinity-labeled with benzodiazepine and the larger one ($Mr = 57$ kDa), termed β, with the GABA agonist, muscimol (Barnard et al., 1988; Olsen et al., 1988).

However, the primary structures deduced from molecular cloning of cDNAs encoding the GABA-A receptor have shown the existence of subtypes of α ($\alpha1$, $\alpha2$, $\alpha3$, $\alpha4$) and β ($\beta1$, $\beta2$, $\beta3$), as well as other classes of subunit, γ and δ. GABA-A/BZ receptors expessed in Xenopus oocytes or mammalian cells with different subunit and subtype combinations displayed different pharmacological properties (reviewed in Olsen and Tobin, 1990).

Autoradiographic comparison of ligand binding has visualized the existence of at least four types of GABA/BZ receptors with different pharmacological properties in rat brain. The distribution of mRNAs for the different subunits and subtypes showed variation of brain regional distributions. It is likely that the GABA-A/BZ receptor is a hetero-oligometric complex whose subunit and subtype combinations vary in different brain regions, resulting in different physiological and pharmacological properties (see chapters by Olsen et al. and Tobin et al., this volume).

Subunit and subtype-specific antibodies have been produced to the specific sequences deduced from cDNAs of the GABA-A/BZ receptor to: (i) demonstrate the existence of gene products of subunit subtypes; (ii) determine the regional distribution of subunit subtypes; and (iii) correlate the distribution and combination of the subtypes with pharmacologically defined subtypes of GABA-A/BZ receptors.

METHODS

Production of Anti-Peptide Antibodies

Subunit and subtype-specific antibodies to the GABA-A/BZ receptor complex were raised against synthetic peptides which correspond to amino acid sequences of α1, α2, and β1 subunit subtypes deduced from cDNAs encoding GABA-A receptor subunits from bovine brain. The sequences for peptide synthesis were selected from non-homologous sequences among subunits and subtypes, and evaluated by hydropathy analysis of Kyte and Doolittle (1982) and secondary structure prediction of Chou and Fasman (1978). The former gives hydrophilic sequences that are most likely to be exposed on the surface of the native molecule and which are therefore potential sites of interaction with antibodies. The latter gives β-turns or loops that tend to be regions of high mobility and are likely to cross-react with corresponding anti-peptide antibodies (Tainer et al., 1984). The sequences chosen for peptide synthesis are shown in Table 1.

Table 1. Synthetic Peptides and Comparison of Their Sequences Among Subtypes

Bovine/Rat	α1	TIEPKEVKPEC (394-403)
Bovine/Rat	α2	TTPEPNKKPEC (392-401)
Bovine	α3	AAPSTSSTPT
Rat	α4	VGTASIRASE
Bovine	β1	ALDRHGAHSKGRIRRRASQLKVKC (416-438) [P]
Rat	β1	GLDRHGVPGKGRIRRRASQLKVK
Bovine	β2	ALERHVAQKKSRLRERASQLKIT
Rat	β2	ALERHVAQKKSRLRRRASQLKIT
Bovine	β3	M DRSVPHKKTHLRRRSSQLKIK
Rat	β3	MGDRSIPHKKTHLRRRSSQLKIK

-: Probable sites of β-turn.
*: Cysteine was included to facilitate coupling to carrier protein.
[P]: Consensus phosphorylation site for the cyclic-AMP dependent protein kinase.
Numbers in parentheses are sequence positions starting from signal sequence (Levitan et al., 1988; Ymer et al., 1989; and Tobin et al., this volume, for sequences).

Peptides used in this study were synthesized with the solid-phase method by Dr. Janis D. Young, Department of Biological Chemistry of UCLA School of Medicine, purified with reverse-phase HPLC, and characterized by amino acid analysis. Peptides were coupled to the carrier protein keyhole limpet hemocyanin (KLH) through the sulfhydryl group of cysteine of the peptide with the bifunctional reagent m-maleimidobenzoyl-N-hydroxysuccinimide ester (Liu et at., 1979; Lerner et al., 1981).

New Zealand White rabbits were immunized with either KLH-peptide conjugates or carrier-free peptide (β1 subunit peptide). Typical immunization schedule was as follows: 200 μg of either KLH-peptide conjugates or carrier-free peptide (0.5 ml in PBS pH 7.2) were emulsified with complete Freund's adjuvant (1.5 ml) and injected subcutaneously. Additional injections of antigens (100 μg of conjugate or 200 μg of carrier-free peptide) in incomplete Freund's adjuvant were made after 3 weeks and then every 2 to 3 week-intervals. Animals were bled 1 to 2 weeks after each injection except the first injection.

Titers of the anti-peptide antisera were determined in indirect immuno-peroxidase enzyme-linked immunosorbent assay (ELISA). Briefly, microtiter plates (Immulon 2; Dynatech) were coated with either BSA-peptide conjugates (0.2 μg/ml in PBS) prepared as KLH-peptide conjugates or carrier-free peptide (β1 subunit peptide, 1 μg/ml in PBS). Antibody bound to the immobilized antigen was reacted with horseradish peroxidase-labeled goat anti-rabbit IgG (Vector laboratories; 1:2000 dilution in 1% BSA/0.05% Tween-20/PBS) and the complex was detected by addition of enzyme substrate (0.4 mg/ml 2,2'-Azino-di-[3-ethylbenzthiazoline sulfonate/0.015% H_2O_2/citrate-phosphate buffer pH 4.0).

For immunohistochemical studies, sequence specific antibodies to α1 and α2 subtypes were purified from anti-α1 peptide and anti-α2 peptide antisera with affinity columns in which each peptide was attached to activated CH-Sepharose 4B.

Purification of GABA-A/BZ Receptor

GABA-A/BZ receptor was purified from rat brain and bovine brain by BZ affinity chromatography essentially as described by Stauber et al. (1987) with some modifications. Solubilization was done in 0.5% (w/v) Triton X-100, 1 M KCL, 20% (w/v) glycerol, 0.5 mM DTT in 50 mM potassium phosphate buffer pH 7.4, including the following protease inhibitors: 1 mM EDTA, 2 mM benzamidine hydrochloride, 0.1 mM benzethonium chloride, 100 μg/ml bacitracin, 10 μg/ml chicken egg white ovomucoid containing trypsin inhibitor, 10 μg/ml soy bean trypsin inhibitor, 1 μM leupeptin, 1 μM pepstatin, 0.3 mM PMSF. The final yield from several preparations was 5 - 30 pmol of [^{3}H]muscimol binding sites/g tissue.

Western Blotting

Purified receptor preparations were subjected to SDS-polyacrylamide gel electrophoresis essentially as described by Laemmli (1970) using 8% acrylamide gels, and electrophoretically transferred to nitrocellulose membrane as described by Haid and Suissa (1983) using the semi-dry transfer method with LKB Multiphor II Electrophoresis apparatus at room temperature at 225 mA for 4.5 hr. The blots were quenched with 5% (w/v) non-fat dry milk/PBS-0.1% (w/v) thimerosal at 4°C overnight and washed 3x15 min with PBS-0.1% (w/v) thimerosal. Subsequent incubations and washings were done at room temperature in the following solutions: [1] Antisera diluted in 5% (w/v) BSA-PBS-0.1% (w/v) thimerosal overnight; [2] 0.05% (v/v) Tween 20-PBS-0.1% (w/v) thimerosal 3x15 min; [3] Biotinylated anti-rabbit IgG (Vector laboratories) diluted 1:200 in 1% (w/v) BSA-0.05% (v/v) Tween 20-PBS-0.1% (w/v) thimerosal for at least 6 hr; [4] the same as step 2; [5] preformed streptavidin-biotinylated horseradish peroxidase complex (Amersham) diluted 1:100 in 1% (w/v) BSA-0.05% (v/v) Tween 20-PBS-0.1% (w/v) thimerosal for at least 4 hr; [6] 3,3'-Diaminobenzidine-H_2O_2-$NiCl_2$ in 100 mM Tris-HCl pH 7.5 as described in Vector laboratories instruction manual; [7] rinse in H_2O to stop the reaction.

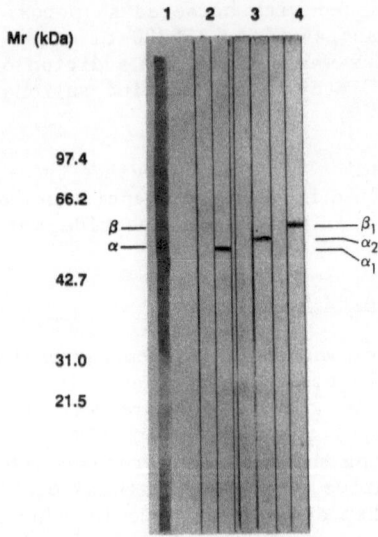

Figure 1. Western Blots of Subtype-Specific Antisera Reacting with Purified Rat Brain GABA/BZ Receptor. The numbers at the left indicate molecular weight (Mr) standards in kiloDaltons (kDa). Lane 1 is stained (Amido Schwartz) affinity-purified rat brain GABA/BZ receptor after SDS-PAGE, indicating α (53 kDa) and β (57 kDa) subunits. Lanes 2, 3, and 4 indicate reaction with subtype-specific antisera prepared against $\alpha1$, $\alpha2$, and $\beta1$ synthetic peptides, respectively, as described in the text. The first lane of each pair is pre-immune serum; the second lane includes antisera at 1:5000 (2 and 3) or 1:1000 (4) dilutions.

RESULTS AND CONCLUSIONS

Anti-α1 peptide, anti-α2 peptide, and anti-β1 peptide antisera recognized distinct molecular weight species in both bovine and rat purified GABA-A receptor on Western blot (α1 = 53 kDa, α2 = 54 kDa, β1 = 57 kDa). Figure 1 shows a Western blot of purified rat receptor recognized by the three antisera. Specificity of the anti-α1 peptide and anti-α2 peptide antisera were also supported by inhibition assays using synthetic peptides as inhibitors in ELISA. Furthermore α1 and α2 peptides sequences do not have homology to those of bovine α3 or rat α4 subunit in the corresponding region (Table 1). Therefore, we conclude that anti-α1 peptide and α2 peptide antisera obtained in this study are specific to the α1 and α2 subtypes, respectively.

The β1-peptide used in this study was chosen from the bovine β1 subunit sequence and it encompasses a consensus phosphorylation sequence for the cyclic-AMP dependent protein kinase which is highly conserved between rat and bovine β subtypes. Therefore, anti-β1 peptide antiserum in this study might be able to recognize other β subtypes. However, anti-β1 peptide antiserum seems to have less reactivity to rat purified receptor than it has to the bovine one, which is probably due to several amino acid substitutions between bovine and rat β1 subunit sequences adjacent to the consensus phosphorylation site.

GABA-A receptors purified from bovine cortex and hippocampus were compared on SDS-polyacrylamide gel electrophoresis and Western blots, by protein staining and immunostaining with subtype-specific antisera. Cortex contained more α1 subtype than α2, and hippocampus contained more α2 subtype than α1. Anti-β1 peptide serum-reactive band was found in both cortex and hippocampus preparations by protein staining as well as immunostaining.

Immunocytochemical staining of rat brain sections was performed by Dr. Carolyn Houser, UCLA Department of Anatomy, using anti-α1 peptide and anti-α2 peptide specific antibodies purified from respective sera. Anti-α1 antibody labeling was observed in cerebral cortex, hippocampus, striatum, hypothalamus and thalamus, whereas anti-α2 antibody labeling was found highly localized to hippocampus. The α1 staining in hippocampus was very dense in the dendritic layers of CA1, and moderate in CA3 and dentate gyrus, while α2 was dense in the pyramidal cell layer of CA1.

These observations agree with a recent report (Pritchett et al., 1989) that the α1β1γ2 subunit combination gives type I GABA-A benzodiazepine receptor, which is widely distributed in brain, and α2β1γ2 or α3β1γ2 combinations give type II receptors, which are localized in a few brain areas including hippocampus.

These antisera and additional ones raised against other subtype and functional domains will be useful in identifying which cDNA sequences correspond with which polypeptides, where they are localized in the central nervous system, and in determining the subunit composition of the various types of oligomeric receptors.

REFERENCES

Barnard, E.A., Darlison, M.G., Fujita, N., Glencorse, T.A., Levitan, E.S., Reale, V., Schofield, P.R., Seeburg, P.H., Squire, M.D., and Stephenson, F.A., 1988, Molecular biology of the GABA_A receptor, in: "Neuroreceptors and Signal Transduction", S. Kito, T. Segawa, K. Kuriyama, M. Tohyama, and R.W. Olsen, eds., Plenum Press, New York.

Chou, P.Y., and Fasman, G.D., 1978, Empirical predictions of protein conformation, Annu. Rev. Biochem., 47:251-276.

Haid, A., and Suissa, M., 1983, "Immunochemical Identification of Membrane Proteins After Sodium Dodecyl Sulfate-Polyacrylamide Gel Electrophoresis", Methods in Enzymology, Volume 96, Academic Press, New York.

Kyte, J., and Doolittle, R.F., 1982, A simple method for displaying hydropathic character of a protein, J. Mol. Biol., 157:105-132.

Laemmli, U.K., 1970, Cleavage of structural proteins during the assembly of the head of bacteriophage T4, Nature, 227:680-685.

Lerner, R.A., Green, N., Alexander, H., Liu, F.-T., Sutcliffe J.G., and Schinnick, T.M., 1981, Chemically synthesized peptides predicted from the nucleotide sequence of the hepatitis B virus genome elicit antibodies reactive with the native envelope protein of Dane particles, Proc. Natl. Acad. Sci. USA, 78:3403-3407.

Levitan, E.S., Schofield, P.R., Burt, D.R., Rhee, L.M., Wisden, W., Köhler, M., Fujita, N., Rodriguez, H.F., Stephenson, F.A., Darlison, M.G., Barnard, E.A., and Seeburg, P.H., 1988, Structural and functional basis for GABA$_A$ receptor heterogeneity, Nature, 335:76-79.

Liu, F.-T., Zinnecker, M., Hamaoka, T., and Katz D.H., 1979, New procedures for preparation and isolation of conjugates of proteins and a synthetic copolymer of D-amino acids and immunochemical characterization of such conjugates, Biochemistry, 18:690-697.

Olsen, R.W., Bureau, M., Ransom, R.W., Deng, L., Dilber, A., Smith, G., Khrestchatisky, M., and Tobin, A.J., 1988, The GABA receptor-chloride ion channel protein complex, in: "Neuroreceptors and Signal Transduction", S. Kito, T. Segawa, K. Kuriyama, M. Tohyama, and R.W. Olsen, eds., Plenum Press, New York.

Olsen, R.W., and Tobin, A.J., 1990, Molecular biology of GABA$_A$ receptors, FASEB J., 4:1469-1480.

Pritchett, D.B., Lüddens, H., and Seeburg, P.H., 1989, Type I and type II GABA$_A$-receptors produced in transfected cells, Science, 245:1389-1392.

Stauber, G.B., Ransom, R.W., Dilber A.I., and Olsen, R.W., 1987, The γ-aminobutyric acid-benzodiazepine receptor protein from rat brain: Large-scale purification and preparation of antibodies, Eur. J. Biochem., 167:125-133.

Tainer, J.A., Getzoff, E.D., Alexander, H., Houghten, R.A., Olson, A.J., Lerner, R.A., and Hendrickson, W.A., 1984, The reactivity of anti-peptide antibodies is a function of the atomic mobility of sites in a protein, Nature, 312:127-134.

Ymer, S., Schofield, P.R., Draguhn, A., Werner, P., Köhler, M., and Seeburg, P.H., 1989, GABA$_A$ receptor β subunit heterogeneity: Functional expression of cloned cDNAs, EMBO J., 8:1665-1670.

THE DIFFERENTIAL EXPRESSION PATTERN OF THE mRNAS ENCODING β SUBUNITS (β_1, β_2, AND β_3) OF GABA$_A$ RECEPTOR IN RAT BRAIN

Jian-Hua Zhang, Makoto Sato, Koichi Noguchi and
Masaya Tohyama

Department of Anatomy II, Osaka University Medical
School, Nakanoshima 4-3-57, Kitaku, Osaka 530
Japan

INTRODUCTION

Gamma-aminobutyric acid (GABA) is a well known major inhibitory neurotransmitter in the vertebrate brain. Its inhibitory action is mediated through the activation of the specific receptors: the GABA$_A$ and GABA$_B$ receptors. Binding of GABA with GADA$_A$ receptor (GABA$_A$-R) causes the chloride channel to open (for reviews see Roberts, 1986; Stephenson, 1988). Recent molecular cloning analysis has revealed that GABA$_A$-R in the brain has at least five subunits and each subunit is consisted of several variants: α_{1-5}, β_{1-3}, γ_{1-2}, δ, and ε (Schofield et al., 1987; Levitan et al., 1988; Richards et al., 1989; Schofield, 1989; Shivers et al., 1989; Ymer et al., 1989a, b). On the other hand, there is much evidence suggesting the heterogeneity of GABA$_A$-R in the brain. For example, using specific oligonucleotide probes for α_1, α_2 or α_3 subunit mRNAs, Wisden et al. (1988) have revealed that the neurons expressing α_1 and α_2 subunit mRNAs are most abundant in layers II-IV of the bovine frontal cortex, while those expressing α_3 subunit mRNA are most abundant in layers V and VI. In addition, recent investigation by Ymer et al. (1989a) has shown that the relative amount of β_2 (8 Kb) and β_3 (6 Kb and 2.5 Kb) subunit mRNAs is higher than that of β_1 (12 Kb) subunit mRNA in the rat brain, although the relative content of β_2 and β_3 mRNAs is also different in several regions, suggesting that heterogeneity of GABA$_A$-R may also occur among the β subunit variants. To examine this possibility, using in situ hybridization histochemistry, we have examined the location of the neurons expressing each β subunit variant (β_1, β_2, and β_3) in some brain areas and spinal cord.

MATERIALS AND METHODS

Animal and tissue preparation: Six male Wistar rats weighing about 150 g each were used. Four of them were perfused transcardially with 4% paraformaldehyde in 0.1 M phosphate buffer (pH 7.2) under sodium pentobarbital anesthesia (50 mg/kg, i.p.). The brain and upper cervical spinal cord were removed and postfixed in the same fresh fixative for another 1 hour at 4℃, and then immersed in 20% sucrose in 0.1 M phosphate buffer at 4℃ for two days. After the tissue were frozen with powdered dry ice, serial sections of 10-15 μm in thickness were cut on a cryostat and thaw-mounted on gelatin-coated slides. These slides were then stored at -80℃ until use (up to 2 months). The other two rats were decapitated. Their fresh brains were removed quickly and immediately frozen on powdered dry ice. The subsequent processing was the same as for fixed brain tissues. No appreciable difference in hybridization signals for each probe was found between the fixed and fresh frozen brain tissues.

In situ hybridization histochemistry: The procedure for in situ hybridization was essentially the same as that reported by Bloch et al. (1986). In brief, after being warmed to room temperature, the slide-mounted sections were fixed in 4% paraformaldehyde in 0.1 M phosphate buffer (pH 7.2) for 5 minutes (the temperature for all steps was room temperature unless otherwise indicated), and rinsed three times (5 minutes each) in 4 X SSC (pH 7.2) (1 X SSC contained 0.15 M sodium chloride and 0.015 M sodium citrate). The slides were then immersed in 4 X SSC containing 1 X Denhardt's solution [1 X Denhardt's solution contained 0.02% bovine serum albumin (fraction V, Sigma), 0.02% Ficoll 400 (Pharmacia), and 0.02 % polyvinylpyrrolidone K-30 (Nakarai tesque Inc.)] for 1 hour and dehydrated through a graded ethanol series (70%-100%). The sections were treated with chloroform for 5 minutes to remove fat from the tissue and immersed in 100% ethanol twice (5 minutes each) before being subjected to hybridization. The hybridization was performed by incubating sections with a buffer [4 X SSC, 50% deionized formamide, 1% N-lauroylsarcosine sodium salt, 0.12 M phosphate buffer (pH 7.2), 1 x Denhardt's solution, 2.5% tRNA, 10% dextran sulfate] containing [α-^{35}S] dATP [1000-15000 Ci/mmol (37-55.5 TB$_q$/mmol), NEN] labeled probes (6-9 X 10^6dpm/ml) for 24-48 hours at 41℃. After hybridization, the sections were rinsed in 1 X SSC (pH 7.2) for 30 minutes, followed by rinsing three to four times in 1 X SSC at 55℃ for 20 minutes each time. Then the sections were dehydrated through a graded ethanol series (70-100%) and coated with Ilford K-5 emulsion (diluted 1:1 with water). These sections were exposed for 4-6 weeks in a tightly sealed dark box at 4℃. After being developed in D-19 (Kodak) developer, fixed with photographic fixer, and washed with taper water, the sections were counter-stained with thionin solution to allow morphological identification.

Oligonucleotide probes: The oligonucleotide probes (48 mer) for each β subunit mRNA were synthesized in an Applied Biosystem 381A DNA synthesizer and then purified using Hitachi high pressure liquid chromatography (ODS column chromatography). The β_1, β_2 and β_3 subunit probes were complementary to bases 1175-1222, 1286-1333 and 1106-1153 of the respective rat cDNAs of the GABA$_A$-R. These segments were located between M3 and M4 of the cDNA and proved to be most dissimilar between each subunit (Ymer et al., 1989). Since all the ligand-gated ion channel receptors (e.g. nicotinic acetylcholine receptor, GABA$_A$ receptor and glycine receptor) belong to the same superfamily and their subunits share high sequence identity with each other. The homology between each probe and all the subunit cDNAs of the ligand-gated ion channel receptors derived from the nervous system of the rat was checked with a computer using DNASIS program (Hitachi Ltd. Japan). The results showed that the homology was about or less than 50% and only the β_1 probe sheared the 64.04% identity with one part of the GABA$_A$-R β_2 subunit cDNA sequence.

For the Northern blot analysis, the probes were labeled at the 3' end with [α-^{32}P]dATP [3000 Ci/mmol (111 TBq/mmol), NEN], resulting in a specific activity of 4.0-9.0 X 10^8 cpm/μg for each probe. Alternatively, for in situ hybridization histochemistry, the probes were labeled at the 3' end using [α-^{35}S]dATP [1000-1500 Ci/mmol (37-55.5 TBq/mmol, NEN] to obtain a specific activity of about 1.4-2.0 X 10^9 dpm/μg.

<u>Northern blot analysis and control experiment</u>: Total RNA of the rat brain was extracted using guanidinim thiocyanate according to the published method (Sambrook et al., 1989). Poly (A)$^+$ RNA was selected from the total RNA by using oligotex-dT30 kit (JSR) following the recommended protocol by the manufacturer (JSR). After electrophoresis in a 1% (w/v) formaldehyde-agarose gel, the poly (A)$^+$ RNA was transferred to GeneScreenPlus (NEN) and then baked for 2 hours at 80℃. The following preincubation (at 42℃), hybridization (at 42℃) and washing (at 60℃) were performed as recommended by the manufacturer (NEN). Finally, the filter was developed onto Kodak X-Omat film for one week at -80℃.

For the competition control experiment, several sections were preincubated with hybridization buffer containing a 10-fold excess of each unlabeled probe. After washing, the same sections were incubated again with hybridization buffer containing each labeled probe.

<u>Analysis of the data</u>: Bright-field and dark-field optics were used to examine the autoradiograms and determine the distributions of labeled cells. The grain density of the unit area (grains/100 μm^2) over the individual cells and the background was determined by the grain counting method reported elsewhere (Noguchi et al., 1988) and indicated as an index of the signal vs noise ratio (S/N). The lowest grain density for positivity was determined to be at least three times more than the mean value of the background density (S/N >3).

RESULTS

<u>Northern blot analysis and competition control experiment</u>
Each probe revealed a single band in Northern blot hybridization. The size of the β_1, β_2 and β_3 subunit transcripts was about 12 kb, 8 kb, and 6 kb and 2.5 kb, respectively.

In the competition experiment, the specific hybridization signals became undetectable on autoradiograms when the same unlabeled and labeled probes were used sequentially, but the signals could still be detected when different unlabeled and labeled probes were used.

All these results established the specificity of the oligonucleotide probes employed in the present study.

Table 1. Summary of the Distribution of Cells Labeled by Different Subunit Variant Probes (β_1, β_2 and β_3) in Some Brain Regions of the Rats

Brain Regions	Subunits		
	β_1	β_2	β_3
OBM	+	+++	+++
OBG	-	-	+++
IsoC	+	++	+++
Pir LI	-	-	-
LII	-	+	+++
LIII	-	-	++
Ammon's horn	++	+	+++
CPu	-	-	+++
GP	-	+++	-
Thalamus	-	+++	-
Hypothalamus	+	-	++
GiA	++	++	++
RMg	++	++	++
Purkinje cells	+	++	-
GC	+	++	+++
VL	-	+++	+++
Vsp	-	-	++
VII	++	+	++
V	-	+	++

+++, high; ++ moderate ; +, low; -, undetectable expression; Regions not listed should not be considered as nonexpressing. Abbreviations: CPu, nucleus caudatus putamen; GC, granule cells of the cerebellum; GiA, α part of gigantocellular reticular nucleus; GP, globus pallidus; IsoC, isocortex; OBG, granule cells of the olfactory bulb (OB); OBM, mitral cells of the OB; Pir, piriform cortex; LI, layer I; LII, layer II; LIII, layer III; RMG, nucleus raphe magnus; VSp, spinal vestibular nucleus; VL, lateral vestibular nucleus; V, trigeminal motor nucleus; VII, facial motor nucleus. The average background levels for probes β_1, β_2 and β_3 were 2.8, 3.5 and 3.0 grain/100 μm^2, respectively.

In Situ Hybridization Histochemistry

Transcripts of each subunit showed region-specific localization. Some areas contained all three variant mRNAs, while in other areas, there were marked differences in their location pattern and level of expression (Table. 1) Neurons in the α part of the gigantocellular reticular nucleus, raphe pallidus nucleus and raphe magnus nucleus were moderately labeled by each probes (Table. 1; Fig. 1). A number of neurons in the isocortex were also labeled by each probe (Table. 1). However, distribution pattern and labeling intensity were different among these probes. The β_3 probe resulted in a moderate to strong labeling of numerous cortical neurons. they were situated evenly from layers II to VI. β_1 and β_2 probes labeled cortical neurons with a weak and moderate intensity, respectively. The neurons labeled by β_2 probe were less numerous than those by β_3 probe, but outernumbered those by β_3 probe. The neurons labeled by β_2 probe in the layers II-IV were more numerous than those seen in layers V and VI. The neurons labeled by β_1 probe in layers IV and VI outernumbered those found in other layers.

Expression pattern of each subunit mRNA in Ammon's horn and dentate gyrus was different from that in the areas mentioned above (Table. 1). Most of the pyramidal cells in the Ammon's horn and the granule cells in the granular layer of the dentate gyrus were very strongly labeled by β_3 probe, they were labeled moderately by β_1 probe but weakly labeled by β_2 probe (Table. 1). In the piriform cortex, no labeled cells were detected in layer I by each probe, whereas a number of neurons in layers II and III were labeled by β_3 probe with a moderate to strong intensity (Table. 1). The β_2 probe resulted in a weak labeling of neurons in layer II only. However, no neurons were labeled by β_1 probe in the piriform cortex (Table. 1).

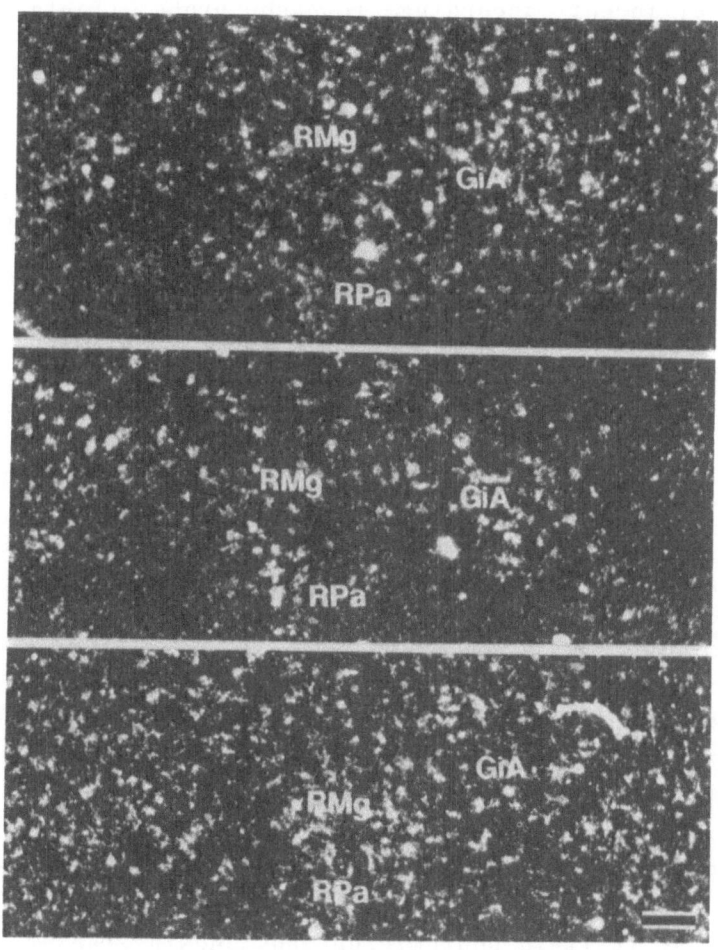

Fig. 1 Dark-field photomicrographs showing the neurons containing mRNAs for β_1 (Uppermost), β_2 (middle) and β_3 (lowest) subunits of GABA$_A$-R in the nucleus raphe magnus (RMg), nucleus raphe pallidus (RPa) and part of the gigantocellular reticular nucleus (GiA). Note most of the neurons in these nuclei are moderately to strongly labeled by each probe. Frontal sections. Bar= 100 μm.

Neurons in the caudate-putamen (Table. 1; Fig. 2) and nucleus accumbens were strongly labeled by β_3 probe but hardly labeled by β_1 and β_2 probes. On the contrary, neurons in the globus pallidus and many thalamic nuclei were strongly labeled by β_2 probes. However, neurons in these nuclei were hardly labeled by other subunit probes (Table. 1; Fig. 2).

The expression pattern of each subunit mRNA in the hypothalamus was opposite to that of the globus pallidus and thalamic nuclei. Most of the neurons in each subnuclei expressed β_3 subunit mRNA strongly or moderately, but the labeling by β_2 probe here was very weak (Table. 1). Expression of β_1 subunit mRNA was not identified in the hypothalamus except the dorsomedial and ventromedial hypothalamic nuclei.

Neurons labeled by each of three subunit probes appeared in the cerebellar cortex (Table. 1), while the localization and intensity of the silver grains over the labeled neurons were different according to each probe (Table. 1). The neurons labeled by β_1 probe were scattered in the granular layer and Purkinje layer and the intensity of the labeling was weak and moderate, respectively (Table. 1). β_2 probe labeled many granule cells and Purkinje cells. The intensity of the labeling of these cells by β_2 probe was stronger than that by β_1 probe (Table. 1). On the other hand, β_3 probe labeled most of the granule cells with a moderate to strong intensity, but failed to label Purkinje cells (Table. 1). In

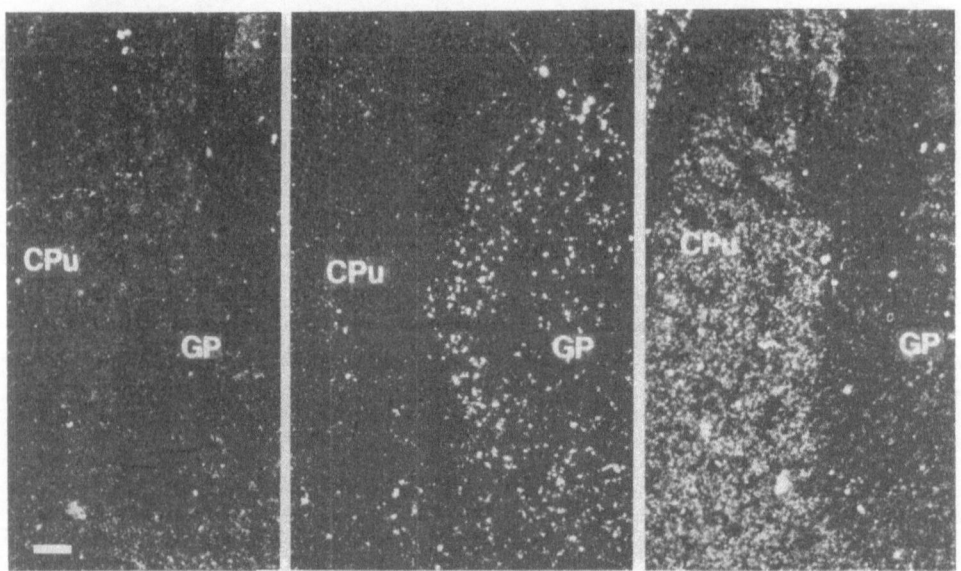

Fig. 2 Dark-field photomicrographs showing the location of the neurons containing mRNAs for β_1 (left), β_2 (middle) and β_3 (right) subunits of GABA$_A$-R in the caudatus putamen (CPu) and globus pallidus (GP). the neurons in CPu are strongly to moderately labeled by β_3 probe, while those in the GP are strongly labeled by β_2 probe. the β_1 labeled neurons in CPu and GP are hardly detected. Frontal sections. Bar= 100 μm.

the cerebellar nuclei, β_2 subunit mRNA was preferentially expressed, having a moderate to strong intensity . Expression of β_3 and β_1 subunit mRNAs was very weak in the cerebellar nuclei.

In the vestibular nucleus, each subnucleus (spinal, lateral, medial and superior) contained many neurons labeled moderately to strongly by β_3 probe, whereas the neurons with a moderate to strong labeling by β_2 probe were only found in the lateral and spinal vestibular nuclei (Fig. 3). On the other hand, β_1 probe resulted in labeling a few neurons with weak intensity in each of these subnuclei. (Fig. 3).

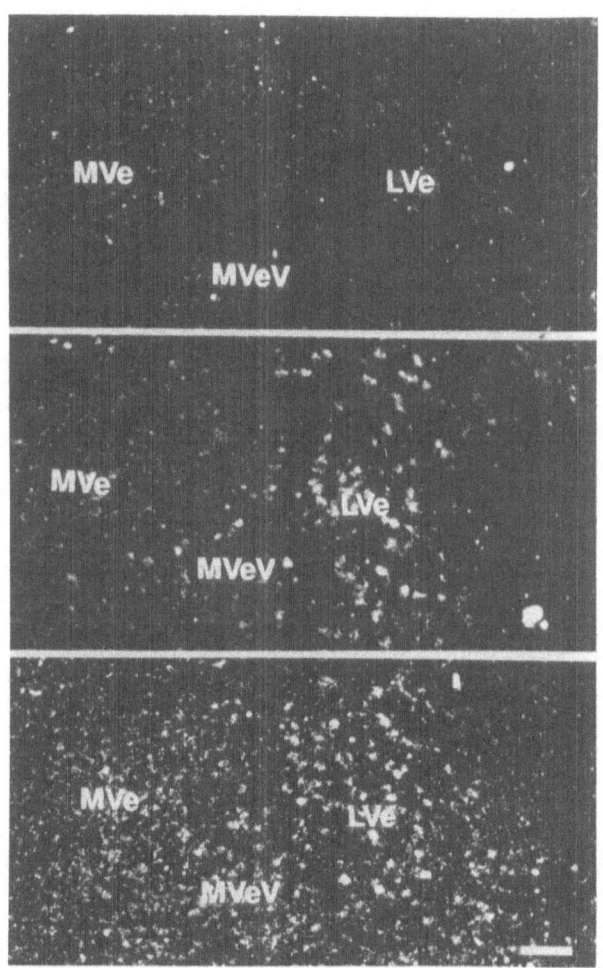

Fig. 3 Dark-field photomicrographs showing the neurons expressing the mRNAs for β_1 (uppermost), β_2 (middle) and β_3 (lowest) subunits of GABA$_A$-R in the lateral vestibular nucleus (LVe), mediodorsal part of the medial vestibular nucleus (MVe) and ventrolateral part of the medial vestibular nucleus (MVeV), respectively. Neurons in these nuclei were labeled moderately to strongly by β_3 probe. Meanwhile β_2 probe labeled neurons in LVe only. Vestibular nucleus lacked the neurons labeled by β_1 probe. Frontal sections. Bar= 100 μ m.

Most of motor neurons in the facial nucleus were labeled by β_1 and β_3 probes moderately. Some neurons in the same nucleus was also weakly labeled by β_2 probe. The same labeling pattern occurred in the hypoglossal nucleus. Weakly labeled neurons by β_3 probe were seen in other cranial motor nuclei such as abducens nucleus and trochlear nucleus, although occulomotor nucleus contained a few neurons labeled by each probe. On the other hand, trigeminal motor nucleus contained a large number of the neurons which were labeled moderately by β_3 probe and several neurons which were weakly labeled by β_2 probe, respectively. However, no labeled neurons by β_1 probe were seen in this nucleus. In the ambiguus nucleus, some neurons which were labeled moderately or weakly by β_2 and β_3 probes, respectively were seen. No labeled cells by β_1 probe were found in this nucleus.

DISCUSSION

Previously, using in situ hybridization histochemistry, Lolait et al. (1989) showed that β_2 subunit mRNA was highly expressed in the globus pallidus, thalamus, substantia nigra and cerebellum, while the expression of β_3 subunit mRNA was high in the caudate putamen, amygdala and cerebellum. Present study confirmed these findings and further revealed much more detailed localization of the neurons expressing β_2 and β_3 subunit mRNAs in the entire rat brain together with overall distribution of neurons containing β_1 subunit mRNA which has not been elucidated. As shown clearly in this study, β_1, β_2 and β_3 subunit mRNAs were differentially and region-specifically expressed in the rat brain. these findings revealed the complicated heterogeneity of GABA$_A$-R from a view point of the location of the neurons containing β subunit variant mRNAs.
Wisden et al. (1988) have revealed that in the cerebral cortex of the bovine, the neurons expressing α_1 and α_2 subunit mRNAs are most abundant in layers II-IV, while those expressing α_3 subunit mRNA are most abundant in layers V and VI. Therefore, heterogeneity of the GABA$_A$-R occurs not only among β subunit variants but also in the α subunit ones. These differential localization of mRNAs of the α and β subunits may indicate the different subtypes of the GABA$_A$-R in different brain regions (see also Olsen et al., 1990; Schofieled et al., 1989)
One of the characteristic observation in this study is that profiles suggesting the possibility of the colocalization of different β subunits in the single neuron are obtained. Most of the neurons in the facial nucleus, raphe magnus and some reticular nuclei, were labeled by all three probes. Most of the pyramidal cells in Ammon's horn and granular cells in the dentate gyrus were strongly labeled by β_3 probe and moderately by β_1 probe, respectively. Similar findings are obtained in many other brain regions. These findings suggest the possibility that GABA$_A$-R contains more than two variants of one subunit.
Localization of the neurons expressing β_2 subunit mRNA in the forebrain shown in this study is very similar to that of neurons containing α_1 subunit mRNA (Hironaka et al., 1990; Zhang et al., unpublished data). According, it is likely in the forebrain that α_1 and β_2 subunit may often form a pair for composition of GABA$_A$-R.

ACKNOWLEDGMENTS

This work was supported in part by grants from the Ministry of Education, Japan, and Naito scientific Foundation.

REFERENCES

Bloch, B., Popovici, T., Le Guellec, D., Normand, E., Chouham, S., Guitteny, A. F., and Bohlen, P., 1986, In situ hybridization histochemistry for the analysis of gene expression in the endocrine and central nervous system tissues: a 3-year experience. J. Neuronsci. Res. 16, 183-200.

Hironaka, T., Morita, Y., Hagihira, S., Tateno, E., Kita, H., and Tohyama, M., 1990, Localization of $GABA_A$-receptor α_1 subunit mRNA-containing neurons in the lower brainstem of the rat. Mol. Brain Res. In press.

Levitan, E.S., Schofield, P. R., Burt, D. R., Rhee, L. W., Wisden, W., Kohler, M., Fujita, N., Rodriguez, H. F., Stephenson, A., Darlison, M. G., Barnard, E. A., and Seeburg , P. G., 1988, Structural and functional basis for $GABA_A$ receptor heterogeneity. Nature 335:76-79.

Lolait, S., O'Carroll, A. -M., Kusano, K., and Mahan, L. C., 1989, Pharmacological characterization and region-specific expression in brain of the β_2- and β_3-subunits of the rat $GABA_A$ receptor. FEBS Letter 258: 17-21.

Noguchi, K., Morita, Y., Kiyama, H., Ono, K., and Tohyama, M., 1988, Noxious stimulus induces the preprotachykinin-A gene expression in the rat dorsal root ganglion: a quantitative study using in situ hybridization histochemistry. Mol. Brain Res. 4:31-35.

Olsen, R. W., McCabe, R. T., and Wamsley, J. K., 1990, $GABA_A$ receptor subtypes: autoradiographic comparison of GABA, benzodiazepine, and convulsant binding sites in the rat central nervous system. J. Chem. Neuroanat. 3:59-76.

Richards, J. G., Sequier, J. M., Malherbe, P., Giller, T., and Mohler, H., 1989, GABA receptor heterogeneity: new insights from receptor radioautography and in situ hybridization histochemistry. Abstract present in the 19th annual meeting of society of neuroscience. p 642.

Roberts, E., 1986, GABA: The road to neurotransmitter status. In R. W. Olsen and J. C. Venter (eds): Receptor Biochemistry and Methodology, Vol. 5. New York: Alan R. Liss, INC, pp. 1-39.

Sambrook, J., Fritsch, E. F., and Maniatis, T., 1989 Molecular cloning: a laboratory manual. Cold Spring Harbor Lab. Press.

Schofield, P.R., Darlison, M. G., Fujita, N., Burt, D. R., Stephenson, F. A., Rodriguez, H., Rhee, L. M., Ramachandran, J., Reale, V., Glencorse, T. A., Seeburg, P. H., and Barnard, E. A., 1987, Sequence and functional expression of the $GABA_A$ receptor shows a ligand-gated receptor super-family. Nature 328:221-227.

Schofield, P. R., 1989, The $GABA_A$ receptor: molecular biology reveals a complex picture. TIPS, 10: 476-478.

Shivers, B.D., Killisch, I., Sprengel, R., Sontheimer, H., Kohler, M. Schofield, P. R., and Seeburg, P. H., 1989, Two novel $GABA_A$ receptor subunits exist in distinct neuronal subpopulations. Neuron 3:327-337.

Stephenson, F.A., 1988, Understanding the GABA$_A$ receptor: a chemically gated ion channel. Biochem. J. 249: 21-32.

Wisden, W., Morris, B. J., Darlison, M. G., Hunt, S. P., and Barnard, E. A., 1988, Distinct GABA$_A$ receptor α subunit mRNAs show differential patterns of expression in bovine brain. Neuron 1:937-947.

Ymer, S., Schofield, P. R., Draguhn, A., Werner, P., Kohler, M., and Seeburg, P. H., 1989a, GABA$_A$ receptor β subunit heterogeneity: functional expression of cloned cDNAs. EMBO J. 8:1665-1670.

Ymer, S., Draguhn, A., Kohler, M., Schofield, P. R., and Seeburg, P. H., 1989b, Sequence and expression of a novel GABA$_A$ receptor α subunit. FEBS Lett. 258:119-122.

"IN VIVO" INHIBITION OF GABAERGIC TRANSMISSION INCREASES ^{35}S-TBPS BINDING IN THE RAT BRAIN

G. Biggio, E. Sanna, M. Serra, G.P. Serra*
and A. Concas

Department of Experimental Biology, Chair of
Pharmacology, University of Cagliari, Italy

INTRODUCTION

Several lines of evidence have well established that the
cage convulsant t-butylbicyclophosphorothionate (TBPS) inhibits the function of the central GABAergic transmission by
binding to specific recognition sites present at the level of
the chloride ionophore coupled to the $GABA_A$/benzodiazepine
receptor complex (Squires et al., 1983; Van Renterghem et
al., 1987). This finding has given a rather unique tool to
study biochemically the function of the GABA-dependent chloride channel. In fact, the specific binding of ^{35}S-TBPS to
the recognition sites associated to the $GABA_A$ receptor complex is modulated in an opposite manner by different compounds which specifically enhance (GABA agonists, benzodiazepines, imidazopyridines, anxiolytic and anticonvulsant ß-
carbolines etc.) and inhibit (GABA antagonists, anxiogenic
and convulsant ß-carbolines etc.) the function of the GABA-
dependent chloride channel, respectively (Squires et al.,
1983; Gee et al., 1986; Concas et al., 1988; Biggio et al.,
1989; Serra et al., 1989; Sanna et al., 1990).
Most of these studies have been performed "in vitro" by the
addition of the above drugs to membranes prepared from cell
culture or tissue of the rat and mouse brain. Thus, we considered it very interesting to verify whether the binding of
^{35}S-TBPS to such membrane preparation could also reveal changes in the function of GABAergic synapses elicited "in vivo"
by treatments which are known to alter the function of the
central GABAergic transmission. Accordingly we decided to evaluate whether the "in vivo" changes in the GABAergic transmission induced either by a specific degeneration of a GA-
BAergic pathway or by drugs which "in vivo" but not "in
vitro" reduce and enhance the function of GABAergic synapses
respectively, could be evaluated by measuring ^{35}S-TBPS binding "ex vivo". The results of this study should indicate
whether the "in vivo" decrease or increase in the amount of

---

*Institute of Zoology, University of Cagliari

Neuroreceptor Mechanisms in Brain, Edited by S. Kito *et al.*
Plenum Press, New York, 1991

GABA available at the GABA$_A$ recognition site causes a conformational changes of the GABA$_A$ receptor subunits, a phenomenon that persists in the "in vitro" membrane preparation. If this is so our experimental model can be considered a suitable biochemical index to study the functional state of the GABA$_A$ receptors "in vivo".

"In vivo" inhibition of GABA synthesis enhances ^{35}S-TBPS binding in the cerebral cortex

It has been clearly shown that isoniazid, an inhibitor of GABA synthesis (Biggio et al., 1977; Horton et al., 1979) produces a functional inhibition of GABAergic transmission in the rat brain. Accordingly, this drug elicits in rats proconvulsant, convulsant and proconflict effects and markedly potentiate the action of drugs which selectively inhibit the central GABAergic synapses (Horton, 1980; Corda et al., 1983).
On the basis of these data we concluded that isoniazid-treated rats might be a very useful model to evaluate biochemically the changes in the function of the GABA-dependent chloride channel elicited by the decreased availability of GABA at the GABA$_A$ recognition site. Since ^{35}S-TBPS has specific binding sites at the level of the chloride ionophore coupled to the GABA$_A$ receptor, the "ex vivo" measurement of ^{35}S-TBPS binding to unwashed membrane preparations from the cerebral cortex of rats treated with isoniazid was used to detect the molecular events related to the inhibition of the GABA dependent chloride channel. All the experiments were performed using male Sprague Dawley rats weighing 200-225 g. ^{35}S-TBPS binding was assayed in cortical unwashed membrane preparations as previously described (Concas et al., 1988).
As reported in Fig. 1 isoniazid (150-600 mg/kg s.c.) induced in 60 min. a dose related increase of ^{35}S-TBPS binding measured "ex vivo" in unwashed membrane of the rat cerebral cortex. The maximal enhancement (+ 90%) of ^{35}S-TBPS binding was obtained with the highest dose (600 mg/kg s.c.) of isoniazid. Time course studies performed with the dose of 300 mg/kg s.c., indicated that isoniazid enhanced significantly (30%) the binding activity of ^{35}S-TBPS as early as 45 min. after the administration. The effect of the drug reached the peak after 2 hours and returned to normal in 5 hours (Fig. 1). Scatchard analysis performed in homogenate of cerebral cortex revealed that the increase of ^{35}S-TBPS binding elicited by isoniazid (300 mg/kg s.c.) was exclusively due to a change in the total number of binding sites with no significant changes in their affinity (K_D) for the ligand (Fig. 2).
To further clarify the functional meaning of the isoniazid-induced increase of ^{35}S-TBPS binding in the rat brain, we studied whether diazepam, a drug known to facilitate the interaction of GABA with its recognition site (Biggio et al., 1977; Haefely, 1985), was able to antagonize the effect of this drug on ^{35}S-TBPS binding.
As shown in Fig. 2, diazepam (3 mg/kg i.p.), administered 30 min. after isoniazid completely prevented the increase of ^{35}S-TBPS binding induced by the latter. Consistent with previous data (Sanna et al., 1989; Serra et al., 1989), diazepam induced per se, a significant decrease in the density of ^{35}S-TBPS binding sites but failed to change the affinity (K_D) for their ligand (result not shown).

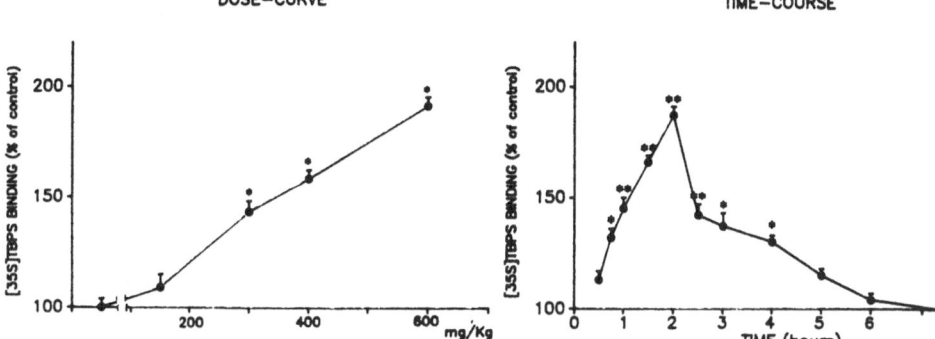

DOSE-CURVE

TIME-COURSE

Fig. 1 Effect of isoniazid administration on ^{35}S-TBPS binding to rat cerebral cortex. For the dose-curve experiments rats were killed 60 min after the administration of isoniazid or saline. Time-course experiments were performed using 300 mg/kg s.c. of isoniazid. Values are the average of six separate experiments each run in triplicate $\pm$ S.E.M. *p < 0.05; **p < 0.01 vs saline-treated rats.

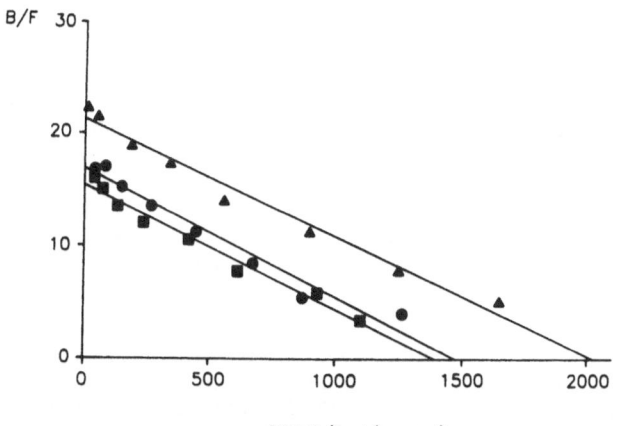

Fig. 2 Increase in ^{35}S-TBPS binding to rat cerebral cortex elicited by administration of isoniazid: antagonism by diazepam. Data are from a representative experiment of five separate experiments each run in triplicate. Linear regression analysis of the individual experiments yielded the following parameters: saline-treated rats (●): Bmax 1491; K_D = 72; isoniazid-treated rats (▲): Bmax 2033, K_D = 75; isoniazid + diazepam treated rats (■): Bmax 1403, K_D = 62. Bmax: fmoles fmoles/mg prot; K_D: nM.

Finally, the intraperitoneal injection of valproic acid, a drug that increases the GABA content in the rat brain (Godyn et al., 1969), produced, like diazepam, a significant dose related and time dependent decrease of ^{35}S-TBPS binding in the rat cerebral cortex, an effect markedly potentiated by the subsequent administration of diazepam (Table 1, 2).

Table 1. Effect of "in vivo" administration of valproic acid on ^{35}S-TBPS binding in the rat cerebral cortex

	^{35}S-TBPS binding (% of control)
Dose curve	
Control	100 ± 7 .
Valproic acid 100 mg/kg	85 ± 9
Valproic acid 200 mg/kg	75 ± 5*
Valproic acid 400 mg/kg	65 ± 5*
Valproc acid 800 mg/kg	64 ± 4*
Time course	
0 min (control)	100 ± 5
30 min	84 ± 4
60 min	78 ± 3*
120 min	74 ± 4*
240 min	74 ± 5*
16 hours	95 ± 3

For the dose-curve experiments rats were killed 60 min after the administration of valproic acid or saline.
The time-course was performed using 200 mg/kg i.p. of valproic acid.
Values are the average of 5 separate experiments each run in triplicate $\pm$ S.E.M. *P < 0.05 vs control.

Our present finding that "in vivo" administration of isoniazid produces an enhancement in the density of ^{35}S-TBPS binding sites measured "ex vivo" in fresh unwashed cortical membranes indicates that this drug inhibits the function of GABA$_A$ receptors. This conclusion is consistent with the previous "in vitro" studies showing that negative modulators (anxiogenic and convulsant ligands and antagonists of the GABA$_A$ receptor complex) of the GABAergic synapses increase ^{35}S-TBPS binding to unwashed membrane preparation from the rat brain (Gee et al., 1986; Concas et al., 1988). Since, isoniazid inhibits the GABAergic transmission by reducing the GABA synthesis and consequently the amount of GABA present at the presynaptic level, our results suggest that the "in vivo" decrease in the availability of GABA at the receptor site causes a conformational modification of the postsynaptic molecular components (α-β-γ subunits) (Barnard and Seeburg, 1988; Pritchett et al., 1989; Shivers et al., 1989) of the GABA$_A$ receptor complex where ^{35}S-TBPS binding sites are located, a

Table 2. Diazepam enhances the decrease of ^{35}S-TBPS binding induced by valproic acid in the rat cerebral cortex

Treatment	^{35}S-TBPS binding (% of control)
Control	100 ± 7
Valproic acid (200 mg/kg i.p.)	79 ± 5*
Diazepam (0.5 mg/kg i.p.)	86 ± 8
Valproic acid + Diazepam	53 ± 6**

Rats were killed 30 and 60 min after diazepam and valproic acid administration, respectively.
Values are the average of 5 separate experiment each run in triplicate ± S.E.M. *P < 0.05 vs control. **P < 0.05 vs valproic acid and diazepam.

phenomenon that persists "in vitro" in the membrane preparation.
This conclusion is also supported by the finding that the "in vivo" administration of valproic acid and diazepam, two drugs that with a different mechanism facilitate the interaction of GABA with the GABA$_A$ recognition site (Godyn et al., 1969; Biggio et al., 1977; Haefely et al., 1985), markedly inhibits ^{35}S-TBPS binding and completely antagonizes the action of isoniazid on this parameter.
In conclusion, our results are biochemical evidence that "in vivo" reduction or enhancement in the interaction of GABA with its recognition site results in opposite changes of ^{35}S-TBPS binding measured "ex vivo" in unwashed cortical membrane preparations. These data suggest that the "ex vivo" measurement of ^{35}S-TBPS binding is a suitable biochemical index of the functional state of GABA$_A$ receptor "in vivo".

"In vivo" administration of ethanol reduces ^{35}S-TBPS binding in the cerebral cortex

Several authors have recently suggested that most of the pharmacological actions induced by ethanol are mediated by the activation of the GABA-dependent chloride channel (Suzdak et al., 1986 a, b; Allan et al., 1987).
This conclusion has been mainly reached on the basis of "in vitro" experiments showing that ethanol at pharmacologically relevant concentrations stimulates GABA receptor-mediated uptake of ^{36}Cl$^-$ into isolated brain vescicles.
In order to further verify the action of ethanol on the function of the GABA-dependent chloride channel we decided to measure ^{35}S-TBPS binding in the cerebral cortex of ethanol treated rats. In fact, we have recently shown (Serra et al., 1989; Sanna et al., 1989; 1990) that the "ex vivo" measurement of ^{35}S-TBPS binding is a very sensitive biochemical index of the "in vivo" function of the GABAergic synapses in the rat brain.

As reported in Table 3, the acute intragastric administration of ethanol induced in 40 min. a dose-dependent (0.5 - 5 g/kg) decrease of ^{35}S-TBPS binding measured "ex vivo" in unwashed membrane preparations from rat cerebral cortex. The effect of ethanol was maximal at the dose of 2 g/kg and increasing doses failed to further inhibit ^{35}S-TBPS binding. The time-course study of ethanol-induced decrease of ^{35}S-TBPS binding showed that the maximal effect (-35%) was reached 60 min. after the administration of 1 g/kg of the drug, persisted for at least 3 hours and returned to the control value after 5-6 hrs (Table 3).

Table 3.   Effect of "in vivo" administration of ethanol on ^{35}S-TBPS binding in the rat cerebral cortex

		^{35}S-TBPS binding (% of control)
Dose curve		
Control		100 ± 5
Ethanol	0.5 g/kg	85 ± 6
Ethanol	1.0 g/kg	72 ± 5*
Ethanol	2.0 g/kg	65 ± 4*
Ethanol	3.0 g/kg	68 ± 5*
Ethanol	5.0 g/kg	67 ± 4*
Time course		
0 min (control)		100 ± 6
30 min		85 ± 6
60 min		65 ± 5*
90 min		67 ± 4*
180 min		68 ± 7*
360 min		96 ± 8

For the dose-response experiment rats were killed 40 min after the administration of ethanol or saline. Time-course experiment was performed using 1 g/kg p.o. of ethanol.
Values are the average of 5 separate experiments each run in triplicate ± S.E.M. *p < 0.05 vs control.

The decrease in ^{35}S-TBPS binding elicited by ethanol in the rat cerebral cortex was shared by the intraperitoneal administration of diazepam (result not shown). However, when the effect of ethanol and diazepam was studied on the kinetic characteristics of ^{35}S-TBPS binding, the Scatchard plot analysis of saturation isotherms of ^{35}S-TBPS binding revealed that diazepam markedly reduced the maximal number (Bmax) of ^{35}S-TBPS recognition sites (Table 4) but failed to affect the affinity constant (K_D). Vice versa, ethanol decreased the apparent affinity of the binding sites for the ligand without affecting the Bmax value (Table 4).

The previous administration to rats of the specific benzodia-zepine receptor antagonist Ro 15-1788 failed to antagonize the inhibitory action of ethanol on ^{35}S-TBPS binding while completely preventing the effect of diazepam (results not shown). On the contrary, the administration of Ro 15-4513 a benzodiazepine receptor partial inverse agonist known to pre-vent some of the behavioural, electrophysiological and bio-chemical effects of ethanol (Bonetti et al., 1985; Mereu et al., 1987; Suzdak et al., 1986b), completely antagonized the effect of both ethanol and diazepam on ^{35}S-TBPS binding.

Table 4. Kinetic characteristics of ^{35}S-TBPS binding in the rat cerebral cortex after the "in vi-vo" administration of ethanol and diazepam

| | ^{35}S-TBPS binding | |
	Bmax (fmol/mg prot)	K_D (nM)
Control	1981 ± 87	70 ± 4
Ethanol	1963 ± 135	91 ± 6*
Diazepam	1327 ± 98*	65 ± 9

Rats were sacrificed 40 and 30 min after administration of etha-nol (1 g/kg, p.o.) and diazepam (3 mg/kg, i.p.), respectively. Unwashed cortical membranes were incubated in presence of ^{35}S-TBPS (2.5 - 500 nM) at 25 °C for 90 min. Each value is the mean ± S.E.M. of five separate experiments. *p < 0.05 vs control.

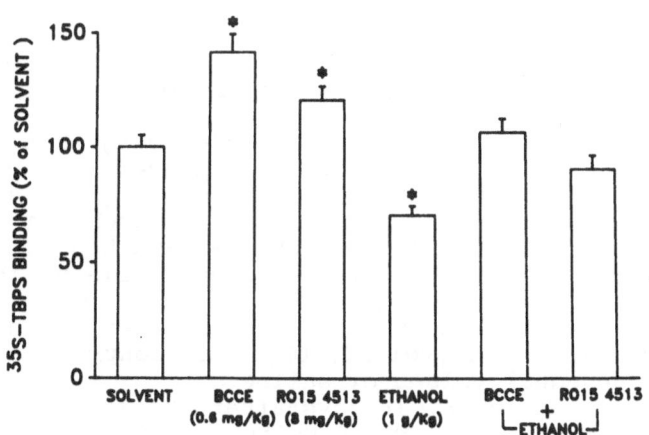

Fig. 3 Ro 15-4513, like ß-carbolines, reverts ethanol-induced de-crease of ^{35}S-TBPS binding in the rat cerebral cortex. Rats were sacrificed 70, 60 and 15 min after administra-tion of Ro 15-4513 (8 mg/kg, i.p.), ethanol (1 g/kg, p.o.) and BCCE (0.6 mg/kg, i.v.), respectively. Values are the average of 4 separate experiments each run in triplicate ± S.E.M. *p < 0.05 vs solvent-treated rats.

Moreover, the ß-carboline derivative ßCCE, like Ro 15-4513, abolished the effect of ethanol on ^{35}S-TBPS binding. Finally, Ro 15-4513 and ßCCE "per se" significantly increased ^{35}S-TBPS binding (Fig. 3).

The data reported here demonstrate for the first time that "in vivo" administration of ethanol enhances the function of the GABA-coupled chloride channel and this effect is reverted by the benzodiazepine receptor inverse agonists. Our finding is in line with several reports suggesting that ethanol has actions similar to those elicited by benzodiazepines and GABA-mimetics (Nestoros, 1980; Liljequist and Engel, 1982; Mereu and Gessa, 1985; Suzdak, 1986 a,b; Sanna et al., 1990). Accordingly, ethanol, like GABA-mimetics and benzodiazepines, reduces the binding of ^{35}S-TBPS on specific recognition sites located at the level of the GABA-coupled chloride channel.

It is worth noting that the "in vivo" administration of ethanol reduces the apparent affinity of these binding sites. This finding is at apparent variance respect to the effect (decrease in the density of ^{35}S-TBPS binding sites) elicited "in vitro" and "in vivo" by GABA mimetics and benzodiazepines (Squires et al., 1983; Gee et al., 1986; Concas et al., 1988; Serra et al., 1989). Thus, we may suggest that the molecular events involved in the enhancement of GABAergic transmission elicited by ethanol are different from those involved in the effects of other GABA mimetics and benzodiazepines.

Striato-nigral denervation enhances ^{35}S-TBPS binding in the substantia nigra

The loss of striato-nigral GABAergic afferents induces a marked decrease of GABA content in the substantia nigra homolateral to the lesioned striatum and this in turn should result in an almost complete functional inhibition of the denervated nigral GABAergic synapses.

In order to prove the above conclusion we studied whether the "in vivo" inhibition in the function of striato-nigral GABAergic transmission elicited by the striato-nigral denervation can be detected by measuring ^{35}S-TBPS binding "ex vivo" in a fresh unwashed membrane preparation from the denervated substantia nigra.

Male Sprague Dawley rats weighing 280-300 g were anesthetized with Equithesin (3 ml/kg, i.p.) and placed in a David Kopf stereotaxic apparatus. Kainic acid (0.75 µg/0.5 µl) was injected at a rate of 0.5 µl/3 min unilaterally into the head (coordinates AP 2.9; LAT 2.7; DV 4.9) and body (coordinates AP 0.8; LAT 3.7; DV 4.3) of the corpus striatum through an injection needle (30 gauge stainless steel tubing) connected through a polyethylene tube to a microburet syringe (Micrometric Instruments, Cleveland, OH). The controlateral striatum received an equal volume of saline.

As reported in Fig. 4, 14 days after the intrastriatal injection of kainic acid, the density of ^{35}S-TBPS binding sites measured in the substantia nigra homolateral to the lesioned striatum was dramatically increased (225%) compared to the contralateral side. Moreover, the intrastriatal injection of kainic acid resulted in a marked (65%) but not complete loss of the GAD activity and GABA content in the substantia nigra homolateral to the injected side (results not shown). Thus, in the denervated substantia nigra a fraction of striato-

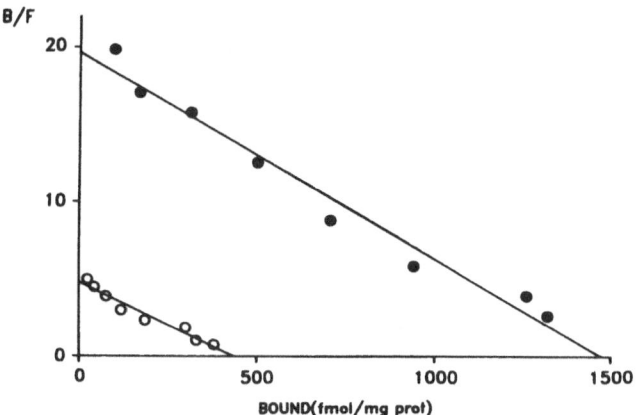

Fig. 4   Increase in ^{35}S-TBPS binding to unwashed nigral membrane preparation elicited by the degeneration of the striato-nigral GABAergic pathway. Scatchard analysis of saturation data from ^{35}S-TBPS binding was carried out on the rat s. nigra from sham-operated side (O) and from intrastriatal kainic acid-injected side (●). Maximum binding (Bmax, fmol/mg prot.) and dissociation constants (K_D, nM) in the s. nigra in the following groups of rats were: sham-operated side, Bmax 436, K_D 92; intrastriatal kainic acid, Bmax 1431, K_D 71. (Sanna et al., 1989).

nigral GABAergic afferent terminals and GABA content was still available after the lesion.

This finding is consistent with the data reported in Table 5 showing that the "in vitro" addition of the GABA receptor antagonist bicuculline, to fresh unwashed membranes prepared from the denervated substantia nigras, produced a marked (63%) further increase of the ^{35}S-TBPS binding. This result suggests that the kainic acid-induced striato-nigral denervation does not completely abolish the inhibitory action of GABA on nigral neurons.

To further clarify whether the increase of ^{35}S-TBPS binding induced by the degeneration of the striato-nigral pathway was related to a reduced availability of GABA at the level of its receptor site, we studied whether diazepam by enhancing the action of the GABA content still present in the denervated substantia nigra antagonized the increase of ^{35}S-TBPS binding induced by the loss of striato-nigral GABAergic pathway. As shown in Fig. 5 the administration of diazepam to kainic acid-lesioned rats almost completely antagonized the increase in the binding of ^{35}S-TBPS in the denervated substantia nigra and produced a significant reduction (30%) of this parameter in the contralateral substantia nigra.

The present results suggest that the loss of the striato-nigral GABAergic pathway elicited by intrastriatal kainic acid produced a dramatic increase in the density of ^{35}S-TBPS binding sites in the denervated substantia nigra, an effect reversed by the "in vivo" administration of diazepam.

Table 5. The addition of bicuculline to nigral homogenates enhances the increase of ^{35}S-TBPS binding elicited by the degeneration of striatonigral GABAergic pathway

	Specific ^{35}S-TBPS (fmoles/mg prot.)	% of sham-operated
Sham-operated	23.7 ± 2	100
Kainic ac.-denervated s. nigra	77.0 ± 5*	325
Kainic ac.-denervated s. nigra + bicuculline (1 μM)	125.5 ± 12**	529

Each value is the mean ± S.E.M. of 4 separate experiments, each done in triplicate. *P < 0.01 vs sham-operated. **P < 0.01 vs kainic ac.-denervated s. nigra.

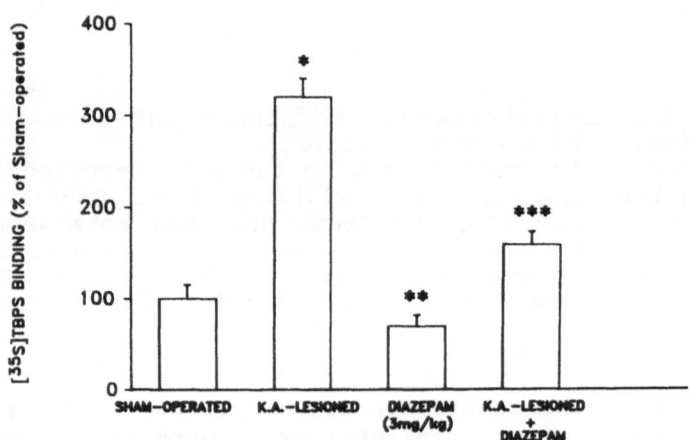

Fig. 5 The "in vivo" administration of diazepam antagonizes the increase in nigral ^{35}S-TBPS binding elicited by the degeneration of the striato-nigral GABAergic pathway. *P < 0.01 with respect to sham-operated side. **P < 0.05 with respect to sham-operated side. ***P < 0.01 with respect to kainic acid-lesioned side. The results are the mean ± S.E.M. of 5 separate experiments. Rats were killed 14 days and 60 min after the intrastriatal injection of kainic acid and the intraperitoneal administration of diazepam, respectively (Sanna et al., 1989).

Our finding is consistent with the general concept that the loss of presynaptic fibers results in an up regulation of the denervated receptors (Burt et al., 1977; Guidotti et al., 1979; Biggio et al., 1981; Porceddu et al., 1985). However it is unlikely that the increase in the density of ^{35}S-TBPS binding sites reflects an increased function of the nigral GABA-dependent chloride channel. In fact the increase of ^{35}S-TBPS binding following the striato-nigral GABAergic denervation is almost completely abolished by the "in vivo" administration of diazepam, a drug known to increase the GABAergic transmission by enhancing the interaction of GABA with its recognition site (Biggio et al., 1977; Haefely et al., 1985; Biggio et al., 1989) while is further enhanced by the "in vitro" addition of the GABA receptor antagonist bicuculline.

It is worth nothing that the action of diazepam in decreasing ^{35}S-TBPS binding is much more potent in the denervated substantia nigra than in the controlateral side.

This finding is in agreement with the knowledge that the benzodiazepines are very potent in enhancing the GABAergic transmission when it is decreased while failing to further increase the maximal effect of GABA (Polc and Haefely, 1976; Biggio et al., 1977; Gallager, 1978). This concept implies that in the rat substantia nigra the GABA$_A$/benzodiazepine/ionophore receptor complex receives a potent tonic modulatory input from the endogenous GABA released by striatal afferent terminals, an effect dramatically reduced by the loss of striato-nigral GABAergic pathway. Accordingly previous studies have shown that intrastriatal kainic acid administration results in a compensatory increase in the density of both nigral GABA (Guidotti et al., 1979) and benzodiazepine recognition sites (Biggio et al., 1981).

Thus, the greater efficacy of diazepam in reducing ^{35}S-TBPS binding in the denervated substantia nigra than in the controlateral intact side can be due to the combination of these two mechanisms: a) a reduced modulatory action of the endogenous GABA; b) an increased availability of GABA and benzodiazepine recognition sites.

The capability of diazepam and bicuculline to decrease and enhance ^{35}S-TBPS binding in the denervated substantia nigra, as well as in other brain areas (Squires et al., 1983; Gee et al., 1986; Concas et al., 1988), respectively, strongly suggests that the increase of nigral ^{35}S-TBPS binding sites following the loss of GABAergic innervation reflects a functional inhibition of the striato-nigral GABAergic system at the level of the GABA-dependent chloride channel. Taken together these results allow us to conclude that the binding of ^{35}S-TBPS to unwashed membrane preparations from the rat substantia nigra may be used as a suitable index of the "in vivo" changes in the function of the striato-nigral GABAergic transmission.

CONCLUSIONS

Our present data support the conclusion that an "in vivo" inhibition or enhancement in the function of the GABA-coupled chloride channel results in a parallel increase or decrease in ^{35}S-TBPS binding measured "ex vivo" in unwashed membrane preparations. This conclusion is consistent with the "in vitro" data showing that the inhibitors and the enhancers of GABAergic transmission markedly increase and decrease ^{35}S-

TBPS binding to unwashed membrane preparations, respectively. Since the opening state of the chloride channel should reflect an enhanced interaction of GABA with its recognition site while the closing state is related to the reduced activation of the $GABA_A$ receptor (Bormann et al., 1987; Bormann, 1988), the increase in nigral ^{35}S-TBPS binding elicited by striato-nigral denervation as well as the effect of isoniazid in the cerebral cortex might be related to the closing state of the chloride channel. In fact, it should reflect a negative allosteric conformation assumed by the portion of the subunits involved in the formation of the chloride channel as a consequence of the loss or reduced functional interaction of $GABA_A$ and its own recognition site. Vice versa the reduction of ^{35}S-TBPS binding elicited by valproic acid, diazepam and ethanol should be consistent with the enhanced activation of $GABA_A$ receptors elicited by these drugs.

All together these findings suggest that GABA plays a crucial role in the physiological and pharmacological modulation of ^{35}S-TBPS binding to brain tissue. Moreover, our results indicate that the "in vivo" conformational changes of the $GABA_A$ receptor subunits elicited by the different availability of GABA at the receptor site persist in the "in vitro" membrane preparations.

In conclusion our data indicate that ^{35}S-TBPS binding to membrane preparations from the rat brain represents a very sensitive biochemical index by which it is possible to evaluate the changes in the function of the GABA-dependent chloride channel induced by the "in vivo" administration of drugs, by the degeneration of specific neuronal populations and by other physiological and pathological conditions. Accordingly, it has been recently found that different stressful conditions enhance ^{35}S-TBPS binding in the rat brain (Trullas et al., 1987; Concas et al., 1988).

Finally, the very high sensitivity of ^{35}S-TBPS binding to the changes in the content of GABA at the synaptic level, suggests that the availability of specific ligands for TBPS binding sites detectable with positron emission tomography will probably allow to get information, directly from the human brain, on the involvement of the $GABA_A$/benzodiazepine/ionophore receptor complex in different neurological and/or psychiatric diseases.

REFERENCES

Allan, A.M., Huidobro-Toro, J.P., Bleck, V. and Harris, R.A., 1987, Alcohol and the GABA receptor-chloride channel complex of brain, Alcohol, 1: 643-646.

Barnard, E.A., and Seeburg, P.H., 1988, Structural basis of the GABA-activated chloride channel: molecular biology and molecular electrophysiology, in: "Chloride Channel and Their Modulation by Neurotransmitters and Drugs", G. Biggio and E. Costa, eds., Raven Press, New York, 1-18.

Biggio, G., Brodie, B.B., Guidotti, A. and Costa, E., 1977, Mechanism by which diazepam, muscimol and other drugs change the content of cyclic GMP in cerebellar cortex, Proc. Natl. Acad. Sci. U.S.A., 74: 3592-3595.

Biggio, G., Concas, A., Corda, M.G. and Serra, M., 1989, Enhancement of GABAergic transmission by zolpidem, an imidazopyridine with preferential affinity for Type I benzodiazepine receptors, Eur. J. Pharmacol., 161: 173-180.

Biggio, G., Corda, M.G., Concas, A. and Gessa, G.L., 1981, Denervation supersensitivity for benzodiazepine recep tors in the rat substantia nigra, Brain Res., 220: 344-349.

Bonetti, E.P., Burkard, W.P., Galb, M. and Mohler, H., 1985, The partial inverse benzodiazepine agonist Ro 15-4513 antagonizes acute ethanol effects in mice and rats, Br. J. Pharmacol., 86: 463p.

Bormann, J., Hamill, O.P. and Sakmann, B., 1987, Mechanism of anion permeation through channels gated by glycine and γ-aminobutyric acid in mouse cultured spinal neurons, J. Physiol., 385: 243-286.

Bormann, J., 1988, Electrophysiology of $GABA_A$ and $GABA_B$ receptor subtypes, TINS, 11: 112-116.

Burt, D.R., Creese, I. and Snyder, S.H., 1977, Antischizophre-nic drugs: chronic treatment elevates dopamine receptor binding in brain, Science, 196: 326-328.

Concas, A., Serra, M., Atsoggiu, T. and Biggio, G., 1988, Foot-shock stress and anxiogenic ß-carbolines increase [^{35}S]t-butylbicyclophosphorothionate binding in the rat cerebral cortex, an effect opposite to anxiolytics and γ-aminobutyric acid mimetics, J. Neurochem., 51: 1868-1876.

Corda, M.G., Costa, E. and Guidotti, A., 1983, Involvement of GABA in the facilitation of punishment-suppressed beha-viour induced by ß-carbolines in rat, in: "Benzodiazepi-ne Recognition Site Ligands: Biochemistry and Pharmaco-logy", G. Biggio and E. Costa, eds., Raven Press, New York, 121-128.

Gallager, D.W., 1978, Benzodiazepines: potentiation of a GABA inhibitory response in the dorsal raphe nucleus, Eur. J. Pharmacol., 49: 133-143.

Gee, K.W., Lawrence, L.J. and Yamamura, H.I., 1986, Modula-tion of the chloride ionophore by benzodiazepine recep-tor ligands: influence of γ-aminobutyric acid and ligand efficacy, Mol. Pharmacol., 30: 218-225.

Godyn, Y., Heiner, L., Mark, J. and Mandel, P., 1969, Effect of d-n-propylacetate, an anticonvulsant compound on GABA metabolism, J. Neurochem., 16: 869-873.

Guidotti, A., Gale, K., Suria, A. and Toffano, G., 1979, Bio-chemical evidence for two classes of GABA receptors in rat brain, Brain Research, 172: 566-571.

Haefely, W., Kyburz, E., Gerecke, M. and Mohler, H., 1985, Recent advances in the molecular pharmacology of benzo-diazepine receptors and in the structure-activity rela-tionships of their agonists and antagonists, Adv. Drug Res., 14: 165-322.

Horton, W.R., 1980, GABA and seizures induced by inhibitors of glutamic acid decarboxylase, Brain Res. Bull., 5: 605-608.

Horton, W.R., Chapman, A.G. and Meldrum, B.S., 1979, Isonia-zid, as a glutamic acid decarboxylase inhibitor, J. Neurochem., 33: 745-750.

Liljequist, S. and Engel, J., 1982, Effects of diazepam and ethanol on punished responding of rats in a conflict test situation, Acta Pharmacol. Toxicol., Suppl. 1, 51: 5.

Mereu, G. and Gessa, G.L., 1985, Low doses of ethanol inhibit the firing of neurons in the substantia nigra pars reti-culata: a GABAergic effect?, Brain Res., 360: 325-330.

Mereu, G., Passino, N., Carcangiu, G.P., Boi, V. and Gessa,

G.L., 1987, Electrophysiological evidence that Ro 15-4513 is a benzodiazepine receptor inverse agonist, European J. Pharmacol., 135: 453-454.

Nestoros, J.N.., 1980, Ethanol specifically potentiates GABA-mediated neurotransmission in feline cerebral cortex, Science, 209: 708-710.

Polc, P. and Haefely, W., 1976, Effects of two benzodiazepines, phenobarbitone, and baclofen on synaptic transmission in the cat cuneate nucleus, Naunyn-Schmiedebergs Arch. Pharmacol., 294: 121-131.

Porceddu, M.L., Ongini, E. and Biggio, G., 1985, ^{3}H-SCH 23390 binding sites increase after chronic blockade of D-1 dopamine receptors, Eur. J. Pharmacol., 118: 367-370.

Pritchett, D.B., Luddens, H. and Seeburg, P.H., 1989, Type I and Type II GABA$_A$-benzodiazepine receptors produced in transfected cells, Science, 245: 1389-1392.

Sanna, E., Concas, A., Serra, M. and Biggio, G., 1990, In vivo administration of ethanol enhances the function of the γ-aminobutyric acid-dependent chloride channel in the rat cerebral cortex, J. Neurochem., 54: 696-698.

Sanna, E., Serra, M., Pepitoni, S. and Biggio, G., 1989, Dramatic increase in nigral t-[^{35}S]butylbicyclophosphorothionate binding sites elicited by the degeneration of the striato-nigral GABAergic pathway: reversal by diazepam, Brain Res., 501: 144-149.

Serra, M., Sanna, E. and Biggio, G., 1989, Isoniazid, an inhibitor of GABAergic transmission, enhances [^{35}S]TBPS binding in rat cerebral cortex, Eur. J. Pharmacol., 164: 385-388.

Shivers, B.D., Killisch, I., Sprengel, R., Sontheimer, H., Kohler, M., Schofield, P.R. and Seeburg, P.H., 1989, Two novel GABA$_A$ receptor subunits exist in distinct neuronal subpopulations, Neuron., 3: 327-337.

Squires, R.F., Casida, J.E., Richardson, M. and Saederup, E., 1983, ^{35}S-t-butylbicyclophosphorothionate binds with high affinity to brain-specific sites coupled to .-aminobutyric acid-A and ion recognition sites, Mol. Pharmacol., 23: 326-336.

Suzdak, P.D., Schwartz, R.D., Skolnick, P. and Paul, S.M., 1986a, Ethanol stimulates γ-aminobutyric acid receptor-mediated chloride transport in rat brain synaptoneurosomes, Proc. Natl. Acad. Sci. U.S.A., 83: 4071-4075.

Suzdak, P.D., Glowa, J.R., Crawley, J.N., Schwartz, R.D., Skolnick, P. and Paul, S.M., 1986b, A selective imidazobenzodiazepine antagonist of ethanol in the rat, Science, 234: 1243-1247.

Trullas, R., Havoundjian, H. and Skolnick, P., 1987, Stress-induced change in t-[^{35}S]butylbicyclophosphorothionate binding to γ-aminobutyric acid-gated chloride channels are mimicked by in vitro occupation of benzodiazepine receptors, J. Neurochem., 49: 968-974.

Van Renterghem, C., Bilbe, G., Moss, S., Smart, T.G., Constanti, A., Brown, D.A. and Barnard, E.A., 1987, GABA receptors induced in xenopus oocytes by chick brain mRNA: evaluatation of TBPS as a use-dependent channel blocker, Mol. Brain Res., 2: 21-31.

123-I-IOMAZENIL-SPECT IN PATIENTS WITH FOCAL EPILEPSIES -
A COMPARATIVE STUDY WITH 99mTc-HMPAO-SPECT, CT and MR

F.J. Ferstl [1], M. Cordes [1], I. Cordes [3], H. Henkes [1], W. Christe [2], H. Eichstädt [1], G. Barzen [1], B. Schmitz [2], P.H. Hasler [4], P.A. Schubiger [4], D. Schmidt [2], R. Felix [1]

Departments of Radiology [1], Neurology [2] and Pediatrics [3]
Universitätsklinikum Rudolf Virchow
Standort Charlottenburg, Berlin, FRG
Paul-Scherrer-Institute [4], Villigen, Switzerland

I. Introduction

Despite optimal pharmacotherapy, about 30% of patients with focal epilepsies arising from the temporal lobe are frequently refractory to anticonvulsant medication (Schmidt, 1981). Up to 50 % of these patients can be cured of their seizures by surgical removal of the epileptogenic focus (Glaser, 1980). Prerequisite for successful surgical intervention is the accurate localisation and lateralisation of the epileptogenic focus, which often can only be achieved through such invasive procedures as EEG examination with subdural or depth electrodes (Wieser, 1983).
As a non-invasive method SPECT studies with blood flow tracers, eg. 99mTc-HMPAO, 123-I-Amphetamine, play an increasingly important role in localisation of the epileptogenic focus. It is well recognised, that there is increased cerebral blood flow (CBF) in the region of the focus during and immediately after partial seizures and decreased CBF interictally. In combination with MR HMPAO-SPECT is reported to be able to detect pathological findings in up to 90% of these patients (Biersack, 1987).
A new approach to this diagnostic problem represents the functional imaging of the cerebral benzodiazepine (BZ) receptor distribution. As part of the GABA/BZ-chloride ionophore complex, which is located at the postsynaptic membrane, the BZ receptor can be used as a marker of the GABA receptor, the main mediator of inhibition in brain (Olsen, 1981). This receptor has been reported to be selectively reduced in epileptic foci from animals and human beings (Ribak, 1979, Roy, 1980). PET examinations with the ^{11}C labeled BZ receptor antagonist Flumazenil have shown that the BZ receptor density in epileptic foci is diminished by 15-35% compared to the nonaffected, contralateral region (Savic, 1988).
In an European Multicenter Trial organised by Paul-Scherrer-Institute SPECT studies with the 123-Iodine labeled BZ receptor antagonist Iomazenil, a derivative of Flumazenil, were performed in patients with

focal epilepsies and Alzheimer's disease. The aim and purpose of the trial was to determine the diagnostic value of 123-I-Iomazenil in assessing these patients in comparison with ^{99m}Tc-HMPAO-SPECT, a validated method for cerebral blood flow mapping.

Iomazenil is a BZ receptor antagonist with a high affinity for the central type BZ receptor and no major pharmacological activity (Richards, 1984). After intravenous application the radioligand is reported to be distributed proportional to the rCBF in the early phase and retained in relation to the BZ receptor density on delayed images.

II. Patients

10 patients, 3 males and 7 females, between the ages of 19 and 47, were investigated interictally. On the course of the seizure and supported by the results of the EEG examinations 9 patients were diagnosed of having temporal lobe epilepsy, in one patient frontal lobe epilepsy was suspected. In all cases the seizures were progressive and considered to be medically intractable. 3 months prior to Iomazenil-SPECT, the anticonvulsive therapy was restricted in all patients to medication, which does not interact with the BZ receptor. At the time of the investigation 4 patients were either treated with phenytoin, carbamazepine or vigabitrin, while 6 patients received a combined drug therapy consisting of phenytoin, primidone, phenobarbital, valproinate or carbamazepine.

All patients gave informed written consent to participate in the trial and received 300 mg potassium perchlorate for thyroid blocking two hours before the study was started.

III. Methods

The application of 200-300 MBq 123-I-Iomazenil was performed slowly by an intravenous route. All patients were resting quietly with eyes closed and were exposed to minimal background noise. No adverse effects of the BZ receptor antagonist were observed.

The SPECT studies were carried out on a rotating gamma camera (APEX 409-ECT, Elscint) with electronic data processing (APEX 415, Elscint).

The subject was placed in the supine position on a couch with the head unfixed. Head-gated scanning was carried out with a rotation radius of 30 cm. A parallel hole, high resolution collimator (APC-4, Elscint) was used with full width at half maximum (FWHM) of 7 mm. Sixty projections were obtained within 1.200 sec (20 sec per projection) on a 64x64 matrix, using Zoom 1. The data acquired were initially processed by filtered back projection with a Hanning-Filter. Without linear attenuation correction reconstructions were then performed in the transverse, sagittal and coronal plane with 13 mm thick slices.

Following the meeting of the Multicenter Trial participants in November 1989 a revised protocol was designed for standardised data processing and quantitative analysis.

An overview of the protocol is outlined by Table 1.

Quantification of side to side differences in brain isotope uptake was achieved by the Regions-of-Interest (ROI) technique. ROI's were positioned in the areas of the suspected cortical lesions and on the contralateral, non-affected side. A reference ROI was fixed 36-48 mm above the basis of the brain in paramedian localisation.

The SPECT examinations of cerebral blood flow were carried out 20 min after i.v. injection of 555 MBq ^{99m}Tc-HMPAO (Ceretec, Amersham, U.K.) using the same apparative equipment and data processing mode as described above for 123-I-Iomazenil-SPECT.

The CT scans were regularly obtained on a Somatom DR2 unit and conducted in an axial imaging plane. All patients had contrast-enhanced and unenhanced CT studies.

Magnetic resonance imaging was performed on 8 patients using a 0.5 Tesla supraconducting unit. MR scans were produced in the axial and coronal plane (10 mm thick slices) by using a mulitple-slice spin echo (SE) technique. SPIN echo sequences with TR 315, TE 14 and TR 1600 TE 70 ms revealed T_1-weighted and T_2-weighted images respectively.

Table 1. Data acquisition and processing of the revised protocol

1. Data Acquisition:
    - Start: 20 min and 120 min p.i
    - Rotation   360 °, continuous mode
    - Frames: 60 images
    - Acquisition time: 1200 sec
    - Zoom: 1.5
    - Matrix: 64x64
    - Final Resolution: 12.7 mm

2. Data Processing:
    - Base Line: cantho-meatal line
    - Filter: Butterworth-Filter
                            (0.5 Nyquist, Order=2)
    - Coefficient of 0.12 cm $^{-1}$ for linear attenuation correction
    - Reconstructions: transverse, sagittal and coronal
    - Slice thickness: 8 mm

IV. Results

1. Preliminary results of the Multicenter Trial

In the course of the Multicenter Trial 92 patients with focal epilepsies were examined by 123-I-Iomazenil-SPECT and ^{99m}Tc-HMPAO-SPECT within 48 hours. Table 2 summarizes the results reported at the meeting of the Multicenter Trial participants. Pathologic findings were defined as focal reduced BZ receptor binding or altered rCBF.

Table 2. Preliminary results of the Multicenter Trial

N	Iomazenil	HMPAO
12	+	-
9	-	+
54	+	+
17	-	-

2. Own Results

123-I-Iomazenil-SPECT showed abnormal patterns of BZ receptor binding in the presumed epileptic focus of all 10 patients. In 4 patients the reduced BZ receptor binding was unifocal, in 6 patients multifocal areas were observed. In comparison with HMPAO-SPECT, which demonstrated pathologic findings in form of decreased rCBF in 8 patients, the reduced BZ receptor binding of 3 patients correlated well in extension and topographic site with the corresponding hypoperfusion. In 3 patients the regions of diminished BZ receptor binding exceeded the altered rCBF, while in one patient the localised hypoperfusion reached major extent than the region of decreased BZ receptor binding.

The quantitative analysis of the epileptogenic focus compared with the contralateral homotopic region by means of ROI revealed a range from 0.49 to 0.95 with median of 0.9 for 123-I-Iomazenil-SPECT and a range from 0.4 to 0.95 with a median of 0.89 for ^{99m}Tc-HMPAO-SPECT. In contrast, left-right ratios of non-affected cortical areas recorded in identical ROI's over the left and right hemisphere differed by only 1-3%.

On CT scans pathologic findings were demonstrated in 4 patients. One patient showed a hypodense tumourlike lesion located in the right temporal lobe. In two other patients a mild cortical atrophy and an extensive right hemisphere atrophy following an infectious encephalitis were found. The CT scan of another patient displayed a right temporal lobe substance defect as a result of an epileptic focus resection.

MR images showed abnormal findings in 5 patients. In 4 patients the structural lesions detected were in the same localisation and of equal size as seen on CT scans. In one patient with normal findings in the CT examination a focal lesion of high signal intensity located in the anterior right temporal lobe was detected on T$_2$-weighted images.

The Iomazenil-SPECT study of this patient displayed a markedly reduced BZ receptor binding surrounding the anterior part of the right temporal lobe, while HMPAO-SPECT demonstrated only a slightly decreased rCBF in the region of the tumour. In the other patients with morphological brain lesions HMPAO- and Iomazenil-SPECT depicted similiar results regarding extension and localisation of reduced BZ receptor binding and decreased rCBF.

Case Report

30 year old female patient with a 12 years history of intractable focal epilepy was admitted to hospital with progressive seizures. A CT scan displayed no abnormal findings. On EEG examination an epileptogenic

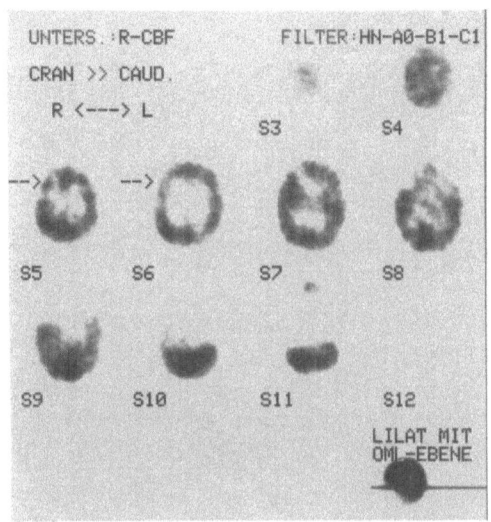

Figure 1a.   HMPAO-SPECT examination showing a slightly decreased rCBF in the right temporal lobe.

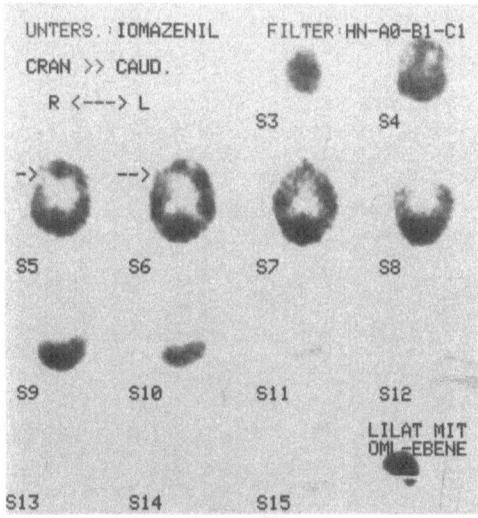

Figure 1b.   Iomazenil-SPECT study demonstrating a markedly reduced BZ receptor binding in the anterior part of the right temporal lobe.

Figure 1.   30 year old female patient with an intractable focal epilepsy. A CT scan was normal. On EEG an epileptogenic focus was suspected in the right temporal lobe. (continued →)

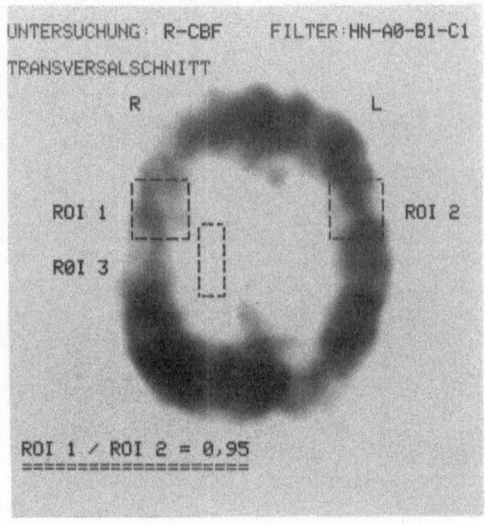

Figure 1c. Quantitative calculation of the suspected area compared to the non-affected contralateral region by means of ROI for HMPAO-SPECT

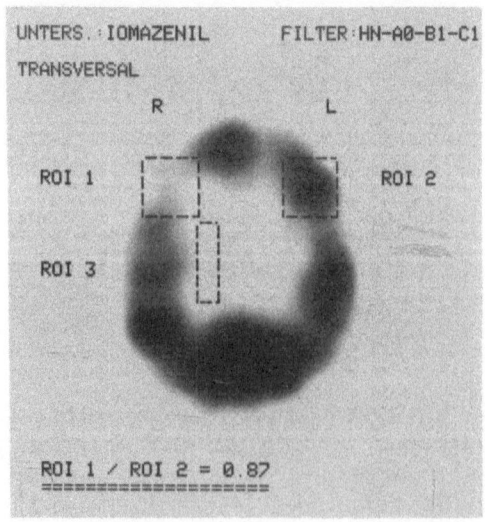

Figure 1d. Quantitative calculation of the suspected area compared to the non-affected contralateral region by means of ROI for Iomazenil-SPECT

focus was suspected in the right temporal lobe. HMPAO-SPECT (fig.1a) demonstrated only a slightly decreased rCBF in the territory of the anterior right temporal lobe. On Iomazenil-SPECT, however, the BZ receptor binding (fig. 1b) was markedly reduced in this area. The quantitative analysis of the presumed epileptogenic focus compared to the contralateral, non-affected region by means of ROI's revealed a right-left ratio of 0.95 for HMPAO-SPECT (fig. 1c) and 0.87 for Iomazenil-SPECT (fig. 1d).

V. Discussion

The cerebral distribution of the BZ receptor antagonist Iomazenil using SPECT scanning was similar to the known BZ receptor distribution in brain (Breastrup, 1977, Richards, 1984, Samson, 1985, Shinotoh, 1986). The SPECT images showed high tracer uptake in the cerebral cortex, mainly frontal and occipital, while the radioligand was less concentrated in cerebellum and basal ganglia. Although these regions were found to possess low BZ receptor density (Richards, 1984, Shinotoh, 1986), they were moderately defined on most of our SPECT images. According to the results published by Bartenstein et al (Bartenstein, 1989), this may relate to the fact, that SPECT scans 30 min after Iomazenil application still register blood flow activity to some extent, thus overlapping specific tracer accumulation in the areas of high BZ receptor density. Following these findings the cerebral BZ receptor distribution is best visualised on SPECT images recorded 60 to 120 min after tracer injection.
In correlation with PET (Savic, 1988) all patients of our study group showed reduction of BZ receptor binding in the presumed epileptic focus. In 80% the diminished BZ receptor density was also associated with a decreased rCBF. Our results, therefore, correspond well to the data of the Multicenter Trial, suggesting that the sensitivity of Iomazenil-SPECT in detecting epileptic disturbances is comparable to that of HMPAO-SPECT.
The quantification of Iomazenil- and HMPAO-SPECT images by computation of left-right ratios of activities recorded in identical ROI's over the left and right hemisphere yielded nearly identical value ranges. While reference values of left-right ratios has already been established for HMPAO-SPECT (Cordes, 1990, Perani, 1988), these have still to be determined for Iomazenil-SPECT. Our results imply, that there is no significant left-right asymmetry of BZ receptor density in normal brain regions.
In evaluating patients with focal epilepsies, MR imaging seems to be qualitatively better than CT scanning. The reasons for this are the superior imaging quality of MR in the region of the temporal lobe close to the base of the skull and the higher detection rate of brain gliosis, which cannot be demonstrated by CT scanning. MR imaging should, therefore, be considered the preferred imaging procedure for studying patients with focal epilepsies (Schörner, 1987).
In summary, 123-I-Iomazenil-SPECT is an efficient method for cerebral BZ receptor mapping. The BZ receptor distribution by SPECT is in accordance with the results of PET and is best visualised on late images. In detecting epileptic foci the image quality and sensitivity of Iomazenil-SPECT are comparable to that of HMPAO-SPECT. At present, the diagnostic utility of Iomazenil-SPECT in patients with focal epilepsies cannot be finally established from the results of the relatively small study groups.

Prospective studies with drug displacement experiments need to be carried out in order to validate this promising method.

VI. References

1. Bartenstein, P., Ludolph, A., Schober, O., Lottes, G., Böttger, I., Beer, H.F., 1989, Vergleich von Blutfluß und Benzodiazepin - Rezeptorverteilung bei fokaler Epilepsie: Vorläufige Ergebnisse einer SPECT-Studie, Nucl. Med., 28: 181-186
2. Biersack, H.J., Stefan, H., Reichmann, K., Linke, D., Kurthen, M., Knopp, J., Penin, H., 1987, HMPAO brain SPECT and epilepsy, Nuc. Med. Com., 8: 513-518
3. Braestrup, C., Albrechtsen, R., Squires, R.F., 1977, High densities of benzodiazepine receptors in human cortical areas, Nature, 269: 702-704
4. Cordes, M., Christe, W., Henkes, H., Delavier, U., Eichstädt, H., Schörner, W., Langer, R., Felix, R., 1990, Focal epilepsies: HMPAO-SPECT compared with CT, MRI and EEG, J. Comp. Assist. Tomogr. (in press)
5. Glaser, G.H., 1980, Treatment of intractable temporal lobe-limbic epilepsy (complex partial seizures) by temporal lobectomy, Ann. Neurol., 8: 455-459
6. Olsen, R.W., 1981, The GABA postsynaptic membrane receptor-ionophore complex, Molec. Cell Biochem., 39:261-279
7. Perani, D., 1988, Tc-99m HMPAO-SPCET study of regional cerebral perfusion in early Alzheimer's disease, J. Nucl. Med., 29:1507-1514
8. Ribak, C.E., Harris, A.B., Andersen, L., 1979, Inhibitory GABAergic nerve terminals decrease at sites of focal epilepsy, Sience, 205:211-214
9. Richards, J.G., Möhler, H., 1984, Benzodiazepine receptors, Neuropharmacol., 23:233-242
10. Roy, A.E., Bakay, A.B., 1980, Neurotransmitter, receptor and biochemical changes in monkey cortical epileptic foci, Brain Res., 206:387-404
11. Samson, Y., Hantrage, R., Baron, J.C., 1985, Kinetics and displacement of (11 C) Ro 15-1788, a benzodiazepine antagonist, studied in human brain in vivo by positron tomography, Eur. J. Pharmacol., 110:247-251
12. Savic, I., Roland, P., Sedvall, G., Persson, A., Pauli, S., Widden, L., 1988, In-vivo demonstration of reduced benzodiazepine receptor binding in human epileptic foci, Lancet II:863-866
13. Schmidt, D., 1981, Die Behandlung der Epilepsien, Thieme, Stuttgart
14. Schörner, W., Meencke, H.J., Felix, R., 1987, Temporal-lobe epilepsy: comparison of CT and MR imaging, AJR, 149:1231-1239
15. Shinotoh, A., Yamasaki, T., Inoue, O., Itoh, T., Suzuki, K., Hashimoto, K., Tateno, Y., Ikehira, H., 1986, Visualization of specific binding sites of benzodiazepine in human brain, J. Nucl. Med., 27:1593-1599
16. Wieser, H.G., 1983, Electroclinical features of the psychomotor seizure, Fischer-Butterworth, Stuttgart

EXPRESSION OF THE GABAB RECEPTOR IN *XENOPUS* OOCYTES AND DESENSITIZATION BY ACTIVATION OF PROTEIN KINASE C

Kohtaro Taniyama, Koichiro Takeda, Hiroshi Ando, Chikako Tanaka

Department of Pharmacology
Kobe University School of Medicine
Kobe 650, Japan

INTRODUCTION

The γ-aminobutyric acid (GABA) receptors have been classified into two subtypes, termed GABAA and GABAB receptors, in the basis of their pharmacological properties (Bowery et al., 1984; 1989; Bormann, 1988). The GABAA receptor with its integrated Cl⁻ channel is now well characterized. Studies revealed the amino acid sequence of the GABAA receptor (Schofield et al., 1987; Levitan et al., 1988). Stimulation of the GABAB receptor has been shown to induce two membrane effects, namely reduction in Ca^{2+} conductance and increase in K^+ conductance (Bormann, 1988; Bowery, 1989), however, the GABAB receptor has not been purified. The expression of GABAB receptor in the *Xenopus* oocytes may provide information on molecular mechanisms of GABAB receptor-mediated responses.

Many neurotransmitter receptors and ion channels are modulated through phosphorylation by protein kinases, such as cyclic AMP-dependent, cyclic GMP-dependent and calcium/calmodulin-dependent protein kinases (Nestler and Greengard, 1984) and protein kinase C (Nishizuka, 1984; Kaczmarek, 1987). The GABAB receptor-mediated response is modulated by activation of protein kinase C present in the hippocampus (Andrade et al., 1986; Dutar and Nicoll, 1988b). We discuss here the expression of GABAB receptor in *Xenopus* oocytes and modulation by protein kinase C of the GABAB receptor-mediated response.

EXPRESSION OF GABAB RECEPTOR IN *XENOPUS* OOCYTES AFTER INJECTION OF RAT CEREBELLAR mRNA

Recent studies have shown that the injection of mRNA isolated from neural tissues into *Xenopus* oocytes results in the functional expression of neurotransmitter receptors and voltage-dependent ion channels (Dascal, 1987; Snutch 1988). In the follicular oocyte, receptors for norepinephrine (NE), acetylcholine (ACh),

Neuroreceptor Mechanisms in Brain, Edited by S. Kito *et al.*
Plenum Press, New York, 1991

5-hydroxytryptamine (5HT) and purinergic agonists, and GTP-binding proteins and adenylate cyclase are localized to the follicle cell layer. Defolliculated oocytes do not respond to these agonists. To eliminate the endogenous receptor, the outer follicular layers of oocytes are removed either manually or by treatment with collagenase, before the injection of mRNA. Injection of rat brain mRNA into the defolliculated oocytes induces synthesis of functional amino acid receptors coupled directly to ion channels, including excitatory (NMDA, kainate) and inhibitory (GABA$_A$, glycine) receptors. The stimulation of either NMDA or kainate receptor activates the currents carried by K^+ and Na^+ (Gundersen et al., 1984; Houamed et al., 1984; Dascal et al., 1986; Verdoorn et al., 1987). The subtype of glutamate receptor, quisqualate receptor-mediated response is mediated by an IP3/Ca^{2+} second messenger pathway (Sugiyama et al., 1987), which differs from the NMDA and kainate receptors. Another group of receptors is coupled to a common messenger pathway. The response to stimulation of this group of receptors is mimicked by the injection of GTP-γ-S or IP3, and is blocked by treatment with pertussis toxin (Dascal et al., 1986; Sugiyama et al., 1987; Hirono et al., 1987; Kato et al., 1988). Thus, these receptors appear to be coupled to a GTP-binding protein - phospholipase C pathway. These include receptors for the neuropeptides, such as neurotensin (Parker et al., 1986; Hirono et al., 1987), substance P (Parker et al., 1986; Harada et al., 1987), tachykinin (Harada et al., 1987), endothelin and angiotensin (Lory et al., 1989), and 5HT (5HT1C subtype) (Gundersen et al., 1983 Dascal et al., 1986; Lubbert et al., 1987; Kato et al., 1988), and ACh (muscarinic M1 subtype) (Sugiyama et al., 1985; Dascal et al., 1986; Hirono et al., 1987).

We succeeded in expression of the functional GABA$_B$ receptor in *Xenopus* oocytes by injection of mRNA from rat cerebellum. Under current-clamp conditions, the application of GABA usually induced a depolarization in RNA injected oocytes via the GABA$_A$ receptor/Cl$^-$ channel complex, however some oocytes of the same donor showed a hyperpolarizing response. There was no response in untreated oocytes from the same donors. To isolate the GABA$_B$ response from that of GABA$_A$, bicuculline was perfused before applying GABA. In the presence of bicuculline, GABA (10^{-6} M to 10^{-4} M) induced a hyperpolarization, in a concentration dependent manner (Fig. 1A). Under voltage-clamp conditions, when the holding potential was -40 mV, GABA induced an outward current (Fig. 1B). When the holding potential was switched to -70 mM, GABA failed to induce any current. Baclofen, a GABA$_B$ agonist mimicked the effect of GABA in the presence of bicuculline, and the response to baclofen was antagonized by phaclofen, a GABA$_B$ antagonist.

In dorsal root ganglion cells, selective stimulation of the GABA$_B$ receptor decreases Ca^{2+} conductance (Deisz and Lux, 1985; Robertson and Taylor, 1986). Voltage-clamp studies confirmed the inhibitory action of baclofen on Ca^{2+} channels, however there were no effects on the voltage-dependent K^+ channels (Robertson and Taylor, 1986). The GABA$_B$ receptor-mediated inhibition of Ca^{2+} current was abolished either by treatment with GDP-β-S, a non-hydrolysable analog of GDP, or by pertussis toxin which ADP-ribosylates the α-subunit of certain GTP-binding proteins (Holz et al., 1986; Dolphin and Scott, 1987), thereby indicating the involvement of GTP-binding proteins in the GABA$_B$ receptor-mediated response of dorsal root ganglion cells.

In the hippocampus, GABAB receptors are located both presynaptically and postsynaptically. The stimulation of GABAB receptor located on the CA1 pyramidal cell increases K^+ conductance, and induces a hyperpolarization of pyramidal cell membranes (Newberry and Nicoll, 1984; 1985; Gahwiler and Brown, 1985; Inoue et al., 1985; Dutar and Nicoll, 1988a; 1988b), without affecting Ca^{2+} conductance (Gahwiler and Brown, 1985). On the other hand, the excitatory postsynaptic potential elicited by stimulation of CA1 afferents (Schaffer collateral-commissural pathway) is suppressed with the stimulation of the presynaptic GABAB receptor (Newberry and Nicoll, 1984; 1985; Dutar and Nicoll,1988b; Colmers and Pittman, 1989). The postsynaptic GABAB receptor-mediated response is antagonized by phaclofen, a GABAB antagonist (Dutar and Nicoll,1988a; 1988b) and is blocked by pertussis toxin (Andrade et al., 1986; Dutar and Nicoll, 1988b; Colmers and Pittman, 1989), however the presynaptic

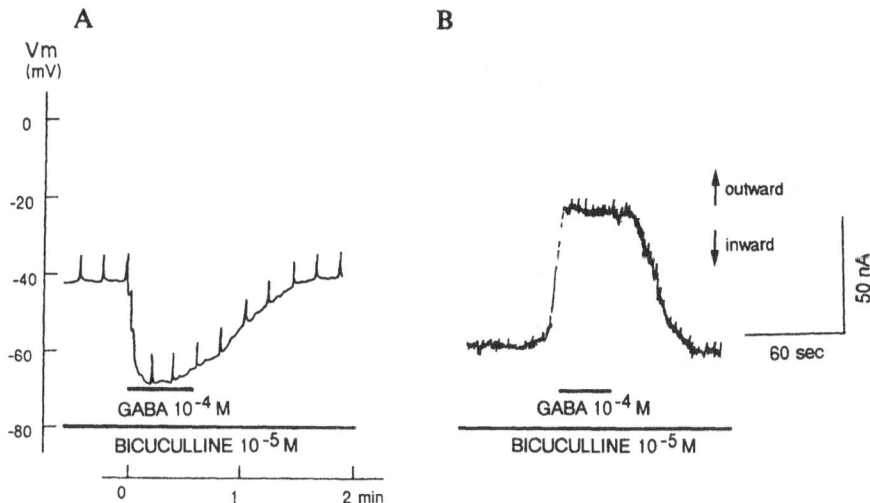

Fig. 1. GABAB receptor-mediated hyperpolarization and outward current in *Xenopus* oocytes. (A) Under current-clamp conditions, (B) Under voltage-clamp conditions. The holding potential was -40 mV.

GABAB receptor-mediated response is phaclofen-resistant and pertussis toxin-resistant (Dutar and Nicoll,1988b; Colmers and Pittman, 1989). GTP-binding proteins appear to be involved in the response mediated by the stimulation of postsynaptic, but not presynaptic GABAB receptors. There are differences between the coupling of GABAB receptors to ion channels in the dorsal root ganglion cells and that in the hippocampal pyramidal cells and the pharmacological property differs in the pre- and postsynaptic receptors of the hippocampus.

The GABAB receptor expressed in *Xenopus* oocytes following the injection of RNA from the rat cerebellum appears to be similar to the postsynaptic GABAB receptor of the hippocampus.

DESENSITIZATION OF THE GABAB RECEPTOR-MEDIATED RESPONSE IN *XENOPUS* OOCYTES BY ACTIVATION OF PROTEIN KINASE C

Various ion channels of various tissues are modulated by the activation of protein kinase C, an event which modulates the native ion channels of folliculated oocytes and the expressed ion channels of oocytes injected with mRNA. In the RNA injected oocytes, the activation of protein kinase C enhances the voltage-dependent Ca^{2+} current (Leonard et al., 1987; Sigel and Baur, 1988), however decreases the voltage-dependent Na^+ current (Sigel and Baur, 1988) and the GABAA receptor- (Sigel and Baur, 1988) and the 5HT receptor- (Kato et al., 1988) mediated Cl^- current. In the folliculated oocytes, activation of protein kinase C decreases the adenosine receptor-mediated K^+ current (Dascal et al., 1985). 5HT-, GABA- and ACh-induced inward currents are also suppressed by the activation of protein kinase C (Moran and Dascal, 1989).

The response to stimulation of the GABAB receptor is mimicked by the activation of protein kinase C in dorsal root ganglia, in which stimulation of the receptor decreases Ca^{2+} conductance (Rane and Dunlap, 1986). In the hippocampus, however, the activation of protein kinase C by phorbol ester suppresses the responses mediated by the stimulation of both pre- (Dutar and Nicoll, 1988b) and postsynaptic (Andrade et al., 1986; Dutar and Nicoll, 1988b) GABAB receptors.

The GABAB receptor-mediated hyperpolarization of the oocyte membrane was suppressed by activation of protein kinase C following treatment with phorbol ester, 1 2 - *O* -tetradecanoylphorbol-13-acetate (TPA), but not with 4α-phorbol-12,13-didecanoate (4α-PDD), a non-activating analog of TPA. The suppressing effect of TPA on the GABAB receptor-mediated response was antagonized by sphingosine, an inhibitor of protein kinase C (Fig. 2). These results indicate that the GABAB receptor-mediated response is suppressed by the activation of protein kinase C. The property of the GABAB receptor expressed in the

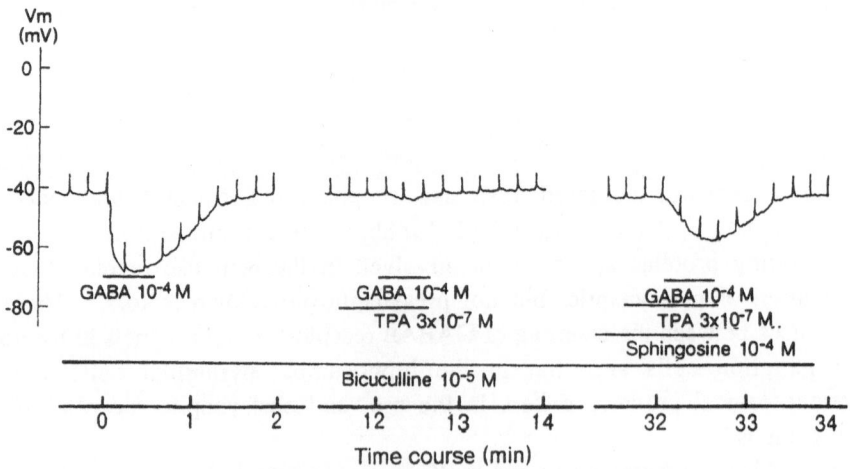

Fig. 2. Suppression of GABAB receptor-mediated hyperpolarization of oocyte membrane by activation of protein kinase C.

Xenopus oocytes by injection of mRNA from rat cerebellum may correspond to that in the postsynaptic GABAB receptor of the hippocampus.

Protein kinase C may possibly phosphorylate the GABAB receptor, GTP-binding proteins and ion channels. Activation of protein kinase C has been shown to phosphorylate the nicotinic ACh receptor (Huganir et al., 1984), β-adreneceptor (Bouvier et al., 1987) and α_1-adrenoceptor (Leeb-Lundberg et al., 1985; Bouvier et al., 1987), and α-subunit of GTP-binding protein (Gi) (Katada et al., 1985), and the Na^+ channel (Costa and Catterall, 1985). It is not clear at present whether the target proteins of protein kinase C are GABAB receptor or ion channels. GTP-binding protein do not seem to be a target protein because the hippocampal presynaptic GABAB receptor-mediated response is resistant to pertussis toxin and yet is suppressed by the activation of protein kinase C (Dutar and Nicoll, 1988b).

CONCLUSION

We described herein the expression of the GABAB receptor in oocytes injected with mRNA from the rat cerebellum. The property was similar to the hippocampal postsynaptic GABAB receptor. Desensitization of the GABAB receptor by activation of protein kinase C was noted in the oocytes. At least 4 subspecies of protein kinase C, α, βI, βII and γ are present in the brain (Nishizuka, 1988). It is still not clear which subspecies of protein kinase C participates in the desensitization of GABAB receptor-mediated response, since the mRNA from the rat cerebellum used in our study may contain mRNA encoding the multiple subspecies of protein kinase C (Tanaka et al., 1988; Nishizuka, 1988) together with that encoding GABAB receptor.

ACKNOWLEDGEMENTS

This work was supported in part by grants from the Ministry of Education, Science and Culture, Japan.

REFERENCES

Andrade, R., Malenka, R.C. and Nicoll, R.A., 1986, A G protein couples serotonin and GABAB receptors to the same channels in hippocampus, Science 234: 1261-1265.

Bormann, J., 1988, Electrophysiology of GABAA and GABAB receptor subtypes, Trend. Neurosci. 11: 112-116.

Bouvier, M., Leeb-Lundberg, L.M.F., Benovic, J.L., Caron, M.G. and Lefkowitz, R.J., 1987, Regulation of adrenergic receptor function by phosphorylation. II. effects of agonist occupancy on phosphorylation of α_1- and β_2-adrenergic receptors by protein kinase C and the cyclic AMP-dependent protein kinase. J. Biol. Chem. 262: 3106-3113.

Bowery, N.G., 1989, GABAB receptors and their significance in mammalian pharmacology, Trends Pharmacol. Sci. 10: 401-407.

Bowery, N. G., Price, G. W., Hudson, A. L., Hill, D. R., Wilkin, G. P. and Turnbull, M. J., 1984, GABA receptor multiplicity, Neuropharmacol. 23: 219-231.

Colmers, W.F. and Pittman, Q.J., 1989, Presynaptic inhibition by neuropeptide Y and baclofen in hippocampus: insensitivity to pertussis toxin treatment, Brain Res. 498: 99-104.

Costa, M.R.C. and Catterall, W.A., 1985, Phosphorylation of the α subunit of the sodium channel by protein kinase C, Cell. Mol. Neurobiol. 4: 291-297.

Dascal, N., 1987, The use of Xenopus oocytes for the study of ion channels, CRC Crit. Rev. Biochem. 22: 317-387.

Dascal, N., Ifune, C., Hopkins, R., Snutch, T. P., Lubbert, H., Davidson, N., Simon, M. I. and Lester, H. A., 1986, Involvement of a GTP-binding protein in mediation of serotonin and acetylcholine responses in Xenopus oocytes injected with rat brain messenger RNA, Mol. Brain Res. 4: 97-105.

Dascal, N., Lotan, I., Gillo, B., Lester, H.A. and Lass, Y., 1985, Acetylcholine and phorbol esters inhibit potassium currents evoked by adenosine and cAMP in Xenopus oocytes, Proc. Natl. Acad. Sci. USA 82: 6001-6005.

Deisz, R.A. and Lux, H.D., 1985, γ-Aminobutyric acid induced depression of calcium currents of chick sensory neurones, Neurosci. Lett. 56, 205-210.

Dolphin, A.C. and Scott, R.H., 1987, Calcium channel currents and their inhibition by (-)baclofen in rat sensory neurones: modulation by guanine nucleotides, J. Physiol. 386: 1-17.

Dutar, P. and Nicoll, R.A., 1988a, A physiological role for GABAB receptors in the central nervous system, Nature 332, 156-158.

Dutar, P. and Nicoll, R.A., 1988b, Pre- and postsynaptic GABAB receptors in the hippocampus have different pharmacological properties, Neuron 1: 585-591.

Gahwiler, B.H. and Brown, D.A., 1985, GABAB-receptor-activated K^+ current in voltage-clamped CA3 pyramidal cells in hippocampal cultures, Proc. natl. Acad. Sci. USA 82, 1558-1562.

Gundersen, C. B., Miledi, R. and Parker, I., 1983, Serotonin receptors induced by exogenous messenger RNA in Xenopus oocytes, Proc. R. Soc. Lond. B219: 103-109.

Gundersen, C. B., Miledi, R. and Parker, I., 1984, Glutamate and kainate receptors induced by rat brain messenger RNA in Xenopus oocytes, Proc. R. Soc. Lond. B 221: 127-143.

Harada, Y., Takahashi, T., Kuno, M., Nakayama, K., Masu, Y. and Nakanishi, S., 1987, Expression of two different tachykinin receptors in Xenopus oocytes by exogenous mRNAs, J. Neurosci. 7: 3265-3273.

Hirono, C., Ito, H. and Sugiyama, H., 1987, Neurotensin and acetylcholine evoke common responses in frog oocytes injected with rat brain messenger ribonucleic acid, J. Physiol. 382: 523-535.

Holz, G.G., Rane, S.G. and Dunlap, K., 1986, GTP binding proteins mediate transmitter inhibition of volatage-dependent calcium channels, Nature 319, 670-672.

Houamed, K.M., Bilbe, G., Smart, T.G., Constanti, A., Brown, D.A., Barnard, E.A. and Richards, B.M., 1984, Expression of functional GABA, glycine and glutamate receptors in Xenopus oocytes injected with rat brain mRNA, Nature 310: 318-321.

Huganir, R.L., Miles, K. and Greengard, P., 1984, Phosphorylation of the nicotinic acetylcholine receptor by an endogenous tyrosine-specific protein kinase, Proc. Natl. Acad. Sci. USA 81, 6968-6972.

Inoue, M., Matsuo, T. and Ogata, N., 1985, Baclofen activates voltage-dependent and 4-aminopyridine sensitive K^+ conductance in guinea-pig hippocampal pyramidal cells maintained in vitro, Br. J. Pharmacol. 84: 833-841.

Kaczmarek, L. K., 1987, The role of protein kinase C in the regulation of ion channels and neurotransmitter release, Trend. neurosci., 10: 30-34.

Katada, T., Gilman, A.G., Watanabe, Y., Bauer, S. and Jacobs, K.H., 1985, Protein kinase C phosphorylates the inhibitory guanin-nucleotide-binding regulatory component and apparently suppresses its function in hormonal inhibition of adenylate cyclase, Eur. J. Biochem. 151: 431-437.

Kato, K., Kaneko, S. and Nomura, Y., 1988, Phorbol ester inhibition of current responses and simultaneous protein phosphorylation in *Xenopus* oocyte injected with brain mRNA, J. Neurochem. 50: 766-773.

Leeb-Lundberg, L.M.F., Cotecchia, S., Lomosney, J.W., DeBernardis, L.F., Lefkowitz, R.J. and Caron, M.G., 1985, Phorbol esters promote α_1-adrenergic receptor phosphorylation and receptor uncoupling from inositol phospholipid metabolism, Proc. Natl. Acad. Sci. 82: 5651-5655.

Leonard, J.P., Nargeot, J., Snutch, T.P., Davidson, N. and Lester, H.A., 1987, Ca channels induced in *Xenopus* oocytes by rat brain mRNA, J. Neurosci. 7: 875-881.

Levitan, E. S., Schofield, P. R., Burt, D. R., Rhee, L. M., Wisden, W., Kohler, M., Fujita, N., Rodriguez, H. F., Stephenson, A., Darlison, M. G., Barnard, E. A.,and Seeburg, P. H., 1988, Structural and functional basis for GABAA receptor heterogeneity, Nature 335: 76-79.

Lory, P., Richard, S., Rassendren, F. A., Tiaho, F. and Nargeot, J., 1989, Electrophysiological expression of endothelin and angiotensin receptors in Xenopus oocytes injected with rat heart mRNA, FEBS Lett. 258: 289-292.

Lubbert, H., Snutch, T. P., Dascal, N., Lester, H. A. and Davidson, N., 1987, Rat brain $5HT_{1C}$ receptors are encoded by a 5-6 kbase mRNA size class and are functionally expressed in injected Xenopus oocytes, J. Neurosci. 7: 1159-1165.

Moran, O. and Dascal, N.,1989, Protein kinase C modulates neurotransmitter responses in Xenopus oocytes injected with rat brain RNA, Mol. Brain Res. 5: 193-202.

Nestler, E. J., Walaas, S. I. and Greengard, P., 1984, Neuronal phosphoproteins: Physiological and clinical implications, Science 225: 1357-1364.

Newberry, N.R. and Nicoll, R.A., 1984, Direct hyperpolarizing action of baclofen on hippocampal pyramidal cells, Nature 308: 450-452.

Newberry, N.R. and Nicoll, R.A., 1985, Comparison of the action of baclofen with γ-aminobutyric acid on rat hippocampal pyramidal cells in vitro, J. Physiol. 360: 161-185.

Nishizuka, Y., 1984, Turnover of inositol phospholipid and signal transduction, Science 225: 1365-1370.

Nishizuka, Y., 1988, The molecular heterogeneity of protein kinase C and its implications for cellular regulation, Nature 334: 661-665.

Parker, I., Sumikawa, K. and Miledi, F. R. S., 1986, Neurotensin and substance P receptors expressed in *Xenopus* oocytes by messenger RNA from rat brain, Proc. R. Soc. Lond. B 229: 151-159.

Rane, S.G. and Dunlap, K., 1986, Kinase C activator 1,2-oleoylacetylglycerol attenuates voltage-dependent calcium current in sensory neurons, Proc. Natl. Acad. Sci. USA 83, 184-188.

Robertson, B. and Taylor, W.R., 1986, Effects of γ-aminobutyric acid and (-)-baclofen on calcium and potassium currents in cat dorsal root ganglion neurones in vitro, Br. J. Pharmacol. 89: 661-672.

Schofield, P. R., Darlison, M. G., Fujita, N., Burt, D. R., Stephenson, F. A., Rodriguez, H., Rhee, L. M., Ramachandran, J., Reale, V., Glencorse, T. A., Seeburg, P. H. and Barnard, E. A., 1987, Sequence and functional expression of the GABAA receptor shows a ligand-gated receptor superfamily, Nature 328: 221-227.

Sigel, E. and Baur, R., 1988, Activation of protein kinase C differentially modulates neuronal Na^+, Ca^{2+}, and γ-aminobutyrate type A channels, Proc. Natl. Acad. Sci. USA 85: 6192-6196.

Snutch, T. P., 1988, The use of Xenopus oocytes to probe synaptic communication, Trend. Neurosci. 11: 250-256.

Sugiyama, H., Hisanaga, Y. and Hirono, C., 1985, Induction of muscarinic cholinergic responsiveness in xenopus oocytes by mRNA isolated from rat brain, Brain Res. 338: 346-350.

Sugiyama, H., Ito, I. and Hirono, C., 1987, A new type of glutamate receptor linked to inositol phospholipid metabolism, Nature 325: 531-533.

Tanaka, C., Saito, N., Kose, A., Hosoda, K., Sakaue, M., Shuntoh, H., Nishino, N and Taniyama, K., 1988, Possible roles of protein kinase C in neurotransmission, p 277-285, in: "Neuroreceptors and signal transduction", S. Kito, T. Segawa, K. Kuriyama, M. Tohyama and R. W. Olsen, eds., Plenum Publishing Corp., New York.

Verdoorn, T. A., Kleckner, N. and Dingledine, R., 1987, Rat brain N-methyl-D-aspartate receptors expressed in *xenopus* oocytes, Science 238: 1114-1116

STRUCTURE AND EXPRESSION OF INHIBITORY GLYCINE RECEPTORS

H. Betz, D. Langosch, W. Hoch, P. Prior, I. Pribilla,
J. Kuhse, V. Schmieden, M.-L. Malosio, B. Matzenbach,
F. Holzinger, A. Kuryatov, B. Schmitt, Y. Maulet, and
C.-M. Becker

Zentrum für Molekulare Biologie
Universität Heidelberg
Im Neuenheimer Feld 282
6900 Heidelberg, FRG

INTRODUCTION

Signal transmission at chemical synapses involves specific receptors that transduce neurotransmitter binding into alterations of membrane potential. Receptors containing integral ion channels mediate rapid (in the $\leq$ msec range) transduction events, whereas receptors activating G-protein coupled channels operate at slower time scales (in the msec to sec range). At resting membrane potential, excitation is generated by cation influx, but inhibition of neuronal firing results from increased chloride permeability.

The nicotinic acetylcholine receptor at the neuromuscular junction initiates muscle contraction. Due to its abundance in the electric organ of certain fish species, it is the best characterized ion channel protein known (reviewed in Changeux et al., 1984). The primary structures of its subunits have been determined in different species, and homologous cDNAs have been isolated from vertebrate and _Drosophila_ brain. The major inhibitory neurotransmitters at central synapses, glycine and γ-aminobutyric acid (GABA), gate chloride channel-forming receptors of similar conductance properties (Bormann et al., 1987), but distinct pharmacology (Betz and Becker, 1988). For example, the convulsive alkaloid strychnine antagonizes postsynaptic inhibition by glycine, the predominant inhibitory neurotransmitter in brain stem and spinal cord, whereas benzodiazepines and barbiturates modify inhibitory GABA$_A$ receptor responses in many regions of the central nervous system.

Keywords: synaptic inhibition / glycine / neurotransmitter receptor /
strychnine / subtype heterogeneity / Xenopus oocyte /
spinal cord / development / chloride channel / transfection.

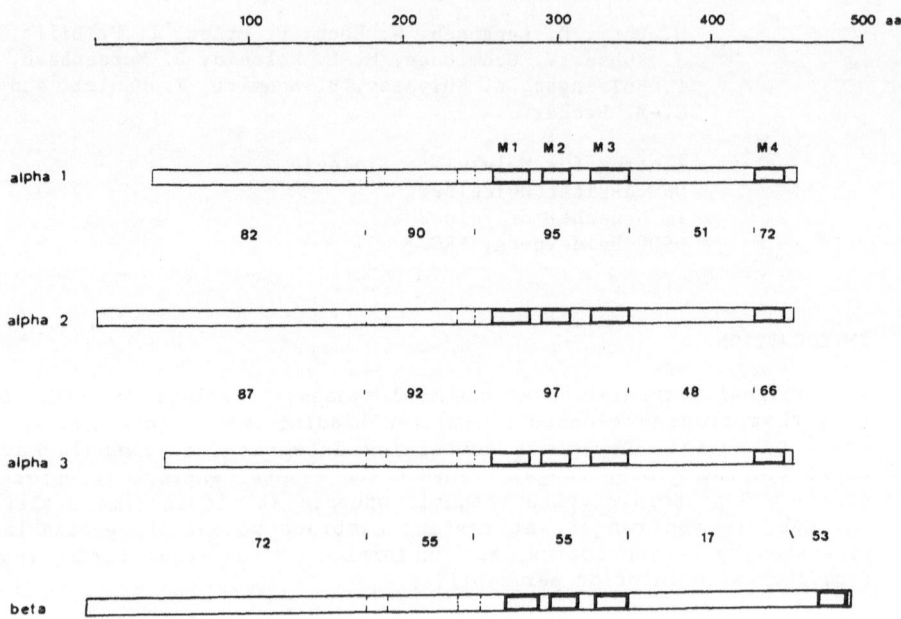

Fig. 1. Structure and homology of mature rat GlyR subunits. The
schematic drawing represents a linear map of subunit organization; amino acid identities (in %) of different domains to the
rat α_1 subunit are indicated. Transmembrane regions are boxed,
and positions of conserved cysteines in the extracellular
domain marked by a weak line. aa, number of amino acids.

THE GLYCINE RECEPTOR, A MEMBER OF THE LIGAND-GATED CHANNEL PROTEIN FAMILY

The glycine receptor (GlyR) was the first receptor protein to be isolated from the mammalian central nervous system (Pfeiffer et al., 1982). Affinity-purified preparations of the GlyR contain two glycosylated integral membrane proteins of 48 kDa (α), and 58 kDa (β) (Pfeiffer et al., 1982; Graham et al., 1985; Becker et al., 1986) which are thought to form the chloride channel of the receptor (Langosch et al., 1988; Betz and Becker, 1988). A copurifying peripheral membrane protein of 93 kDa associated with cytoplasmic domains of the GlyR (Schmitt et al., 1987; Becker et al., 1989) has been implicated in its synaptic localization and/or anchoring to cytoskeletal elements. Indeed, immunocytochemical data have revealed its co-distribution with other GlyR subunits at central synapses (Triller et al., 1985; Altschuler et al., 1986; Araki et al., 1988).

Peptide mapping (Pfeiffer et al.,1984) and cDNA sequencing (Grenningloh et al., 1987a, and in preparation) of α and β GlyR subunits revealed a high homology between these proteins (Fig. 1). Both possess a cleavable signal sequence and, in the C-terminal half of the polypeptide, four hydrophobic segments (M1 to M4) long enough to form transmembrane α-helices. This arrangement resembles that of nicotinic acetylcholine receptor (Noda et al., 1983; Changeux et al., 1984) and GABA$_A$ receptor proteins (Schofield et al., 1987) suggesting that all channel-forming receptors are composed of subunits sharing a common transmembrane topology. Furthermore, significant amino acid sequence homology exists between the subunits of different ligand-gated ion channels (Grenningloh et al., 1987a and b; Schofield et al., 1987). Thus, these receptors constitute a protein superfamily that evolved by gene duplication from a common ancestor early in phylogeny.

PUTATIVE CHLORIDE CHANNEL DOMAINS

The transmembrane segments M1 to M3 are highly conserved between GlyR and GABA$_A$ receptor subunits, pointing to their potential importance in chloride channel function (Grenningloh et al., 1987b). Segment M2 contains a high content of uncharged polar amino acid residues and therefore is thought to provide the hydrophilic inner lining of the chloride channel. Here, eight consecutive amino acid residues are identical in most GABA$_A$ receptor and GlyR subunits. Interestingly, transmembrane segment M2 of nicotinic acetylcholine receptor proteins is known to be involved in cation transport and channel blocker binding (Giraudat et al., 1986; Hucho et al., 1986; Imoto et al., 1986). Segment M2 thus may be a common structural determinant of ligand-gated ion channel function. Indeed, comparison of both cation and anion channel forming M2 sequences unravels some common periodicity where a bulky hydrophobic residue at roughly every fourth position, i.e. on the same side of a potential transmembrane α-helix, is followed by a small or polar side chain (Betz, 1990). Having the latters arranged towards the channel lumen, a pore sufficiently large and polar for passage of permeating ions may be created in the center of only five symmetrically arranged α-helices (Bormann et al., 1987).

The M2 segments of anion-selective GlyR and GABA$_A$ receptor proteins usually terminate with positively charged residues both intra- and extracellularly. Furthermore, positively charged residues are abundant in close vicinity to the M1 and M3 sequences. Cation-conducting nicotinic acetylcholine receptors in contrast have negatively charged side chains close to positive charges bordering transmembrane segment M2

(Noda et al., 1983). Patch clamp data indicate two sequentially occupied anion binding sites in both GlyR and GABA_A receptor channels (Bormann et al., 1987). The charged residues at the termini of the M2 segments may be the structural correlate of these sites at the presumptive inner and outer mouths of receptor ion channels, and thus provide their ion selectivity filter. Indeed, mutation of the negatively charged residues bordering the M2 regions of nicotinic acetylcholine subunits has been found to modulate the conductivity of this cation channel (Imoto et al., 1988). From single channel analysis of the expressed mutant receptors, three rings of negatively charged and glutamine residues neighboring the M2 segments have been identified which serve as major determinants of cation transport through the nicotinic receptor. Also, a synthetic peptide corresponding to segment M2 of the GlyR α subunit has been shown to produce randomly gated "channels" upon incorporation into planar lipid bilayers (Langosch et al., 1990). Interestingly, the ion selectivity of these channels was modulated upon inversing the terminal charges of the peptide. As outlined above, hydroxylated residues are abundant within the M2 regions of anion channel-forming receptor subunits. The high positive ion potential of the hydroxyl terminus may stabilize permeating anions and thus contribute to ion selection in these channels.

THE EXTRACELLULAR LIGAND-BINDING REGION

Besides considerable conservation of transmembrane sequences, high homology exists also in the extracellular N-terminal domain (Fig. 1). Remarkable are two precisely conserved cystein residues which are also present in nicotinic acetylcholine and GABA_A receptor polypeptides. For the acetylcholine receptor, these cysteines have been proposed to form a disulfide bridge essential for receptor tertiary structure (Mishina et al., 1985; Stroud and Finer-Moore, 1985). Similar folding patterns thus may exist in the extracellular portion of channel forming receptor proteins.

Photoaffinity labelling experiments using [³H]strychnine have localized the ligand binding site of the GlyR on the α subunit (Graham et al., 1981 and 1983). From theoretical considerations, a stretch of charged residues preceeding the first transmembrane segment has been postulated to be part of the binding pocket (Grenningloh et al., 1987a). Interestingly, a corresponding region containing two neighboring cysteine residues is known to be important for acetylcholine binding to the α-subunits of the nicotinic acetylcholine receptor (Kao et al., 1984). Within the latter, however, other extracellular residues have in addition been shown to react upon covalent incorporation of acetylcholine derivatives (Dennis et al., 1988). These data suggest that agonist binding to ligand-gated channel proteins usually involves multiple interactions with an extended extracellular domain of the respective ligand binding subunits.

THE GlyR IS A PENTAMERIC PROTEIN

Analysis of the subunit composition of purified GlyR preparations by sedimentation and crosslinking techniques has indicated a pentameric channel core composed of three α and two β subunits, respectively (Langosch et al., 1988). This subunit stoichiometry resembles that of the nicotinic acetylcholine receptor which is known to contain five membrane-spanning subunits (Changeux et al., 1984; Stroud and Finer-Moore, 1985). In view of the above discussed sequence homology and common transmembrane topology of different channel forming receptor proteins, a quasisymmetrical pentameric complex of transmembrane polypeptides around a central ion pore is proposed as the common quaternary structure of different members of the ligand-gated ion channel super-family (Langosch et al., 1988).

GLYCINE RECEPTOR SUBTYPES

Recent biochemical and cDNA sequence data have established subtype diversity as a general phenomenon for brain nicotinic acetylcholine and GABA$_A$ receptor subunits. In case of these receptors, heterologous expression of different subtype combinations has been shown to produce functionally and/or pharmacologically distinct channel entities. For the GlyR, subtype heterogeneity has first been detected during spinal cord development (Becker et al., 1988). There, a neonatal receptor isoform is prevalent at birth that differs in strychnine binding affinity, immunological properties and molecular weight (49 kDa) of its ligand binding subunit from the adult receptor protein (Becker et al., 1988). This neonatal GlyR isoform is abundantly expressed in primary cultures of embryonic spinal cord, a condition that facilitated its biochemical analysis (Hoch et al., 1989). Pulse-chase experiments indicate that the neonatal GlyR is a metabolically stable protein (t$_{1/2}$ ≈ 2 days) that contains only α subunits.

Evidence for glycine receptor heterogeneity also comes from DNA sequencing data. By screening cDNA and genomic libraries under conditions of low stringency, variants of the originally isolated GlyR α subunit (now termed α$_1$) have been isolated (Fig. 1). The novel α$_2$ subunit sequences predicted from human and rat cDNAs display about 80% amino acid identity to their α$_1$ counterparts and probably correspond to the ligand binding subunits of the neonatal GlyR as α$_2$ transcripts are abundant in spinal cord of newborn rats (Grenningloh et al., 1990; Kuhse et al., 1990). In addition, an α$_3$ sequence has been isolated from rat (Fig. 1), and clones encoding exons of a fourth variant, α$_4$, have been identified in mouse genomic libraries (J. Kuhse and Y. Maulet, unpublished data). Thus, considerable diversity of ligand binding subunits exists for the GlyR. So far, however, no variants of the β subunit could be detected (A. Kuryatov, unpublished). In conclusion, not only one, but several GlyRs appear to exist in the vertebrate CNS.

Alternative splicing further contributes to GlyR heterogeneity in spinal cord. For the rat α$_1$ subunit, a variant cDNA has been identified that originates from alternate splice acceptor site selection at an exon encoding the cytoplasmic domain adjacent to transmembrane segment M3 (Malosio et al., 1990). The insertion resulting contains eight additional amino acid residues and may create novel sites for protein modification and/or association to the cytoskeleton. S1 nuclease mapping and PCR experiments indicate that this splice variant is expressed at all stages of postnatal development.

FUNCTIONAL EXPRESSION OF GLYCINE RECEPTOR SUBUNITS

Heterologous expression of individual agonist binding subunits of glycine and GABA$_A$ receptors in <u>Xenopus</u> oocytes and mammalian cell lines generates functional ligand-gated channels which display most of the typical pharmacology of their natural counterparts (Pritchett et al., 1988; Schmieden et al., 1989; Sontheimer et al., 1989). Although the efficiency of assembly of such homooligomeric receptors may be low

(Sontheimer et al., 1989), their formation clearly indicates that individual receptor subunits must have similar, exchangeable oligomerisation sites. This may be exploited in vivo for generating functional diversity from a limited set of subunit subtypes.

Expression of individual rat GlyR α_1, or human α_1 and α_2, subunit RNAs in Xenopus oocytes (Schmieden et al., 1989; Grenningloh et al., 1990), or transfection of corresponding expression constructs into mammalian cells (Sontheimer et al., 1989; and F. Holzinger, unpublished data) results in generation of glycine-gated chloride channels. These homooligomeric receptors are blocked by nanomolar concentrations of strychnine, as is the GlyR in spinal neurons (Betz and Becker, 1988). A marked difference, however, was found upon rat α_2 subunit expression. The receptor resulting displayed reduced glycine affinity combined with low strychnine sensitivity (IC50 $\approx$ 50μM), whereas gating by the amino acids β-alanine and taurine was comparable to that found for α_1 subunit receptors (Kuhse et al., 1990). This closely resembles the pharmacological profile of neonatal GlyR in biochemical studies and demonstrates the functional importance of GlyR subunit heterogeneity in the mammalian CNS.

CONCLUSIONS AND PERSPECTIVES

The present biochemical and sequence data on GlyRs provide considerable structural details of these ion channel proteins. Furthermore, they underline the importance of glycinergic synapses in the control of neuronal activity. The crucial role of GlyRs in regulating diverse motor and sensory functions is also supported by studies on animal mutants. Spastic mice and myoclonic Poll Hereford cattle display severe motor deficits resulting from GlyR deficiencies which strongly reduce the life span of the affected animals (White and Heller, 1982; Becker et al., 1986; Gundlach et al., 1988). GlyR anomalies may also be implicated in human neurological diseases. Interestingly, the gene of one GlyR subunit in man, α_2, has been localized in close vicinity to the Duchenne-Becker muscular dystrophy locus (Grenningloh et al., 1990). Thus, unraveling the biochemistry and molecular biology of inhibitory GlyRs may not only contribute to understanding the functioning and pharmacology of this neuronal channel protein, but elucidate pathogenic mechanisms in humans.

REFERENCES

Altschuler, R., Betz, H., Parakkal, M. H., Reeks, K. A., and Wenthold, R. J., 1986, Identification of glycinergic synapses in the cochlear nucleus through immunocytochemical localization of the postsynaptic receptor, Brain Res., 369:316-320.

Araki, T., Yamano, M., Murakami, T., Wanaka, A., Betz, H., and Tohyama, M., 1988, Localization of glycine receptors in the rat central nervous system: an immunocytochemical analysis using monoclonal antibody, Neuroscience, 25:613-624.

Becker, C.-M., Hermans-Borgmeyer, I., Schmitt, B., and Betz, H., 1986, The glycine receptor deficiency of the mutant mouse spastic: evidence for normal glycine receptor structure and localization, J. Neurosci. 6:1358-1364.

Becker, C.-M., Hoch, W., and Betz, H., 1988, Glycine receptor heterogeneity in rat spinal cord during postnatal development, EMBO J., 7: 3717-3726.

Becker, C.-M., Hoch, W., and Betz, H., 1989, Sensitive immunoassay shows selective association of peripheral and integral membrane proteins of the inhibitory glycine receptor complex, J. Neurochem., 53:125-131.

Betz, H., 1990, Homology and analogy in transmembrane channel design: lessons from synaptic membrane proteins, Biochemistry, in press.

Betz, H. and Becker, C.-M., 1988, The mammalian glycine receptor: Biology and structure of a neuronal chloride channel protein, Neurochem. Int., 13:137-146

Bormann, J., Hamill, O. P., and Sakmann, B., 1987, Mechanism of anion permeation through channels gated by glycine and γ-aminobutyric acid in mouse cultured spinal neurones, J. Physiol., 385:243-286.

Changeux, J.-P., Devilliers-Thiéry, A., and Chenouilli, P., 1984, Acetylcholine receptor: an allosteric protein, Science, 225:1335-1345.

Dennis, M., Giraudat, J., Kotzyba-Hibert, F., Goeldner, M., Hirth, C., Chang, J.-Y., Lazure, C., Chrétien, M., and Changeux, J.-P., 1988, Amino acids of the Torpedo marmorata acetylcholine receptor α subunit labeled by a photoaffinity ligand for the acetylcholine binding site, Biochemistry, 27:2345-2351.

Giraudat, J., Dennis, M., Heidmann, T., Chang, J.-Y., and Changeux, J.-P., 1986, Structure of the high-affinity binding site for noncompetitive blockers of the acetylcholine receptor: serine-262 of the γ subunit is labeled by [³H] chlorpromazine, Proc. Natl. Acad. Sci. U.S.A., 83:2719-2723.

Graham, D., Pfeiffer, F., and Betz, H., 1981, UV light-induced cross-linking of strychnine to the glycine receptor of rat spinal cord membranes, Biochem. Biophys. Res. Commun., 102:1330-1335.

Graham, D., Pfeiffer, F., and Betz, H., 1983, Photoaffinity-labelling of the glycine receptor of rat spinal cord, Eur. J. Biochem., 131:519-525.

Graham, D., Pfeiffer, F., Simler, R., and Betz, H., 1985, Purification and characterization of the glycine receptor of pig spinal cord, Biochemistry, 24:990-994.

Grenningloh, G., Rienitz, A., Schmitt, B., Methfessel, C., Zensen, M., Beyreuther, K., Gundelfinger, E. D., and Betz, H., 1987a, The strychnine-binding subunit of the glycine receptor shows homology with nicotinic acetycholine receptors, Nature (London), 328:215-220.

Grenningloh, G., Gundelfinger, E., Schmitt, B., Betz, H., Darlison, M. G., Barnard, E. A., Schofield, P. R., and Seeburg, P. H., 1987b, Glycine vs GABA receptors, Nature, 330:25-26.

Grenningloh, G., Schmieden V., Schofield, P. R., Seeburg, P. H., Siddique, T., Mohandas, T. K., Becker, C.-M., and Betz, H., 1990, Alpha subunit variants of the human glycine receptor: primary structures, functional expression and chromosomal localization of the corresponding genes, EMBO J., in press.

Gundlach, A. L., Dodd, P. R., Grabara, C. S. G., Watson, W. E. J., Johnston, G. A. R., Harper, P. A. W., Dennis, J. A., and Healy, P. J., 1988, Deficit of spinal cord glycine/strychnine receptors in inherited myoclonus of Poll Hereford calves, Science, 241:1807-1810.

Hoch, W., Betz, H. and Becker, C.-M., 1989, Primary cultures of mouse spinal cord express the neonatal isoform of the inhibitory glycine receptor, Neuron, 3:339-348.

Hucho, F., 1986, The nicotinic acetylcholine receptor and its ion channel, Eur. J. Biochem., 158:211-226.

Imoto, K., Methfessel, C., Sakmann, B., Mishina, M., Mori, Y., Konno, T., Fukuda, K., Kurasaki, M., Bujot, H., Fujita, Y., and Numa, S., 1986, Location of a δ subunit region determining ion transport through the acetylcholine receptor channel, Nature (London), 324:670-674.

Imoto, K., Busch, C., Sakmann, B., Mishina, M., Konno, T., Nakai, J., Bujo, H., Mori, Y., Fukuda, K., and Numa, S., 1988, Rings of negatively charged amino acids determine the acetylcholine receptor channel conductance, Nature (London), 335:645-648.

Kao, P., Dwork, A., Kaldany, R., Silver, M., Wideman, J., Stein, S., and Karlin, A., 1984, Identification of the α-subunit half-cystine specifically labelled by an affinity reagent for the acetylcholine receptor binding site, J. Biol. Chem., 259:11662-11665.

Kuhse, J., Schmieden, V., and Betz, H., 1990, A neonatally expressed subunit of the rat glycine receptor forms chloride channels of low strychnine sensitivity, submitted.

Langosch, D., Thomas, L., and Betz, H., 1988, Conserved quaternary structure of ligand-gated ion channels: the postsynaptic glycine receptor is a pentamer, Proc. Natl. Acad. Sci. U.S.A., 85:7394-7398.

Langosch, D., Hartung, K., Grell, E., Bamberg, E. and Betz, H., 1990, Synthetic transmembrane segments of the inhibitory glycine receptor form ion-selective channels in lipid bilayers, submitted.

Malosio, M.-L., Grenningloh, G., Kuhse, J., Schmieden, V., Schmitt, B., Prior, P., and Betz, H., 1990, Glycine receptor heterogeneity in rat spinal cord: evidence for alternative splicing of the α₁ subunit, submitted.

Mishina, M., Takai, T., Imoto, K., Noda, M., Takahashi, T., Numa, S., Methfessel, C., and Sakmann, B., 1986, Molecular distinction between fetal and adult forms of muscle acetylcholine receptor, Nature (London), 321:406-410.

Noda, M., Takahashi, H., Tanabe, T., Toyosato, M., Kikyotani, S., Furutani, Y., Hirose, T., Takashima, H., Inayama, S., Miyata, T., and Numa, S., 1983, Structural homology of Torpedo californica acetylcholine receptor subunits, Nature, 302:528-532.

Pfeiffer, F., Graham, D., and Betz, H., 1982, Purification by affinity chromatography of the glycine receptor of rat spinal cord, J. Biol. Chem., 257:818-823.

Pfeiffer, F., Simler, R., Grenningloh, G., and Betz, H., 1984, Monoclonal antibodies and peptide mapping reveal structural similarities between the subunits of the glycine receptor of rat spinal cord, Proc. Natl. Acad. Sci. U.S.A., 81:7224-7227.

Pritchett, D. B., Sontheimer, H., Shivers, B. D., Ymer, S., Kettenmann, H., Schofield, P. R., and Seeburg, P. H., 1989, Importance of a novel GABA_A receptor subunit for benzodiazepine pharmacology, Nature, 8:695-700.

Schmieden, V., Grenningloh, G., Schofield, P. R., and Betz, H., 1987, Functional expression in Xenopus oocytes of the strychnine binding 48 kd subunit of the glycine receptor, EMBO J., 338:582-585.

Schmitt, B., Knaus, P., Becker, C.-M., and Betz, H., 1987, The M_r 93,000 polypeptide of the postsynaptic glycine receptor is a peripheral membrane protein, Biochemistry, 26:805-811.

Schofield, P. R., Darlison, M. G., Fujita, N., Rodriguez, H., Burt, D. R., Stephenson, F. A., Rhee, L., M., Ramachandran, J., Glencorse, T. A., Reale, V., Seeburg, P. H., and Barnard, E. A., 1987, Sequence and functional expression of the GABA_A receptor shows a ligand-gated receptor superfamily, Nature, 328:221-227.

Sontheimer, H., Becker, C.-M., Pritchett, D. B., Schofield, P. R., Grenningloh, G., Kettenmann, H., Betz, H., and Seeburg, P. H., 1989, Functional chloride channels by mammalian cell expression of rat glycine receptor subunit, Neuron, 2:1491-1497.

Stroud, R. M. and Finer-Moore, J., 1985, Acetylcholine receptor structure, function, and evolution, Ann. Rev. Cell. Biol., 1:317-351.

Triller, A., Cluzeaud, F., Pfeiffer, F., Betz, H., and Korn, H., 1985, Distribution of glycine receptors at central synapses: an immunoelectron microscopy study, J. Cell Biol., 101:683-688.

White, F. and Heller, A. H., 1982, Glycine receptor alteration in the mutant mouse spastic, Nature, 298:655-657.

HETEROGENEOUS DISTRIBUTION OF GLUTAMATE RECEPTOR SUBTYPES IN

HIPPOCAMPUS AS REVEALED BY CALCIUM FLUOROMETRY

Yoshihisa Kudo[1], Etsuro Ito[2] and Akihiko Ogura[1]

[1]Department of Neuroscience, Mitsubishi Kasei Institute of Life
Sciences, Minamiooya, Machida, Tokyo 194 and [2]Advanced
Research Center for Human Sciences, Waseda University
Tokorozawa, Saitama 359, JAPAN

INTRODUCTION

Recent advances in the fluorometry of intracellular Ca^{2+} concentration ($[Ca^{2+}]_i$) enabled detecting the time course and the two-demensional distribution of $[Ca^{2+}]_i$ in single isolated cells as well as the tissue preparations (Kudo and Ogura, 1986; Nohmi et al., 1989). Since the role of Ca^{2+} on the modulation of the efficacy of synaptic transmission (synaptic plasticity) has been suggested from physiological and biochemical studies, the methods have been applied to neuronal cells in culture and brain slices, and elevations of $[Ca^{2+}]_i$ during neuronal activity have directly been shown (Connor et al., 1988; Kudo et al., 1987).

In the present study we examined the distributions of glutamate receptor subtypes in the single neuron and in the slice preparation from the rat hippocampus using Ca^{2+} fluorometry. This is based on the recent knowledge about the mobilization of Ca^{2+} by the activation of on glutamate receptor subtypes; these are (1) activation of N-methyl-D-aspartate (NMDA)-subtype of L-glutamate receptor leads to an influx of Ca^{2+} into the cell (Collingridge and Bliss, 1987), (2) quisqualate (QUIS)-subtype mobilizes intracellularly stored Ca^{2+} by facilitating phosphoinositide metabolism (Sugiyama et al., 1987), (3) kainate (KAI)-subtype of elevates $[Ca^{2+}]_i$ via unspecified pathway(s) (Murphy and Miller, 1989; Ogura et al., 1989), and (4) the diffusion of Ca^{2+} in the neuronal cytoplasm is so slow that the state of elevated $[Ca^{2+}]_i$ remains in a limited volume after the activation of receptors for neurotransmitters or of voltage-operated Ca^{2+} channels (Connor et al., 1988). Combining these facts, we expect that the measurement of $[Ca^{2+}]_i$ of high spatial resolution should reveal the distribution of each subspecies of L-glutamate receptor. Digital analysis of fluoromicroscopic images of cultured neurons and slices of rat hippocampus loaded with a Ca^{2+}-indicator, fura-2, is of sufficiently high temporal and spatial resolutions.

This type of study has conventionally been done by autoradiography using a radioactive ligand for each receptor (Cotman et al., 1987). But the binding sites for ligands do not necessarily mean that functional receptors are present . Moreover it is difficult in autoradiography to reveal the distributions of multiple species of receptors in a single preparation. Development of silver grains needs considerable time.

The digital Ca^{2+} fluorometry should be free of those problems, since a single living preparation can be stimulated by different agonists in a repeated manner. Moreover, comparison of the magnitudes of $[Ca^{2+}]_i$ elevation induced by agonists should become possible by Ca^{2+} fluorography, which is important in elucidating the role of Ca^{2+} in the establishment of long term potentiation of synaptic efficacy(LTP).

MATERIALS AND METHODS

<u>Dissociation cell culture of rat hippocampal neurons</u> Hippocampal neurons were isolated from rat embryos of gestational day 18 and maintained for more than 7 days on a thin glass coverslip attached to a silicon rubber wall. The neurons were treated with $5\,\mu M$ fura-2 acetoxymethylester (fura-2/AM) for *ca.* 60 min (37°C) and then placed on an inverted fluorescence microscope. The culture was continuously perfused (2 ml/min) with warmed (32 °C) balanced salt solution (BSS) composed of (mM): NaCl 130, KCl 5.4, $CaCl_2$ 1.8, glucose 5.5, tetrodotoxin 0.001, and HEPES-NaOH 20 (pH 7.3). For detailed method see previous report[10].

<u>Preparation of rat hippocampal slice</u> Hippocampal slices of *ca.* $300\,\mu m$ in thickness were obtained from young adult rats. The slice was treated with fura-2/AM (7.5 - 10 μM) for 45 min (37°C) and placed in a chamber made of a thin glass coverslip and a leucite wall. The chamber was continuously perfused with oxygenated artificial cerebrospinal fluid (32°C; 2 ml/min) composed of (mM): NaCl 124,KCl 5.0, NaH_2PO_4 1.25, $NaHCO_3$ 22 and glucose 10. Population spikes were recorded from the pyramidal layer of CA1 or CA3 region on stimulating Schaffer-collateral or mossy fiber, respectively. LTP in the population spike was induced by the tetanization (50 Hz, 5 s) of each pathway.

<u>Device for measurement of $[Ca^{2+}]_i$</u> The device is composed of an inverted fluorescence microscope (Olympus IMT-2), a silicon-intensified-target video camera (Hamamatsu photonics; C2400-8) and an image processor (Hamamatsu photonics; ARGUS 100). For details see previous description (Ogura et at., 1988). Briefly, a fluorescence image of a fura-2-loaded specimen obtained under the illumination light of 340 nm in wavelength was divided by that obtained under the light of 360 nm. In some cases we calculated the ratio of the fluorescence images obtained both by 340 nm but taken before and after treatment with agonists for subtypes of glutamate receptor. This ratio gives the relative magnitude of $[Ca^{2+}]_i$ changes in response to the applied agonists. The specific software used for this pixel-to-pixel division includes correction for photo-bleaching of fura-2 and provides a series of processed (rationized) pictures at an arbitrary (> 1/30 s) sampling interval.

Objective lens used for the epifluorescence observation of cultured neurons was x40 or x100, while that for viewing slices was x4 or x10. Conversion of the data of fluorescence ratio to the absolute levels of $[Ca^{2+}]_i$ (according to a calibration curve previously obtained by illuminating the standard solutions of known Ca^{2+} concentrations) was applied to the cultured neurons but not to the slices. This is because the autofluorescence of the slice (due presumably to respiratory coenzymes) diluted the Ca^{2+}-dependent signal of the fluoroprobe in a variable proportion. To overcome the

432

autofluorescence, we had to load the slice with fura-2 sufficiently heavily until the fluorescence signals under excitation lights of 340 and 380 nm behaved oppositely (autofluorescence moves in the same direction). In some preparations, population spikes were recorded simultaneously.

RESULTS

Characteristics of the effects of glutamate receptor subtype agonists. When agonists for glutamate receptor-subtypes were applied on the cultured hippocampal neurons, the $[Ca^{2+}]_i$ increased dose-dependently. The effect of QUIS was always more obvious than the others. As has been reported in the electrophysiological technique (Johnson and Ascher, 1987), NMDA-induced $[Ca^{2+}]_i$ was markedly augmented, under the treatment with glycine (1 μM), while the effect of QUIS remained unchanged (Fig. 1). On the other hand, the effect of NMDA was abolished in medium containing 3 mM Mg^{2+}, while the effects of QUIS and KAI were not affected. These results indicate that the elevation of $[Ca^{2+}]_i$ induced by QUIS and KAI is independent from the NMDA receptor coupled Ca^{2+} channel and that these amino acids can elevate the level of $[Ca^{2+}]_i$ by their own mechanisms. To examine if the elevations in $[Ca^{2+}]_i$ induced by these amino acids were dependent on the voltage-operated Ca^{2+} channels, we blocked pharmacologically the voltage-operated Ca^{2+} channel. From electrophysiological examination La^{3+} was effective for blocking the channel (Ozawa et al., 1989). Actually the treatment with La^{3+} (50 μM) completely blocked the KCl (50 mM)-induced increase in $[Ca^{2+}]_i$. Although La^{3+} partially blocked the response to NMDA as well as QUIS and KAI, these agonists still produced obvious increases in $[Ca^{2+}]_i$. It is noted that the elevation of fluorescence did not recover fully to the initial level during the observation of more than 10 min. Since we confirmed that the fluorescence of fura-2 is increased when the dye makes a chelate with La^{3+}, the long "tail response" in the fluorescence was concluded to be due to the influx of La^{3+} through the ionic entry pathways linked to these amino acid receptors.

Since the influx of Ca^{2+} was induced by the activation of all subtypes of glutamate receptor, it should be possible to detect the regions sensitive to each agonist by two-dimensional image analysis of La^{3+} treated neurons. Fig.3 exemplifies the results of such two dimensinal analysis of a fura-2 loaded neuron on the exposure to NMDA (10 μM) and QUIS (10 μM) in the presence of La^{3+}. Although there is some overlap of the area sensitive to NMDA and QUIS, a noticeable difference was observed between the areas susceptible to these agonists. We also observed the heterogeneous distribution of high $[Ca^{2+}]$ area on the activations of KAI and NMDA receptors. The pattern of spatial heterogeneity in the sensitive sites differed from cell to cell.

The results suggest that the receptor subtypes are expressed separately in neuronal membranes, casting a question to the theory of "functional pair" which hypothesizes co-localization of NMDA-receptor and QUIS- or KAI-receptor (Collingridge and Bliss, 1987). There remains a possibility, however, that the receptors are expressed in the brain *in vivo* (where inherent synaptic connections were formed in the restricted regions of dendrites) differently from those in the neurons grown *in vitro* (where synaptic connections would be formed randomly). So we applied our fluoromicroscopic image analysis to the slice of rat hippocampus.

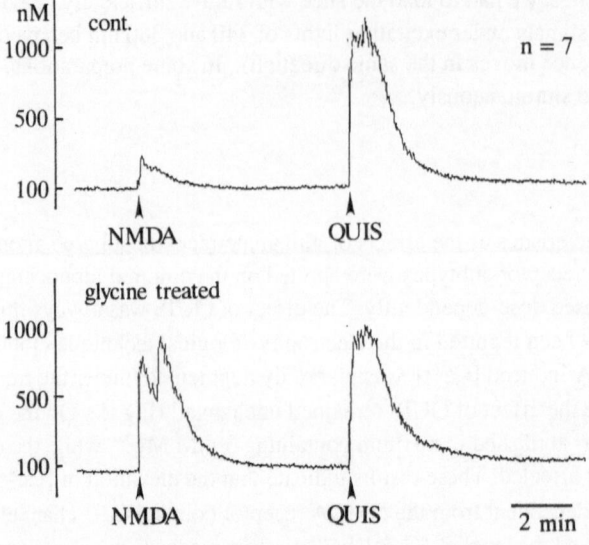

Fig. 1. Effect of glycine on the elevation of $[Ca^{2+}]_i$ induced by NMDA and QUIS.
Upper trace: Effects of NMDA and QUIS in the control recording medium.
Lower trace: Effects of NMDA and QUIS under glycine (1 μM) exposure.
Traces are the digitally averaged recordings from 8 separate neurons in the
same culture dish.

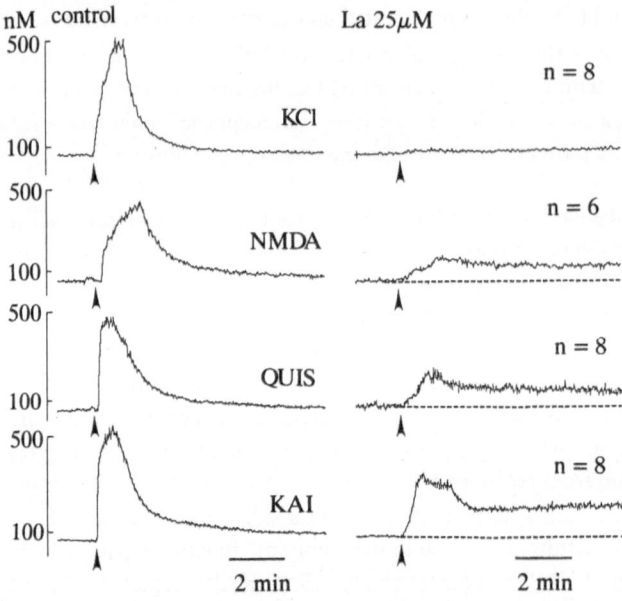

Fig. 2. Effects of glutamate receptor-subtype agonists on the hippocampal
neurons in culture after La^{3+} treatment. The effects of KCl, NMDA, QUIS
and KAI on the neurons cultured in separate dishes before (life column) and
after the La^{3+} treatment (right column). Each trace represents the
averaged response over 6 - 8 neurons.

Heterogeneous distributions of L-glutamate receptor subtypes in hippocampal. In a low power magnification , an image of a whole hippocampal slice was shown Fig. 4. The ratio images (F340/F340) were obtained after the application of NMDA, QUIS or KAI (50 $\dagger$M) by the same experimental setup as in the case of single neurons. We applied multiple agonists on the same preparation one after another placing the recovery period of more than 30 min between the applications to reveal the distribution of the receptor subtypes in the slice. Fig. 4 shows a representative result. Among the agonists, NMDA was the most effective on the *st. radiatum* of CA1 region and dentate gyrus (DG), but was less effective on the CA3 region. On the other hand, QUIS and KAI showed a rather specific effect in the CA3 region, whereas they were less effective in the *st.radiatum* of CA1 region.

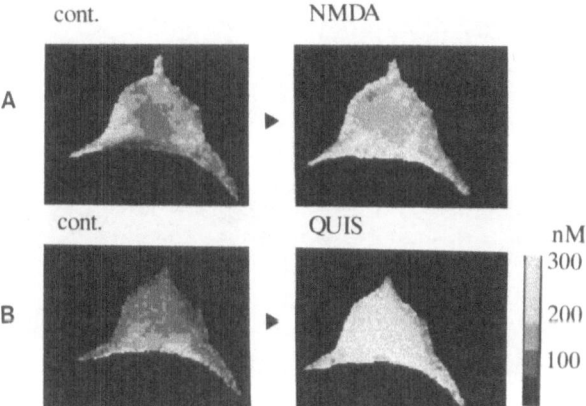

Fig.3 Two demensional distribution of NMDA- and QUIS- sensitive regions in a cultured hippocampal neuron treated with La^{3+} as detected by the $[Ca^{2+}]_i$ rises. The same neuron was exposed to NMDA (10 µM)(A) and QUIS (10 µM)(B). More than 20 min interval was taken between two applications to minimize a possible desensitization.

Detailed heterogeneity of the sensitive regions was examined in a higher magnification (pseudo-color images as shown in Fig. 5). NMDA-sensitive regions seemed to be localized around *st. radiatum* proximal to pyramidal cell layer of CA1, whereas those sensitive to QUIS and KAI seemed to be distributed in more distal regions (Fig. 5, upper images). Such heterogeneity was observed in CA3 region (Fig. 5 lower images), where NMDA-and KAI-sensitive sites are localized in the distal dendritic region, and QUIS sensitive sites are distributed in the proximal region. Although the heterogeneous distribution of the subtypes of glutamate receptors have been shown by autoradiography using radioactive ligands (Cotman, et al. 1987), the present results demonstrate the distribution of functional receptors.

Regional difference of the change in [Ca^{2+}]$_i$ during tetanic stimulation to cause long term potentiation. Fig. 6 (upper panels) show the changes in [Ca^{2+}]$_i$ during a tetanic stimulation given to the orthodromic synaptic pathways. When Schaffer-collaterals were stimulated at 50 Hz for 5 s, the [Ca^{2+}]$_i$ level in the proximal layer of *st. radiatum* increased markedly, but the increase in *st. oriens* was less prominent (Fig. 6). The increase lasted for about 2 min, during which time LTP in the population spikes developed markedly.

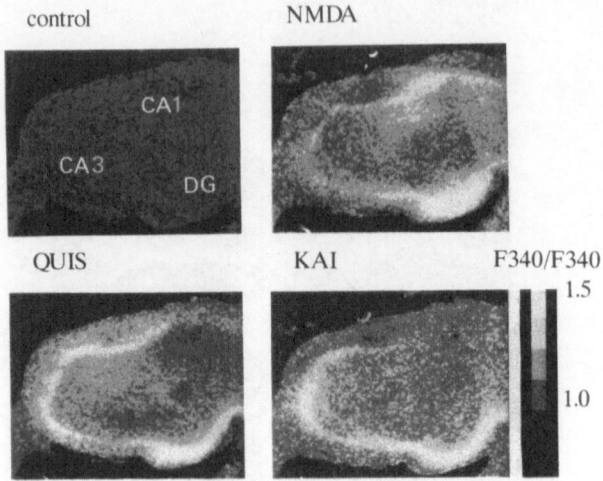

Fig. 4. Heterogeneous changes in [Ca^{2+}]$_i$ 60 s after exposures to glutamate receptor agonists (50 µM) in a rat hippocampal slice. Displayed images are the ratio of fluorescence intensities under an excitation light of 340 nm before and after the applications of the agonists indicating the distribution patterns of the respective receptor subtypes.

When mossy fibers were stimulated tetanically at 50 Hz for 5 s, a [Ca^{2+}]$_i$ increase in the distal layer of *st. radiatum* in CA3 region was evoked, which lasted for about 2 min (Fig. 6, lower panels), and LTP in the population spikes recorded from the CA3 pyramidal cell layer was established. The increases in the proximal layer of *st. radiatum* and in the *st. lucidum* were little observed.

The increase of [Ca^{2+}]$_i$ in CA1 region induced by stimulation of Schaffer-collaterals was greatly reduced by the treatment with 2-amino-5-phosphonovalerate(APV; 1µM) which is known to block the establishment of LTP (Malenka et al., 1988), but it was resistant to the treatment with 6-cyano-7-nitroquinoxaline-2,3-dione (CNQX; 10 µM). To the contrary, the increase in [Ca^{2+}]$_i$ in the distal layer of *st. radiatum* induced by the stimulation of mossy fibers was little affected by APV (10 µM), but was markedly blocked by CNQX (10 µM).

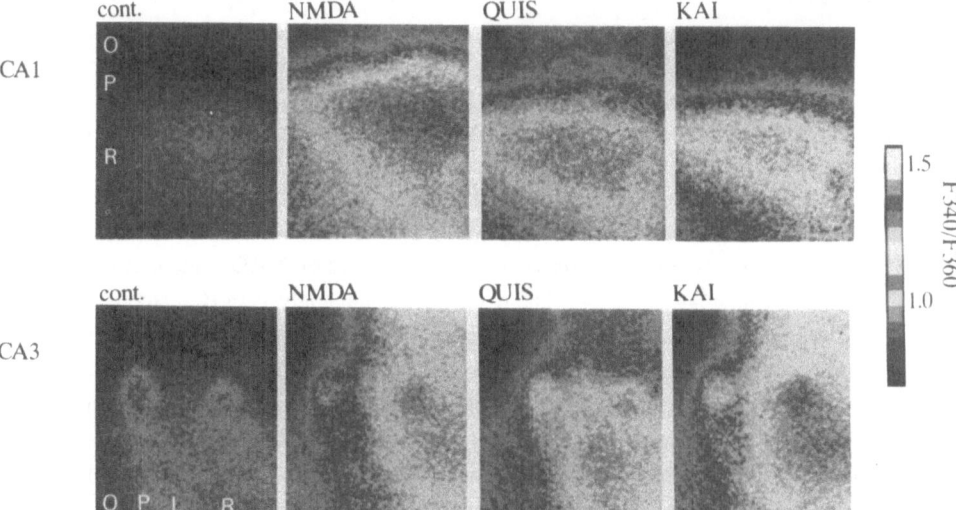

Fig.5 Effects of agonists for glutamate receptor subtypes on $[Ca^{2+}]_i$ changes in CA1 and CA3 region. In each experiment, each agoninst (50 μM) was applied for 2 min to the same preparation sequentially with over 30 min interval between the applications. Pseudocolor presentations of a ratio of F340/F360 obtained 60 sec after the application of the agents.

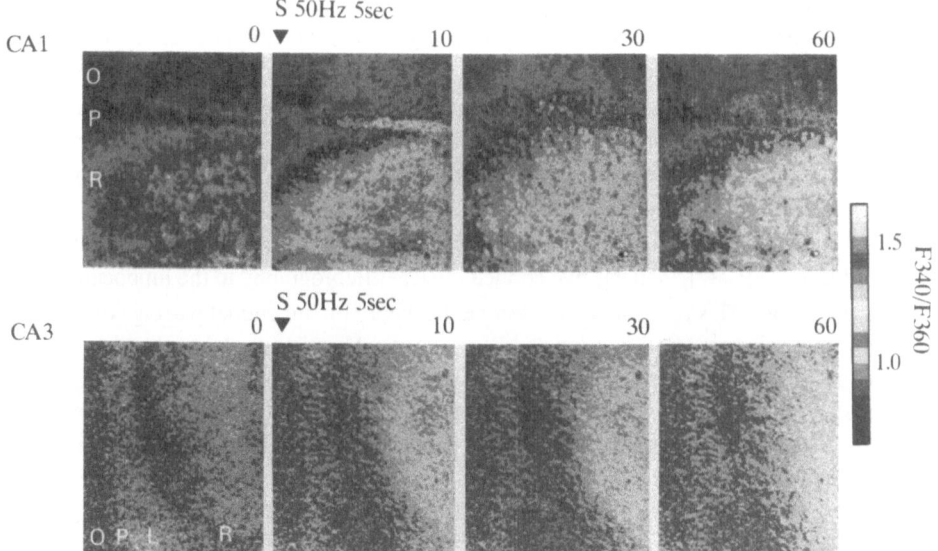

Fig. 6 Changes in $[Ca^{2+}]_i$ in CA1 and CA3 regions after tetanic stimulations to cause LTP. A, Change in $[Ca^{2+}]_i$ in CA1 region after the stimulation of Schaffer collaterals at a rate of 50 Hz for 5 sec; B, Change in $[Ca^{2+}]_i$ in CA3 region after the tentanization of mossy fibers.

DISCUSSION

In the present study we monitored $[Ca^{2+}]_i$ by an image analysis of fura-2 fluorescence. In the method for measuring temporal change in $[Ca^{2+}]_i$, we can measure the concentration of more than 20 separate neurons in a single microscopic viewfield. Ordinarily we measured 6 - 10 neurons in a single examination. This is helpful to know the response to a given stimultation of a heterogeneous population of the hippocampal neurons in culture. Since glycine augmented the effect of NMDA preferentially, we confirmed the distinct mechanisms underlying the increases in $[Ca^{2+}]_i$ induced by NMDA and non-NMDA subtypes of glutamate receptors. The two-dimensional analysis of $[Ca^{2+}]_i$ changes in cultured hippocampal neurons after the stimulation with agonists for glutamate receptor subtypes suggests that the receptor subtypes are expressed independently in the neuronal membrane. The Ca^{2+} influx through voltage-dependent Na^+ channels, which might be significant in some occasions, was blocked by tetrodotoxin. The Ca^{2+} influx via voltage-dependent Ca^{2+} channels was blocked by the treatment with La^{3+} ($50\,\mu M$), which was ascertained by a total suppression of a $[Ca^{2+}]_i$ elevation triggered by high $K^+(50\,mM)$-depolarization (Fig. 2). Under the La^{3+} treatment the applications of NMDA, QUIS and KAI resulted in the long lasting increase in fura-2 fluorescence. The observed increase in the fura-2 fluorescence in the neurons treated with La^{3+} should be partly due to the invaded La^{3+}. The entry of polyvalent cation through the receptor linked ionic pathway was further confirmed by the quenching of the fura-2 fluorescence during the stimulation in the presence of the Mn^{2+} in BSS (Ogura et al., 1989). The results strongly support that the activation of receptor-coupled cationic channels are permeable not only to Ca^{2+} but also to a broad spectrum of cations. Image processing of the fura-2 fluorescence showed that NMDA and KAI evoked the $[Ca^{2+}]_i$ increase preferentially in the submembraneous compartments, but QUIS tended to produce $[Ca^{2+}]_i$ increases in both submembraneous and deep cytoplasmic compartments (Fig. 3). This difference may reflect distinct sources of Ca^{2+} according to the species of agonists. Since a micrograph of a neuron is the two-dimensional projection of a sphere, the shell-formed distribution of high $[Ca^{2+}]_i$ volume should result in an image of ring-formed distribution of high $[Ca^{2+}]_i$ area.

Since the activation not only of NMDA receptors but also of non-NMDA receptors induced the $[Ca^{2+}]_i$ increase, we applied the Ca^{2+} fluorography to the hippocampal slice preparation. This is to prove or disprove the theory of "functional pairing" of glutamate receptor subtypes (Collingridge and Bliss, 1978). This theory claims that the activation of non-NMDA receptor produces depolarization of postsynaptic spine membranes so as to make NMDA-receptor present in the same membrane functional. An extreme opinion insists that the NMDA-receptor and non-NMDA receptor share the same molecular entity activated differentially by different species of agonist (Mayer, 1987).

We have shown above that the theory does not stand in the case of cultured neurons. Molecules of NMDA-, QUIS- and KAI-receptors should be distinct. However, the receptors expressed in the cultured neuron are not necessarily arranged in the same way as those in the brain *in situ*, since synaptic activity should be crucial in determination of expressed receptor subtypes and of their distribution, as in the well-known case of the neuromuscular synapse. Since the La^{3+} treatment to suppress the voltage-sensitive Ca^{2+} channels, which was effective in cultured neurons, was found to be invalid in slices, we were afraid that it would be difficult to detect the regional difference in slice preparation. In fact we can detect the heterogeneous $[Ca^{2+}]_i$ rise in the slice preparation under TTX treatment as shown in Fig.4 and 5. The difference in layers of the large $[Ca^{2+}]_i$ rise in response to distinct agonist species should represent the differential distribution of receptors, be that the participation of voltage-sensitive Ca^{2+} channels, if any, might have diluted the topographical characteristics of the $[Ca^{2+}]_i$ rise. With some overlap, NMDA receptors are dense in the proximal dendrite of CA1 and dentate gyrus regions, while the non-NMDA receptors are distributed dominantly in the CA3 region. These results are apparently not in favor of the theory of 'functional pair' in a strict sense. It is interesting that the tetanic stimulation of the Schaffer collaterals to cause LTP resulted in the increase in the fura-2 fluorescence in the same layer as in the case of NMDA administration. We do not insist that the non-NMDA receptors are totally absent in this layer. The 'functional pairing' of L-glutamate receptor subtypes during the establishment of LTP in a looser sense might be accepted.

The topographical analysis of the $[Ca^{2+}]_i$ rise in the CA3 region showed that NMDA- and KAI- receptors are distributed in the layer of distal dendrites in *st. radiatum*. QUIS-receptors are found to be distributed in the layer of proximal dendrite in *st. radiatum* and *st. lucidum*. Tetanic stimulation of mossy fibers induced an increase in fluorescence signal in the distal dendritic layer. A treatment with CNQX, a non-NMDA receptor antagonist, blocked the fluorescence increase due to the tetanic stimulation. But a treatment with APV, an NMDA receptor antagonist, had little effect on the tetanus-induced $[Ca^{2+}]_i$ increase. These results suggest the receptor responsible for the LTP at mossy fiber/CA3 pyramidal cell synapse should be subtype, assuming that Ca^{2+} plays a key role in LTP in this synapse. Recently Williams and Johnston (1989) have suggested the requirement of an increase in postsynaptic Ca^{2+} for induction of mossy fiber LTP by injecting Ca^{2+} chelators. The possibility has been discussed that LTP at the mossy fiber/CA3 pyramidal cell synapse would be independent of $[Ca^{2+}]_i$ unlike LTP at the Schaffer collateral/CA1 pyramidal cell synapse (Nicoll et al., 1988). This discussion has been based firstly on the autoradiographic observation that NMDA-receptor is scarcely distributed in the mossy fiber/CA3 pyramidal cell synapse and secondly on the assumption that NMDA-receptor is the sole species of L-glutamate receptor that allows the $[Ca^{2+}]_i$ increase. Apparently, this discussion was not supported by the present study and by the finding of Williams and Johnston.

Discrimination between the signals related to presynaptic activity and those of postsynaptic activity is difficult. From the relative volume and from the agonists' and antagonists' effects, we tentatively assumed that the majority of the dynamic fluorescence signal originates from the postsynaptic compartment. But if presynaptic terminals possess glutamate receptors with their pharmacological nature common to the postsynaptic ones, and if the $[Ca^{2+}]_i$ rise in presynaptic cytoplasm much surpasses that in postsynaptic space to compensate for its small volume, our assumption will collapse. In this context, comparison of the results obtained by the present method with the data from an independent technique, such as a direct injection of Ca^{2+} fluoroprobe into postsynaptic cytoplasm (Tank et al., 1988), is required.

In spite of the above-mentioned methodological limitation, we conclude that the subtypes of glutamate receptors are expressed and distributed separately in the neuronal membranes in the hippocampus. The activation of not only the NMDA-subtype receptor, but also of non-NMDA receptors induces the $[Ca^{2+}]_i$ elevation. The $[Ca^{2+}]_i$ rise owing to non-NMDA receptors should not be neglected in elucidating the mechanism underlying LTP. In other words, the absence of an effect of APV on the establishment of LTP does not explicitly mean the absence of the role of Ca^{2+}.

ACKNOWLEDGMENTS   This work was supported by grant-in-aid from Japanese Ministry of Education, Science and Culture. Technical assistance by staff of Hamamatsu Photonics Co. and by Ms. K. Akita is acknowledged.

REFERENCES

Collingridge, G.L. and Bliss, T.V.P. 1987, NMDA receptors - their role in long-term potentiation. Tr. Neurosci., 10 : 288 - 293.
Connor, J. A., Wadman, W. J., Hockberger, P. E. and Wong, R. K. S. 1988, Sustained dendritic gradients of Ca^{2+} induced by excitatory amino acids in CA1 hippocampal neurons. Science 240: 649 - 653 .
Cotman, C.W., Monaghan, D.T., Ottersen, O.P., Storm-Mathisen, J. 1987, Anatomical organization of excitatory amino acid receptors and their pathways. Tr. Neurosci., 10: 273-280 .
Johnson, J. W. and Ascher, P. 1987, Glycine potentiates the NMDA response in cultured brain neurons. Nature 325: 529 - 531.
Kudo, Y. and Ogura, A. 1986, Glutamate-induced increase in intracellular Ca^{2+} concentration in isolated hippocampal neurones. Br. J. Pharmacol., 89: 191-198.
Kudo, Y., Ito, K., Miyakawa, H., Izumi, Y., Ogura, A. and Kato, H. 1987, Cytoplasmic calcium elevation in hippocampal granule cell induced by perforant path stimulation and L-glutamate application. Brain Res., 407: 168-172 .
Malenka, R.C., Kauer, J.A., Zucker, R.S. and Nicoll, R.A. 1988, Postsynaptic calcium is sufficient for potentiation of hippocampal synaptic transmission. Science (Wash.), 242: 81-84 .
Mayer, M. 1987, Two channels reduced to one. Nature (Lond.), 325: 480-481.
Murphy, S.N. and Miller, R.J. 1989, Regulation of Ca^{2+} influx into striatal neurons by kainic acid. J. Pharmacol. Exp. Therap., 249: 184-193 .
Nicoll, R.A., Kauer, J.A. and Malenka, R.C. 1988, The current excitement in long-term potentiation. Neuron, 1: 97-103.
Nohmi, M., Kuba, K., Ogura, A. and Kudo, Y., 1988 Measurement of intracellular Ca^{2+} in the fullfrog Sympathetic ganglion Cells Using fura-2 fluorescence. Brain Res. 438: 175-181.
Ogura, A., Miyamoto, M. and Kudo, Y. 1988, Neuronal death in vitro. Exp. Brain Res., 73: 447-458.
Ogura, A., Akita, K. and Kudo, Y. 1989, Cytosolic calcium rise mediated by non-NMDA receptors in cultured rat hippocamapl neurons. Soc. Neurosci. Abs. 15: p1161 .
Ozawa, S., Tsuzuki, K., Iino, M., Ogura, A. and Kudo, Y., 1989, Three types of voltage-independent clcium current in cultured rat hippocampal neurons. Brain Res. 495: 329-336.
Sugiyama, H., Ito, I, and Hirono, C. 1987, A new type of glutamate receptor linked to inositol phopholipid metabolism. Nature (Lond.), 325: 531-533 .
Tank, D.W., Sugimori, M., Connor, J.A. and Llinas, R.R. 1988,Spatially resolved calcium dynamics of mammalian Purkinje cells in cerebellar slice. Science (Wash.), 242: 773-776 .
Williams, S. and Johnston, D. 1989,Long-term potentiation of hippocampal mossy fiber synapses is blocked by postsynaptic injection of calcium chelators. Neuron 3: 583 - 588

EXCITATORY AMINO ACID RECEPTORS IN

THE XENOPUS OOCYTE EXPRESSION SYSTEM

Raymond Dingledine, Nancy W. Kleckner, and Christopher J. McBain

Department of Pharmacology
University of North Carolina School of Medicine
Chapel Hill, NC 27599
U.S.A.

EXCITATORY AMINO ACID RECEPTOR FAMILY

The excitatory amino acid (EAA) receptors are neurotransmitter receptors responsive to glutamate, aspartate, and perhaps also to homocysteate or related compounds (Fonnum, 1984; Do et al., 1986). Although the nomenclature is still somewhat unsettled, electrophysiologists have identified at least two EAA receptors, namely the well-known N-methyl-D-aspartate (NMDA) receptor and a less well characterized "non-NMDA" receptor that can be powerfully activated by kainic acid and (RS)-α-amino-3-hydroxy-5-methyl-isoxazole-4-propionate (AMPA). A third receptor, stimulated by glutamate, quisqualate and ibotenate, is linked via a G-protein to the activation of phospholipase C and consequent generation of the second messenger, inositol trisphosphate (Sugiyama et al., 1987). Due to their speed of activation (within millisec of applying agonist, e.g., Benveniste et al., 1990), the NMDA and non-NMDA receptors are thought to be receptor-channel complexes similar in architecture to the nicotinic acetylcholine receptor; i.e., all ligand binding sites and ion permeation pathways are thought to reside within the same macromolecular complex. The discussion that follows focuses on these two receptor-channel complexes.

Interest in NMDA receptors stems largely from the variety of important brain phenomena that are influenced by NMDA receptor blockers. Antagonists of this receptor are, in some animal models, effective anticonvulsants (e.g., Croucher et al., 1982, reviewed in Dingledine et al., 1990), and are also able to ameliorate ischemic or hypoxic neuron injury (e.g., Simon et al., 1984). Moreover, NMDA antagonists can disrupt the development of normal neuronal connectivity in the visual and olfactory systems (Kleinschmidt et al., 1987; Lincoln et al., 1988) and can impair learning or memory in tasks that require spatial orientation (Morris, 1989). The drive to create therapeutically useful drugs targeted to the NMDA receptor is strong.

Ligand binding studies have identified a bewildering number of binding sites for L-glutamate (Foster and Fagg, 1984), not all of which have yet been assigned functions as synaptic receptors. The most intriguing example of a binding site in search of a function is the high affinity kainate (Km 5-10 nM) recognition site that is present in high density at the termination zone of granule cell mossy fibers onto CA3 pyramidal neurons (Foster et al., 1981; Monaghan et al., 1983; Unnerstall and Wamsley, 1983). This binding site is

Neuroreceptor Mechanisms in Brain, Edited by S. Kito *et al.*
Plenum Press, New York, 1991

highly localized at identified synapses yet it is unclear whether it is coupled to an ion channel. Several lines of evidence indicate this binding site is unlikely to correspond to the electrophysiologists' "non-NMDA" receptor. The relative potencies of agonists that open non-NMDA channels (L-glutamate = domoate > kainate) (Verdoorn and Dingledine, 1988) do not match those of the high affinity kainate recognition site (domoate > kainate > L-glutamate) (Foster and Fagg, 1984). Moreover, the affinity of CNQX for the high affinity binding site is 1700 nM (Honore et al., 1988), which is higher than the K_B estimated from Schild analaysis of kainate-activated ionic currents in oocytes (295 nM) (Verdoorn et al., 1989). The available evidence points to the depolarizing action of kainate being mediated via activation of the [3H]-AMPA binding site rather than the high affinity kainate binding site (Fletcher et al., 1988; Verdoorn et al., 1989).

The structural basis of the pharmacological differences among EAA receptors will only be resolved once these receptors are purified and their molecular cloning has been achieved. From three observations one expects some structural similarities to be found among the EAA receptor subunits. First, both NMDA and non-NMDA receptors are cation channels; second, glutamate is a strong agonist at all EAA receptors (e.g., Mayer and Westbrook, 1987); and third, antagonists of the glycine site on NMDA receptors are also antagonists at non-NMDA receptors (Kleckner and Dingledine, 1989). However, whether these functional similarities will be reflected in a high degree of sequence homology is uncertain. The recent molecular cloning of three kainate recognition proteins (Hollman et al., 1989; Gregor et al., 1989; Wada et al., 1989) reveals little sequence homology to known ligand-gated ion channels. The cDNA encoding a functional rat brain kainate-activated receptor (or receptor subunit) (Hollman et al., 1989) is of particular interest, since it appears to have some pharmacological properties expected of the "non-NMDA" receptor. Thus domoate is a more potent agonist than kainate in the Xenopus oocyte system, and the absolute potencies of these agonists are similar to those measured for receptors translated in Xenopus oocytes injected with rat brain mRNA (Verdoorn and Dingledine, 1988). The molecular cloning of other members of the EAA receptor family is awaited.

THE XENOPUS OOCYTE RECEPTOR EXPRESSION SYSTEM

The EAA receptors are normally synthesized principally, if not exclusively, by neurons of the central nervous system. The surprising absence of permanent cell lines that express EAA receptor-channel complexes has hampered the detailed study of EAA receptor properties. Until recently the only preparations that could be used were intact central neurons, which are fragile and difficult to voltage clamp adequately. Tortuous diffusion pathways and uptake systems in intact tissues further complicate the interpretation of pharmacological experiments since the concentration of drug at its receptors cannot generally be known with precision. The Xenopus oocyte expression system (Gurdon et al., 1971; Snutch, 1988) can serve as an alternative preparation that offers some advantages over the study of receptor properties expressed natively.

The isolation, microinjection and culture of oocytes are described in detail elsewhere (Verdoorn and Dingledine, 1988). Briefly, oocytes dissected from ovarian lobes of Xenopus laevis are treated with 1.5-2 mg/ml neutral dispase for 45-75 min to loosen connective tissue, after which individual oocytes are shrunk in hypertonic saline so that the follicle cell layer can be removed manually with fine forceps. Oocytes are then individually injected with mRNA solution prepared from brain or a brain region, such that each oocyte

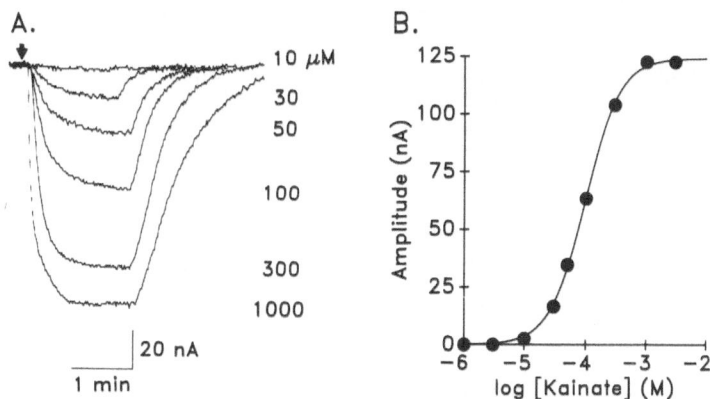

Fig 1. A complete concentration-response curve for kainate in an individual oocyte that had been microinjected with 50 ng rat brain mRNA 3 days previously. A, a series of six superimposed currents at a holding potential of -60 mV evoked by perfusion with the concentrations of kainate shown to the right. B, current amplitudes are plotted as a function of log kainate concentration. From Verdoorn and Dingledine (1988).

receives a fixed amount of mRNA usually ranging from 10 to 50 ng. The injected cells are placed individually in wells of a 96 well culture plate together with approximately 100 μl of antibiotic-containing saline and held at 19° for two to seven days. During this incubation time the cells translate the mRNA, assemble multisubunit receptors and insert them into the plasma membrane where they are accessible to study under voltage clamp. The frog oocyte is a hardy cell that easily survives prolonged experiments involving many solution changes.

Oocytes are voltage clamped with one or two microelectrodes and agonists are applied by perfusion at known concentration. An example of kainate-evoked inward cation currents and the concentration-response curve derived from these records are shown in figure 1. Concentration-response curves obtained in this manner are well behaved and quite reproducible; EC50s for kainate determined over a period of about 18 months were normally distributed about their mean (87 μM) and fell within a narrow range (95% confidence interval 79-96 μM, n=38) (Verdoorn et al., 1989).

Although the oocyte is convenient for the quantitative study of EAA receptors, this preparation has several drawbacks that must be borne in mind, one of which is obvious from the slow timecourse of the kainate-evoked currents shown in figure 1. The oocyte is such a large cell (approximately 1 mm in diameter) that rapid and even perfusion of its whole surface has proven elusive. Thus quantitative study of responses that desensitize within tens or even hundreds of millisec (e.g., Trussel et al., 1988; Mayer et al., 1989) is difficult if not impossible with present perfusion systems. Other limitations of the oocyte system include the presence of an endogenous calcium-dependent chloride conductance that can interfere with studies of receptors which have substantial calcium conductance (e.g., the NMDA receptor), and uncertainty whether the oocyte carries out subunit assembly, posttranslational processing and modification of receptors exactly as would occur in the native neuronal membrane.

NMDA RECEPTOR EXPRESSED IN OOCYTES

The native NMDA receptor has several distinguishing characteristics. This receptor is activated by glutamate, aspartate and NMDA, potentiated by submicromolar glycine (Johnson and Ascher, 1987), blocked by Zn^{+2} (Westbrook and Mayer, 1987) and blocked in a voltage-dependent manner by micromolar Mg^{+2} (Nowak et al., 1984; Mayer et al., 1984), ketamine and phencyclidine (MacDonald et al., 1987). Binding of ligands to the NMDA receptor has also been shown to be modulated by certain polyamines (Ransom and Stec, 1988).

Oocytes injected with rat brain mRNA were first shown to respond to high (mM) concentrations of NMDA by Gunderson et al. (1984). Micromolar concentrations of NMDA were later shown to elicit an inward current that could be blocked by D-2-amino-5-phosphonovaleric acid (D-APV), potentiated by glycine and blocked in a voltage-dependent manner by Mg^{+2} (Verdoorn et al., 1987). NMDA-induced currents are also blocked by phencyclidine (Kushner et al., 1988). Schild analysis (Arunlakshana and Schild, 1959) was used to demonstrate that the block by D-APV was competitive over a wide range of antagonist concentrations (Verdoorn et al., 1989) (figure 2); the slope of the Schild plot was 1.03 ± 0.02 and the K_B estimated from these experiments was $1.35 \mu M$ (95% confidence interval, 1.11-$1.64 \mu M$, n=9), near the dissociation constant of D-APV for NMDA receptors in brain membranes (Olverman et al., 1984). Aspartate also activates the NMDA receptor expressed in oocytes; as expected block of aspartate currents by D-APV was competitive (Schild slope 0.97 ± 0.04) and the K_B was identical ($1.35 \mu M$) to that for D-APV block of NMDA-induced currents. These results provide confidence that NMDA, L-aspartate and D-APV all act at the same recognition site in oocytes, and that this recognition site is an NMDA receptor.

GLYCINE RECOGNITION SITE ON NMDA RECEPTORS

In 1986 the distribution of strychnine-insensitive [3H]-glycine binding in the forebrain (Bristow et al., 1986) was found to match that of NMDA-sensitive [3H]-glutamate binding (Monaghan and Cotman, 1985), two observations that foreshadowed the discovery of a new modulatory site on NMDA receptors. Glycine was first shown to potentiate NMDA-evoked currents markedly in cultured mouse brain neurons by Johnson and Ascher (1987). The glycine recognition site on NMDA receptors is distinct from the conventional inhibitory glycine receptor, since the glycine response was not blocked by micromolar strychnine. (Results from a later study suggested that higher concentrations of strychnine, 20-60 μM, blocked NMDA receptor activation but at a different site (Bertolino and Vicini, 1988)). The effect of glycine was selective in that kainate- and quisqualate-evoked currents were unaffected by glycine. In studies of single NMDA-gated channels on murine neurons, glycine was shown to increase channel opening frequency without affecting mean open time or channel conductance (Johnston and Ascher, 1987). There followed a plethora of experiments with neurons and neuronal membranes confirming the distinction of this glycine site from the strychnine-sensitive inhibitory glycine receptor. Among the more provocative of these reports was an autoradiographic study showing that glycine preferentially potentiates agonist binding at NMDA receptor of the rat thalamus and cerebral cortex compared with striatum (Monaghan et al., 1988); these authors interpreted their data as pointing to the existence of multiple NMDA receptors.

Mechanism of Action of Glycine

The original report by Johnson and Ascher (1987) indicated that a saturating concentration (1 μM) of glycine could potentiate NMDA-gated currents

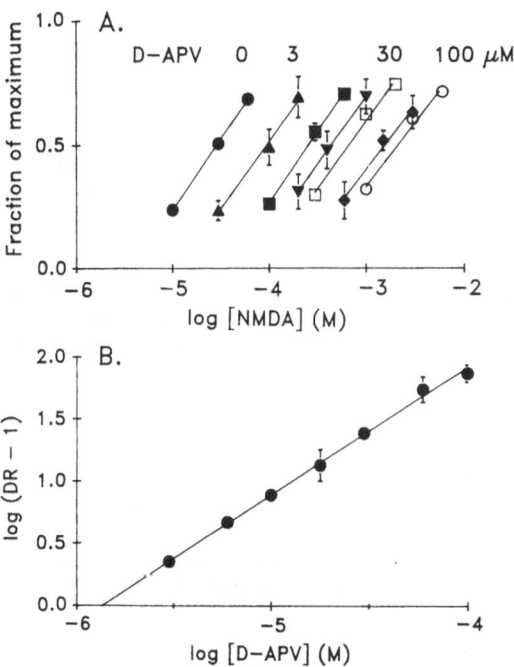

Fig. 2. Schild analysis of the antagonism by D-APV of currents evoked by NMDA in frog oocytes injected with rat brain mRNA. A, NMDA concentration-response curves in the presence of different concentrations of D-APV shown above each curve. Each point represents the mean ($\pm$ SEM) of measurements from three to nine cells. The maximum response (to 300 μM NMDA) was determined before and after the construction of each curve and the NMDA-gated currents expressed as a fraction of the average maximum responses. B, Schild regression from the data illustrated in A. The points represent the mean $\pm$ SEM of three to six determinations (with the exception of the point at 6 μM, which was from a single cell). The solid line is the regression over the entire data set (n=28). The average Schild slope determined individually in nine cells was 1.03$\pm$0.02, consistent with competitive block by D-APV, and the mean pA2 was 5.87$\pm$0.043. From Verdoorn et al., 1989.

by 17 fold. Later studies carried out in <u>Xenopus</u> oocytes injected with rat brain mRNA showed that when precautions were taken to reduce ambient glycine concentration as much as possible the effect was even more pronounced, glycine imparting an 80-100 fold or more potentiation of NMDA-gated currents (Kleckner and Dingledine, 1988). Indeed, in the absence of added glycine the NMDA-gated current was not statistically different from zero, even with high (EC95) concentrations of NMDA (figure 3). Thus it appears that, at least at equilibrium, glycine is required for the activation of NMDA receptors by NMDA. This finding has recently been confirmed in patch clamp studies of NMDA-gated currents in cultured neurons of the rat spinal cord, visual cortex and hippocampus (Huettner, 1989; Lester et al., 1989). Studies with antagonists also support this view, since several competitive antagonists at the glycine recognition site can abolish NMDA-evoked responses (Kemp et al., 1988; Huettner, 1989; Lester et al., 1989; Kleckner and Dingledine, 1989). Because neither glycine nor NMDA acting alone can gate NMDA channels, we refer to these compounds as "coagonists". The NMDA receptor is to our knowledge the first neurotransmitter receptor to require two different agonists for activation.

Although glycine would seem to be required for gating of the channel by NMDA, the molecular mechanism of action of glycine is unknown. Several possible modes of action of glycine can be mentioned. First, glycine may simply induce a conformational change in the receptor similar to the manner in which conventional agonists are thought to act. Depending on the nature of this conformational change, glycine would modulate (Johnson and Ascher, 1987) or enable (Kleckner and Dingledine, 1988) NMDA-gated currents. One prominent action of glycine may be to increase the rate of recovery of the NMDA receptor from desensitization (Mayer et al., 1989). Finally, glycine could act as an allosteric effector by increasing the affinity of NMDA for its recognition site. Some evidence for allosteric interactions between glycine and NMDA agonists has been obtained in ligand binding studies to rat forebrain membranes. Glycine and D-serine increased the binding of [3H]-glutamate for NMDA-sensitive binding sites in rat brain (Monaghan et al., 1988; Fadda et al., 1988). Conversely, glutamate was reported to increase the binding of [3H]-glycine to rat brain membranes (Kessler et al., 1989). Another study, however, failed to demonstrate reciprocal effects of the binding of the two ligands under conditions in which glycine markedly potentiated the access of a channel ligand, [3H]-TCP to its binding site (Bonhaus et al., 1989). Likewise, in voltage clamped hippocampal neurons (Chizhamakov et al., 1989) and mRNA-injected oocytes (Verdoorn et al., 1987; McBain et al., 1989), the EC50 of NMDA agonists was not affected by glycine; glycine simply increased the maximum agonist-evoked current. These conflicting findings have not been entirely reconciled. It would not be unexpected if two ligands that were each required for channel gating influenced each other's binding; however there is no evidence yet that shifts in NMDA binding affinity underlie the potentiation of NMDA-gated currents by glycine.

Glycine Site Agonists

The original report of Johnson and Ascher (1987) pointed out that alanine and serine but not other amino acids tested could substitute for glycine in NMDA receptor activation. Glycine itself is the simplest amino acid, having only two protons for α-carbon substituents. The characterization of agonists at the glycine recognition site has proceeded in numerous laboratories (e.g., Reynolds et al., 1987; Snell et al., 1988; Nadler et al., 1988; Hood et al., 1989; Marvizon et al., 1989). From these studies and a detailed investigation of more than 60 glycine analogs (McBain et al., 1989), five structural features of an effective agonist emerge. First, sterically unhindered and ionized

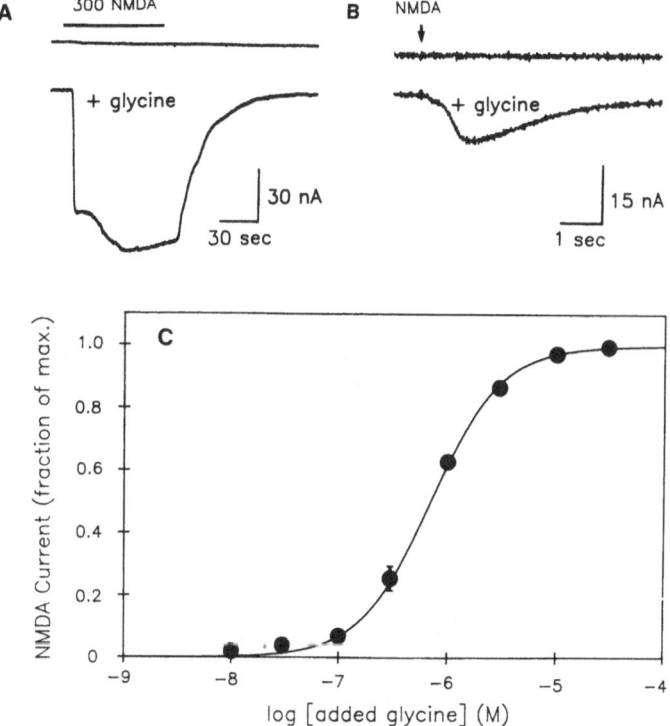

Fig. 3. Control of NMDA-gated currents by glycine
in oocytes injected with rat brain mRNA. A,
currents produced at a holding potential of -60 mV
by perfusion with 300 μM NMDA in the absence (top),
and in the presence (bottom) of 3 μM added glycine.
B, currents induced by the brief (30 msec) pressure
application of NMDA in the absence (top) and
presence (bottom) of 30 μM glycine. In the absence
of added glycine neither method of NMDA application
elicited a measureable ionic current. C, a
concentration-response curve to glycine was
generated as in A by applying 100 μM NMDA in the
presence of sequentially increasing glycine
concentrations, with washout periods between
agonist applications. Each point represents the
mean current ($\pm$ SEM), as a fraction of the maximum
response, from five cells. The curve is the least
squares fit of the data to the logistic equation
assuming the response to NMDA in the absence of
added glycine is zero. The EC50 of glycine was 670
nM in these cells. From Kleckner and Dingledine, 1988.

carboxyl and amino termini appear essential for activity; N-methylglycine and glycinamide, for example, possess 6% or less the activity of glycine itself. The ionized hydroxyl group of the carboxyl terminus appears to be the active constituent, since (R)-(+)-cycloserine, which has only a free hydroxyl group, was active whereas several compounds with only carbonyl groups (e.g., glycinehydroxymate) were inactive. Second, it is important that the terminal carbons be separated by only a single carbon; β-alanine, isoserine and 3-aminoisobutyric acid are all without activity compared with the parent compounds (glycine, serine and alanine, respectively). Third, although glycine itself does not possess an asymmetric carbon, study of analogues that are chiral shows that activity usually resides in the D-isomer. For example, D-serine is 30 times as potent as L-serine in the oocyte bioassay. Longer-chain compounds with weak agonist activity (e.g., valine, phenylglycine), however, exhibit reversed stereoselectivity.

The fourth and fifth requirements for good agonist activity at this site have to do with the size and properties of the α-carbon substituent. Replacement of the proton on the α-carbon with more bulky aliphatic or aromatic moities usually reduces activity substantially. For example, D-alanine (methyl substituent) is about six times less potent than glycine, and D-2-aminobutyric acid (ethyl group on the α-carbon) is completely inactive. α-Carbon substitution with a substituent of equal size but capable of hydrogen bonding, on the other hand (e.g., the hydroxymethyl of D-serine, the fluoromethyl of β-fluoro-D-alanine), restores activity. These findings imply that glycine analogues like D-serine owe their activity to the ability of the β-substituent to hydrogen bond at an additional site of the receptor not normally occupied by glycine.

The structure-activity information summarized above suggests a picture of the glycine recognition site on the NMDA receptor. The active region of this site appears to be a small pocket containing both positively and negatively charged groups capable of binding the carboxyl and amino termini of glycine, and an additional site capable of hydrogen bonding.

Glycine Site Antagonists

Highly selective and competitive antagonists are needed to study the role of the glycine recognition site in NMDA receptor mediated synaptic transmission. A number of blockers of the glycine recognition site are now known. The most potent antagonists are derivatives of kynurenic acid or have a quinoxaline ring structure (figure 4), although indole-2-carboxylic acid (Huettner, 1989), cycloleucine (Snell and Johnson, 1988), 1-aminocyclobutane-1-carboxylate (Hood et al., 1989) and 1-hydroxy-3-aminopyrrolidone (HA-966, Fletcher and Lodge, 1988; Foster and Kemp, 1989) possess weaker antagonist activity. HA-966 appears to be a partial agonist at the glycine site (e.g., Foster and Kemp, 1989).

Interestingly, the kynurenine and quinoxaline compounds are not only glycine site blockers but also possess significant blocking activity at the kainate-activated non-NMDA receptors (Frey et al., 1988; Kleckner and Dingledine, 1989; Lester et al., 1989). The most selective competitive glycine site antagonist known to date is 7-chlorokynurenic acid (Kemp et al., 1988; Kleckner and Dingledine, 1989). However, by Schild analysis of the block of EAA-gated currents in the Xenopus oocyte system, 7-chlorokynurenic acid was only 40-fold more potent at the glycine site than at the kainate-activated non-NMDA receptor (Kleckner and Dingledine, 1989). The K_B for block of ionic currents gated by glycine in the presence of NMDA was 350 nM, whereas the K_B for block of kainate-gated currents was 14.1 μM. 7-Chlorokynurenic acid has been shown to reduce NMDA receptor-mediated epileptic bursts in hippocampal slices selectively without affecting the non-NMDA receptor-mediated evoked potential (Kleckner and Dingledine, 1989). This result supports earlier findings that less selective glycine blockers could reduce excitatory events evoked by NMDA application (Kemp et al., 1988; Birch et al., 1988; Fletcher and Lodge, 1988), all of which together confirm the idea that the glycine site is functional in intact tissues.

Glycine site antagonists may find eventual uses to treat seizures or neuronal injury associated with hypoxia or ischemia. As discussed above 7-chlorokynurenic acid is an effective anticonvulsant in an in vitro model of epileptic activity, however, tests of the in vivo efficacy of glycine site antagonists against seizures have yet to emerge.

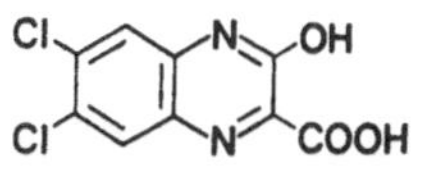

6,7–dichloro–3–hydroxy–
2–quinoxalinecarboxylic acid

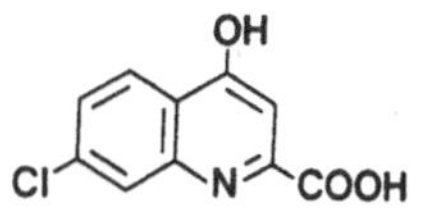

7–chlorokynurenate

Fig. 4. Structures of two competitive antagonists at the glycine recognition site of NMDA receptors.

REFERENCES

Arunlakshana, O. and Schild, H.O., 1959, Some quantitative uses of drug antagonists, Br. J. Pharmacol., 14:48-57.

Benveniste, M., Mienville, J.-M., Sernagor, E. and Mayer, M.L., 1990, Concentration-jump experiments with NMDA antagonists in mouse cultured hippocampal neurons, J. Neurophysiol., in press.

Bertolino, M. and Vicini, S., 1988, Voltage-dependent block by strychnine of N-methyl-D-aspartic acid-activated cationic channels in rat cortical neurons in culture, Molec. Pharmacol., 34:98-103.

Bonhaus, D.W., Yeh, G.-C., Skaryak, L. and McNamara, J.O., 1989, Glycine regulation of the N-methyl-D-aspartate receptor-gated ion channel in hippocampal membranes, Molec. Pharmacol., 36:273-279.

Birch, P.J., Grossman, C.J. and Hayes, A.G., 1988, 6,7-Dinitroquinoxaline-2,3-dion and 6-nitro-7-cyanoquinoxaline-2,3-dion antagonize responses to NMDA in the rat spinal cord via an action at the strychnine-insensitive glycine receptor, Eur. J. Pharmacol., 156:177-180.

Bristow, D.R., Bowery, N.G. and Woodruff, G.N., 1986, Light microscopic autoradiographic localization of [3H]glycine and [3H]strychnine binding sites in the brain, Eur. J. Pharmacol., 126:303-307.

Chizhamakov, I.V., Kiskin, N.I., Krishtal, O.A. and Tsyndrenko, A.Ya., 1989, Glycine action on N-methyl-D-aspartate receptors in rat hippocampal neurons, Neurosci. Lett., 99:131-136.

Croucher, M.J., Collins, J.F. and Meldrum, B.S., 1982, Anticonvulsant activity of excitatory amino acid antagonists, Science, 216:899-901.

449

Dingledine, R., McBain, C.J. and McNamara, J.O., 1990, Excitatory amino acid receptors in epilepsy, Trends in Pharmacol. Sciences, in press.

Do, K.Q., Mattenberger, M., Streit, P. and Cuenod, M., 1986, In vitro release of endogeneous excitatory sulfur-containing amino acids from various rat brain regions, J. Neurochem., 46:779-786.

Fadda, E., Danysz, W., Wroblewski, J.T. and Costa, E., 1988, Glycine and d-serine increase the affinity of N-methyl-D-aspartate sensitive glutamate binding sites in rat brain synaptic plasma membranes, Neuropharmacol., 27:1183-1185.

Fletcher, E.J. and Lodge, D., 1988, Glycine reverses antagonism of N-methyl-D-aspartate (NMDA) by 1-hydroxy-2-aminopyrrolidone (HA-966) but not by D-2-amino-5-phosphonovalerate (D-APV) on rat cortical slices, Eur. J. Pharmacol., 151:161-162.

Fletcher, E.J., Martin, D., Aram, J.A., Lodge, D. and Honore, T., 1988, Quinoxalinediones selectively block quisqualate and kainate receptors and synaptic events in rat neocortex and hippocampus and frog spinal cord in vitro, Br. J. Pharmacol., 95:585-587.

Fonnum, F., 1984, Glutamate: a neurotransmitter in mammalian brain, J. Neurochem., 42:1-16.

Foster, A.C. and Fagg, G.E., 1984, Acidic amino acid binding sites in neuronal membranes: their characteristics and relationship to synaptic receptors, Brain Res. Rev., 7:103-164.

Foster, A.C. and Kemp, J.A., 1989, HA-966 antagonizes N-methyl-D-aspartate receptors through a selective interaction with the glycine modulatory site, J. Neurochem., 9:2191-2196.

Foster, A.C., Mena, E.E., Monaghan, D.T. and Cotman, C.W., 1981, Synaptic localization of kainic acid binding sites, Nature, 298:73-75.

Frey, P., Berney, C., Herrling, P.L., Mueller, W. and Urwyler, S., 1988, 6,7-Dichloro-3-hydroxy-2-quinoxalinecarboxylic acid is a relatively potent antagonist of NMDA and kainate receptors, Neurosci. Lett., 91:194-198.

Gregor, P., Mano, I., Maoz, I., McKeown, M. and Teichberg, V.I., 1989, Molecular structure of the chick cerebellar kainate-binding subunit of a putative glutamate receptor, Nature, 342:689-692.

Gunderson, C.B., Miledi, R. and Parker, I., 1984, Glutamate and kainate receptors induced by rat brain messenger RNA in Xenopus oocytes, Proc. Roy. Soc. London Ser. B, 221:127.

Gurdon, J.B., Lane, C.D., Woodland, H.R. and Marbaix, G., 1971, Use of frog eggs and oocytes for the study of messenger RNA and its translation in living cells, Nature, 233:173-182.

Hollman, M., O"Shea-Greenfield, A., Rogers, S.W. and Heinemann, S., 1989, Cloning by functional expression of a member of the glutamate receptor family, Nature, 342:643-648.

Honore, T., Davies, S.N., Drejer, J., Fletcher, E.J., Jacobsen, P., Lodge, D. and Nielsen, F.E., 1988, Quinoxalinediones: potent competitive non-NMDA glutamate receptor antagonists, Science, 241:701-703.

Hood, W.F., Sun, E.T., Compton, R.P. and Monahan, J.B., 1989, 1-aminocyclobutane-1-carboxylate (ACBC): a specific antagonist of the N-methyl-D-aspartate receptor coupled glycine receptor, Eur. J. Pharmacol., 161:281-282.

Huettner, J.E., 1989, Indole-2-carboxylic acid: a competitive antagonist of potentiation by glycine at the NMDA receptor, Science, 243:1611-1613.

Johnson, J.W. and Ascher, P., 1987, Glycine potentiates the NMDA response in cultured mouse brain neurones, Nature, 325:529-531.

Kemp, J.A., Foster, A.C., Leeson, P.D., Priestley, T., Tridgett, R., Iversen, L.L. and Woodruff, G.N., 1988, 7-Chlorokynurenic acid is a selective antagonist at the glycine site of the N-methyl-D-aspartate receptor complex, Proc. Nat. Acad. Sci. USA, 85:6547-6550.

Kessler, M., Terramani, T., Lynch, G. and Baudry, M., 1989, A glycine site associated with N-methyl-D-aspartic acid receptors: characterization and identification of a new class of antagonists, J. Neurochem., 52:1319-1328.

Kleckner, N.W. and Dingledine, R., 1988, Requirement for glycine in activation of NMDA receptors expressed in Xenopus oocytes, Science, 241:835-837.

Kleckner, N.W. and Dingledine, R., 1989, Selectivity of quinoxalines and kynurenines as antagonists of the glycine site on N-methyl-D-aspartate receptors, Molec. Pharmacol., 36:430-436.

Kleinschmidt, A., Bear, M.F. and Singer, W., 1987, Blockade of "NMDA" receptors disrupts experience-dependent plasticity of kitten striate cortex, Science, 238:355-358.

Kushner, L., Lerma, J., Zukin, R.S. and Bennett, M.V.L., 1988, Coexpression of N-methyl-D-aspartate and phencyclidine receptors in Xenopus oocytes injected with rat brain mRNA, Proc. Nat. Acad. Sci. USA, 85:3250-3254.

Lester, R.A.J., Quarum, M.L., Parker, J.D., Weber, E. and Jahr, C.E., 1989, Interaction of 6-cyano-7-nitroquinoxaline-2,3-dione with the N-methyl-D-aspartate receptor-associated glycine binding site, Molec. Pharmacol., 35:565-570.

Lincoln, J., Coopersmith, R., Harris, E.W., Cotman, C.W. and Leon, M., 1988, NMDA receptor activation and early olfactory learning, Dev. Brain Res., 39:309-312.

MacDonald, J.F., Miljkovic, Z. and Pennefather, P., 1987, Use-dependent block of excitatory amino acid currents in cultured neurons by ketamine, J. Neurophysiol., 58:251-266.

Marvizon, J.C.G., Lewin, A.H. and Skolnick, P., 1989, 1-Aminocyclopropanecarboxylic acid: a potent and selective ligand for the glycine modulatory site of the N-methyl-D-aspartate receptor complex, J. Neurochem., 52:992-994.

Mayer, M.L. and Westbrook, G.L., 1987, The physiology of excitatory amino acids in the vertebrate central nervous system, Prog. Neurobiol., 28:197-291.

Mayer, M.L., Westbrook, G.L. and Guthrie, P.B., 1984, Voltage-dependent block by Mg of NMDA responses in spinal cord neurones, Nature, 309, 261-263.

Mayer, M.L., Vyklicky, L. and Clements, J., 1989, Regulation of NMDA receptor desensitization in mouse hippocampal neurons by glycine, Nature, 338:425-427.

McBain, C.J., Kleckner, N.W., Wyrick, S. and Dingledine, R., 1989, Structural requirements for activation of the glycine coagonist site of N-methyl-D-aspartate receptors expressed in Xenopus oocytes, Molec. Pharmacol., 36:556-565.

Monaghan, D.T. and Cotman, C.W., 1985, Distribution of N-methyl-D-aspartate-sensitive L-[3H]-glutamate binding sites in rat brain, J. Neuroscience, 5:2909-2919.

Monaghan, D.T., Holets, V.R., Toy, D.W. and Cotman, C.W., 1983, Anatomical distribution of four pharmacologically distinct 3H-L-glutamate binding sites, Nature, 306:176-178.

Monaghan, D.T., Olverman, H.J., Nguyen, L., Watkins, J.C and Cotman, C.W., 1988, Two classes of N-methyl-D-aspartate recognition sites: differential distribution and differential regulation by glycine, Proc. Nat. Acad. Sci. USA, 84:9836-9840.

Morris, R.G.M., 1989, Synaptic plasticity and learning: selective impairment of learning in rats and blockade of long-term potentiation in vivo by the N-methyl-D-aspartate antagonist AP5, J. Neurosci., 9:3040-3057.

Nadler, V., Kloog, Y. and Sokolovsky, M., 1988, 1-Aminocyclopropane-1-carboxylic acid (ACC) mimics the effects of glycine on the NMDA receptor ion channel, Eur. J. Pharmacol., 157:115-116.

Nowak, L., Bregestovski, P., Ascher, P., Herbet, A. and Prochiantz, A., 1984, Magnesium gates glutamate-activated channels in mouse central neurones, Nature, 307:462-465.

Olverman, H.J., Jones, A.W. and Watkins, J.C., 1984, L-Glutamate has a higher affinity than other amino acids for [3H]-D-APV binding sites in rat brain, Nature, 307:460-462.

Ransom, R.W. and Stec, N.L., 1988, Cooperative modulation of [3H]-MK-801 binding to the N-methyl-D-aspartate receptor-ion channel complex by L-glutamate, glycine and polyamines, J. Neurochem., 51:830-836.

Reynolds, I.J., Murphy, S.N. and Miller, R.J., 1987, 3H-labeled MK-801 binding to the excitatory amino acid receptor complex from rat brain is enhanced by glycine, Proc. Nat. Acad. Sci. USA, 84:7744-7748.

Simon, R.P., Swan, J.H., Griffiths, T. and Meldrum, B.S., 1984, Blockade of N-methyl-D-aspartate receptors may protect against ischemic damage in the brain, Science, 226:850-852.

Snell, L.D. and Johnson, K.M., 1988, Cycloleucine competitively antagonizes the strychnine-insensitive glycine receptor, Eur. J. Pharmacol., 151:165-166.

Snell, L.D., Morter, R.S. and Johnson, K.M., 1988, Structural requirements for activation of the glycine receptor that modulates the N-methyl-D-aspartate operated ion channel, Eur. J. Pharmacol., 156:105-110.

Snutch, T.P., 1988, The use of Xenopus oocytes to probe synaptic communication, Tr. in Neuroscience, 11:250-256.

Sugiyama, H., Ito, I. and Hirono, C., 1987, A new type of glutamate receptor linked to inositol phosphate metabolism, Nature, 325:7180-7182.

Trussel, L.O., Thio, L.L., Zorumski, C.F. and Fischbach, G.D., 1988, Rapid desensitization of glutamate receptors in vertebrate central neurons, Proc. Nat. Acad. Sci USA, 85:2834-2838.

Unnerstall, J.R. and Wamsley, J.K., 1983, Autoradiographic localization of high affinity [3H]-kainic acid binding sites in the rat forebrain, Eur. J. Pharmacol., 86:361-364.

Verdoorn, T.A. and Dingledine, R., 1988, Excitatory amino acid receptors expressed in Xenopus oocytes: agonist pharmacology, Molec. Pharmacol., 34:298-307.

Verdoorn, T.A., Kleckner, N.W. and Dingledine, R., 1987, Expresion of rat brain N-methyl-D-aspartate receptors in Xenopus oocytes. Science, 238:1114-1116.

Verdoorn, T.A., Kleckner, N.W. and Dingledine, R., 1989, N-methyl-D-aspartate/glycine and quisqualate/kainate receptors expressed in Xenopus oocytes: antagonist pharmacology, Molec. Pharmacol., 35:360-368.

Wada, K., Dechesne, C.J., Shimasaki, S., King, R.G., Kusano, K., Buonanno, A., Hampson, D.R., Banner, C., Wenthold, R.J. and Nakatani, Y., 1989, Sequence and expression of a frog brain complementary DNA encoding a kainate-binding protein, Nature, 342: 684-689.

Westbrook, G.L. and Mayer, M.L., 1987, Micromolar concentrations of Zn^{++} antagonize NMDA and GABA responses of hippocampal neurones, Nature, 328:640-643.

NOVEL FOURTH BINDING SITES OF [³H]SPERMIDINE WITHIN THE NMDA RECEPTOR COMPLEX

Yukio Yoneda and Kiyokazu Ogita

Department of Pharmacology, Faculty of Pharmaceutical Sciences
Setsunan University, Hirakata, Osaka 573-01, Japan

INTRODUCTION

N-Methyl-D-aspartic acid (NMDA)-sensitive subclass of the synaptic receptors for central excitatory amino acid neurotransmitters is supposed to form a receptor ion channel complex with modulatory glycine (Gly) binding sites which are insensitive to a classical Gly antagonist strychnine (for example, see review by Robinson and Coyle, 1987), based on the electrophysiological findings that Gly drastically potentiates in a strychnine-insensitive manner currents mediated by the NMDA-sensitive subclass without affecting those mediated by the other subclasses in cultured neuronal cells (Johnson and Ascher, 1987). The complex consists of at least three distinctly different sites; 1) NMDA recognition sites, 2) ion channel sites and 3) Gly recognition sites. The latter strychnine-insensitive Gly sites are referred to as GlyB sites to differentiate them from the strychnine-sensitive GlyA sites just for convenience' sake (Ogita et al., 1989).

Biochemical radioligand labeling of the NMDA recognition sites is rather difficult (Ferkany and Coyle, 1983; Foster and Fagg, 1987; Murphy et al., 1987a). Selective antagonists for the subclass, D-2-amino-5-phosphonovaleric acid (AP5) (Davies and Watkins, 1982) and D-2-amino-7-phosphonoheptanoic acid (AP7) (Perkins et al., 1981), as well as NMDA itself are shown to be unsuitable as a radioligand to label these sites. Recent studies have introduced novel competitive antagonists highly selective to the NMDA-sensitive receptors, including (±)-3-(2-carboxypiperazin-4-yl)propyl-1-phosphonic acid (CPP) (Davies et al., 1986) and cis-4-phosphonomethyl-2-piperidine carboxylic acid (CGS 19755) (Lehmann et al., 1988), and made it possible to label the NMDA recognition sites by these two novel antagonists (Murphy et al., 1987a; 1988). In addition to the labeling by these competitive antagonists, NMDA recognition sites are labeled by a radioactive endogenous ligand L-glutamic acid (Glu) in several synaptic membrane preparations of the rodent brain (Fagg and Matus, 1984; Monaghan and Cotman, 1986; Monahan and Michel, 1987). We have demonstrated a drastic disclosure of this NMDA-sensitive [³H]Glu binding by treatment of synaptic membrane preparations with a low concentration of Triton X-100 (Ogita and Yoneda, 1988a; Yoneda et al., 1989), possibly by way of reducing the endogenous levels of acidic amino acids as well as solubilizing membrane-bound proteins, other than NMDA-sensitive receptors, with a relatively high affinity for Glu. In addition, Triton-treated membranes also exhibit an inversely temperature-dependent binding of [^{3}H]CPP which is sensitive to the displacement by NMDA (Yoneda et al., 1990a; Ogita and Yoneda, 1990a).

On the other hand, the ion channel sites are radiolabeled by noncompetitive antagonists for the NMDA-sensitive receptors. Several dissociative anesthetics, such as ketamine (Anis et al.,, 1983) and phencyclidine (PCP) (Davies et al., 1988), block the NMDA-mediated currents in a noncompetitive manner. An anticonvulsant (+)-5-methyl-10,11-dihydro-5H-dibenzo-[a,d]cyclohepten-5,10-imine (MK-801) elicits its pharmacological action through noncompetitively antagonizing the NMDA-sensitive receptors (Wong et al., 1986). These noncompetitive antagonists are proposed to associate with the activated but not inactivated state of ion channels linked to NMDA recognition sites within the NMDA receptor complex (Loo et al., 1986; Fagg, 1987; Bonhaus et al., 1987; Snell et al., 1987; Foster and Wong, 1987; Reynolds et al, 1987; Wong et al., 1988). Triton-treated membranes are again superior for evaluating these NMDA channels to the membranes not treated with a detergent, in terms of the freedom from the confounding effects by the endogenous effectors including L-Glu, L-aspartic acid (Asp) and Gly (Yoneda et al., 1988; Yoneda and Ogita, 1989a).

The third GlyB sites are radiolabeled by [^{3}H]Gly in brain sections (Bristow et al., 1986) and synaptic membrane preparations (Kishimoto et al., 1981; Kemp et al., 1988; Marvizon and Skolnick, 1988; Snell and Johnson, 1988). The Triton treatment again potentiates this strychnine-insensitive [^{3}H]Gly binding in brain synaptic membranes through lowering the Kd value with a concomitant increase in the Bmax value (Yoneda, et al., 1988; Ogita and Yoneda, 1989). However, the binding is not potentiated in the spinal cord synaptic membranes by Triton treatment (Yoneda et al., 1990b). D-Stereoisomers of serine and alanine more potently displace the binding in Triton-treated membranes than the respective L-isomers (Ogita and Yoneda, 1989).

Furthermore, a recent binding study has raised the possibility that one polyamine spermidine (SPD) may potentiate the NMDA-mediated responses through associating with its own binding sites on the complex, which is based on the findings that SPD markedly potentiates both [^{3}H]MK-801 and [^{3}H]N-(1-[2-thienyl]cyclohexyl)piperidine (TCP) bindings in rat cortical membranes (Ransom and Stec, 1988). The latter compound is another noncompetitive NMDA antagonist used to label the activated state of ion channels linked to NMDA recognition sites on the NMDA receptor complex (Loo et al., 1986; Fagg, 1987; Bonhaus et al., 1987; Snell et al., 1987). The endogenous factor extracted from rat brains that enhances [^{3}H]MK-801 binding found in the presence of maximally effective concentrations of both Glu and Gly is identified as the polyamine SPD (Williams et al., 1989). The aforementioned findings are all favorable to the idea that brain synaptic membranes may contain novel fourth binding sites of polyamines which are located within the NMDA receptor complex. Indeed, our preliminary experiments have demonstrated the existence of novel binding sites of [^{3}H]SPD in brain synaptic membranes treated with a low concentration of Triton X-100 (Yoneda and Ogita, 1989b). In this study, therefore, the putative 4 different sites within the whole macromolecular NMDA receptor complex have been biochemically analyzed using a radioligand binding technique for the respective sites.

MATERIALS AND METHODS

Materials

[^{3}H]Glu (L-[2,3,4-^{3}H]glutamic acid, 2.22 TBq/mmol) and [^{3}H]Gly ([2-^{3}H]glycine, 1.11 TBq/mmol) were purchased from American Radiolabeled

Chemicals (St. Louis, MO, U.S.A.). [^{3}H]MK-801 ([2-^{3}H]MK-801, 832.5 GBq/mmol), [^{3}H]CPP ([propyl-1,2-^{3}H]CPP, 1.14 TBq/mmol), [^{3}H]DL-α-amino-3-hydroxy-5-methylisoxazole-4-propionic acid (AMPA; DL-α-[5-methyl-^{3}H]AMPA, 1.02 TBq/mmol), [^{3}H]kainic acid (KA; [vinylidene-^{3}H]KA, 2.22 TBq/mmol) and [^{3}H]SPD (spermidine hydrochloride, [terminal methylenes-^{3}H]SPD, 1.10 TBq/mmol) were all purchased from NEN/DuPont (Boston, MA, U.S.A.). AMPA, 6-cyano-7-nitroquinoxaline-2,3-dione (CNQX), CPP, 6-chloroquinoxaline-2,3-dione (CQX), D-AP5, 6,7-dichloroquinoxaline-2,3-dione (DCQX), 6,7-dichloro-3-hydroxy-2-quinoxaline-carboxylic acid (DHQXC), 6,7-dinitro-quinoxaline-2,3-dione (DNQX), 3-amino-1-hydroxy-2-pyrrolidone (HA-966) and 7-chlorokynurenic acid (7-ClKYNA) were all supplied by Tocris Neuramin (Buckhurst Hill, U.K.). SPD, spermine (SPN), 1,3-diaminopropane (DA3), 1,4-diaminobutane (DA4; putrescine), 1,5-diaminopentane (DA5; cadaverine), 1,6-diaminohexane (DA6), 1,7-diaminoheptane (DA7), 1,8-diaminooctane (DA8), putreanine (PTA), bacitracin, trypsin, chymotrypsin, papain, pronase E, β-galactosidase, β-glucosidase, hyaluronidase, neuraminidase and phospholipases A$_2$, C and D were obtained from Sigma Chemical (St. Louis, MO, U.S.A.). Bis-(3-aminopropyl)amine (BAA) was provided by Fluka AG (CH-9470 Buchs, Switzerland). Other chemicals used were all of the highest purity commercially available.

Membrane Preparation

Crude synaptic membrane fractions were obtained from the brains of Male Wistar rats weighing 200-250 g and washed by the suspension in 40 vol 50 mM Tris-acetate buffer (pH 7.4) followed by the centrifugation at 50,000 g for 20 min (Ogita and Yoneda, 1986). Buffers and any other solutions used in the present study were all sterilized immediately before each use by the filtration through a nitrocellulose membrane filter with a pore size of 450 nm to avoid possible microbial contamination (Yoneda and Ogita, 1989c). These washing procedures were repeated 3 times and the final pellets were suspended in 0.32 M sucrose (Yoneda and Ogita, 1986). The suspensions were frozen at -80°C until use. On the day of experiments, frozen suspensions were thawed at room temperature and the thawed suspensions were treated with 0.08% Triton X-100 at an approximate protein concentration of 0.32 mg/ml at 2°C for 10 min with a gentle stirring (Ogita and Yoneda, 1988a). The treatment was terminated by centrifuging at 50,000 g for 20 min and pellets thus obtained were washed once more as described above to obtain "Triton-treated membranes" (Ogita and Yoneda, 1988b). The thawed suspensions were also subjected to the same washing procedures to obtain "untreated membranes" as needed.

[^{3}H]MK-801 Binding

An aliquot of Triton-treated membranes was incubated with 5 nM [^{3}H]MK-801 in 0.5 ml 50 mM Tris-acetate buffer (pH 7.4) at a protein concentration of 0.2-0.3 mg/ml at 30°C for 30 min or 16 hr unless indicated otherwise. Incubation was terminated by the addition of 3 ml ice-cold buffer (2°C) and subsequent filtration through a Whatman GF/B glass fiber filter under a constant vacuum of 15 mm Hg (Yoneda and Ogita, 1987). The filter was rinsed with cold buffer (2°C) 4 times within 10 sec, and radioactivity retained on the filter was measured by a liquid scintillation spectrometer using 5 ml modified Triton-toluene scintillant (Ogita et al., 1986) at a counting efficiency of 40-42%. Nonspecific binding was defined by 0.1 mM (+)-MK-801 and accounted for 10-15% of the total binding found in the absence of unlabeled MK-801. For Scatchard analysis, membranes were incubated with 5 nM [^{3}H]MK-801 at 30°C for 16 hr in the presence of at least 6 varying concentrations of unlabeled MK-801 to cover a concentration range of 5 to 50 nM.

[³H]SPD Binding

An aliquot of Triton-treated or untreated synaptic membranes was incubated with 20 nM [³H]SPD in 0.5 ml 50 mM Tris-acetate buffer (pH 7.4) at a protein concentration range between 0.2 to 0.3 mg/ml at 2°C for 10 min unless indicated otherwise. Incubation was terminated by the filtration method as mentioned above with a minor modification. Whatman GF/B glass fiber filters were soaked in 0.3% polyethyleneimine solution at least 1 hr prior to use in order to reduce the nonspecific binding found in the presence of 10 mM unlabeled SPD. For Scatchard analysis, membranes were incubated with 20 nM [³H]SPD at 2°C for 10 min in the presence of at least 6 varying concentrations of unlabeled SPD to cover a concentration range of 20 to 400 µM.

Other Bindings

Other bindings including NMDA-sensitive [³H]Glu binding (Ogita and Yoneda, 1988a), temperature-independent [³H]CPP binding (Ogita and Yoneda, 1990a) and strychnine-insensitive [³H]Gly binding (Ogita et al., 1989) were all determined as described elsewhere. [³H]AMPA binding was determined by incubating Triton-treated membranes with 10 nM [³H]AMPA at 2°C for 30 min in 0.5 ml 50 mM Tris-acetate buffer (pH 7.4) in the presence of 100 mM KSCN as reported previously (Murphy et al., 1987b; Terramani et al., 1988). Triton-treated membranes were also incubated with 10 nM [³H]KA in 0.5 ml buffer at 2°C for 30 min (Murphy et al., 1987b). All assays were terminated by the rapid filtration method.

Regional Distribution

Brain regions were dissected on an ice-cold plastic plate according to the procedures described by Glowinski and Iversen (1966). Each brain structure was homogenized in 40 vol (40 ml/g wet wt.) 5 mM Tris-acetate buffer (pH 7.4) containing 0.01% bacitracin, 0.1 mM phenylmethylsulfonyl fluoride (PMSF) and 1 mM EDTA. Homogenates were then centrifuged at 50,000 g for 30 min and the pellets were suspended in the same volume of 50 mM Tris-acetate buffer (pH 7.4). After repeating these washing procedures 3 times, the final pellets were suspended in 8 vol 0.32 M sucrose to the original wet weight. The suspensions were frozen at -80°C for no longer than 1 week. On the day of experiments, frozen suspensions were thawed at room temperature and the thawed suspensions were subjected to the Triton treatment as described above.

Analyses

Binding assays were all carried out at an interval of 20 sec in triplicate with a variation of less than 10%. Results were expressed as the mean ± S.E. and the statistical significance was determined by the Student's $\underline{t}$-test or ANOVA analysis. Saturation isotherms were analyzed by the LIGAND computer program (Munson and Rodbard, 1980). The concentration of each test drug at which it inhibited (IC50) or potentiated (EC50) the binding by 50% was calculated according to the Hill plot analysis by a computer with a LOTUS 1-2-3 program.

RESULTS

Triton Treatment

Treatment of brain synaptic membranes with 0.08% Triton X-100 resulted in an almost complete abolition of both [³H]MK-801 (Yoneda and Ogita, 1989a) and [³H]TCP bindings (Ogita et al., 1990). However, Glu drastically potentiated both bindings in Triton-treated membranes through associating with the NMDA recognition sites. Further addition of Gly completely restored [³H]MK-801 as well as [³H]TCP binding in the presence of Glu alone in Triton-treated membranes to the level found in untreated membranes. Pharmacological profiles for the potentiation and displacement of both [³H]MK-801 and [³H]TCP bindings were not altered after the treatment of synaptic membranes with a low concentration of Triton X-100 (Yoneda and Ogita, 1989a; Ogita et al., 1990). Accordingly, it is likely that Triton-treated membranes have kept their regulatory properties unchanged.

Potentiation by Polyamines

Figure 1 shows chemical structures of numerous polyamines tested in this study. Each polyamine at different concentrations was included in the incubation mixture for [³H]MK-801 binding at 2 °C or 30 °C in either the presence or absence of Glu and Gly. At 2 °C, the binding was significantly potentiated by the addition of both Glu and Gly, but not by the individual addition of Glu or Gly. Further addition of SPD potentiated the binding at 2 °C in the presence of Glu alone in a concentration dependent manner at concentrations above 0.01 mM, with a concomitant slight enhancement of the binding in the presence of both Glu and Gly (Table 1). In addition, SPD alone markedly potentiated the binding in the absence of added amino acids in Triton-treated membranes at the concentrations used. SPD also elicited a drastic potentiation of the binding in Triton-treated membranes at 30 °C to the extent similar to that at 2 °C. Although SPD alone markedly potentiated the binding determined at 30 °C as seen at 2 °C, SPD was not only effective in markedly potentiating the enhanced binding induced by Glu alone at concentrations above 0.01 mM, but also active as an additional enhancer of the potentiated binding caused by both Glu and Gly. No marked potentiation was induced by SPD at the concentrations used in the presence of Gly alone, irrespective of the incubation temperature. These results suggest that Gly may elicit an inhibitory property against the potentiation of [³H]MK-801 binding by SPD in Triton-treated membranes.

Spermidine	[SPD]	$H_2N(CH_2)_3NH(CH_2)_4NH_2$
Spermine	[SPN]	$H_2N(CH_2)_3NH(CH_2)_4NH(CH_2)_3NH_2$
Bis-(3-aminopropyl)amine	[BAA]	$H_2N(CH_2)_3NH(CH_2)_3NH_2$
1,3-Diaminopropane	[DA3]	$H_2N(CH_2)_3NH_2$
1,4-Diaminobutane(Putrescine)	[DA4]	$H_2N(CH_2)_4NH_2$
1,5-Diaminopentane(Cadaverine)	[DA5]	$H_2N(CH_2)_5NH_2$
1,6-Diaminohexane	[DA6]	$H_2N(CH_2)_6NH_2$
1,7-Diaminoheptane	[DA7]	$H_2N(CH_2)_7NH_2$
1,8-Diaminooctane	[DA8]	$H_2N(CH_2)_8NH_2$
Putreanine	[PTA]	$H_2N(CH_2)_4NHCH_2CH_2COOH$

Fig. 1. Chemical structures of numerous polyamines.

Table 1. Effect of spermidine (SPD) on [³H]MK-801 binding

Addition (-log M)	[³H]MK-801 binding (% of control)		Addition (-log M)		
	2 °C	30 °C		2 °C	30 °C
NONE	100 ± 6	100 ± 4	GLY (5)	100 ± 10	100 ± 4
+SPD (7)	89 ± 12	168 ± 27	+SPD (7)	92 ± 4	109 ± 13
(6)	125 ± 26	136 ± 9	(6)	89 ± 1	97 ± 11
(5)	118 ± 11	128 ± 10	(5)	110 ± 6	97 ± 2
(4)	156 ± 30	159 ± 27	(4)	114 ± 17	122 ± 7
(3)	294 ± 49*	373 ± 62*	(3)	132 ± 17	173 ± 13*
(2)	411 ± 43#	404 ± 39#	(2)	169 ± 20	144 ± 13
GLU (5)	100 ± 7	100 ± 4	GLU + GLY	100 ± 9	100 ± 3
+SPD (7)	94 ± 1	84 ± 4	+SPD (7)	104 ± 4	98 ± 4
(6)	89 ± 14	89 ± 5	(6)	118 ± 5	98 ± 1
(5)	101 ± 11	90 ± 6	(5)	124 ± 5	109 ± 6
(4)	219 ± 15#	236 ± 9#	(4)	190 ± 7#	201 ± 5#
(3)	395 ± 73#	370 ± 18#	(3)	229 ± 18#	239 ± 15#
(2)	368 ± 56#	355 ± 17#	(2)	239 ± 16#	212 ± 19#

An aliquot of Triton-treated membranes was incubated with [³H]MK-801 in the buffer containing varying concentrations of SPD at a concentration range of 100 nM to 10 mM in either the presence or absence of 10 µM Glu and 10 µM Gly. *P<0.05, #P<0.01, compared with each control value. Control binding (fmol/mg protein): NONE, 11.7 ± 0.7 (2°C), 9.2 ± 0.4 (30°C); GLU, 23.6 ± 1.6 (2°C), 102.2 ± 4.2 (30°C); GLY, 27.2 ± 2.8 (2°C), 27.4 ± 1.1 (30°C); GLU + GLY, 38.8 ± 3.2 (2°C), 168.8 ± 4.4 (30°C).

Table 2. Effect of various polyamines on [³H]MK-801 binding

Addition (-log M)	[³H]MK-801 binding (% of control)			
	(5)	(4)	(3)	(2)
SPD	109 ± 6	201 ± 5#	239 ± 14#	212 ± 19#
SPN	153 ± 1*	216 ± 2#	184 ± 10#	175 ± 7#
BAA	112 ± 2	195 ± 5#	237 ± 16#	194 ± 10*
DA3	104 ± 1	114 ± 2	133 ± 11	169 ± 11*
DA4	102 ± 1	101 ± 2	97 ± 10	100 ± 9
DA5	100 ± 1	90 ± 2	87 ± 7	81 ± 10
DA6	102 ± 1	101 ± 1	103 ± 13	97 ± 4
DA7	99 ± 1	105 ± 2	99 ± 9	92 ± 6
DA8	99 ± 1	94 ± 3	81 ± 6	47 ± 2*
PTA	109 ± 4	105 ± 4	95 ± 7	12 ± 3*

Triton-treated membranes were incubated with [³H]MK-801 in the presence of varying concentrations of numerous polyamines at a concentration range of 0.01 to 10 mM. *P<0.05, #P<0.01, compared with the control value obtained in the absence of added polyamine.

In addition to SPD, several polyamines significantly potentiated the binding at 30 °C found in the presence of both Glu and Gly in a concentration-dependent fashion (Table 2). Among various polyamines tested, SPN most potently potentiated the binding at concentrations higher than 1 µM followed by BAA and SPD in a rank order of potency. These 3 active polyamines were rather inhibitory at 10 mM as compared with the respective levels found at 1 mM. DA3 also potentiated the binding at a concentration range of 0.1 to 10 mM, whereas none of DA4, DA5, DA6 and DA7 was an active effector of the binding in the presence of both Glu and Gly in Triton-treated membranes. In contrast, both DA8 and PTA markedly inhibited the binding at a concentration of 10 mM. Therefore, it is evident that the potentiation by polyamines is a structure selective phenomenon.

Kinetics

In the absence of added amino acids, [³H]MK-801 binding gradually increased with incubation time up to 24 hr at 30 °C in Triton-treated membranes. Addition of 10 µM Glu markedly facilitated the association rate with a concomitant slight potentiation of the steady-state level determined at 16 hr after the initiation of incubation at 30 °C [observed association rate (K obs); 0.088 ± 0.018 vs. 0.65 ± 0.07 min^{-1}]. Further addition of 10 µM Gly in addition to Glu supplementarily accelerated the initial association rate without affecting the steady-state level found in the presence of Glu alone (K obs = 1.00 ± 0.06 min^{-1}) (Fig. 2). Further addition of Gly did not significantly elevate the steady-state level found in the presence of Glu alone. SPD at 1 mM also accelerated the association rate in the presence of Glu alone to that found in the presence of both Glu and Gly, while being inactive in additionally facilitating the initial association rate found in the presence of both Glu and Gly. The initial association rate found in the presence of Glu/Gly was not supplementarily facilitated by the additional inclusion of SPD (K obs: Glu alone, 0.65 ± 0.07; Glu/Gly, 1.00 ± 0.06; Glu/Gly/SPD, 1.10 ± 0.13). However, SPD induced about 2-fold potentiation of the binding at equilibrium in the presence of Glu alone as well as that in the presence of both Glu and Gly (Fig. 2). The steady-state level in the presence of Glu/SPD was similar to that found in the presence of Glu/Gly/SPD. These results suggest that SPD may potentiate [³H]MK-801 binding through a molecular mechanism distinctly different from that underlying the potentiation by Glu and/or Glu/Gly.

Novel Antagonists

Since it is impossible to completely deplete these endogenous effectors from the membrane preparations while keeping the binding sites intact at any cost, the search for competitive antagonists highly selective to the respective sites is crucial for the analysis of molecular mechanisms underlying the potentiation of [³H]MK-801 binding. For this purpose, several quinoxaline derivatives (Yoneda and Ogita, 1989d) and the known antagonists for GlyB sites such as HA-966 (Fletcher and Lodge, 1988), KYNA and 7-ClKYNA (Kemp et al., 1988) were tested for the inhibitory potency on [³H]MK-801 binding in the presence of Glu alone in Triton-treated membranes (Table 3). Among the test compounds, 7-ClKYNA was most effective in inhibiting the binding followed by DHQXC, DCQX, DNQX, CNQX, CQX, KYNA, QX and HA-966 in a rank order of decreasing potency. Further inclusion of Gly in addition to Glu obviously attenuated the inhibitory activities of all 9 inhibitors. Only 3 compounds out of the positive 9 antagonists, including DHQXC, 7-ClKYNA and DNQX, were still active as inhibitors of the binding found in the presence of both Glu and Gly at 0.1 mM. These results imply that Gly is able to reverse or antagonize the potencies of all 9 test compounds to inhibit the binding in the presence of Glu alone and thereby that all substances may have some affinity for the GlyB sites.

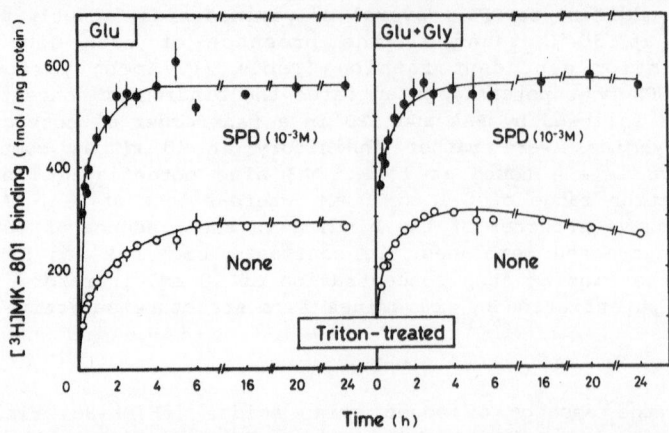

Fig. 2. Effect of incubation time on [³H]MK-801 binding. An aliquot of
Triton-treated membranes was incubated with [³H]MK-801 in the
buffer containing either 10 μM Glu alone or 10 μM Glu plus 10 μM
Gly, in either the presence or absence of 1 mM SPD.

Table 3. Inhibition of various bindings by numerous antagonists

Antagonists	IC50 values (μM)				
	[³H]MK-801 binding		[³H]Glu	[³H]CPP	[³H]Gly
	Glu	Glu + Gly	binding	binding	binding
CNQX	7.48± 1.39	100	73.3±11.2	16.4± 2.1	5.31±1.02
DNQX	1.90± 0.13	97.3±11.7	98.9±16.1	31.5± 1.0	1.42±0.27
DCQX	1.12± 0.08	>100	>100	NI	0.62±0.11
CQX	9.88± 1.30	>100	>100	80.7± 1.4	6.13±0.89
QX	75.9 ±39.2	NI	>100	98.6±11.0	44.1 ±3.3
DHQXC	1.03± 0.22	55.5±12.1	18.9± 3.9	21.6± 4.2	0.49±0.07
HA-966	>100	NI	NI	NI	8.25±2.15
KYNA	40.0 ±12.0	NI	>100	58.9±15.9	13.7 ±1.1
7-ClKYNA	0.94± 0.13	60.7±5.2	47.7± 1.8	53.6± 6.9	0.36±0.06

An aliquot of Triton-treated membranes was incubated with [³H]MK-801,
[³H]Glu, [³H]CPP or [³H]Gly under the respective incubation conditions as
described in the text, in the presence of at least 6 different concentrations of the individual antagonists at a concentration range of 10 nM to
0.1 mM. NI; no inhibition.

To test such a possibility, an attempt was made to determine whether
or not these compounds inhibit strychnine-insensitive [³H]Gly binding in
Triton-treated membranes. As shown in Table 3, 7-ClKYNA potently displaced
the binding in a concentration-dependent manner at concentrations above 10
nM. Both DCQX and DHQXC were as potent inhibitors of [³H]Gly binding as 7-
ClKYNA, whereas DNQX exhibited a relatively intensive displacement of the
binding with progressively less potent inhibitions by CNQX, CQX, HA-966,

KYNA and QX. Most of these test compounds but DCQX and HA-966 also displaced NMDA-sensitive [³H]Glu binding, with decreasing potencies by DHQXC, 7-ClKYNA, CNQX, DNQX, KYNA, CQX and QX. Both DCQX and HA-966 at 0.1 mM did not inhibit [³H]Glu binding. The above 7 active inhibitors were effective in displacing the temperature-independent binding of [³H]CPP (Ogita and Yoneda, 1990b). Although DCQX did not significantly affect [³H]CPP binding at concentrations below 0.1 mM, HA-966 enhanced [³H]CPP binding at concentrations above 1 μM. These results clearly indicate that DCQX is a competitive antagonist highly selective to the strychnine-insensitive Gly binding sites within the NMDA receptor complex.

Requirement for Gly

In the absence of any antagonists, Glu potentiated the binding of [³H]MK-801 in a concentration-dependent manner at concentrations above 10 nM with an EC_{50} of $0.18 \pm 0.01 \mu M$, and further inclusion of Gly supplementarily enhanced the Glu-potentiated binding with an EC_{50} of $0.10 \pm 0.02 \mu M$ (Table 4). Addition of competitive NMDA antagonists, such as CPP and D-AP5, resulted in a concentration-dependent reduction of the EC_{50} value of Glu without affecting that of Gly, while the Gly antagonists including

Table 4. Effects of both NMDA and Gly antagonists on the potentiation of [³H]MK-801 binding by Glu and Gly

Antagonists (-log M)	Glu		Gly	
	EC50 (μM)	Hill coefficient	EC50 (μM)	Hill coefficient
None	0.18 ± 0.01	1.12 ± 0.29	0.10 ± 0.02	1.01 ± 0.27
CPP				
(6)	0.71 ± 0.07#	1.05 ± 0.10	0.11 ± 0.02	0.80 ± 0.06
(5)	4.16 ± 0.64#	1.00 ± 0.11	0.18 ± 0.03	0.95 ± 0.07
D-AP5				
(6)	0.78 ± 0.04#	0.95 ± 0.07	0.16 ± 0.04	0.80 ± 0.06
(5)	4.63 ± 0.65#	1.04 ± 0.04	0.14 ± 0.03	1.05 ± 0.24
DCQX				
(6)	0.41 ± 0.15	1.36 ± 0.41	0.23 ± 0.04*	0.98 ± 0.15
(5)	NP	-	0.87 ± 0.06#	1.07 ± 0.04
7-ClKYNA				
(6)	0.39 ± 0.19	1.14 ± 0.31	0.31 ± 0.03*	1.07 ± 0.11
(5)	NP	-	1.53 ± 0.12#	1.17 ± 0.03
HA-966				
(6)	0.17 ± 0.02	0.93 ± 0.12	0.17 ± 0.04	1.14 ± 0.18
(5)	0.20 ± 0.02	1.21 ± 0.19	0.39 ± 0.08*	0.85 ± 0.11

An aliquot of Triton-treated membranes was incubated with [³H]MK-801 in the buffer containing at least 6 different concentrations of Glu at a concentration range of 10 nM to 0.1 mM in either the presence or absence of the indicated concentration of each antagonist, to determine the EC_{50} values for Glu. To calculate the EC_{50} value for Gly, membranes were incubated with [³H]MK-801 in the buffer containing at least 6 different concentrations of Gly at a concentration range of 10 nM to 0.1 mM in addition to 10 μM Glu in either the presence or absence of the antagonist. *P<0.05, #P<0.01, compared with each control value obtained in the absence of added antagonist.

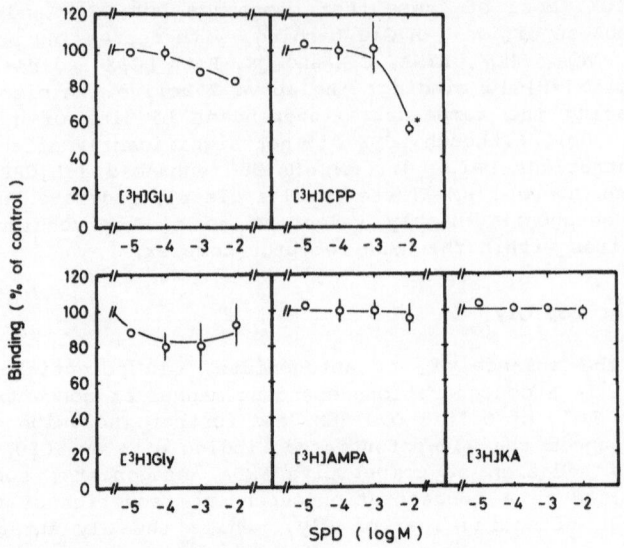

Fig. 3. Effect of SPD on various radioligand bindings. An aliquot of
Triton-treated membranes was incubated with [³H]Glu, [³H]CPP,
[³H]Gly, [³H]AMPA or [³H]KA under the respective routine experimental conditions in the presence of varying concentrations of
SPD. *P<0.05, compared with the control value obtained in the absence of added SPD.

DCQX, 7-ClKYNA and HA-966 significantly reduced the EC50 value of Gly.
However, the former 2 Gly antagonists also abolished the ability of Glu to
potentiate the binding. In the presence of DCQX or 7-ClKYNA at 10 µM, Glu
was unable to markedly potentiate the binding even at 0.1 mM, whereas HA-
966 failed to affect the potency of Glu at the concentrations used. Hill
coefficients of both Glu and Gly were not significantly different from
unity under any experimental conditions. These results raise the possibility that the activation of GlyB sites may be essential for the Glu-
induced potentiation of [³H]MK-801 binding in the brain.

The potentiation by SPD in the presence of Glu was not only
protected by the NMDA antagonists, but also antagonized by the Gly
antagonists. The Gly antagonists prevented the potentiation by SPD to a
much greater extent than the NMDA antagonists, while CPP at a concentration of 0.1 mM drastically abolished the potentiation by SPD. Inclusion of
Gly in addition to Glu significantly reversed the preventive action of the
Gly antagonists against the potentiation by SPD, without affecting that of
the NMDA antagonists. CPP at 0.1 mM again potently prevented the potentiation by SPD even in the presence of added Gly. Therefore, SPD seems to be
unable to potentiate [³H]MK-801 binding in brain synaptic membranes unless
both NMDA recognition and GlyB sites are under the activated state by
respective agonists.

Other Bindings

Since the potentiation by SPD was dependent on the activated states
of the NMDA recognition sites and GlyB sites, the effect of SPD on both
bindings was next examined. As shown in Fig. 3, however, neither NMDA-
sensitive [³H]Glu binding nor strychnine-insensitive [³H]Gly binding was
significantly affected by the <u>in vitro</u> addition of SPD at a concentration
range of 10 µM to 10 mM. In contrast, SPD at 10 mM markedly inhibited the

Table 5. Effect of rinsing times of filters on [³H]SPD binding

Rinsing (times)	[³H]SPD binding (fmol/mg protein)		
	Total	Nonspecific	Specific
0	4240 ± 451	4877 ± 196	–
1	1784 ± 148	1197 ± 142	587 ± 148
2	1028 ± 94	284 ± 17	744 ± 94
3	829 ± 78	199 ± 18	630 ± 78
4	915 ± 129	160 ± 23	754 ± 129
5	795 ± 112	146 ± 6	649 ± 112
7	472 ± 75	102 ± 10	369 ± 74
10	339 ± 30	79 ± 8	260 ± 29

An aliquot of Triton-treated membranes was incubated with 20 nM [³H]SPD in either the presence (Nonspecific) or absence (Total) of 10 mM unlabeled SPD to determine the specific binding. Incubation was terminated by the rapid filtration through a glass fiber filter, and the filter was rinsed with cold buffer various times indicated.

binding of [³H]CPP to the NMDA recognition sites. SPD also failed to alter both the binding of [³H]AMPA to the quisqualate (QA)-sensitive receptors and the binding of [³H]KA to the KA-sensitive receptors at the concentrations used. These results raise the possibility that SPD may potentiate [³H]MK-801 binding through associating with its own binding sites located on the NMDA receptor complex, rather than interacting with the NMDA recognition as well as GlyB sites within the complex.

[³H]SPD Binding

Conditions. To evaluate the presence of [³H]SPD binding sites in brain synaptic membranes treated with Triton X-100, an attempt was made to determine the binding by a filtration method using Whatman GF/B filters to separate the bound ligand. Triton-treated membranes were incubated with 20 nM [³H]SPD and the incubation was terminated by the filtration under vacuum. Then, the filter was rinsed with cold buffer several times. Table 5 shows the effect of varying rinsing times on the total, nonspecific and specific bindings of [³H]SPD. No significant specific binding was detected on the filters without rinsing. After rinsing the filters with buffer once, the specific binding accounted for about 30% of the total binding. The specific binding was rather constant when the filters were rinsed with cold buffer from 2 to 5 times, whereas further rinsing up to 10 times decreased the specific binding with a concomitant decrease in both the total and nonspecific bindings. Accordingly, the specific binding accounted for more than 70% of the total binding on the filters rinsed more than 2 times up to 10 times. These results clearly indicate that [³H]SPD bound specifically is not easily dissociated from the binding sites during the rinsing of filters to a significant extent. Subsequent experiments were carried out using the 4 times-rinsing procedures.

Properties. The specific binding of [³H]SPD increased linearly with increasing protein concentrations in Triton-treated membranes up to 0.4 mg protein (Fig. 4). In untreated membranes, the binding increased in proportion to the incubation time up to 10 min and reached a plateau thereafter at 2 °C. Elevation of the incubation temperature from 2°C to 30°C resulted in about 2-fold reduction of the binding, while the time required to attain equilibrium at 30°C was within 2 min. The bindings at 2°C and 30°C were rapidly dissociated by the addition of an excess unlabeled SPD (10

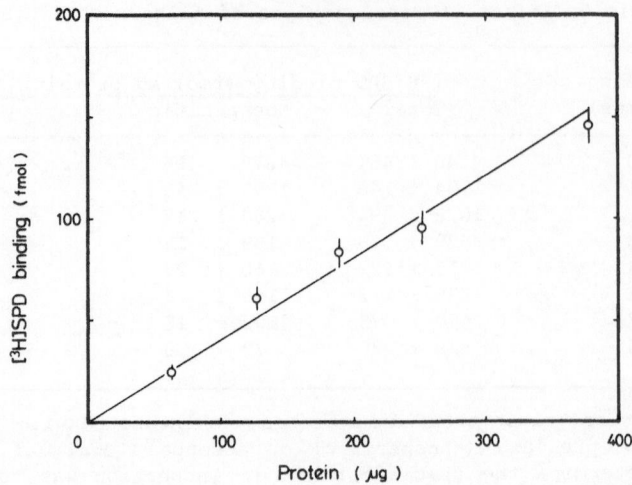

Fig. 4. Effect of increasing protein concentrations on [³H]SPD binding.
The routine binding assays were carried out using Triton-treated
membranes at varying protein concentrations indicated.

Table 6. Binding parameters for [³H]SPD binding

Membranes	Kd (µM)	Bmax (nmol/mg protein)
Untreated	161.8 ± 14.7	5.89 ± 0.53
Triton-treated	161.5 ± 12.8	4.99 ± 0.20

An aliquot of untreated or Triton-treated membranes was
incubated with [³H]SPD under the routine conditions.

mM) 10 min after the initiation of incubation within 10 min to the level
at most 3 times lower than that at equilibrium, respectively. In contrast,
the binding reached a steady-state level within 2 min after the initiation
of incubation at 2 ℃ in Triton-treated membranes. In these Triton-treated
membranes, elevation of the incubation temperature from 2°C to 30°C again
reduced the binding which reached a transient maximal value 40 sec after
the initiation of incubation with a gradual decline thereafter up to 30
min. The steady-state level in Triton-treated membranes was invariably approximately two-third of that in untreated membranes irrespective of the
incubation temperature employed. However, the binding was completely
abolished by the addition of excess SPD 10 min after the initiation of incubation to the nonspecific binding level within 5 min after the addition,
whereas the time required to attain a complete dissociation at 30°C was
less than 1 min. Therefore, [³H]SPD binding is evidently a time-dependent,
inversely temperature-dependent and reversible process.

Scatchard analysis. The binding was also saturable with increasing
concentrations of the ligand in both membrane preparations. The binding
increased with increasing ligand concentrations up to 500 µM and reached a
steady-state level at concentrations higher than 500 µM. Scatchard
analysis of these data revealed that the binding sites consisted of a

single component with a Kd of 161.8 ± 14.7 μM and a Bmax of 5.89 ± 0.53 nmol/mg protein in untreated membranes, while the Kd and Bmax were 161.5 ± 12.8 μM and 4.99 ± 0.20 nmol/mg protein in Triton-treated membranes, respectively (Table 6). These results clearly indicate that [³H]SPD binding is a saturable phenomenon with a relatively low affinity and a high density.

Pharmacology. Table 7 shows a correlation between the abilities of numerous polyamines to potentiate [³H]MK-801 binding and to displace [³H]SPD binding in Triton-treated membranes. Among various polyamines tested, SPN was the most potent displacer of [³H]SPD binding, accompanied by BAA, SPD and DA3 in a rank order of potency. Similar rank order was seen with the potency of polyamines to enhance [³H]MK-801 binding. However, the polyamines examined were all effective as displacers of [³H]SPD binding at 10 mM, in contrast to the inefficiency to potentiate [³H]MK-801 binding at the same concentration. These results suggest that some polyamines might potentiate [³H]MK-801 binding by way of associating with [³H]SPD binding sites located on the NMDA receptor complex.

Preincubation. Since the affinity of [³H]SPD binding sites was considerably low as specific receptor binding sites with protein structures, an attempt was made to determine whether or not [³H]SPD binding is significantly inhibited by the preincubation with different proteases (Table 8). In untreated membranes, only pronase E at 0.1 mg/ml significantly inhibited the binding, with the other 3 proteases employed including papain, trypsin and chymotrypsin being inactive at the highest concentration used. In Triton-treated membranes, however, the proteases were all ineffective in inhibiting the binding at a concentration range of 1 to 100 μg/ml. In addition, only neuraminidase potentiated the binding in untreated but not in Triton-treated membranes at 10 mU/ml among 4 different glycosidases. The other 3 negative glycosidases included β-galactosidase, β-glucosidase and hyaluronidase. The binding was markedly potentiated by the preincubation with phospholipases A_2 and C at 10 U/ml in untreated as well as Triton-treated membranes, while phospholipase D being inactive at the same concentration. These results imply that [³H]SPD binding sites may be

Table 7. Effects of various polyamines on [³H]MK-801 binding and [³H]SPD binding

Polyamines	[³H]MK-801 binding EC50 (μM)	[³H]SPD binding IC50 (μM)
SPD	59 ± 8	238 ± 55
SPN	12 ± 1	22 ± 1
BAA	49 ± 8	173 ± 41
DA3	> 1000	> 1000
DA4	NP	> 1000
DA5	NP	> 1000
DA6	NP	> 1000
DA7	NP	> 1000
DA8	NP	> 1000
PTA	NP	> 1000

Triton-treated membranes were incubated with [³H]MK-801 or [³H]SPD under the respective routine conditions in the presence of varying concentrations of each polyamine. NP; no potentiation.

Table 8. Effects of preincubation with several enzymes
on [³H]SPD binding

| | [³H]SPD binding (% of control) | |
	Untreated	Triton-treated
[PROTEASES] (100 μg/ml)		
Papain	90 ± 7	173 ± 19
Pronase E	62 ± 5#	120 ± 11
Trypsin	87 ± 7	153 ± 18
Chymotrypsin	97 ± 9	153 ± 17
[PHOSPHOLIPASES] (10 U/ml)		
A2	265 ± 14#	786 ± 61#
C	260 ± 32#	313 ± 16#
D	142 ± 10*	124 ± 11
[GLYCOSIDASES] (10 U/ml)		
β-Glucosidase	97 ± 7	94 ± 10
β-Galactosidase	106 ± 10	91 ± 8
Hyaluronidase	104 ± 9	94 ± 9
Neuraminidase	149 ± 11	129 ± 11
(10 mU/ml)		

An aliquot of untreated or Triton-treated membranes was
preincubated with the indicated concentration of each
enzyme at 30°C for 30 min, and subsequently [³H]SPD binding was determined under the routine conditions.

Table 9. Regional variations of several bindings in rat brain

| Regions | Binding (fmol/mg protein) | | | |
| | [³H]Glu | [³H]SPD | [³H]MK-801 binding | |
			(Glu/Gly)	(Glu/Gly/SPD)
CC	475.4 ± 10.5	281.7 ± 18.0	467.4 ± 52.5	1100.1 ± 18.3
ST	339.6 ± 19.4	509.6 ± 27.5	221.3 ± 31.0	545.2 ± 18.3
MB	122.2 ± 6.7	671.3 ± 40.3	115.1 ± 15.8	279.1 ± 6.6
HT	200.1 ± 10.5	322.4 ± 18.2	131.0 ± 22.0	232.1 ± 42.2
HPC	493.8 ± 14.8	348.5 ± 15.4	456.4 ± 58.5	1003.3 ± 55.3
CB	80.2 ± 4.8	367.3 ± 26.2	36.9 ± 3.4	50.3 ± 2.7
MP	38.3 ± 4.7	940.6 ± 45.8	39.6 ± 5.0	68.4 ± 3.8

Membranous homogenates of each brain structure were treated with Triton X-
100 before subjecting to the determination of respective bindings under
the routine conditions. Abbreviations: CB, cerebellum; CC, cerebral
cortex; HPC, hippocampus; HT, hypothalamus; MB, midbrain; MP, medulla-
pons; ST, striatum.

membranous constituents insensitive to proteases but sensitive to phospholipases in brain synaptic membranes.

Regional Variation

[³H]SPD binding. In order to evaluate the functional significance of the novel binding sites of [³H]SPD, an attempt was made to determine whether or not the binding is distributed unevenly in the brain. Among the rat brain structures examined, unexpectedly, the highest binding of [³H]SPD measured at a fixed ligand concentration was found in the medulla-pons followed by the midbrain, striatum, cerebellum, hippocampus, hypothalamus and cerebral cortex in a rank order of decreasing activity (Table 9). Scatchard analysis using ligand concentrations between 20 and 400 μM revealed that the binding sites invariably consisted of a single component with Kd values ranging from 32 to 48 μM and Bmax values ranging between 837 and 2661 pmol/mg protein in all brain structures examined. The medulla-pons again had the highest density with the cerebral cortex exhibiting the lowest density, whereas the affinities for [³H]SPD were not different each other in the brain structures evaluated.

NMDA receptor complex. In contrast, NMDA-sensitive [³H]Glu binding was highest in the hippocampus among all brain regions examined. Progressively lower bindings were found in the cerebral cortex, striatum, hypothalamus, midbrain, cerebellum and medulla-pons. The affinities for [³H]Glu among these structures were not significantly different one another, while having different densities. Similar distribution profiles were observed with [³H]MK-801 binding determined in the presence of Glu/Gly or Glu/Gly/SPD (Table 9). All brain regions tested had the same affinities for [³H]MK-801 with different densities when determined in the presence of both Glu and Gly, whereas further addition of SPD significantly increased those affinities without affecting the densities except the cerebellum and medulla-pons. SPD at 1 mM failed to significantly potentiate [³H]MK-801 binding in the presence of both Glu and Gly in the latter 2 brain regions. A clear correlation was found between the distribution profiles in the brain structures of NMDA-sensitive [³H]Glu binding and [³H]MK-801 binding determined in the presence of Glu/Gly and Glu/Gly/SPD. It thus appears that the distribution of [³H]SPD binding is not consistent with that of the NMDA receptor complex in the brain although all the bindings exhibit an uneven distribution profile.

DISCUSSION

The data provided here clearly indicate that several polyamines potentiate [³H]MK-801 binding in brain synaptic membranes in a structure-selective manner. The propyl moiety between the terminal amino and imino groups seems to be essential for the potentiation. Since substitution of the imino group with amino group reduced but did not eliminate the ability of a polyamine to potentiate the binding, the imino residue is not necessarily crucial but important for the potentiation by polyamines. However, the molecular mechanism underlying the potentiation by SPD is evidently different from that underlying the potentiation by Gly. For example, Gly facilitated the accelerated association rate without affecting the increased steady-state level induced by Glu, whereas SPD stimulated both the accelerated association rate and increased steady-state level induced by Glu.The Gly-induced facilitation of the association rate may agree with the previous electrophysiological data that Gly dramatically potentiates the NMDA-mediated currents through increasing the frequency of opening of the NMDA channels, rather than through altering either the amplitude of a single NMDA current or the mean open time of the channels (Johnson and Ascher, 1987). Therefore, polyamines could be maximally effective in increasing the probability of opening of the NMDA channels, with a concomitant enhancement of the amplitude and/or the mean open time.

However, it is still open to be evaluated that polyamines in fact potentiate various physiological responses mediated by the NMDA-sensitive receptors in the mammalian brain. One possible discouraging interpretation of the potentiation of [³H]MK-801 binding by SPD is that SPD may only potentiate the association of an open channel blocker MK-801 with the Glu-activated NMDA channels, while being ineffective as a stimulant of the channels. The polyamine would potentiate the noncompetitive inhibition by MK-801 of the NMDA-mediated responses through accelerating its association with the channels. If this is the case, the potentiation of [³H]MK-801 binding by SPD could be a type of experimental artifacts with no significant physiological meanings. Further confirmation of the putative stimulatory effect of polyamines on the NMDA-mediated responses should be made using electrophysiological techniques before drawing any conclusion. However, much attention has to be paid to electrophysiological analysis on the exact action sites of polyamines on the NMDA-mediated responses in the brain. Polyamines may potentiate the NMDA responses by way of associating with its own binding sites located on the intracellular surfaces of plasma membranes.

From this standpoint, it is noteworthy that brain synaptic membrane preparations indeed contain structure-selective, reversible and saturable binding sites of [³H]SPD with low affinity and high density, irrespective of the Triton treatment. Similar rank order was seen between the abilities of various polyamines to potentiate [³H]MK-801 binding and to displace [³H]SPD binding. In contrast, several independent lines of evidence have shown the possible binding of SPD to membrane constituents such as ATPases and nucleases (Tabor and Tabor, 1984; Pegg, 1986). Nevertheless, the virtual ineffectiveness of several proteases on [³H]SPD binding is unfavorable to the idea that some membranous proteins may be responsible for the specific binding of [³H]SPD in brain synaptic membranes treated and not treated with a low concentration of Triton X-100. Possible interaction of SPD with some nucleic acids (Tabor and Tabor, 1984) does not appear to participate in the observed [³H]SPD binding by taking into consideration the permeability of polyamine-nucleic acid conjugate across a glass fiber filter.

Although the molecular mechanism underlying the drastic potentiation of [³H]SPD binding by the 2 phospholipases is not clarified at present, these data raise the possibility that membrane phospholipids may be at least in part responsible for the saturable binding of [³H]SPD with low affinity and high density. In fact, SPD as well as SPN is shown to inhibit the activity of phospholipase C possibly through interacting with phosphatidylinositol in human platelets (Nahas and Graff, 1982). Both DA5 and DA6 at 10 mM are effective as inhibitors of both phospholipase C and [³H]SPD binding, while being inactive in potentiating [³H]MK-801 binding. These findings along with the present results maintain the aforementioned phospholipids hypothesis. The differentiation between the abilities of both DA5 and DA6 at 10 mM to potentiate [³H]MK-801 binding and to displace [³H]SPD binding is suggestive of the proposal that SPD may potentiate [³H]MK-801 binding through associating with its own binding domains different from the specific binding sites of [³H]SPD demonstrated in this study. This proposal is also supported by the differential distribution profiles between [³H]MK-801 binding and [³H]SPD binding in the rat brain.

Consequently, there is no doubt that brain synaptic membrane preparations contain specific binding sites of [³H]SPD, whereas the functional significance of these binding sites as well as their potential role in the NMDA receptor complex still remains to be elucidated in future studies. Some unidentified endogenous factor would facilitate the signal transfer from [³H]SPD binding sites to ion channel sites within the NMDA receptor complex. Furthermore, possible heterogeneity of the NMDA receptor complex in the mammalian brain is not ruled out at present. In fact, SPD markedly potentiated [³H]MK-801 binding in the hippocampus and cerebral cortex due to an increase in the affinity, without significantly affecting the binding in the cerebellum and medulla-pons. The heterogeneity of NMDA receptors has been indeed postulated by Perkins and Stone (1983). The cerebellum and spinal cord could have quinolinate-resistant "NMDA1" receptors, whereas the cerebral cortex and hippocampus might contain quinolinate-sensitive "NMDA2" receptors (Perkins and Stone, 1983; McLennan, 1984). Therefore, it seems reasonable to speculate that the former NMDA1 receptors may have no functional communication with [³H]SPD binding sites, while the latter NMDA2 receptors having the communication. The potentiation by SPD would be a useful tool to differentiate these subpopulations in addition to the sensitivity to quinolinate. The differential potentiation by SPD among the brain structures is also against the aforementioned idea that SPD may only artifactually potentiate the association of open channel blockers with the NMDA channels with no physiological significance.

Since DCQX is a competitive antagonist highly selective to the GlyB sites on the NMDA receptor complex (Yoneda and Ogita, 1989d; Ogita and Yoneda, 1990b), complete abolition of the Glu-induced potentiation of [³H]MK-801 binding by DCQX is demonstrative of the absolute requirement for Gly in the signal transduction between NMDA recognition sites and ion channel sites within the NMDA receptor complex. In the voltage-clamped Xenopus oocytes injected with rat brain mRNA, NMDA elicits a negligible inward current in the absence of added Gly (Kleckner and Dingledine, 1988). A competitive antagonist with a relatively low affinity for the GlyB sites, indole-2-carboxylic acid, completely blocks the response to NMDA in solutions containing a low concentration of Gly in primary cultures of the rat visual cortex (Huettner, 1989). Since it is impossible to completely deplete the endogenous Gly from any biological materials while keeping the cellular functions intact, the use of a competitive antagonist with high affinity and specificity is crucial for elucidating the molecular mechanism underlying the Gly-induced potentiation of NMDA-mediated responses. Under a complete obstruction of the GlyB sites, NMDA agonists would be unable to open ion channels associated with the NMDA recognition sites. Occupation of the GlyB sites by Gly could enable a very small amount of Glu released from nerve terminals to activate the channels. Accordingly, the extracellular Gly seems to be a type of "sensitizer" rather than a "co-agonist" (Kleckner and Dingledine, 1988), since it occurs in cerebrospinal fluids at concentrations that already saturate the GlyB sites (Ferraro and Hare, 1985).

CONCLUSION

It is evident that the rat brain contains specific binding sites of [³H]SPD with low affinity and high density which exhibit an uneven distribution profile. However, the functional significance as well as possible role in the NMDA receptor complex is still uncertain at present. The NMDA receptor complex may be further classified into at least 2 subpopulations according to the difference in sensitivity to the potentiation by SPD. At any rate, Gly plays a key role in physiological responses mediated by the NMDA receptors.

Acknowledgement

This work was supported in part by a Grant-in-Aid for Scientific Research 01304037 from the Ministry of Education, Science and Culture, Japan.

REFERENCES

Anis, N.A., Berry, S.C., Burton, N.R., and Lodge, D., 1983, The dissociative anesthetics, ketamine and phencyclidine, selectively reduce excitation of central mammalian neurones by N-methyl-aspartate, Br. J. Pharmacol., 79: 565-575.

Bonhaus, D.W., Burge, B.C., and McNamara, J.O., 1987, Biochemical evidence that glycine allosterically regulates an NMDA receptor-coupled ion channels, Eur. J. Pharmacol., 142: 489-490.

Bristow, D.R., Bowery, N.G., and Woodruff, G.N., 1986, Light microscopic autoradiographic localisation of [³H]glycine and [³H]strychnine binding sites in rat brain, Eur. J. Pharmacol., 126: 303-307.

Davies, J., and Watkins, J.C., 1982, Action of the D- and L-forms of 2-amino-5-phosphonovalerate and 2-amino-4-phosphonobutyrate in the cat spinal cord, Brain Res., 235: 378-386.

Davies, J., Evans, R.H., Herrling, P.L., Jones, A.W., Olverman, H.J., Pook, P., and Watkins, J.C., 1986, CPP, a new potent and selective NMDA antagonists. Depression of central neuron responses, affinity for D-[³H]AP5 binding sites on brain membranes and anticonvulsant activity, Brain Res., 382: 169-173.

Davies, S.N., Martin, D., Millar, J.D., Aram, J.A., Church, J., and Lodge, D., 1988, Differences in results from in vivo and in vitro studies on the use-dependency of N-methylaspartate antagonism by MK-801 and other phencyclidine receptor ligands, Eur. J. Pharmacol., 145: 141-151.

Fagg, G.E., 1987, Phencyclidine and related drugs bind to the activated N-methyl-D-aspartate receptor-channel complex in rat brain membranes, Neurosci. Lett., 76: 221-227.

Fagg, G.E., and Matus, A., 1984, Selective association of N-methyl aspartate and quisqualate types of L-glutamate receptor with brain postsynaptic densities, Proc. Natl. Acad. Sci. U.S.A., 81: 6876-6880.

Ferkany, J.M., and Coyle, J.T., 1983, Specific binding of [³H](±)-2-amino-7-phosphonoheptanoic acid to rat brain membranes in vitro, Life Sci., 33: 1295-1306.

Ferraro, T.N., and Hare, T.A., 1985, Free and conjugated amino acids in human CSF: influence of age and sex, Brain Res., 338: 53-60.

Fletcher, E.J., and Lodge, D., 1988, Glycine reverses antagonism of N-methyl-D-aspartate (NMDA) by 1-hydroxy-3-aminopyrrolidone-2 (HA-966) but not by D-2-amino-5-phosphonovalerate (D-AP5) on rat cortical slices, Eur. J. Pharmacol., 151: 161-162.

Foster, A.C., and Fagg, G.E., 1987, Comparison of L-[³H]glutamate, D-[³H]aspartate, DL-[³H]AP5 and [³H]NMDA as ligands for NMDA receptors in crude postsynaptic densities from rat brain, Eur. J. Pharmacol., 133: 291-300.

Foster, A. C., and Wong, E.H.F., 1987, The novel anticonvulsant MK-801 binds to the activated state of the N-methyl-D-aspartate receptor in rat brain, Br. J. Pharmacol., 91: 403-409.

Glowinski, J., and Iversen, L.L., 1966, Regional studies of catecholamines in the rat brain. I. The disposition of [³H]norepinephrine, [³H]-dopamine and [³H]DOPA in various regions of the brain, J. Neurochem., 13: 655-669.

Huettner, J.E., 1989, Indole-2-carboxylic acid: A competitive antagonist of potentiation by glycine at the NMDA receptor, Science, 243: 1611-1613.

Johnson, J.W., and Ascher, P., 1987, Glycine potentiates the NMDA responses in cultured mouse brain neurons, Nature, 325: 529-531.

Kemp, J.A., Foster, A.C., Leeson, P.D., Priestley, T., Tridgett, R., Iversen, L.L., and Woodruff, G.N., 1988, 7-Chlorokynurenic acid is a selective antagonist at the glycine modulatory site of the N-methyl-D-aspartate receptor complex, Proc. Natl. Acad. Sci. U.S.A., 85: 6547-6550.

Kishimoto, H., Simon, J.R., and Aprison, M.H., 1981, Determination of the equilibrium dissociation constants and number of glycine binding sites in several areas of the rat central nervous system, using a sodium-independent system, J. Neurochem., 37: 1015-1024.

Kleckner, N.W., and Dingledine, R., 1988, Requirement for glycine in activation of NMDA-receptors expressed in Xenopus oocytes, Science, 241: 835-837.

Lehmann, J., Hutchinson, A.J., McPherson, S.E., Mondadori, C., Schmutz, M., Sinton, C.M., Tsai, C., Murphy, D.E., Steel, D.J., Williams, M., Cheney, D.L., and Wood, P.L., 1988, CGS 19755, a selective and competitive N-methyl-D-aspartate-type excitatory amino acid receptor antagonist, J. Pharmacol. Exp. Ther., 246: 65-75.

Loo, P., Braunwalder, A., Lehmann, J., and Williams, M., 1986, Radioligand binding to central phencyclidine recognition sites is dependent on excitatory amino acid receptor agonists, Eur. J. Pharmacol., 123: 467-468.

Marvizon, J.C.G., and Skolnick, P., 1988, [^{3}H]Glycine binding is modulated by Mg^{2+} and other ligands of the NMDA receptor-cation channel complex, Eur. J. Pharmacol., 151: 157-158.

McLennan, H., 1984, A comparison of the effects of NMDA and quinolinate on central neurones in the rat, Neurosci. Lett., 46: 157-160.

Monaghan, D.T., and Cotman, C.W., 1986, Identification and properties of N-methyl-D-aspartate receptors in rat brain synaptic membranes, Proc. Natl. Acad. Sci. U.S.A., 83: 7532-7536.

Monahan, J.B., and Michel, J., 1987, Identification and characterization of an N-methyl-D-aspartate-specific L-[^{3}H]glutamate recognition sites in synaptic plasma membranes, J. Neurochem., 48: 1699-1708.

Munson, P.J., and Rodbard, D., 1980, LIGAND: a versatile computerized approach for characterization of ligand-binding system, Anal. Biochem., 107: 220-239.

Murphy, D.E., Schneider, J., Boehm, C., Lehmann, J., and Williams, M., 1987a, Binding of [^{3}H]3-(2-carboxypiperazin-4-yl)propyl-1-phosphonic acid to rat brain membranes: a selective, high affinity ligand for N-methyl-D-aspartate receptors, J. Pharmacol. Exp. Ther., 240, 778-784.

Murphy, D.E., Snowhill, E.W., and Williams, M., 1987b, Characterization of quisqualate recognition sites in rat brain tissue using DL-[^{3}H]-amino-3-hydroxy-5-methylisoxazole-4-propionic acid (AMPA) and a filtration assay, Neurochem. Res., 12: 775-782.

Murphy, D.E., Hutchinson, A.J., Hurt, S.D., Williams, M., and Sills, M.A., 1988, Characterization of the binding of [^{3}H]-CGS 19755: a novel N-methyl-D-aspartate antagonist with nanomolar affinity in rat brain, Br. J. Pharmacol., 95: 932-938.

Nahas, N., and Graff, G., 1982, Inhibitory activity of polyamines on phospholipase C from human platelets, Biochem. Biophys. Res. Comm., 109: 1035-1040.

Ogita, K., and Yoneda, Y., 1986, Characterization of Na^+-dependent binding of [^{3}H]glutamate in synaptic membranes from rat brain, Brain Res., 397: 137-144.

Ogita, K., and Yoneda, Y., 1988a, Disclosure by Triton X-100 of NMDA-sensitive [^{3}H]glutamate binding sites in brain synaptic membranes, Biochem. Biophys. Res. Comm., 153: 510-517.

Ogita, K., and Yoneda, Y., 1988b, Temperature-dependent and -independent apparent binding activities of [^{3}H]glutathione in brain synaptic membranes, Brain Res., 463: 37-46.

Ogita, K., and Yoneda, Y., 1990a, Temperature-independent binding of [^{3}H](±)-3-(2-carboxypiperazin-4-yl)propyl-1-phosphonic acid in brain synaptic membranes treated with Triton X-100, Brain Res., in press.

Ogita, K., and Yoneda, Y., 1990b, 6,7-Dichloroquinoxaline-2,3-dione is a competitive antagonist specific to strychnine-insensitive [^{3}H]glycine binding sites on the N-methyl-D-aspartate receptor complex, J. Neurochem., 54: 699-702.

Ogita, K., Kitago, T., Nakamuta, H., Fukuda, Y., Koida, M., Ogawa, Y., and Yoneda, Y., 1986, Glutathione-induced inhibition of Na^+-independent and -dependent bindings of L-[^{3}H]glutamate in rat brain, Life Sci., 39: 2411-2418.

Ogita, K., Suzuki, T., and Yoneda, Y., 1989, Strychnine-insensitive binding of [^{3}H]glycine to synaptic membranes in rat brain, treated with Triton X-100, Neuropharmacology, 28: 1263-1270.

Ogita, K., Nabeshima, T., and Yoneda, Y., 1990, [^{3}H]Thienylcyclohexyl-piperidine binding activity in brain synaptic membranes treated with Triton X-100, J. Neurochem., in press.

Pegg, A.E., 1986, Recent advances in the biochemistry of polyamines in eukaryotes, Biochem. J., 234: 249-262.

Perkins, M.N., and Stone, T.W., 1983, Quinolinic acid: regional variations in neuronal sensitivity, Brain Res., 259: 172-176.

Perkins, M.N., Stone, T.W., Collins, J.F., and Curry, K., 1981, Phosphonate analogues of carboxylic acids as amino acid antagonists on rat cortical neurones, Neurosci. Lett., 23: 333-336.

Ransom, R.W., and Stec, N.L., 1988, Cooperative modulation of [^{3}H]MK-801 binding to the N-methyl-D-aspartate receptor-ion channel complex by L-glutamate, glycine and polyamines, J. Neurochem., 51: 830-836.

Reynolds, I.J., Murphy, S.N., and Miller, R.J., 1987, ^{3}H-Labeled MK-801 binding to the excitatory amino acid receptor complex from rat brain is enhanced by glycine, Proc. Natl. Acad. Sci. U.S.A., 84: 7744-7748.

Robinson, M.B., and Coyle, J.T., 1987, Glutamate and related acidic neuro-transmitters: from basic science to clinical application, FASEB J., 1: 446-455.

Snell, L.D., and Johnson, K.M., 1988, Cycloleucine competitively an-tagonizes the strychnine-insensitive glycine receptor, Eur. J. Pharmacol., 151: 165-166.

Snell, L.D., Morter, R.S., and Johnson, K.M., 1987, Glycine potentiates N-methyl-D-aspartate-induced [^{3}H]TCP binding to rat cortical mem-branes, Neurosci. Lett., 83: 313-317.

Tabor, C.W., and Tabor, H., 1984, Polyamines, Ann. Rev. Biochem., 53: 749-790.

Terramani, T., Kessler, M., Lynch, G., and Baudry, M., 1988, Effects of thiol-reagents on [^{3}H] -amino-3-hydroxy-5-methylisoxazole-4-propionic acid binding to rat telencephalic membranes, Mol. Pharmacol., 34: 117-123.

Williams, K., Romano, C., and Molinoff, P.B., 1989, Effects of polyamines on the binding of [^{3}H]MK-801 to the N-methyl-D-aspartate receptor: Pharmacological evidence for the existence of a polyamine recogni-tion site, Mol. Pharmacol., 36: 575-581.

Wong, E.H.F., Kemp, J.A., Priestley, T., Knight, A.R., Woodruff, G.N., and Iversen, L.L., 1986, The anticonvulsant MK-801 is a potent N-methyl-D-aspartate antagonist, Proc. Natl. Acad. Sci. U.S.A., 83: 7104-7108.

Yoneda, Y., and Ogita, K., 1986, Localization of [^{3}H]glutamate binding sites in rat adrenal, Brain Res., 383: 387-391.

Yoneda, Y., and Ogita, K., 1987, Are Ca^{2+}-dependent proteases really responsible for Cl^--dependent and Ca^{2+}-stimulated binding of [^{3}H]glutamate in rat brain?, Brain Res., 400: 70-79.

Yoneda, Y., and Ogita, K., 1989a, Labeling of NMDA receptor channels by [^{3}H]MK-801 in brain synaptic membranes treated with Triton X-100, Brain Res., 499: 305-314.

Yoneda, Y., and Ogita, K., 1989b, Possible role of polyamines as modulators of NMDA receptor complex, Bull. Jap. Neurochem. Soc., 28: 136-137. (in Japanese)

Yoneda, Y., and Ogita, K., 1989c, Microbial methodological artifacts in [^{3}H]glutamate receptor binding assays, Anal. Biochem., 177: 250-255.

Yoneda, Y., and Ogita, K., 1989d, Abolition of the NMDA-mediated responses by a glycine antagonist, 6,7-dichloroquinoxaline-2,3-dione (DCQX), Biochem. Biophys. Res. Comm., 164: 841-849.

Yoneda, Y., Ogita, K., and Suzuki, T., 1988, Multiple binding sites on the NMDA receptor/ion channel complex in brain synaptic membranes, _in_: "Neurotransmitters: Focus on Excitatory Amino Acids," I. Kanazawa, ed., pp. 47-65, Excerpta Medica, Tokyo.

Yoneda, Y., Ogita, K., Ohgaki, T., Uchida, S., and Meguri, H., 1989, N-Methyl-D-aspartate-sensitive [^{3}H]glutamate binding sites in brain synaptic membranes treated with Triton X-100, _Biochim. Biophys. Acta,_ 1012: 74-80.

Yoneda, Y., Ogita, K., Kouda, T., and Ogawa, Y., 1990a, Radioligand labeling of N-methyl-D-aspartic acid (NMDA) receptors by [^{3}H]($\pm$)-3-(2-carboxypiperazin-4-yl)propyl-1-phosphonic acid in brain synaptic membranes treated with Triton X-100, _Biochem. Pharmacol.,_ 39: 225-228.

Yoneda, Y., Ogita, K., and Suzuki, T., 1990b, Interaction of strychnine-insensitive glycine binding with MK-801 binding in brain synaptic membranes, _J. Neurochem.,_ in press.

SOLUBILIZATION OF THE NMDA RECEPTOR ION CHANNEL COMPLEX FROM RAT BRAIN

Kiyokazu Ogita and Yukio Yoneda

Department of Pharmacology, Faculty of Pharmaceutical Sciences
Setsunan University, Hirakata, Osaka 573-01, Japan

INTRODUCTION

Several lines of evidence indicate that N-methyl-D-aspartic acid (NMDA)-sensitive receptors form a macromolecular complex which consists of at least three distinct sites (for example, see review by Robinson and Coyle, 1987); 1) NMDA recognition sites labeled by [³H]glutamic acid (Glu) (Ogita and Yoneda, 1988) or [³H](±)-3-(2-carboxypiperazin-4-yl)propyl-1-phosphonic acid (CPP) (Murphy et al., 1987), 2) cation channel sites labeled by radiolabeled noncompetitive NMDA antagonists, (+)-5-methyl-10,11-dihydro-5H-dibenzo[a,d]cyclohepten-5,10-imine (MK-801) (Foster and Wong, 1987) and N-[1-(2-thienyl)cyclohexyl]piperidine (TCP) (Loo et al., 1986), 3) glycine (Gly) recognition sites labeled by [³H]Gly, which are insensitive to strychnine (GlyB sites) (Ogita et al., 1989). In addition, recent binding studies have raised the possibility that modulatory sites by polyamines may also exist on the NMDA receptor ion channel complex in addition to the aforementioned three sites, which is based on the findings that polyamines, such as spermidine (SPD) and spermine, markedly enhance both [³H]MK-801 and [³H]TCP bindings (Ransom and Stec, 1988) and that the potentiation by SPD is antagonized by some other polyamines, including putrescine and cadaverine (Williams et al., 1989).

Recently, solubilization of the NMDA receptor ion channel complex has been carried out using [³H]MK-801 or [³H]TCP binding as a biochemical measure (McKernan et al., 1989; Ambar et al., 1988). These studies, however, failed to demonstrate either the potentiation by SPD or any significant bindings of [³H]Glu and [³H]Gly in solubilized preparations. Therefore, we have attempted here to solubilize the whole macromolecular complex, including NMDA recognition sites, ion channel sites, GlyB sites and SPD sites.

METHODS

Solubilization

Whole brains of Wistar rats were homogenized in 15 vol 50 mM Tris-acetate buffer (pH 7.4) containing 0.1 mM phenylmethylsulfonyl fluoride (PMSF), 0.01% bacitracin and 1 mM EDTA (BUFFER A) using a Physcotron homogenizer at setting #6 for 1 min at 4 °C. The homogenates were centrifuged at 50,000 g for 30 min. Resultant pellets were suspended in BUFFER A and the suspensions were subsequently centrifuged as above. The pellets were resuspended in BUFFER A containing 20% glycerol and the suspensions were stored at -80 °C until use. On the day of experiments, frozen suspensions were thawed at room temperature and washed with BUFFER A by the centrifugation. The pellets were suspended in BUFFER A containing 20% glycerol and the suspensions were treated with 1% deoxycholate (DOC) at an approximate protein concentration of 3-4 mg/ml at 4 °C for 30 min with stirring. This treatment was terminated by centrifuging at 200,000 g for 60 min and the resultant supernatants were applied to a Sephadex G-25 column (3 x 45 cm) equilibrated with 50 mM Tris-acetate buffer (pH 7.4) containing 0.01 mM PMSF, 0.001% bacitracin, 0.1 mM EDTA, 20% glycerol and 0.01% DOC. Each 5 ml fraction eluted with the same buffer was subjected to the binding assay.

Binding assay

An aliquot (about 0.3 mg protein) of solubilized preparations was incubated with 10 nM [^{3}H]MK-801 (832.5 GBq/mmol, NEN/DuPont) in 0.5 ml 50 mM Tris-acetate buffer (pH 7.4) at 30 °C for 2 or 16 hr. Incubation was terminated by the addition of 3 ml ice-cold buffer and subsequent filtration under a constant vacuum of 15 mm Hg (Ogita and Yoneda, 1986) through a Whatman GF/B glass fiber filter soaked in 0.3% polyethyleneimine solution at least 1 hr before use (Bruns et al., 1983). Nonspecific binding was defined by 0.1 mM (+)-MK-801. Binding assays were carried out in triplicate. Results were represented as the mean ± S.E..

RESULTS

Binding activity of [^{3}H]MK-801 was most effectively solubilized by DOC among various detergents used, including Triton X-100, cholate, CHAPS, digitonin, n-octylglucoside and n-octylthioglucoside. Although addition of Glu or Gly did not enhance the binding in supernatant preparations solubilized by DOC before gel filtration on Sephadex G-25 column, the activity was markedly inhibited not only by competitive NMDA antagonists, such as CPP and D-2-amino-5-phosphonovaleric acid (D-AP5), but also by a GlyB antagonist, 7-chlorokynurenic acid (7-ClKYNA). The activity of [^{3}H]MK-801 binding was eluted in void volume fractions of the Sephadex column in parallel with the peak of absorption at 280 nm. Therefore, subsequent experiments were carried out using the void volume fractions as solubilized preparations.

The basal binding, which was determined in the absence of both Glu and Gly added, gradually increased with incubation time up to 8 hr. Addition of Glu and Gly accelerated the initial association rate of the binding, without altering the steady-state level found in the absence of added Glu and Gly. In contrast, addition of SPD did not significantly affect the association rate accelerated by both Glu and Gly, with a concomitant elevation of the steady-state level. Addition of Glu (EC50 = 0.37 ± 0.08 μM) or NMDA (10.8 ± 4.7 μM) significantly potentiated the binding in a concentration-dependent manner, whereas neither DL-α-amino-3-hydroxy-5-methylisoxazole-4-propionic acid (AMPA) nor kainic acid affected the binding (Fig. 1). The potentiation by Glu was markedly antagonized by competitive NMDA antagonists, such as CPP and D-AP5. The potentiated binding induced by Glu was further enhanced by Gly (0.19 ± 0.06 μM) (Fig. 2), and GlyB antagonists, including 7-ClKYNA and 6,7-dichloroquinoxaline-2,3-dione (DCQX), competitively inhibited the enhancement by Gly. Furthermore, SPD

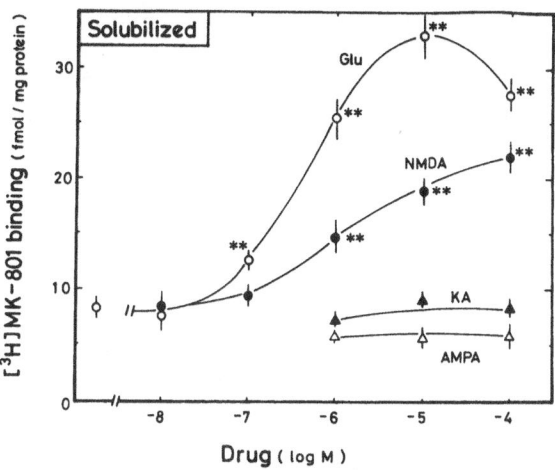

Fig. 1. Effect of some agonists for excitatory amino acid receptors on
[³H]MK-801 binding in solubilized preparations. An aliquot of
solubilized preparations was incubated with [³H]MK-801 at 30°C for
2 hr in the presence of varying concentrations of various compounds indicated. **P<0.01, compared with the control value
obtained in the absence of added agonist.

(19.5 ± 8.3 μM) displayed a marked potentiation of the binding in the
presence of Glu/Gly. EC50 values for Glu and Gly in solubilized preparations were similar to those obtained in Triton-treated membrane preparations (Yoneda and Ogita, 1989). In contrast, EC50 value for SPD in
solubilized preparations was somewhat lower than that obtained in Triton-
treated membrane preparations.

Saturation isotherms were analyzed 16 hr after the initiation of incubation when complete equilibrium was attained. Scatchard analysis of
these data revealed that the binding sites consisted of a single component
with a Kd of 9.7 ± 1.5 nM and a Bmax of 64.9 ± 6.7 fmol/mg protein in the
presence of Glu/Gly. Addition of SPD induced an increase of the affinity
of binding sites for [³H]MK-801, without significantly altering the density (Kd = 4.0 ± 0.5 nM, Bmax = 74.9 ± 6.9 fmol/mg protein). (+)-MK-801
(IC50 = 0.010 ± 0.001 μM) was the most potent displacer of the binding in
the presence of Glu/Gly/SPD among various drugs tested. Progressively
decreasing rank order of potency was observed for (-)-MK-801 (0.041 ±
0.003 μM), TCP (0.055 ± 0.001 μM), PCP (0.38 ± 0.02 μM), SKF 10,047 (3.92
± 0.24 μM) and ketamine (6.66 ± 0.20 μM). On the other hand, both (+)-3-
(3-hydroxyphenyl)-N-(1-propyl)-piperidine and haloperidol, which are
ligands with high affinity for sigma sites, did not markedly displace the
solubilized binding in the presence of Glu/Gly/SPD at concentrations up to
0.1 mM. This rank order of potency was similar to that observed with the
inhibition of [³H]MK-801 binding in the presence of Glu/Gly in Triton-
treated membrane preparations (Yoneda and Ogita, 1989).

Both NMDA-sensitive [³H]Glu binding and strychnine-insensitive
[³H]Gly binding were detected in these supernatant preparations after the
gel filtration. The preparations also exhibited a significant activity of
[³H]SPD binding, in addition to [³H]AMPA binding and [³H]kainate binding
(data not shown).

479

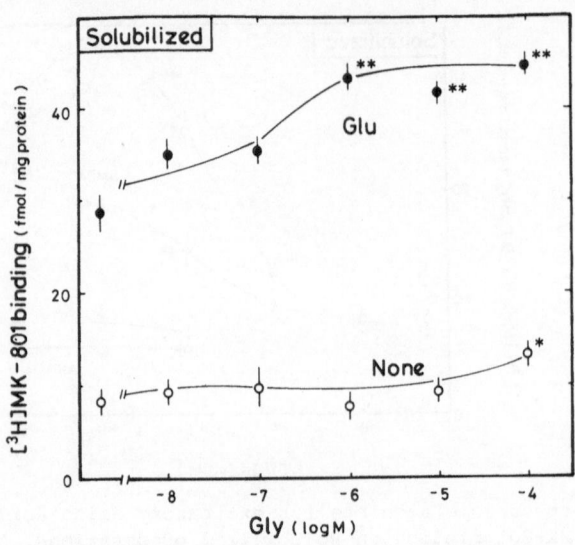

Fig. 2. Effect of Gly on [³H]MK-801 binding in solubilized preparations.
An aliquot of solubilized preparations was incubated at 30°C with
[³H]MK-801 for 2 hr in the buffer containing varying concentrations of Gly, in either the presence or absence of 10 M Glu.
*P<0.05, **P<0.01, compared with each control value obtained in
the absence of added Gly.

DISCUSSION

Low recovery of [³H]MK-801 binding in spite of the high recovery of
protein contents in supernatants solubilized by DOC, may be due to the
following reasons as demonstrated previously (McKernan et al., 1989; Ambar
et al., 1988); (1) The receptor ion channel complex is unstable in solutions at a high concentration of DOC. (2) The binding of [³H]MK-801 to
solubilized receptors is inhibited by DOC. In fact, recovery of the binding was elevated through removing excess detergent by gel filtration immediately after the solubilization. The potentiation by Glu, Gly and SPD
supports the proposal that there may exist macromolecules consisting of
respective recognition sites for NMDA, Gly and SPD in solubilized
preparations. One direct evidence for this proposal is that solubilized
preparations indeed exhibit NMDA-sensitive [³H]Glu, strychnine-insensitive
[³H]Gly and [³H]SPD bindings. Demonstration of the existence of each
recognition site on the NMDA receptor ion channel complex, are quite important for a type of solubilization studies on neurotransmitter receptors.

A sigma ligand haloperidol slightly displaces [³H]MK-801 binding in
the presence of Glu/Gly in Triton-treated membranes (Yoneda and Ogita,
1989), without affecting the binding in the presence of Glu/Gly/SPD in
solubilized preparations. Although the exact molecular mechanism underlying this phenomenon is not clear at present, the experimental conditions
employed here may be convenient to a predominant detection of the NMDA
receptor ion channel complex which distinctly differs from the sigma sites.

Acknowledgment

This work was supported in part by a Grant-in-Aid for Scientific Research 01772048 to K.O. from the Ministry of Education, Science and
Culture, Japan.

REFERENCES

Ambar, I., Kloog, Y., and Sokolovsky M., 1988, Solubilization of rat brain phencyclidine receptors in an active binding form that is sensitive to N-methyl-D-aspartate receptor ligands, J. Neurochem., 51: 133-140.

Bruns, R.F., Lawson-Wendling, K., and Pugsley, T.A., 1983, A rapid filtration assay for soluble receptors using polyethyleneimine-treated filters, Anal. Biochem., 132: 74-81.

Foster, A.C., and Wong, E.H.F., 1987, The novel anticonvulsant MK-801 binds to the activated state of the N-methyl-D-aspartate receptor in rat brain, Br. J. Pharmacol., 91: 403-409.

Loo, P., Braunwalder, A., Lehmann, J., and Williams, M., 1986, Radioligand binding to central phencyclidine recognition sites is dependent on excitatory amino acid receptor agonists, Eur. J. Pharmacol., 123: 467-468.

McKernan, R.M., Castro, S., Poat, J.A., and Wong, E.H.F., 1989, Solubilization of the N-methyl-D-aspartate receptor channel complex from rat and porcine brain, J. Neurochem., 52: 777-785.

Murphy, D.E., Schneider, J., Boehm, C., Lehmann, J., and Williams, M., 1987, Binding of [^{3}H]3-(2-carboxypiperazin-4-yl)propyl-1-phosphonic acid to rat brain membranes: a selective, high affinity ligands for N-methyl-D-aspartate receptors, J. Pharmacol. Exp. Ther., 240: 778-784.

Ogita, K., and Yoneda, Y., 1986, Characterization of Na$^+$-dependent binding of [^{3}H]glutamate in synaptic membranes from rat brain, Brain Res., 397: 137-144.

Ogita, K., and Yoneda, Y., 1988, Disclosure by Triton X-100 of NMDA-sensitive [^{2}H]glutamate binding sites in brain synaptic membranes, Biochem. Biophys. Res. Comm., 153: 510-517.

Ogita, K., Suzuki, T., and Yoneda, Y., 1989, Strychnine-insensitive binding of [^{3}H]glycine to synaptic membranes in rat brain, treated with Triton X-100, Neuropharmacology, 28: 1263-1270.

Ransom, R.W., and Stec, N.L., 1988, Cooperative modulation of [^{3}H]MK-801 binding to the N-methyl-D-aspartate receptor-ion channel complex by L-glutamate, glycine and polyamines, J. Neurochem., 51: 830-836.

Robinson, M.B., and Coyle, J.T., 1987, Glutamate and related acidic neurotransmitters: from basic science to clinical application, FASEB J., 1: 446-455.

Williams, K., Romano, C., and Molinoff, P.B., 1989, Effects of polyamines on the binding of [^{3}H]MK-801 to the N-methyl-D-aspartate receptor: Pharmacological evidence for the existence of a polyamine recognition site, Mol. Pharmacol., 36: 575-581.

Yoneda, Y., and Ogita, K., 1989, Labeling of NMDA receptor channels by [^{3}H]MK-801 in brain synaptic membranes treated with Triton X-100, Brain Res., 499: 305-314.

NMDA RECEPTOR AGONISTS: RELATIONSHIPS BETWEEN

STRUCTURE AND BIOLOGICAL ACTIVITY

Bjarke Ebert, Ulf Madsen, Tommy Nørskov Johansen and
Povl Krogsgaard-Larsen

Department of Organic Chemistry
The Royal Danish School of Pharmacy
DK-2100 Copenhagen, Denmark

NEURODEGENERATIVE DISORDERS - ROLE OF EXCITATORY AMINO ACIDS

The acidic amino acids L-aspartic acid (ASP) and, in particular, L-glu-tamic acid (GLU) are generally accepted to be the major excitatory amino acid (EAA) neurotransmitters in the mammalian central nervous system (CNS). Hyperactivity of central EAA neuronal pathways has been associated with the etiology of certain neurodegenerative diseases, such as status epilepticus, Huntington's chorea and dementia of the Alzheimer type (Lodge, 1988; Rothman and Olney, 1986; Greenamyre 1986; Greenamyre et al., 1988; Bridges et al., 1988). In Alzheimer patients a regional degeneration of cholinergic, nor-adrenergic and serotonergic neurones is observed (Narang and Cutler, 1986).

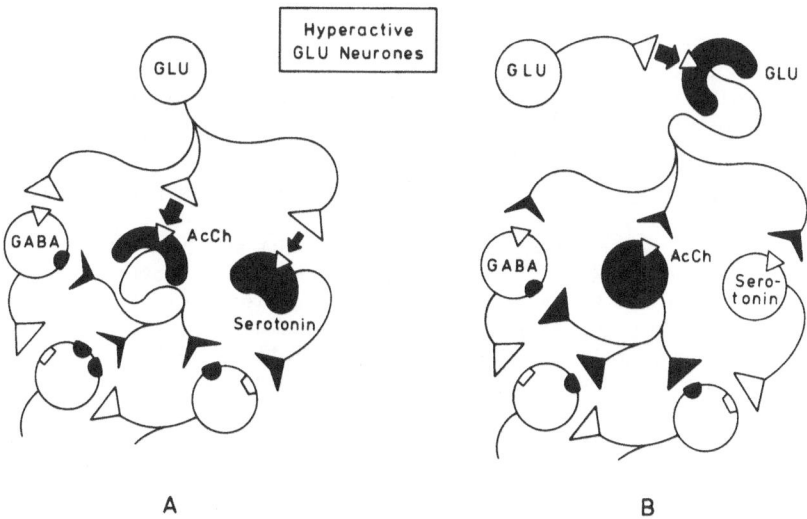

A   B

Fig. 1. Schematic illustration of (A) the degeneration of acetylcholine (AcCh) and serotonin neurones caused by hyperactive L-glutamic acid (GLU) neurones; (B) the neurodegeneration of GLU neurones caused by hyperactive GLU neurones. GABA: 4-aminobutanoic acid.

Neuroreceptor Mechanisms in Brain, Edited by S. Kito *et al.*
Plenum Press, New York, 1991

In addition, loss of glutamatergic neurones is seen in the progression of Alzheimer's disease (Greenamyre et al., 1988) (Fig. 1). Thus, hyperactive as well as hypoactive EAA neuronal mechanisms may be operative in Alzheimer's disease. The hypoactivity of central EAA neuronal pathways has recently been proposed to be associated with learning and memory deficits (Greenamyre et al., 1988). The neuronal degeneration observed after ischemia following stroke, hypoxia, or hypoglycemia may also be due to prolonged and excessive stimulation of EAA receptors (Rothman and Olney, 1986). There is some evidence that hypoactivity of central EAA system(s) may be a key factor in schizophrenia (Deutsch et al., 1989).

Since hyperactivity as well as hypoactivity of EAA neuronal systems may be implicated in for example Alzheimer's disease, drugs capable of both protecting and activating EAA receptors may be of therapeutic interest. An obvious challenge is to design partial agonists with appropriately balanced agonist/antagonist profiles.

EAA receptors are subdivided into four main classes, some, if not all, of which may be heterogeneous (McLennan, 1983; Hansen and Krogsgaard-Larsen, 1990; Watkins et al., 1990): (**1**) N-Methyl-D-aspartic acid (NMDA) receptors at which NMDA and ibotenic acid (Figs. 2 and 4) are selective agonists and (R)-2-amino-5-phosphonopentanoic acid (D-AP5) and (RS)-3-(2-carboxypiperazin-4-yl)propyl-1-phosphonic acid (CPP) selective antagonists; (**2**) AMPA receptors at which quisqualic acid is a non-selective and α-amino-3-hydroxy-5-methyl-4-isoxazolepropionic acid (AMPA) a highly selective agonist (Krogsgaard-Larsen et al., 1980); (**3**) kainic acid receptors, which are selectively activated by kainic acid (Coyle, 1983; Shinozaki, 1988); (**4**) L-AP4 receptors, through which L-2-amino-4-phosphonobutanoic acid (L-AP4) inhibits synaptic excitation (Johnson and Koerner, 1988).

The considerable flexibility of the GLU molecule probably is essential for its synaptic activity. This molecular flexibility makes GLU itself inapplicable for studies of its "receptor active conformations".

In order to gain insight into the structural and conformational requirements for binding to and activation of the different GLU receptor subtypes, a number of conformationally constrained GLU analogues have been designed and tested pharmacologically (Krogsgaard-Larsen et al., 1984, 1985; Hansen and Krogsgaard-Larsen, 1990; Watkins et al., 1990; Madsen et al., 1990). In this paper we describe the relationship between structure and in vitro biological activity of a series of piperidinedicarboxylic acids (PDAs) and a number of analogues of ibotenic acid structurally related to ASP.

PIPERIDINEDICARBOXYLIC ACIDS AS NMDA RECEPTOR AGONISTS OR ANTAGONISTS

The acidic amino carboxylic acids 2,3-PDA and 2,4-PDA are cyclic analogues of ASP/NMDA and GLU, respectively (Fig. 2). Whereas the trans-forms of 2,3-PDA and 2,4-PDA are NMDA receptor agonists, cis-2,3-PDA and cis-2,4-PDA show NMDA antagonist properties (Watkins et al., 1990). Using 1H NMR spectroscopy and molecular mechanics calculations we have studied the relationship between structure, conformational flexibility, and effects on NMDA receptors of these cyclic amino acids. Whilst the NMDA antagonist cis-2,4-PDA was shown to exist almost exclusively in one conformation in aqueous solution, the NMDA agonist trans-2,4-PDA exists in aqueous solution as an equilibrium mixture of two conformations (Fig. 3). Similarly, only one conformation of the NMDA antagonist cis-2,3-PDA could be detected in aqueous solution, whereas the NMDA agonist trans-2,3-PDA, like trans-2,4-PDA, exists as an equilibrium mixture of two conformations. The cis-form of 2,5-PDA, which shows weak NMDA agonist effects, also exists as a mixture of two conformations in water (not illustrated) (Madsen et al., 1990).

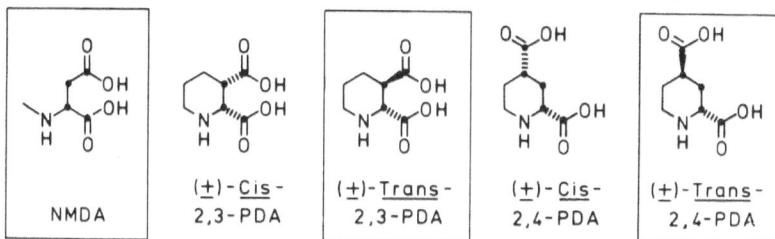

Fig. 2. Structures of NMDA and some piperidinedicarboxylic acids (PDAs) showing NMDA receptor agonist or antagonist profiles. The framed compounds are NMDA agonists.

These observations seem to indicate that a certain degree of conformational flexibility of GLU or ASP/NMDA analogues is a necessary condition for agonist activity at NMDA receptors. Alternatively, the active NMDA agonist conformation(s) are not accessible to, for example, cis-2,3-PDA or cis-2,4-PDA, which show NMDA antagonist profiles.

IBOTENIC ACID ANALOGUES STRUCTURALLY RELATED TO ASPARTIC ACID AND NMDA

Ibotenic acid, which is an analogue of GLU (Fig. 4), is a potent NMDA receptor agonist, but it also interacts with AMPA and kainic acid receptor sites (Krogsgaard-Larsen et al., 1980). Furthermore, ibotenic acid is chemically unstable (Nielsen et al., 1985). As an attempt to develop chemically stable compounds showing potent and specific NMDA agonist effects, we have synthesized and tested a number of ibotenic acid analogues structurally related to ASP/NMDA (Fig. 4), including α-amino-3-hydroxy-5-methyl-4-isoxazoleacetic acid (AMAA) and the bicyclic analogue 3-hydroxy-4,5,6,7-tetrahydroisoxazolo[4,5-c]pyridine-4-carboxylic acid (4-HPCA) (Table 1). None of these compounds showed any detectable affinity for AMPA, high- and low-affinity kainic acid, or strychnine-insensitive glycine receptor sites (see footnote to Table 1).

AMAA was shown to be more potent than NMDA as an NMDA agonist and to interact more effectively with different sites of the NMDA receptor complex (Table 1). N-methylation of AMAA was accompanied by a marked loss of activity, whereas N,N-dimethylation resulted in almost complete loss of activity. The pharmacological profile of 4-HPCA was very similar to that of quinolinic acid. The very weak NMDA agonist activity of 4-HPCA may be attributed to its conformationally restricted structure. In aqueous solution, 4-HPCA does, however, exist in a conformation, which is distinctly different from

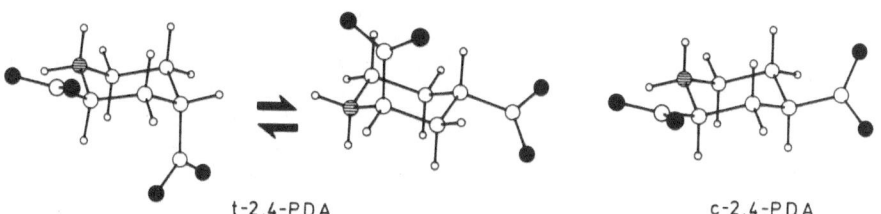

Fig. 3. Illustrations of the conformations in aqueous solution of the fully charged molecules of trans-2,4-PDA (t-2,4-PDA) and cis-2,4-PDA (c-2,4-PDA) as determined by 1H NMR spectroscopy.

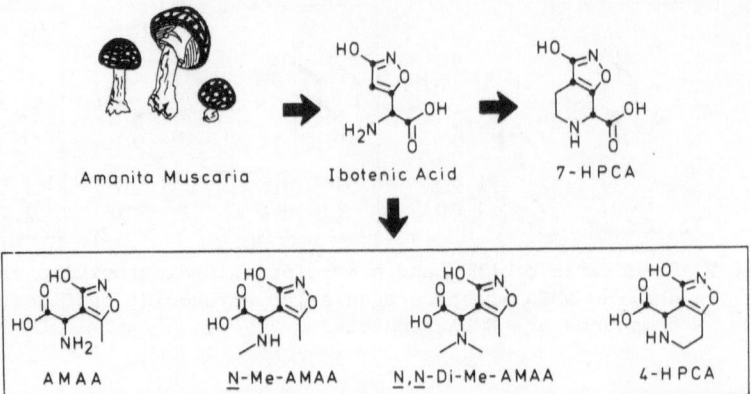

Fig. 4. The structures of the <u>Amanita muscaria</u> constituent ibotenic acid, the specific AMPA receptor agonist 7–HPCA, and a number of ibotenic acid analogues structurally related to ASP/NMDA.

that of <u>cis</u>-2,3–DPA, and this may explain why 4–HPCA is not recognized by the NMDA receptors as an antagonist (Madsen et al., 1990). Comparative studies on the potencies of AMAA, 4–HPCA, and quinolinic acid as neuronal excitants and neurotoxins are in progress.

Acknowledgements – This work was supported by grants from The Danish Medical and Technical Research Councils and The Lundbeck Foundation. The secretarial assistance of Mrs. B. Hare is gratefully acknowledged.

Table 1. In Vitro Pharmacological Profiles of NMDA, Quinolinic Acid and a Number of Ibotenic Acid Analogues

Compounds	Effects of the Binding[a] of radioactive			D–AP5–Sensitive Excitation in the Rat Cortical Slice
	CPP Inhibition (IC_{50}, µM)	MK–801[b] Inhibition (IC_{50}, µM)	MK–801[c] Stimulation (ED_{50}, µM)	(ED_{50}, µM)
NMDA	35	>100	0.45	15
AMAA	4.5	>100	0.20	12
N–Me–AMAA	18	>100	50	70
N,N–Di–Me–AMAA	>100	>100	>100	>10000
4–HPCA	90	>100	>100	1000
Quniolinic Acid	>100	>100	N.t.	1500

[a]None of the compounds tested showed any detectable effects on the receptor binding of AMPA. Similarly, the compounds were inactive as inhibitors of the binding of kainic acid to high–affinity (in the absence of calcium ions) or low–affinity (in the presence of 100 mM calcium chloride) receptor sites and also inactive as inhibitors of the binding of glycine to strychnine–insensitive glycine sites.
[b]Binding to rat brain membranes was tested in the presence of GLU (30 µM) and glycine (1 µM) (fully stimulated).
[c]Binding to rat brain membranes was tested in the presence of glycine (100 nM) and various concentrations of test compound (baseline binding).

References

Bridges, R. J., Geddes, J. W., Monaghan, D. T., and Cotman, C. W., 1988, Excitatory amino acid receptors in Alzheimer's disease, in "Excitatory Amino Acids in Health and Disease," D. Lodge, ed., J. Wiley & Sons, Chichester, pp. 321-335.

Coyle, J. T., 1983, Neurotoxic action of kainic acid, J. Neurochem., 41: 1-11.

Deutsch, S. I., Mastropaolo, J., Schwartz, B. L., Rosse, R. B., and Morihisa, J. M., 1989, A "glutamatergic hypothesis" of schizophrenia: rationale for pharmacotherapy with glycine, Clin. Neuropharmacol., 12:1-13.

Greenamyre, J. T., 1986, The role of glutamate in neurotransmission and in neurologic disease, Arch. Neurol., 43:1058-1063.

Greenamyre, J. T., Maragos, W. F., Albin, R. L., Penney, J. B., and Young, A. B., 1988, Glutamate transmission and toxicity in Alzheimer's disease, Prog. Neuro-Psychopharmacol. Biol. Psychiat., 12:421-430.

Hansen, J. J. and Krogsgaard-Larsen, P., 1990, Structural, conformational, and stereochemical requirements of central excitatory amino acid receptors, Med. Res. Rev., 10:55-94.

Johnson, R. L. and Koerner, J. F., 1988, Excitatory amino acid neurotransmission, J. Med. Chem., 31:2057-2066.

Krogsgaard-Larsen, P., Brehm, L., Johansen, J. S., Vinzents, P., Lauridsen, J., and Curtis, D. R., 1985, Synthesis and structure-activity studies on excitatory amino acids structurally related to ibotenic acid, J. Med. Chem., 28:673-679.

Krogsgaard-Larsen, P., Honoré, T., Hansen, J. J., Curtis, D. R., and Lodge, D., 1980, New class of glutamate agonists structurally related to ibotenic acid, Nature (London), 284:64-66.

Krogsgaard-Larsen, P., Nielsen, E. Ø., and Curtis, D. R., 1984, Ibotenic acid analogues. Synthesis and biological in vitro activity of conformationally restricted agonists at central excitatory amino acid receptors, J. Med. Chem., 27:585-591.

Lodge, D., ed., 1988, "Excitatory Amino Acids in Health and Disease", J. Wiley & Sons, Chichester.

Madsen, U., Brehm, L., Schaumburg, K., Jørgensen, F. S., and Krogsgaard-Larsen, P., 1990, Relationship between structure, conformational flexibility and biological activity of agonists and antagonists at the N-methyl-D-aspartic acid subtype of excitatory amino acid receptors, J. Med. Chem., 33:374-380.

McLennan, H., 1983, Receptors for the excitatory amino acids in the mammalian central nervous system, Prog. Neurobiol., 20:251-271.

Narang, P. K. and Cutler, N. R., 1986, Pharmacotherapy in Alzheimer's disease: basis and rationale, Prog. Neuropsychopharmacol. Biol. Psychiat., 10:519-531.

Nielsen, E. Ø., Schousboe, A., Hansen, S. H., and Krogsgaard-Larsen, P., 1985, Excitatory amino acids: studies on the biochemical and chemical stability of ibotenic acid and related compounds, J. Neurochem., 45:725-731.

Rothman, S. M. and Olney, J. W., 1986, Glutamate and the pathology of hypoxic/ischemic brain damage, Ann. Neurol., 19:105-111.

Shinozaki, H., 1988, Pharmacology of the glutamate receptor, Prog. Neurobiol., 30:399-435.

Watkins, J. C., Krogsgaard-Larsen, P., and Honoré, T., 1990, Structure-activity relations in the development of excitatory amino acid receptor agonists and competitive antagonists, Trends Pharmacol. Sci., 11:25-33.

CONTRIBUTORS

Heinrich Betz
Zentrum für Molekulare Biologie
Universität Heidelberg
Im Neuenheimer Feld 282
D-6900 Heidelberg, F.R.G.

Giovanni Biggio
Department of Experimental Biology
University of Cagliari
Via Palabanda 12
Cagliari 09123, Italy

Juei-Tang Cheng
Department of Pharmacology
College of Medicine
National Cheng Kung University
Tainan City, Taiwan 70101, R.O.C.

Raymond J. Dingledine
Department of Pharmacology
CB #7365 Faculty Laboratory Office Bldg.
University of North Carolina
   School of Medicine
Chapel Hill, NC 27599, U.S.A.

Bjarke Ebert
Department of Organic Chemistry
Royal Danish School of Pharmacy
2 Universitets Parken, DK-2100
Copenhagen, Denmark

Frederick J. Ehlert
Department of Pharmacology
College of Medicine
University of California, Irvine
Irvine, CA 92717, U.S.A.

Shuichi Endo
Department of Pharmacology
UCLA School of Medicine
Center for Health Sciences
Los Angeles, CA 90024, U.S.A.

Salvatore J. Enna
Nova Pharmaceutical Corporation
6200 Freeport Centre, Baltimore
MD 21224, U.S.A.

Franz J. Ferstl
Radiologische Klinik
Universitätsklinikum
Rudolf Virchow Standort Charlottenburg
Spandauer Damm 130
D-1000 Berlin 19, F.R.G.

Gilles Fillion
Department of Pharmacology
Institut Pasteur
28 rue du Docteur-Roux 75724
Paris Cedex 15, France

Motohatsu Fujiwara
First Department of Pharmacology
Faculty of Medicine, Kyoto University
Yoshidakonoemachi, Sakyo-ku
Kyoto 606, Japan

George P. Hess
Section of Biochemistry
Molecular and Cell Biology
217 Biotechnology Building
Cornell University
Ithaca, NY 14853, U.S.A.

Richard L. Huganir
Howard Hughes Medical Institute
Department of Neuroscience
Johns Hopkins University School of Medicine
725 North Wolfe Street
Baltimore, MD 21205, U.S.A.

Hiro-o Kamiya
Department of Pharmacology
Faculty of Pharmaceutical Sciences
Fukuoka University
8-19-1 Nanakuma, Jyonan-ku
Fukuoka 814-01, Japan

Ryuichi Kato
Department of Pharmacology
Keio University School of Medicine
35 Shinanomachi, Shinjuku-ku
Tokyo 160, Japan

John S. Kelly
Department of Pharmacology
University of Edinburgh
1 George Square, Edinburgh
EH8 9JZ, Scotland, U.K.

Shozo Kito
Division of Health Sciences
University of the Air
2-11 Wakaba, Chiba 260, Japan

Yoshihisa Kudo
Department of Neuroscience
Mitsubishi Kasei Institute of Life Sciences
11 Minamioohya, Machida, Tokyo 194, Japan

Kinya Kuriyama
Department of Pharmacology
Kyoto Prefectural University of Medicine
465 Kajiicho Hirokohji Agaru
Kawaramachi-Hirokoji, Kamikyo-ku
Kyoto 602, Japan

Jon M. Lindstrom
Salk Institute for Biological Studies
P.O.Box 85800
San Diego, CA 92138, U.S.A.

Katsuhiko Mikoshiba
Division of Regulation of Macromolecular
   Function
Institute for Protein Research
Osaka University
3-2 Yamadaoka, Suita
Osaka 565, Japan

Hitoshi Nakayama
Department of Pharmacology
Nara Medical University
Shijyou-cho 840
Kashihara 634, Japan

Toshio Narahashi
Department of Pharmacology
Northwestern University Medical School
303 East Chicago Avenue
Chicago, IL 60611, U.S.A.

Yasuyuki Nomura
Department of Pharmacology
Faculty of Pharmaceutical Sciences
Hokkaido University
Kita 12 Nishi 6, Kita-ku
Sapporo 060, Japan

Toshihide Nukada
Department of Biochemistry
Institute of Brain Research
Faculty of Medicine, University of Tokyo
7-3-1 Hongo, Bunkyo-ku
Tokyo 113, Japan

Norio Ogawa
Department of Neurochemistry
Institute for Neurobiology
Okayama University Medical School
2-5-1 Shikata-cho
Okayama 700, Japan

Kiyokazu Ogita
Department of Pharmacology
Faculty of Pharmaceutical Sciences
Setsunan University
45-1 Nagaotohge-cho, Hirakata
Osaka 573-01, Japan

Richard W. Olsen
Department of Pharmacology
UCLA School of Medicine
Center for Health Sciences
Los Angeles, CA 90024, U.S.A.

Elena Porsche-Wiebking
Hoechst AG, Werk Kalle Albert
Pharmacological Research
Adolfsallee 16
D-6200 Wiesbaden 1, F.R.G.

Masamichi Satoh
Department of Pharmacology
Faculty of Pharmaceutical Sciences
Kyoto University
Shimoadachicho, Yoshida, Sakyo-ku
Kyoto 606, Japan

Tomio Segawa
Department of Pharmacology
Institute of Pharmaceutical Sciences
Hiroshima University School of Medicine
1-2-3 Kasumi, Minami-ku
Hiroshima 734, Japan

Hitoshi Shinotoh
Department of Neurology
Chiba University School of Medicine
1-8-1 Inohana-cho
Chiba 280, Japan

Catherine D. Strader
Department of Molecular Pharmacology &
Biochemistry
Merck Sharp & Dohme Research Laboratories
126 E. Lincoln Avenue
Rahway, NJ 07065. U.S.A.

Chikako Tanaka
Department of Pharmacology
Kobe University School of Medicine
7-5-1 Kusunokicho, Chuo-ku
Kobe 650, Japan

Allan J. Tobin
Department of Biology
UCLA, 405 Hilgard Avenue
Los Angeles, CA 90024, U.S.A.

Masaya Tohyama
Department of Anatomy II
Osaka University Medical School
4-3-57 Nakanoshima, Kita-ku
Osaka 530, Japan

Shuji Uchida
Department of Pharmacology I
Osaka University Medical School
4-3-57 Nakanoshima, Kita-ku
Osaka 530, Japan

James K. Wamsley
Neuropsychiatric Research Institute
700 First Avenue South
Fargo, ND 58103, U.S.A.

Hiroshi Watanabe
Department of Anatomy
Yamagata University School of Medicine
2-2-2 Iida-Nishi
Yamagata 990-23, Japan

Donald J. Woodward
Department of Cell Biology and Neuroscience
University of Texas, Southwestern
   Medical Center
5323 Harry Hines Boulevard
Dallas, TX 75235, U.S.A.

Henry I. Yamamura
Department of Pharmacology
College of Medicine
University of Arizona
Health Sciences Center
Tucson, AZ 85724, U.S.A.

Yukio Yoneda
Department of Pharmacology
Faculty of Pharmaceutical Sciences
Setsunan University
45-1 Nagaotohge-cho, Hirakata
Osaka 573-01, Japan

INDEX

498